Fire Inspection and Code Enforcement

Sixth Edition

Edited by
Michael Wieder and Carol Smith

Cover photo courtesy of Michael Wieder, Stillwater, Oklahoma.

Validated by the International Fire Service Training Association
Published by Fire Protection Publications, Oklahoma State University

THE INTERNATIONALFIRE SERVICE TRAINING ASSOCIATION

The International Fire Service Training Association (IFSTA) was established as a "nonprofit educational association of fire fighting personnel who are dedicated to upgrading fire fighting techniques and safety through training." This training association was formed in November 1934, when the Western Actuarial Bureau sponsored a conference in Kansas City, Missouri. The meeting was held to determine how all the agencies interested in publishing fire service training material could coordinate their efforts. Four states were represented at this initial conference. Because the representatives from Oklahoma had done some pioneering in fire training manual development, it was decided that other interested states should join forces with them. This merger made it possible to develop training materials broader in scope than those published by individual agencies. This merger further made possible a reduction in publication costs, because it enabled each state or agency to benefit from the economy of relatively large printing orders. These savings would not be possible if each individual state or department developed and published its own training material.

To carry out the mission of IFSTA, Fire Protection Publications was established as an entity of Oklahoma State University. Fire Protection Publications' primary function is to publish and disseminate training texts as proposed and validated by IFSTA. As a secondary function, Fire Protection Publications researches, acquires, produces, and markets high-quality learning and teaching aids as consistent with IFSTA's mission. The IFSTA Executive Director is officed at Fire Protection Publications.

IFSTA's purpose is to validate training materials for publication, develop training materials for publication, check proposed rough drafts for errors, add new techniques and developments, and delete obsolete and outmoded methods. This work is carried out at the annual Validation Conference.

The IFSTA Validation Conference is held in July in the state of Oklahoma. Fire Protection Publications, the IFSTA publisher, establishes the revision schedule for manuals and introduces new manuscripts. Delegates are selected for technical input by the Delegate Selection Committee. The Delegate Selection Committee consists of three Board members and two conference delegates; the committee is chaired by the Vice-Chair of IFSTA. Applications are reviewed by the committee, and delegates are selected based upon technical expertise and demographics. Committees meet and work at the conference addressing the current standards of the National Fire Protection Association and other standard-making groups as applicable.

Most of the delegates are affiliated with other international fire protection organizations. The Validation Conference brings together individuals from several related and allied fields, such as:

- Key fire department executives and training officers
- Educators from colleges and universities
- Representatives from governmental agencies
- Delegates of firefighter associations and industrial organizations
- Engineers from the fire insurance industry

Delegates are not paid nor are they reimbursed for their expenses by IFSTA or Fire Protection Publications. They come because of commitment to the fire service and its future through training. Being on a committee is prestigious in the fire service community, and delegates are acknowledged leaders in their fields. This unique feature provides a close relationship between the International Fire Service Training Association and other fire protection agencies, which helps to correlate the efforts of all concerned.

IFSTA manuals are now the official teaching texts of most of the states and provinces of North America. Additionally, numerous U.S. and Canadian government agencies as well as other English-speaking countries have officially accepted the IFSTA manuals.

Table of Contents

List of Tables

Preface

This is the Sixth Edition of ***Fire Inspection and Code Enforcement***. The greatly expanded text and illustrations detail all aspects of inspection responsibilities and procedures, plan review, hazardous materials and their associated storage and handling, building construction, exiting systems, and general fire safety practices.

Acknowledgment and special thanks are extended to the members of the IFSTA validating committee who contributed their time, wisdom, and knowledge to this manual.

Chairman:
Ed Steiner
Edmond Fire Department
Edmond, Oklahoma

Howard Boyd
Consultant
Nashville, Tennessee

James Hebert
St. Martin Parish Fire District
Breaux Bridge, Louisiana

James Kater
Butler County Community College
El Dorado, Kansas

Andy King
Brentwood Fire Department
Brentwood, Tennessee

Barbara Koffron
Phoenix Fire Department
Phoenix, Arizona

Brett Lacey
Colorado Springs Fire Department
Colorado Springs, Colorado

J.T. O'Neal
Henderson Fire Department
Henderson, Nevada

Richard Pippinger
MACTEC-ERS
Grand Junction, Colorado

Paul Valentine
Mount Prospect Fire Department
Mount Prospect, Illinois

A book of this scope would be impossible to publish were it not for the assistance of many persons and organizations. We are grateful to the following people and organizations who assisted us in the completion of this manual:

Bilco Company
Professors Pat Brock and Jim Hanson, Oklahoma State University School of Fire Protection and Safety Engineering Technology
Illinois Fire Inspectors Association
Institute of Makers of Explosives
International Conference of Building Officials (ICBO)
Justrite Manufacturing, Inc.
Mount Prospect (Illinois) Fire Department
National Fire Protection Association (NFPA)
Plano (Texas) Fire Department
Edward Prendergast, Chicago (Illinois) Fire Department
Southern Building Code Congress International, Inc. (SSBCC)
Houston (Texas) Fire Department
Conoco Oil Co., Ponca City, Oklahoma Refinery
Air Force Academy Fire Department, Colorado Springs, Colorado
Dan Gross, Maryland Fire & Rescue Institute

C.H. Guernsey & Co. Engineers, Architects, and Consultants, Oklahoma City, Oklahoma
Ron Jeffers, Union City, New Jersey
The Tumbleweed Dance Hall and Concert Arena, Stillwater, Oklahoma
Leslie Miller, Oklahoma State University Environmental Health Services

Gratitude is also extended to the following members of the Fire Protection Publications staff, whose contributions made the final publication of the manual possible.

Lynne Murnane, Senior Publications Editor
Don Davis, Coordinator, Publications Production
Ann Moffat, Graphic Design Analyst
Desa Porter, Senior Graphic Designer
Connie Cook, Senior Graphic Designer
Don Burull, Graphics Assistant
Ben Brock, Graphic Assistant
Tim Frankenberg, Research Technician
Jack Krill, Research Technician

Introduction

Fire suppression activities are not the only way to combat fires: a well-planned and executed fire prevention and inspection program is a less expensive and more effective way to accomplish the goal of the fire service. The goal is, of course, to minimize the risk of life and property loss from fire. By observing, making recommendations, and subsequently controlling or eliminating hazardous conditions, the inspector can make major strides toward accomplishing this goal before a fire occurs.

The common view of the fire inspector's role is one of an "enforcing" or "policing" authority. However, when the job is performed properly, the inspector is just as much an *educator* as anything else. The inspector helps to educate occupants in ways to control hazards, in proper methods of evacuation, and in overall fire safety practices. It is much more effective, in the long run, to educate the public than it is to simply keep enforcing regulations on them.

Historically, courts have ruled that fire departments and/or individuals empowered to perform fire inspections may be held legally accountable for their performance. Chapter 1 addresses the issue of liability, as well as authority and responsibilities of the inspector. A discussion of police power, right of entry, organizational structure, and cooperation among other agencies is also included.

Chapter 2, Inspection Procedures, covers all aspects of conducting an inspection. These include preparation for the inspection, the inspection itself, documentation, follow-up inspections, and taking cases to court.

The fire inspector must know how fires start and spread in order to recognize fire hazards and their potential consequences. Chapter 3, Principles of Combustion and Fire Growth, discusses such fire basics as the combustion process, sources of heat energy, phases of fire, and methods of heat transfer. This information is intended as a review for the firefighter/inspector and as an overview of fire behavior for the civilian inspector. The last part of the chapter concentrates on how fires grow and spread. Understanding fire growth and spread is becoming more important with the increase in use of performance-based fire codes.

Chapter 4, Fire Hazard Recognition, addresses safety considerations that affect many occupancies. These include good housekeeping, regulation smoking, open burning, and the use of flammable decorations. Electrical safety is also critical, so basic electrical theory, electrical hazards, and the dangers of static electricity are covered.

In order to determine that a building meets the appropriate building and fire safety codes, the inspector must possess the ability to determine how the building has been constructed and for what purposes it will be used. Chapter 5, Construction and Occupancy Classifications, describes the various construction and occupancy classifications that are used by all of the model code organizations.

One of the most crucial elements in building safety is the ability of occupants to quickly and efficiently evacuate the structure in the event of a fire or other hazardous situation. Chapter 6 focuses on the inspector's role in evaluating the means of egress for a structure. Also included in this chapter is information for ensuring proper site access for emergency personnel and emergency planning procedures for building occupants.

Extinguishing equipment and fire protection systems are effective ways to increase occupant safety and to control fires during their early stages. Chapter 7 covers water-based fire protection systems, including fire sprinkler systems, standpipe systems, fixed fire pumps, and water supply systems. Fire inspectors must be well-versed in the procedures for inspecting and testing each of these systems. Detailed descriptions of testing procedures are an important part of this chapter.

In addition to water-based fire protection systems, there are a wide variety of other types of fire protection systems and equipment with which inspectors must be very knowledgeable. Chapter 8 provides descriptions and testing information on portable fire extinguishers, special agent fixed extinguishing systems, and fire detection and alarm systems.

Fire departments have come to understand that plans review is a necessary component of overall fire protection and prevention. Chapter 9 introduces the inspector to the many aspects of plans review, including developing a plans review system and evaluating architectural, structural, mechanical, and electrical drawings. Although the information in this chapter will by no means turn the inspector into a trained plans reviewer, it will help provide a better understanding of construction drawings.

Because hazardous materials is such a complex subject, it is treated in two chapters. Chapter 10 begins with a discussion of the common methods for recognizing and identifying hazardous materials that may be found during building inspections. The remainder of the chapter is dedicated to the requirements involving the use, handling, and storage of flammable and combustible liquids and gases. Of all the different hazardous materials in use today, most inspectors will find that flammable and combustible liquids and gases account for the majority of their time. Chapter 11 covers the wide array of other types of hazardous materials that fire inspectors may encounter during their work day.

PURPOSE AND SCOPE

Fire Inspection and Code Enforcement (6th Edition) is designed to educate the inspector in the principles and techniques of fire prevention and life safety inspection and code compliance. It is written to meet the professional standards set forth in NFPA 1031, *Standard for Professional Qualifications for Fire Inspector* (1998 edition), Levels I and II. This manual is designed for use by public or private sector individuals who perform inspections.

NOTICE ON GENDER USAGE

The English language has historically given preference to the male gender. Among many words, the pronouns, "he" and "his" are commonly used to describe both genders. Society evolves faster than language, and the male pronouns still predominate our speech. IFSTA/Fire Protection Publications has made great effort to treat the two genders equally, recognizing that a significant percentage of fire service personnel are female. However, in some instances, male pronouns are used to describe both males and females solely for the purpose of brevity. This is not intended to offend readers of the female gender.

1

Authority, Responsibilities, and Organization

Chapter 1
Authority, Responsibilities, and Organization

Fire prevention inspections are the single most important nonemergency activity performed by the fire service. This chapter presents the general concepts related to the fire inspector's responsibilities and legal authority. Because they vary widely from jurisdiction to jurisdiction, it is the inspector's responsibility to become thoroughly familiar with the statutes, codes, regulations, and permitting processes in his own jurisdiction.

Personnel who perform fire inspections must possess a great deal of knowledge regarding fire safety and building codes. They must also have a versatile personality that allows them to interact favorably with a wide variety of people. The inspector must also have good written and oral communications skills.

Throughout the course of this manual, we will refer to the person performing the outlined functions as the *fire inspector,* or simply the *inspector*. However, keep in mind that personnel other than designated fire inspectors may perform the inspection procedures highlighted in this manual. In many jurisdictions, basic inspections are performed by fire companies (Figure 1.1). In these jurisdictions, the company officer and/or the firefighters on the company may be required to be certified to NFPA 1031, *Standard for Professional Qualifications for Fire Inspector*. The information in this manual pertains to anyone who performs inspections, whether they are in an inspection bureau or a fire company. It may also pertain to employees of a private business who are charged with performing inspections in privately owned facilities.

Figure 1.1 Some jurisdictions use in-service fire companies to perform basic fire inspections.

Any person who performs fire inspections must have a thorough knowledge and understanding of the following:

- The statutes that create their position or designate them to perform inspections
- The statutes that provide the legal basis and requirements for fire prevention activities
- The laws, codes, and ordinances that detail various fire safety requirements and establish a fire inspector's duties and responsibilities
- The statutes that set the limits of authority
- The edition of the code that is being used and when it was adopted
- The appeals process
- The ways that the statutes or laws can be changed

These statutes may include federal regulations, state laws, and municipal ordinances. Statutes, laws, codes, and ordinances vary from jurisdiction to jurisdiction. Few are exactly alike, but most are similar in concept. Inspectors must not assume that statutes in place in their jurisdiction are the same for another jurisdiction, or vice versa, as this is often not the case. Court decisions or local preferences may dictate changes that range from subtle to major, even in neighboring jurisdictions.

AUTHORITY

There are certain key areas that fire inspectors must consider regarding the legal basis of and limits to their authority. These areas include:

- The fire inspector's status as a member of the public sector, including the right, responsibility, and liabilities inherent in being public officers or employees
- The limits and scope of their authority as fire inspectors

The fire inspector must have a clear understanding of the statutes he is responsible for enforcing. This information should be provided to the inspector by fire department management. Management generally obtains this information from the jurisdiction's legal counsel. The legal counsel may also provide information on the inspector's authority and liability. If the legal counsel does not have this information readily available, they may have to obtain it from legal research companies or law libraries.

Public Sector Inspector's Legal Status

Fire inspectors who work for a governmental agency, such as a local fire department or a state fire marshal's office, must be cognizant of their legal status as members of the public sector. In particular, they should know whether their job position classifies them as a public employee (nonsworn position) or a public officer (sworn position). In some jurisdictions, this distinction can affect potential liability, compensation, and other benefits.

In the United States, the legal status for public sector employees is in most cases determined by individual state governments. In addition to legal requirements for inspectors, the states may also have regulations on the organization of fire protection agencies, retirement systems, and civil service requirements. The state may also specify jurisdictional lines between the state fire marshal's office and local agencies. Each state may also have specific statutes concerning the legal liability and responsibility of inspectors.

POLICE POWER

Public sector fire inspectors must clearly understand any authority they may have to arrest, to issue summons, and to issue citations for fire code violations. Fire inspectors may also have the authority to interrupt business operations that contain eminently dangerous conditions. An example of this would be to close down a nightclub that is grossly over its legal occupant load (Figure 1.2). Some jurisdictions give designated fire inspectors police power to act in these roles. Local statutes delegate authority to fire inspectors with specific words such as "to issue summons," "to write a ticket," "to issue a warrant," or "to arrest." When fire inspectors have this authority, they must have appropriate law enforcement training, particularly if they could be involved in prosecuting a fire code violator.

The exact relationship between the police department and the fire inspector must be clearly defined when the fire inspector has limited police

Figure 1.2 Serious code violations may require the fire inspector to close a business, such as this nightclub, until the situation can be corrected.

power. This will avoid conflicts when the fire inspector must exercise police powers that require police department support. An example of this type of situation is the fire inspector needing assistance in making an arrest of a code violator, such as the nightclub owner in the previous example, who is known to be belligerent in these situations. Police backup may be required to safely make the arrest and close the club. The model code adopted by the jurisdiction may contain language that further explains the relationship between the fire inspector and the police department.

Private Sector Inspector's Legal Status

The role of fire inspector is not limited to the public sector. Fire inspectors are an important part of the loss control mission in many private sector organizations. There are two common roles within which private sector fire inspectors operate. The first is an inspector that is employed by a particular company to ensure that appropriate levels of fire safety are maintained within all company facilities (Figure 1.3). The inspector's ability to report and correct deficiencies that are found will vary widely. Each company will have policy and procedures for handling fire safety hazards. The private sector inspector must know the proper chain of command for handling these situations and the limits of relationships with public sector inspectors.

The second common role is that of an inspector who is employed by an insurance company. The inspector evaluates processes and facilities that are underwritten by the inspector's insurance company. The goal of this inspector's work is to identify hazards within the insured's facility so that they can be corrected prior to any loss affecting both the insured and the insurance company. Again, the power that the insurance inspector will have to make corrections or improvements will vary widely. It will be affected by the insurance company's policies and working relationship between the insured and the insurance company. Appropriate corporate procedures for handling noted hazards must be followed at all times.

Often, insurance companies dictate protection requirements for their clients that are more stringent than local codes or ordinances. Fire inspection personnel may encounter facilities that are inspected internally, causing discrepancies between the two parties. As long as the situation meets or exceeds the requirements of the local code, the inspector should be satisfied. Inspectors must also remember that some requirements by insurance companies are targeted more at property protection concerns than life safety concerns. The fire inspector may have to point out these issues and see that the code being enforced is addressed.

Figure 1.3 Some fire inspectors may actually be employees of the facility they are inspecting.

Liability Incurred as a Result of Authority

In today's society, the legal lines of liability change on a regular basis. It is difficult in a manual of this type to give specific, accurate information on liability, as this varies from jurisdiction to jurisdiction and changes with time. There are, however, some general liability considerations that are useful for all fire inspectors to be aware of.

The lines of liability may be established by state and/or local statutes. In general, fire inspectors are not held liable for discretionary acts. *Discretionary acts* involve actions fire inspectors consider necessary to fulfill their responsibilities. Fire inspectors may be held liable for ministerial actions. *Ministerial actions* involve the manner in which the fire inspector carries out or performs an act or policy. An example would be citing a business for a fire code violation and then not following up to make sure the problem is corrected. If the problem that was found to be a violation results in a fire, the occupant may have a cause for action against the inspector and his agency.

Most of the model fire prevention codes contain language that limits the liability to the jurisdiction using the code and its inspectors. In order for this language to be effective, that portion of the code must be specifically adopted by the authority having jurisdiction. Merely adopting a reference to the liability section will be insufficient to provide adequate protection. The following example of a liability clause in a model code is taken from the 1997 edition of the *Uniform Fire Code*™:

> ***101.5 Liability.*** *The chief and other individuals charged by the chief with the control or extinguishment of any fire, the enforcement of this code or any other official duties, acting in good faith and without malice in the discharge of their duties, shall not thereby be rendered personally liable for any damage that may accrue to persons or property as a result of any act or by reason of any act or omission in the discharge of their duties. Any suit brought against the chief or such individuals because of such an act or omission performed in the enforcement of any provision of such codes or other pertinent laws or ordinances implemented through the enforcement of this code or enforced by the code enforcement agency shall be defended by this jurisdiction until final termination of such proceedings, and any judgment resulting therefrom shall be assumed by this jurisdiction.*
>
> *This code shall not be construed to relieve from or lessen the responsibility of any person owning, operating or controlling any building or structure for any damages to persons or property caused by defects, nor shall the code enforcement agency or its parent jurisdiction be held as assuming such liability by reason of the inspections authorized by this code or any permits or certificates issued under this code.*

Although this example was taken from the *Uniform Building Code*™, all of the other model fire and building codes contain similar statements.

Over the years, there have been some court rulings against the immunity provisions of the model codes. The courts have ruled that the immunity provisions conflict with statutes that establish an inspection authority and require the enforcement of codes and regulations. In other words, a community cannot be required to do something and at the same time be immune from liability if it or its officers (fire inspectors) do the job inadequately or negligently.

Most communities indemnify their fire inspection personnel or provide liability insurance to protect them in the areas that they may be held liable. To *indemnify* the fire inspector means to assume the responsibility for any claims against the individual in total. The procedures for indemnification generally depend on prevailing state law. It is most important for the fire inspectors to determine whether they do or do not have indemnity or liability insurance when they are performing their official duties. This will protect the fire inspector from costs involved with providing legal counsel or court judgments.

A 1976 court ruling (Adams v. State of Alaska; see Appendix A) held that fire inspectors, in conducting code inspections, had taken on a duty and must use reasonable care in the exercise of that duty. For example, if fire inspectors inspect a property and determine that violations are present but fail to follow up to ensure that the violations are corrected, they can be held liable if a fire related to the violations occurs. In addition, they can be held liable for the deaths or injuries resulting if they can be attributed foreseeable or in fact to the code violation.

Likewise, fire inspectors taking on a special duty to a person can be held liable. A *special duty* is one in which a person has moved from a position of safety to a position of danger because he relied on the expertise of the inspector. For example, by issuing an occupancy permit, a fire inspector establishes the duty to ensure that the building complies with applicable codes and regulations.

Most model codes contain a *duty to inspect* clause. These clauses are significant for a fire inspector because they normally do not allow selective enforcement; rather, they charge the inspector with total enforcement. This means that fire inspectors may not target only certain buildings within their jurisdiction for enforcement. The code must be applied equally, within reason, to all applicable occupancies in a given jurisdiction. Failure to follow this clause may subject the department or individual fire inspector to personal and professional liability, such as the Adams v. State of Alaska case previously cited.

As with liability clauses, the duty-to-inspect clauses in each model code are very similar in nature. The following example of a duty-to-inspect clause is also from the 1997 edition of the *Uniform Fire Code*™:

> ***103.2 Authority for Inspection and Enforcement.***
>
> ***103.2.1 Authority of the chief and the fire department.***
>
> ***103.2.1.1 General.*** *The chief is authorized to administer and enforce this code. Under the chief's direction, the fire department is authorized to enforce all ordinances of the jurisdiction pertaining to:*
>
> 1. *The prevention of fires,*
> 2. *The suppression or extinguishment of dangerous or hazardous fires,*
> 3. *The storage, use and handling of hazardous materials,*
> 4. *The installation and maintenance of automatic, manual and other private fire alarm systems and fire-extinguishing equipment,*
> 5. *The maintenance and regulation of fire escapes,*
> 6. *The maintenance of fire protection and the elimination of fire hazards on land and in buildings, structures and other property, including those under construction,*
> 7. *The maintenance of means of egress, and*
> 8. *The investigation of the cause, origin and circumstances of fire and unauthorized releases of hazardous materials.*

From the previous example, it should be apparent that the fire inspector must know all of the provisions of the code and the various ways that the codes can be interpreted. Knowing all of this information is no small task. Periodic training is necessary to renew skills and knowledge and to keep abreast of changes.

From the liability standpoint, it is better for fire inspectors to conduct fewer, but more thorough, inspections and to follow up on all violations than to perform more frequent inspections in a haphazard, incomplete, or negligent manner. Failure to inspect a property does not impose a duty on the inspector, unless laws or statutes impose such a duty, or there is a known code violation present. Laws that single out a particular class occupancy for a predetermined number of inspections establish a duty on the inspection staff. Fire inspectors need to be aware of these laws so that they are fully aware of their duties.

Another potential source of major liability that fire inspectors must be concerned about involves the issue of civil rights. If fire inspectors perform inspections in a jurisdiction in a manner that discriminates against a certain group of people (by singling out certain groups of people or businesses without justification, for example), the jurisdiction can be held liable. Civil rights cases tend to be high-priced, high-profile affairs that should be avoided at all costs.

Outside Technical Assistance

Frequently, fire inspectors will be confronted by code issues or conditions in an occupancy that are beyond their level of expertise. Examples of these situations include the use of alternative building materials or fire protection systems to meet the intent of an existing code requirement. These situations generally require the involvement of a more highly trained individual, such as a fire protection engineer. Many medium- to larger-size fire departments and industrial operations have fire protection engineers on staff to handle these issues (Figure 1.4). However, most smaller jurisdictions do not have this luxury. Lacking this technical support on staff, smaller jurisdictions may choose to contract with an engineering or fire protection firm to provide technical assistance when required. Before entering into such an agreement, local officials must make sure that such an agreement is legal and that decisions made by the outside firm can be made binding by the local enforcement official.

Figure 1.4 Inspectors may need to consult with their department's fire protection engineer on complex issues.

Right of Entry

The right to enter a property to inspect for code compliance is essential in order for fire inspectors to fulfill their duties. In most cases, the issue of right to enter is not a problem for fire inspectors. However, the U.S. Supreme Court has ruled that property owners have the right to refuse admittance to an inspector unless the inspector has obtained a warrant based on the belief that a serious fire hazard exists on the property.

The U.S. Supreme Court has held that portions of commercial premises that are not open to the public may only be entered for inspections without consent of the owner through prosecution or physical force within the framework of a warrant procedure. For the actual documentation on this ruling, review *See v. City of Seattle 387 vl. 541, 87 S. Ct 1737*. See Appendix A.

The following recommendations should be used by fire inspectors to operate within the generally accepted legal guidelines:

- Inspectors must be adequately identified (Figure 1.5).
- Inspectors must state the reason for the inspection.
- Inspectors must request permission for the inspection.
- Inspectors should invite a building representative to walk along during the inspection.
- The local electrical, mechanical, plumbing, and building inspectors may also participate in this inspection. It is best to try to do all of these at once to avoid the appearance of harassment of the occupant.
- Inspectors should carry and follow a written inspection procedure.
- Inspectors should request a search warrant if entry is denied.

Figure 1.5 One method that the inspector may use to readily identify himself is to wear a uniform while conducting inspections.

- Inspectors may issue stop orders for extremely hazardous conditions, even if entry is denied, while search warrants are being prepared.
- Inspectors should have guidelines available that define conditions whereby they may stop operations without obtaining either permission to enter or a search warrant.
- Inspectors should be sure that all licenses and permits indicate that compliance inspection can be made throughout the duration of the permit or license.
- Inspectors must be trained in fire hazard recognition and in applicable laws and ordinances.
- Inspectors should develop a reliable record-keeping system of inspections.

Many state and federal courts have made decisions that protect the right of privacy of owners of private dwellings where no known or suspected fire hazards exist. To insist on making an inspection under these conditions is generally considered to be an unreasonable search.

Similar constitutional protection has been extended to owners of commercial property on the grounds that administrative entry without the owner's consent results in warrantless entry and, as such, violates the owner's rights under the Fourth and Fourteenth Amendments to the U.S. Constitution.

Challenges to the right to enter a property are generally rare; however, the inspector must understand explicitly the legal boundaries of his authority. State or local jurisdictions may have their own set of guidelines on right of entry by public personnel. Most of the model codes that a jurisdiction may adopt have language in them regarding right of entry. The fire inspector must know the right-of-entry language in the code under which he operates.

If the fire inspector must obtain a search warrant, he must understand the process required in granting the warrant so that he can provide the necessary supporting documentation. The fire department may wish to develop a form that can be used when a property owner refuses the right to enter a property (Figure 1.6). The form can provide the basis of information for obtaining a warrant to enter the property at a later time. Any form that is developed for this purpose should be approved by the jurisdiction's legal counsel.

RIPLEY FIRE PROTECTION DISTRICT
Fire Marshal's Office
P.O. Box 65
Ripley, Oklahoma 74062
(405) 555-3473

I ______ (BUSINESS REPRESENTATIVE), do hereby refuse ______ (INSPECTOR), authorized representative of the Ripley Fire Protection District, entry into my business, ______ (BUSINESS NAME)), located at ______ (STREET ADDRESS), ______ (INCORPORATED UNINCORPORATED AREA), to conduct a fire inspection.

______ (AUTHORIZED COMPANY OFFICIAL)

______ (TITLE)

______ (DATE)

The following conditions were observed in my attempt to secure entry:
1. ______
2. ______
3. ______

Past inspections have disclosed few/several fire code violations.
Date of last inspection: ______
Occupancy Group ______ Occupant Load ______
Construction Type ______ Life Hazard ______
Distance to adjacent occupancies structures

______ RIPLEY FIRE PROTECTION DISTRICT

Figure 1.6 Fire departments may wish to develop a form that can be filled out when an occupant refuses to admit them to his property.

MODIFICATION OF REQUIREMENTS, APPEALS PROCEDURES, AND JUDICIAL REVIEW

A frequent occurrence in the world of fire inspections and code compliance is the need to modify various provisions of the code that has been adopted in a given jurisdiction. Requests to modify the code generally involve the property owner's desire to use alternative materials, products, or systems to meet the intent of the code. The key issue in these cases is whether or not the substitution will provide an equal or greater level of protection than what is required by the code. Fire inspectors must be aware of what latitude they have, if any, to make judgments on equivalency issues. The inspectors must also know what requirements exist in the code that govern judgment in determining equivalency or the use of alternatives. Most model codes have language built into them that addresses this issue.

In most jurisdictions, fire inspectors themselves have little authority to approve modification of code requirements. The inspector usually processes the requests and receives a formal interpretation from a superior inspection bureau officer, a staff fire protection engineer, a contract fire protection consultant, or an appeals board.

To receive consideration for a code modification, the applicant must generally make a formal written request to the authority having jurisdiction. The appropriate office then analyzes the request to ensure that the general intent of the code is being observed and that public safety is maintained. Following the decision of the official, the applicant receives a signed copy of the decision. Detailed records of the decision are generally kept available in the fire prevention bureau office. An example of a request for the modification of a fire code is provided in Figure 1.7.

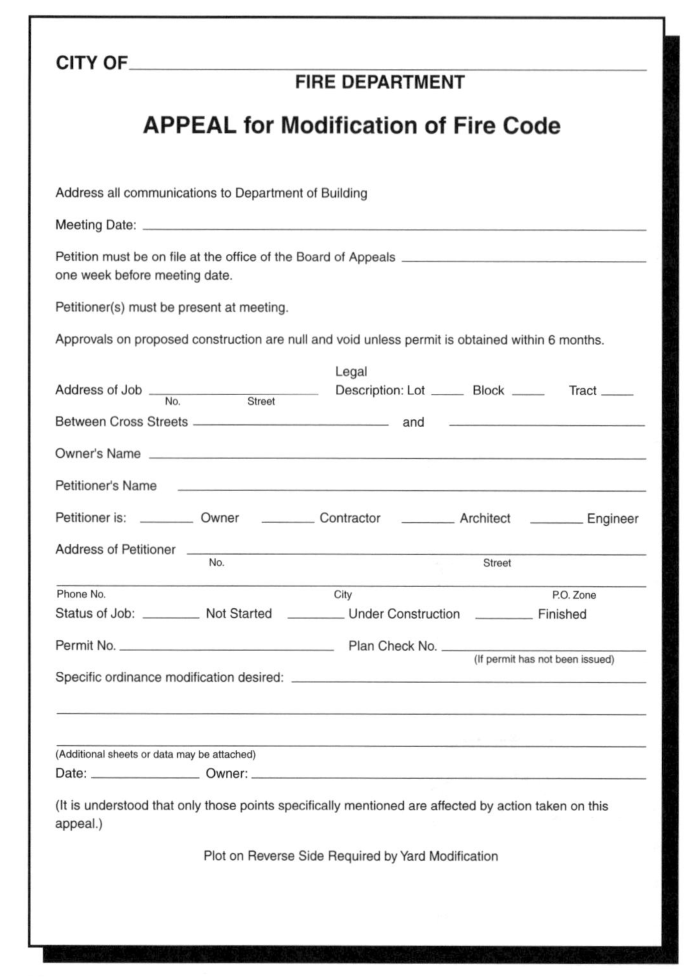

CITY OF ____________________

FIRE DEPARTMENT

APPEAL for Modification of Fire Code

Address all communications to Department of Building

Meeting Date: ____________________

Petition must be on file at the office of the Board of Appeals ____________________
one week before meeting date.

Petitioner(s) must be present at meeting.

Approvals on proposed construction are null and void unless permit is obtained within 6 months.

Address of Job ______ No. Street   Legal Description: Lot ____ Block ____ Tract ____

Between Cross Streets ____________________ and ____________________

Owner's Name ____________________

Petitioner's Name ____________________

Petitioner is: ______ Owner ______ Contractor ______ Architect ______ Engineer

Address of Petitioner ____________________ No. Street

Phone No. City P.O. Zone

Status of Job: ______ Not Started ______ Under Construction ______ Finished

Permit No. ____________________ Plan Check No. ____________________ (If permit has not been issued)

Specific ordinance modification desired: ____________________

(Additional sheets or data may be attached)

Date: ____________ Owner: ____________________

(It is understood that only those points specifically mentioned are affected by action taken on this appeal.)

Plot on Reverse Side Required by Yard Modification

Figure 1.7 Most jurisdictions have a form that can be filled out when a citizen requests a modification of a code regulation in his community.

If the applicant feels that the decision reached was unfair or that the fire prevention code was enforced unfairly or misinterpreted by the enforcement officer, the individual may file a request with the Board of Appeals. Most codes establish an appeals procedure with a Board of Appeals or other body empowered to interpret the code and issue a ruling. The Board of Appeals usually consists of three to seven members who have previous experience in the field of fire prevention or building construction. The exact number of members and their professional qualifications are specified by the adopted code. Fire inspectors need to understand the appeals process and the workings of the Board of Appeals.

Some issues involving the appeals process and the Board of Appeals that the fire inspector must be particularly familiar with include:

- Whether or not the fire inspector can continue to enforce codes on the property during the appeal process.
- Whether or not a victory by the property owner will affect the way that the fire inspector enforces that section of the code or ordinance for other properties in the future.
- Whether or not further action is required. An example of this might be asking the Board to clarify whether it has granted a general or one-time variance.

A one-time variance means that the decision is binding only for this particular circumstance and may not be directly applied to other similar situations. When the Board grants a general variance by reason of equivalency, fire inspectors must implement this ruling the same way with future code enforcement. If the Board of Appeals rules that the code is too vague for enforcement, the fire inspectors must take steps to ensure that the vagueness in the code is removed.

Adopted code regulations usually specify a time limit that the property owner has to submit an appeal. The period of seven days from the time of the inspection is a common figure, but this will vary in different municipalities. The Board will then accept or reject the appeal and file an appropriate notice of denial or acceptance. Rules and regulations used by the Board during its hearing generally are made public. A schematic of a typical appeals process is provided in Figure 1.8.

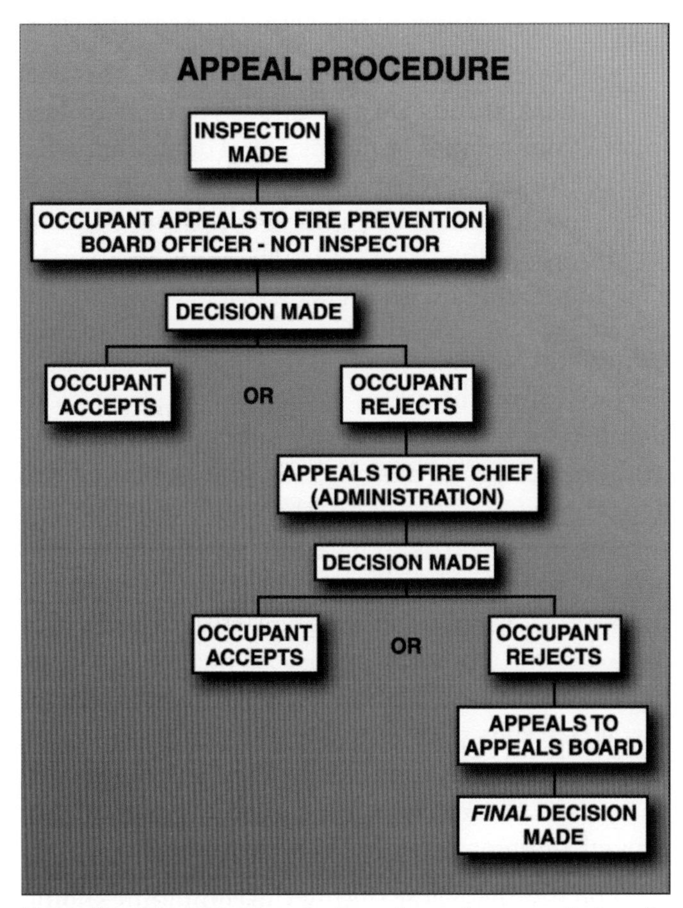

Figure 1.8 This chart shows one type of appeals process that can be followed within a jurisdiction.

Enforcement Procedures

The fire inspector must fully understand the steps in the process of gaining compliance and in taking a person to court if necessary. In assessing the statutes of the local jurisdiction dealing with code compliance, the inspector should know the answers to the following questions:

- Is noncompliance with the code a violation of criminal or civil law?
- What have been the major methods of achieving compliance?
- What inducements are applied for voluntary compliance?
- What is the nature of penalties applied to ensure compliance?

The actions taken to ensure code compliance vary from jurisdiction to jurisdiction and are dictated by local codes. It is important for inspectors to fully understand the enforcement procedures to be followed and to keep complete, accurate records of all actions taken. Notifying the responsible party in writing of the violations found is usually the first step taken in the enforcement procedure. This step is followed by a follow-up inspection after a predetermined amount of time. If the violation has not been corrected within this amount of time, most jurisdictions issue some type of sanction. This sanction may be in the form of a citation, fine, court summons, or stop-work order. The procedures for issuing sanctions differ; however, each jurisdiction should have a written procedure that details the process. Appendix B contains an example of a citation procedure.

Taking Cases to Court

Because of the legal implications involved with fire inspections and code compliance work, on occasion a sanction taken against a property owner will end up as a court case. This usually will occur when the property owner has exhausted all other means of appeal and still is not satisfied with the outcome. As a result, the fire inspector and any other fire department personnel who were involved in this situation may find themselves involved in the legal proceedings needed to adjudicate the complaint to a final outcome.

Fire inspectors may be placed in the position of being either a witness or a courtroom advisor to the prosecution. In most cases, the fire inspector will find himself in the position of doing a little of each of these. As an advisor, the fire inspector can assist the prosecuting attorney with information about the fire ordinances, technical terms, and the facts of this particular case. In order to preserve the case and be as helpful as possible, the fire inspector should heed the suggestions in the following list regarding courtroom procedure and behavior:

- Reinspect the facility the day before the trial. Take new notes and photographs to reinforce the case.
- Go over your testimony with the prosecutor before entering the courtroom. This is perfectly acceptable and recommended. Resist any attempt by superiors or prosecution personnel to encourage you to modify your testimony.
- Appear in proper uniform or be neatly dressed because appearance is important (Figure 1.9).
- Confine testimony to the facts of the case. Avoid hearsay (information from a third party) and irrelevant statements.
- Remain impartial and be especially careful not to give the impression that you have a personal dislike of the defendant.
- Limit the information that you provide to only that which is necessary to answer a given question. *Never volunteer information that has not been requested.*
- Make your response as brief as possible, while conveying all the information necessary. If the answer calls for a simple *yes* or *no* answer, do just that. However, if you feel that you need to explain your answer, you have the right to request the court's permission to do so.
- Do not attempt to give the court the impression that you are a "know-it-all." No one knows the answers to every question. If you are presented a question to which you are not certain of the answer, do not attempt to venture a guess at the answer. Simply state you are unable to answer that question.
- Make sure that all physical evidence, exhibits, photographs, notes, reference materials, and other materials pertinent to the case are brought to court for use as needed. Notes are often invaluable in refreshing your memory of important facts in the case.
- Never become argumentative on the witness stand. Do not allow yourself to lose your temper or professional demeanor with the attorney examining you. Anticipate personal attacks or challenges to diminish your credibility. Remember to simply answer the questions with facts.

Figure 1.9 Inspectors must be prepared to testify in court proceedings that involve code issues.

Complaint Handling

Part of maintaining a successful fire prevention program is allowing the public to be an integral part of it. In many cases, the public or fire department personnel in the field will identify serious fire hazards in locations the fire inspector has yet to have the opportunity to inspect. Many times a member of the public will call the fire department to report a suspected hazard that they have encountered. Fire inspectors must treat concerned citizens with courtesy and express genuine interest in their complaint. This approach will encourage the citizens continued support of fire prevention measures in the community.

The fire inspector taking the complaint should record all pertinent information. Departmental complaint forms are useful in helping to assure that all the data collected is complete and correct. Complaints that do not require immediate attention can be routed through the normal channels for processing and handling. Complaints that appear to be serious in nature and deserving of immediate corrective action should be immediately forwarded to the appropriate fire department officer for further action.

When investigating a complaint, fire inspectors do not need to give the owner advance notice. In fact, there are many situations where it is best not to give advance notice of a complaint inspection. An example would be a citizen calling to complain about chained and padlocked emergency exits at a nightclub (Figure 1.10). If the management of the theater knew that the fire inspector was coming out to check on this, they certainly would have the exits opened before the inspector got there. This would not allow the inspector to prove that the theater had not been complying with the law prior to his arrival.

The inspector should show appropriate identification and explain the purpose of the inspection. If the inspector does find code violations, appropriate departmental actions should be taken to ensure that they are corrected. Inspectors should be prepared to deal with negative attitudes by the property owner or representative when they act on complaints. It is important to maintain a cool, professional demeanor, even when the person you are dealing with has not.

Figure 1.10 Fire inspectors frequently receive calls from citizens regarding locked exits and other perceived fire/life safety hazards.

RESPONSIBILITY

In researching authority and responsibility for fire inspection and code compliance, fire inspection personnel will find that there are federal, state, and local laws that all impact their duty. It is quite common for more than one of these laws to impact the same issue. In cases where there are conflicting requirements for the same issue, generally the most stringent or strict requirement will apply. However, this may not always be the case, and fire inspectors must be aware of when this is true for their particular jurisdiction. Code administration and enforcement must be built upon a foundation of current, consistent standards and codes as well as a clear assignment of code enforcement responsibility.

Current Codes and Standards

Fire inspectors need to know whether the codes and standards in their jurisdiction are up to date. Most of the model codes that jurisdictions adopt as their local code are revised on a regular basis. Three to five years is the typical life cycle of a model code edition. Unless specified by a local statute, a revised edition of the model code in effect in a community ***does not*** automatically take effect when it is released. The previous (existing) edition remains in effect until the authority having jurisdiction chooses to formally and legally adopt the newer edition. For example, a jurisdiction has been enforcing the 1993 edition of a particular model code. In 1996, the model code organization releases a new edition of the code. Even though the new edition is available, the jurisdiction in question will continue to enforce the 1993 version until its city commission chooses to adopt the 1996 edition. It is not uncommon for an old edition to continue to be enforced for several years after the new edition has been released. On the other hand, some jurisdictions have legislation in place that allows the new code to automatically be adopted when it is released. The inspector must be familiar with how his jurisdiction handles this situation.

In addition to keeping current with the jurisdiction's fire code, the inspector must also keep current on codes (building, plumbing, etc.)

that other agencies enforce within the community. Interaction with these other agencies will also allow the fire inspector to be made aware of developments within the jurisdiction that otherwise may have gone unnoticed for some time. Staying aware of changes in occupancies is a challenge in terms of "keeping current" with codes in a community or jurisdiction (Figure 1.11). Having a good exchange of information with other agencies, such as the building department or electrical inspectors, is one of the positive steps that fire inspectors can take to overcome this problem. Other steps include monitoring business license applications, monitoring business and occupancy permits issued, and conducting an annual occupancy inventory survey. If fire inspections are tied to the issuance of licenses and permits, fire inspectors should have a notification mechanism that keeps them aware of business and commercial changes in the community.

Figure 1.11 Inspectors must stay aware of changes in occupancy involving structures in their jurisdiction. An example would be this former clothing factory that has been converted into apartments.

Consistent Codes and Standards

As stated previously, codes and standards enacted at a variety of governmental levels may conflict with each other. Fire codes and standards that include fire-related considerations must be consistent to avoid confusion and duplication of effort. In addition, the interpretation of fire-related codes must be clear and consistent with the purpose and requirements of the fire codes themselves. The following are examples of other codes and regulations that may impact fire code enforcement:

- Housing codes
- Zoning ordinances
- Subdivision regulations
- Building codes
- Electrical codes
- Mechanical codes
- Gas codes
- Transportation regulations
- Health regulations
- Plumbing codes
- Life safety codes
- Insurance codes and regulations
- Handicapped regulations (such as the *Americans With Disabilities Act* [ADA] in the United States)
- Municipal engineering ordinances

The assignment of which agencies will enforce which codes in a certain community must be very clear and consistent. Lack of communication between agencies can undermine public safety, which is the agency's primary goal. Most codes allow the authority having jurisdiction some degree of latitude in enforcing the code provisions, in terms of interpretation, judgment, or both.

The willingness to communicate and share information on the part of fire inspectors becomes a vital element in code enforcement. Just as vital is the ability to produce creative, alternate solutions. Fire inspectors should consider this a major priority in maintaining an effective fire inspection program.

Alternate solutions are the basis for a growing form of code development commonly called *performance-based codes and standards*. Performance-based codes and standards are just as the name implies — requirements based on specific performance outcome objectives and specific installation and construction techniques. These are combined with acceptance testing criteria that must be successfully performed to prove the effectiveness of the assembly.

For example, suppose a traditional code specifies that a roof assembly must use a minimum of 2 x 10-

inch (50 mm by 250 mm) lumber for its main structural support to carry a specific load. A performance-based code would not specify the size of the lumber to be used. The performance-based code would only require that the roof assembly be capable of supporting the required load. The architect/engineer who designs the structure would be responsible for proving that the design he intends to use will meet the load requirement.

The difficulties associated with enforcing performance-based codes should become evident when you consider the previous example. The authority having jurisdiction must develop acceptance testing criteria, methods of evaluating data, and procedures for approving the outcomes of the testing. Significant technical expertise is required to administer and evaluate these various criteria. In most cases, the expertise required to adjudicate these matters is beyond the ability of fire inspectors. Professional engineers are often required to handle these matters. Many jurisdictions lack the personnel on staff to handle these matters.

Another problematic issue with the enforcement of performance-based codes is how the authority having jurisdiction mantains this same performance criteria over the life of the building or facility. In order for the safety of a building or facility to remain constant, more frequent and complex inspections, evaluations, and testing will have to be performed. This places great demands on the inspection agency. Inspectors in jurisdictions using performance-based codes must explicitly understand their role in the enforcement process and know what things they are or are not allowed to do.

TYPES OF LAWS AND PERMITTING PROCESSES

As stated previously in this chapter, personnel who perform fire inspections must be familiar with a myriad of different laws, ordinances, and other legal regulations. This section details the various types of federal, state/provincial, and local laws that fire inspectors may have to deal with. In addition to laws, fire inspectors may be involved in issuing various types of permits for their jurisdictions. The types of permits and permit processes are also covered in this section.

Federal Laws

Many federal agencies have set forth regulations designed to ensure the safety of the public. These regulations cover a broad spectrum of activities and include such matters as employee safety, transportation of hazardous materials, patient safety in health care facilities, access issues for handicapped citizens, and minimum housing standards. The federal agency that sets standards in any particular area typically is also responsible for enforcing them. However, in some cases, such as with workplace safety laws, the state or province may choose to enforce federal regulations, as opposed to the federal agency doing it.

In most cases, the local fire inspector is not responsible for enforcing federal regulations. However, if the local fire inspector becomes aware of hazards or violations within federal properties or jurisdiction, he should know how to report them to the proper authority to see that they are corrected.

Federally (and some state) owned buildings located within the local jurisdiction are not required to comply with local codes (Figure 1.12). In the past, agencies that operated these buildings usually enforced their own fire protection regulations. In recent years, the government has chosen to follow local codes in its facilities. Fire inspectors should get to know the people who are responsible for federal properties and work with them to ensure that fire protection is maintained at a high level, regardless of which code they follow. Even if the fire department does not have the authority to perform code compliance inspections in a federal

Figure 1.12 Almost every community has at least one federally owned or controlled building, such as this post office, within their jurisdiction.

facility, it still should perform prefire planning visits and company walk-throughs so that it is prepared to handle any fire that may occur within the facility.

State/Provincial Laws

In addition to enforcing selected federal laws as previously described, states/provinces are empowered to enforce state laws and statutes. The state/provincial government may also regulate specific fire inspection activities within its jurisdiction, as well. For example, in some states or provinces, the duty to inspect nursing homes and day-care centers is specifically assigned to the state fire marshal's office.

State and provincial laws can define and specify building construction and maintenance details in terms of fire protection, and empower agencies to issue regulations. State labor laws, insurance laws, and health laws also have a bearing on fire safety and sometimes encompass fire inspection responsibilities. It is the fire inspector's responsibility to communicate the authority of the local jurisdiction to other agencies with fire-related concerns. This kind of communication can form the basis for an effective working relationship between local and state or provincial agencies.

Local Laws and Ordinances

Local laws and ordinances, although sometimes based on state laws, are more specific and tailored toward the exact needs of the county, municipality, or fire protection district they are adopted by to protect. Typically, states or provinces have legislation in place that enable local jurisdictions to adopt state/provincial regulations. The regulations may be adopted by the local jurisdiction by reference or in the form of enabling acts. To adopt by reference means that the local jurisdiction will follow the state/provincial laws exactly as drawn. Adopting them in the form of enabling acts gives the local jurisdiction the use of state/provincial laws as their basis but then add or delete regulations or ordinances based on local needs or preference.

Many communities adopt a code from one of the four major model fire code organizations for use in their community. These are:

- *National Fire Codes®,* published by the National Fire Protection Association (NFPA)
- *BOCA® National Fire Prevention Code*, published by the Building Officials & Code Administrators International (BOCA)
- *Uniform Fire Code™,* published by the International Conference of Building Officials (ICBO)
- *Standard Fire Prevention Code*, published by the Southern Building Code Congress International (SBCCI)

Each of these model code organizations has a variety of other codes in addition to their fire codes. These include building, plumbing, electrical, and other codes. Each jurisdiction may actually have adopted several of the different codes from one of these organizations.

Laws and ordinances should be kept current and should meet growth and changes in the community. Fire inspectors should be thoroughly familiar with the requirements of both current and past laws that may apply to their fire inspection duties. They should also be involved in updating those laws. Fire inspectors can use the model codes to help structure more adequate and effective fire prevention and inspection requirements.

Fire safety regulations generally fall into one of three categories:

- Those that govern the construction and occupancy of a building when it is being planned and constructed (Figure 1.13)

Figure 1.13 Some fire and safety regulations are enforced prior to and during the construction of a new building.

- Those that regulate activities that are conducted within a building once it has been constructed (fire prevention code regulations)
- Codes for the maintenance of building components

Fire inspectors should be actively involved in all three categories. It is, of course, easier to gain compliance with safety requirements during a building's planning or construction phase than it is to require retroactive implementation of fire protection measures after the building has been occupied. Gaining compliance before a structure is occupied requires a close working relationship with the members of the local building, zoning, and planning departments. If such a relationship does not exist, the fire inspector should make it a priority to establish and foster these relationships.

Types of Permits

Most local agencies have ordinances that provide for the issuance of permits for special operations and conditions within their jurisdiction. A permit is an official document that grants a property owner or other party permission to perform a specific activity. Typically, permits are used to control:

- Maintenance, storage, or use of hazardous products
- Hazardous operations
- Installation/operation of equipment in connection with the first two activities

Permits should only be issued if the condition being permitted meets applicable code requirements or regulations. In general, permits are not issued to allow the party to disregard or exceed code requirements in any manner.

The purpose of the permitting process is twofold. First, by requiring occupants or property owners to obtain permits when they are seeking to install, store, or use hazardous products or operations, fire inspection personnel will have the opportunity to ensure that these conditions will meet the applicable code requirements. This allows fire personnel to devote specific attention to potential target hazards within their response area. Secondly, and perhaps an even more basic reason for permits, the permitting process should ensure that no hazardous situations develop within the jurisdiction of which fire department personnel are not aware.

Each of the model codes that a jurisdiction may adopt have requirements in them for permits and permitting processes. Each of the codes list specific activities, operations, practices, or conditions that require permits. Local governments may add to or delete from these lists based on local needs. Examples of situations requiring permits according to the model codes include, but are not limited to, the following:

- Aerosol products
- Aircraft refueling vehicles (Figure 1.14)
- Asbestos removal
- Bowling pin or alley refinishing
- Candles or open flames in assembly areas
- Compressed gas operations
- Explosive or blasting operations
- Use of fire hydrants or water system valves
- Fumigation or insecticidal fogging
- Temporary displays in malls
- Open burning operations and bonfires
- Parade floats
- Tents, canopies, and temporary membrane structures (Figure 1.15)

Figure 1.14 Some codes require aircraft refueling vehicles to have special permits.

Figure 1.15 Many communities enact codes that require special permits for large tents or canopies.

- Tire storage
- Welding and cutting operations

Permits are usually issued for a specific condition, at a specific location, and for a specific period of time. They are not transferable beyond the conditions stated on the permit. The permit authorizes, by law, the right of entry for the fire inspector at anytime to ensure compliance with code requirements and the conditions of the permit.

The Permit Process

The permit process begins with the occupant or property owner recognizing the need to get a permit for an operation or condition he wishes to create on the property. Many citizens will not be totally sure of the permit requirements for their municipality. Thus, fire inspectors and other code enforcement personnel will frequently be contacted by citizens to determine whether or not they need a permit. The inspector should advise them whether or not a permit is needed and explain the permit process. The inspector should emphasize that this conversation does not give them permission to begin engaging in their activity.

Beyond inquiring about the need to obtain a permit, the first step in the permit process is for the property owner to obtain a permit application. Most jurisdictions have a specific form that the applicant must complete (Figure 1.16). This form may be obtained from the fire inspector, other code enforcement personnel, or a designated municipal office. Depending on the type of permit being sought and local requirements, the applicant may be required to submit additional documentation along with the application. This additional documentation may include:

- Shop drawings
- Construction documents
- Plot diagrams
- Material safety data sheets (MSDS) or other chemical documentation

Typically, there will also be a permit processing fee that is required to be paid when the application is submitted. The amount of the fee will vary from jurisdiction to jurisdiction. Usually, this fee is nonrefundable whether or not the permit is actually granted. This may vary depending on local practices.

Once all proper documentation has been submitted by the applicant, the fire inspector will need to examine this documentation. The jurisdiction may specify a time frame that the inspector has to review the application. The first step for the inspector is to make sure that the application has been filled out properly and all required supporting documentation has been provided. If this is not the case, the application can be immediately rejected, or the applicant can be requested to submit the necessary additional information. If the paperwork appears to be in order, the fire inspector may issue the permit immediately or perform a site visit to further investigate the request. If the site visit shows everything to be acceptable, the permit is then issued. Once issued, the permit must be kept at the location for which it was issued and be readily available for inspection by the fire inspector.

The permit must explicitly explain the conditions under which it has been issued. This includes what actions are being allowed, guidelines that must be followed, and the time frame for which the permit is applicable. Most jurisdictions have time limit requirements built into the ordinances that give them the authority to issue permits. The time frames for which permits will be effective will vary depending on the circumstance. For example, permits for fuel handling vehicles at an airport may be issued on a yearly basis. However, a permit to burn a brush pile may be given with a one-week time limit to complete the process. Each jurisdiction must also have procedures for granting extensions.

The fire inspector has the authority to revoke a permit if, upon inspection, it is noted that the stipulations within the permit are not being followed. Also cause for revocation is an inspection turning up the fact that false statements or misrepresentation of the actual conditions on the property or documentation were made. Depending on the problem, the permit may be revoked permanently or temporarily until such time that the condition is corrected.

APPLICATION FOR BUILDING PERMIT

VILLAGE OF MOUNT PROSPECT

COMMUNITY DEVELOPMENT - BUILDING DIVISION

All information below must be filled in prior to submission.
(Please Print)

Applicant's Name ______________________ Permit No. ______________

Project Address ______________________ Date Issued ______________

Real Estate Index No. ____________ Lot No. ________ Valuation of Work ______________

☐ Single Family Residential ☐ Multi-Family Residential ☐ Commercial ☐ Fire Alarm ☐ Other ________

☐ Industrial ☐ Garage ☐ Fire Suppression ☐ Institutional

	Name	Address	Phone	Office Use Only		
				B/L	Bond	Reg.
Owner			H W			
Architect			H W			
General Contractor			H W			
Excavating						
Concrete						
Carpenter						
Mason						
Drywall						
Painting						
HVAC						
Fire Suppression						
Electrical						
Plumbing						
Roofing						
Fire Alarm						

No error or omission in either plans or application, whether said plans or application has been approved by the Building Division or not, shall permit or relieve the applicant from constructing the work in any other manner than that provided for in all the ordinances of the Village of Mount Prospect relating thereto. I will submit this work to the required inspections and prohibit the occupancy of any space until a Certificate of Occupancy has been obtained from the Director of Community Development. The applicant having prepared and read this application and fully understanding the intent thereof declares that the statements made are true to the best of his ability, knowledge and belief. Construction to be started within 60 days and completion one (1) year of date of application. (See Section 21.203 (d) (1)(2)(3)(4)(5)(6)(7)(8)(9)).

Signature ______________ Title ______________ Date ______________

Address ______________ City ______________ State ______________ Zip ______________

Permit Authorized ______________ by ______________
Date Community Development Director

WHITE COPY - OFFICE YELLOW COPY - FILE PINK COPY - INSPECTION GREEN COPY - TOWNSHIP GOLD COPY - COUNTY

Figure 1.16 A typical permit application form. *Courtesy of the Mount Prospect (IL) Fire Department.*

ORGANIZATIONAL STRUCTURE

There are two different organizational structures that may affect the inspection activities of fire inspectors:

- The structure of the fire department or code enforcement division
- The structure of other related, outside agencies, such as departments of planning, zoning, streets, and sanitation

Fire/Code Enforcement Department Structure

To function effectively, the fire inspector must understand the structure of the organization he functions within. In most communities, fire code enforcement is performed by the fire department. It may be done by company inspections, inspection/ fire prevention bureau activities, or a combination of the two.

The fire inspection or fire prevention bureau within a fire department is typically managed by a fire marshal. In most fire departments, the fire marshal is a chief officer with battalion, assistant, or deputy chief rank (Figure 1.17). Typically, the fire marshal reports directly to the fire chief. On very small fire departments, the fire marshal may be the only full-time member of the department assigned to fire prevention. In larger fire departments, the fire marshal may have inspectors, plans reviewers, fire protection engineers, public fire education specialists, and other personnel assigned to the bureau. Personnel assigned to a fire prevention bureau may be either sworn members of the fire department or civilians. Many larger fire prevention bureaus have a combination of both.

The use of in-service companies to perform fire inspection activities will require coordination between the fire prevention bureau and the operations division of the fire department. Typically, general directions on required inspection activities will be given from the fire prevention bureau to the officer in charge of the operations division. Delegation of inspection duties to the companies will then be filtered down through the command structure of the operations division.

In some jurisdictions, the task of performing fire inspections and fire code enforcement is relegated to some department other than the fire department. The municipality may have a separate code enforcement department or a building department that enforces the fire code. In these situations, the fire inspectors usually will be civilian employees of these departments. In municipalities where this is the case, the fire department must maintain an excellent liaison with the code enforcement people. This will ensure that all the operational concerns of the fire department are addressed by the code enforcement program.

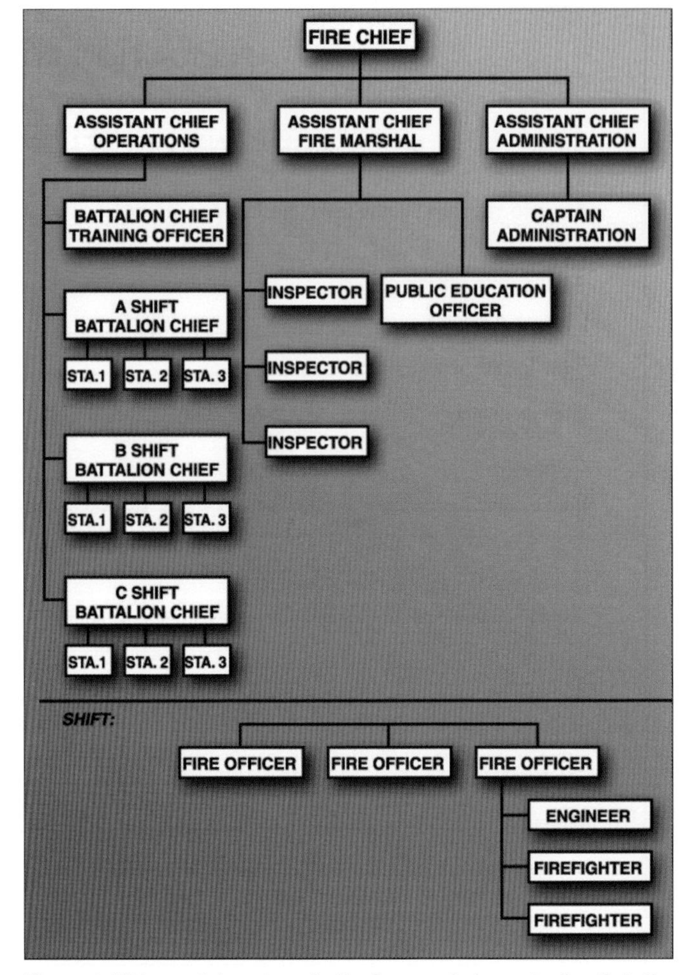

Figure 1.17 In most departments, the fire marshal reports directly to the fire chief.

STRUCTURES OF RELATED AGENCIES

As stated earlier in this chapter, there are a myriad of code enforcement agencies with which the fire inspector may have to interact. These may include state/provincial agencies such as the state/ provincial fire marshal or Department of Labor. The fire inspector will also have to interact with other local agencies such as the building or zoning departments. The fire inspector must have clear understanding of the lines of authority between each agency. The fire inspector must also know the method for reporting important information between his agency and the other agencies.

In cases where there are conflicts on code issues between more than one agency, the fire inspector must understand the process for resolving the issues. Typically, the agency with the most stringent requirement will have jurisdiction. However, this is not always the case. Thus, the fire inspector must understand the procedure for making sure these situations are handled in a proper and legal manner.

2

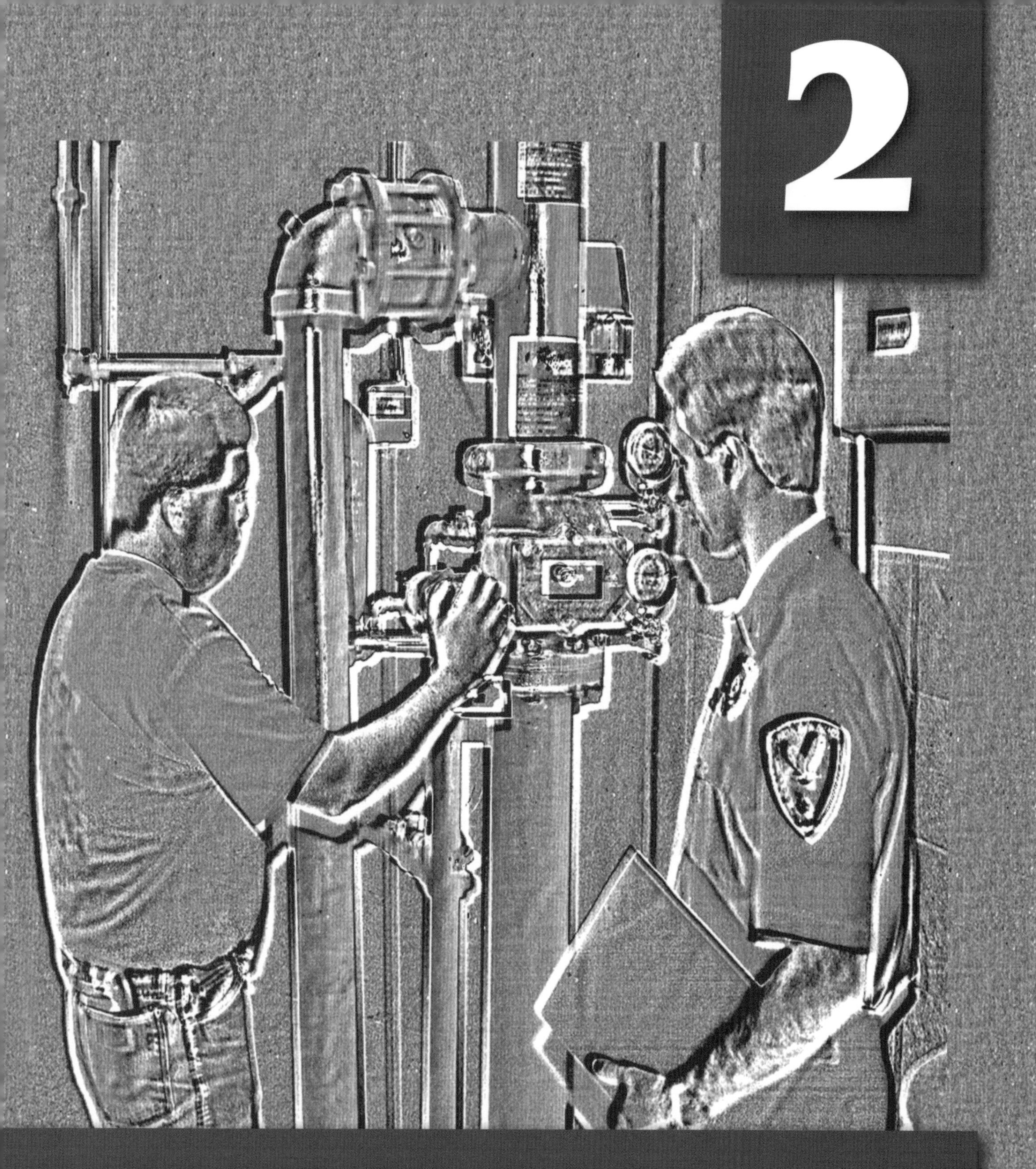

Inspection Procedures

This chapter contains information that will assist the reader in meeting the listed job performance requirements contained in NFPA 1031, ***Standard for Professional Qualifications for Fire Inspector and Plan Examiner*** (proposed 1998 edition).

Chapter 3 Fire Inspector I

3-2.1 Prepare written correspondence to communicate fire protection and prevention practices, given a common fire safety issue, so that the correspondence is concise, accurately reflects applicable codes and standards, and is appropriate for the intended audience.

3-2.2 Prepare inspection reports, given observations from a field inspection, so that the report is clear and concise and accurately reflects the findings of the inspection in accordance with applicable codes and standards.

3-2.6 Maintain files, given inspection reports, complaint investigations, and related documents, so that information can be easily retrieved and is filed in compliance with the record-keeping policies of the organization.

Chapter 4 Fire Inspector II

4-2.1 Develop written correspondence to communicate fire protection and prevention requirements, given a complex fire safety issue, so that the correspondence reflects the research and accurate interpretations of applicable codes and standards and is appropriate for the intended audience.

4-2.2 Create inspection checklists and forms, given applicable codes, standards, and policies and procedures of the jurisdiction, so that the materials developed are clear and concise and key issues are addressed.

4-2.8 Recommend policies and procedures for the delivery of inspection services, given management objectives, so that inspections are carried out in accordance with the policies of the jurisdiction and due process of the law is followed.

Chapter 2
Inspection Procedures

The benefits of pre-incident planning and standard operating procedures have been plainly established in the fire and emergency services. Often, post-incident critiques of routine and major incidents show that one of the major factors in the successful handling of an incident is the preparation done by suppression forces before being faced with an incident. The other major factor commonly credited for the success is the tendency of personnel to follow the established incident plan.

Just as frontline fire suppression personnel must plan and follow standard operating procedures to successfully handle emergency incidents, so must fire inspectors. Preparing for an inspection, following a standard inspection procedure, and using good interpersonal skills are all necessary if the fire inspector is to perform the job consistently and effectively.

This chapter covers the basic principles of performing an effective fire inspection. These basic principles include:

- Interacting with people
- Preparing for the inspection
- Conducting the inspection
- Performing a follow-up inspection
- Maintaining inspection files and records

INTERACTING WITH PEOPLE

There is no doubt that the key to success in virtually every aspect of life is the ability to effectively communicate and deal with people. This is certainly crucial for fire inspectors because often the people being dealt with are not particularly happy about the fire inspector's presence in their occupancy (Figure 2.1). The fire inspector should view the inspection as an opportunity to win over the property owner to ensure that his property is maintained in a safe condition. This may require the fire inspector to be a "salesperson," but in the long run this is a very effective method. It is much easier to sell the occupant on the idea of voluntarily making a change than it is to use legal methods to force the change.

Figure 2.1 Inspectors frequently encounter people who are unhappy with the inspection process.

The fire inspector has an opportunity to educate the occupants on important safety issues while performing the inspection. When conducted correctly, a fire inspection is just as much a public fire education program as it is a code enforcement program. Local officials should look at their fire inspection programs in this manner and make sure that the people assigned to carry out the inspections are capable of taking this approach with the public. Only a very small percentage of the public will ever deal with the fire department during an emergency; their main contact will be through inspection activities. Therefore, the image portrayed by fire department personnel performing an inspection will have a significant effect on the overall public perception of the fire department. The fire inspector should be capable of conveying to the public why the codes are written, what the intent of the codes are, and give historical recounts of tragic fires as examples of what can happen.

Making a positive first impression with the occupant is always an important step in getting the inspection off to a good start. The fire inspector should approach the occupant in a friendly but businesslike manner. If departmental policy requires the fire inspector to wear a uniform, he should wear the complete uniform. If the inspector is allowed to wear plain clothes, he should have a professional appearance. Some sort of official identification should be visible or shown to the occupant on first meeting (Figure 2.2). Regardless of whether the inspector is wearing a uniform or plain clothes, all clothing items should be neat, clean, and in good repair.

As the inspector, you should immediately greet the occupant and identify yourself and the purpose of your visit in a friendly manner. For example, *"Good morning. I am Inspector Jim Heimbach with the Pennsburg Fire Department. I have a two o'clock appointment with your manager to perform our annual fire inspection."* If you are not certain that the first person greeting you at the entrance is the person with whom you made the appointment or the person in charge, ask that the appropriate person meet with you.

It is important to be prompt when making a scheduled inspection. Most routine inspections are scheduled in advance so that the occupant knows when the inspector will be arriving. Make every effort to arrive at the scheduled time. If you are going to be delayed, contact the occupant and advise him of this, even if it is only a 10 or 15 minute difference. If the difference is going to be greater than that, it might be courteous to ask if the delay is too much of an inconvenience and if he would like to reschedule the inspection for a later date.

Figure 2.2 The inspector should be prepared to show a photo identification card to the occupant, even if he is in uniform.

People are more likely to respect you and be courteous to you if you treat them in the same manner. In addition to having a courteous demeanor, you should also conduct your physical actions in a courteous, respectful manner. Little things such as wiping your feet before you enter the building, walking on sidewalks instead of the grass, and carefully closing doors as opposed to slamming them will project a positive image to the occupant(Figure2.3).

Figure 2.3 Always approach the structure using the sidewalks that are provided.

The fire inspector should attempt to promote a helpful attitude toward the occupant, as opposed to a fault-finding attitude. Try not to be critical or demeaning of the occupant, building architect, contractor, or anyone else associated with the property. Even when violations or problems are found, try and convey them to the occupant in a noncritical manner. Convey your findings in a manner that gives the occupant the impression that your goal is to help him become better and safer, not to belittle him or criticize his mistakes.

The fire inspector should not accept any favors from the occupant. The occupant may think that he can improve his chances of not being cited for violations by offering you a soda, cup of coffee, or other "friendly" inducements (Figure 2.4). You

Figure 2.4 Inspectors should never accept gifts of any kind from an occupant.

should kindly refuse such offers. In fact, you must be careful to avoid any behavior that shows favoritism toward an occupant. Do not interpret the code too liberally for one occupant and then be very strict on the same point with another. Business owners do have a tendency to talk to each other and any variations in treatment will soon become public knowledge. You must always be able to justify your actions.

The most difficult aspect of dealing with people in an inspection setting is the need to discuss problems or violations that have been encountered during the inspection. When handled in a tactful, professional manner, the inspector can make these discussions very positive. The first thing that you must be sure of before beginning a discussion of this nature is that you are correct on the points you are about to make. Nothing erodes the occupant's confidence in the inspector more than to catch a mistake that the inspector has made. Do not assume that all occupants are ignorant of the codes and that they can be fooled by any argument you make. You should make an effort to hear and understand the occupant's side of the story. Even if it does not change your decision on the violation, it still gives the occupant a feeling that you are fair and concerned. You should fully explain the problems that have been noted and offer solutions for their correction.

When trying to communicate with an unhappy or difficult occupant, try to make the situation as calm as possible. Never argue with a disgruntled individual. If unable to continue a discussion with the occupant in a rational, calm manner, you should advise the occupant that the matter is going to be passed to a higher authority and then leave the property.

PREPARING FOR AN INSPECTION

It cannot be overemphasized that the preparation phase is one of the most important parts of the inspection process. Inspectors who spend an adequate amount of time preparing for the inspection are much more likely to be successful in the long run than ones who "fly by the seat of their pants."

The preparation phase begins by scheduling the inspection. Most fire inspections are conducted with little or no advance notice to the property owner. You must use common sense in the manner in which inspections are scheduled. The amount of time it takes to inspect a given occupancy will have many variables, including:

- Size of the occupancy (Figure 2.5)
- Complexity of operations within the occupancy
- Whether it is a preliminary inspection or a follow-up inspection
- Location of the occupancy
- The inspector's familiarity and experience with the occupancy

You must also use common sense and professional judgment when determining how many occupancies can be scheduled in a normal work day. Some facilities may take a whole day to inspect, while several small occupancies may be able to be

Figure 2.5 Large occupancies will require considerable time for a thorough inspection.

inspected in the same day. Attempt to rely on experience when scheduling inspections. Try not to schedule more than can be realistically accomplished during the allotted time.

When scheduling multiple occupancies in the same day, give consideration to the geographic location of the inspection sites. Try to schedule occupancies that are in the same general area and that require a minimum travel time between them. Try to schedule the visits in a manner that allows for a logical route from one place to the next (Figure 2.6). Much time can be wasted if you are required to "hopscotch" all over the jurisdiction to complete inspections.

Figure 2.6 Inspectors may be able to inspect multiple small occupancies in the same day if the occupancies are close together.

Once contact has been made with an appropriate representative of the occupancy, the fire inspector should record the appointment time, contact's name, address, and phone number on the inspection schedule. The fire department may have inspection schedule forms to use or the inspector may simply use a date book or time planner.

Once the inspection has been scheduled, the next thing for the fire inspector to do is to mentally prepare for the visit to the particular site. There are a number of things that should be done before making the visit. If the occupancy has been inspected in the past, previous inspection records should be reviewed. This will allow you to become familiar with the occupancy if you have not been there before or refresh your memory if you were there a long time ago. Pay particular attention to information on special hazards or previous violations that were noted in the past. Noting information from previous inspections will also alert you to important changes in occupancy if the same conditions are not found on the current visit.

If the occupancy contains structural elements, manufacturing processes, or operations that are not commonly encountered by the inspector, time should be spent reviewing applicable sections of the local code prior to making the visit. This will aid you in being thorough and accurate when assessing the hazards at the occupancy.

The inspector should also make sure that he has assembled all of the necessary tools and equipment prior to leaving for the inspection. The actual tools and equipment needed will vary with the jurisdiction, type of inspection being conducted, and the inspector's preference. However, the following is a general list of items commonly needed for fire inspections:

- Safety equipment such as a hard hat, eye protection, and hearing protection
- Coveralls for operating in dirty areas
- Clipboard with inspection forms and map symbols
- Electronic data-collection equipment, if used by the jurisdiction
- Materials needed to make sketches
- Measuring tape, at least 50 feet (15 m) in length
- Flashlight
- Camera with flash equipment
- Pitot tube and gauge, hydrant wrench, static pressure cap gauge, if water supply testing will be conducted
- Reference materials such as code books, building plans, and previous reports
- Cloth or wet-wipes for cleaning hands

The inspector may choose to develop a checklist or form that aids in making sure all of the preparation for doing an inspection is completed.

CONDUCTING THE INSPECTION

It is important for the fire inspector to follow an established routine when conducting an inspection. This will ensure that a thorough assessment of the occupancy is made and that all necessary information is gathered. As stated earlier, the first step in conducting an inspection is arriving at the scheduled time. In most cases, occupancies should be inspected during their normal business or operating hours. These hours will vary depending on the type of occupancy, and the inspector must be flexible when scheduling inspections (Figures 2.7 a and b). For example, a hardware store will normally be open during daytime, business hours. However, a nightclub and dance hall normally will be open only during the evening. While it may be possible to inspect many of the necessary building features of the nightclub during nonoperating, daytime hours, there will be a number of factors that can only be properly assessed during evening operations. These factors include number of people in the building and proper illumination and operation of exits.

Inspecting an occupancy during normal operating hours is especially important when making surprise inspections to check on occupant loads, open exits, and to follow up on public complaints. It will be impossible to properly evaluate these situations or cite violations unless the conditions are documented as they actually exist. For example, if the fire department receives complaints about overcrowding conditions in a local nightclub, it would make little sense for inspectors to visit the club during daytime hours. Inspectors should visit at night, when the club is open, to determine if a problem really exists.

You can make some observations regarding the exterior of the premises before entering the structure to begin the inspection. Some of the things that can be noted include:

- Make sure the building name and address are correct, and note if they are visible on the outside of the building.
- Check the exterior condition of the building for any notable maintenance or housekeeping problems (Figure 2.8).
- Record the type of construction and building height.
- Note any conditions that may affect or hinder the fire department access to the building, including parking problems, large trees, overhangs, etc. (Figures 2.9 a and b).
- Record the location of both private and public fire hydrants, fire department sprinkler/standpipe connections, fire escapes, and other items of particular interest to fire companies.

You should then enter the structure through the main public entrance. A building representative should be located as soon as possible, and you

Figure 2.7a A hardware store is usually open during the day and may be inspected during normal business hours.

Figure 2.7b The logical time to inspect movie theaters is during the evening when they are open to the public.

Figure 2.8 Cracks in walls around tension rods are one type of problem that is visible from the outside of the structure.

Figure 2.9a Security gates may provide access problems for emergency vehicles.

Figure 2.9b Streets that have been closed and barricaded to provide pedestrian walkways may limit access to fire apparatus during an emergency.

should identify yourself to that person. If the first person contacted is not the person with whom the appointment was made, ask for that person to be summoned. In the case of a surprise inspection, you should request that the manager or highest-ranking employee of the facility be summoned. That higher-ranking employee may choose to delegate the responsibility for hosting the inspector to a subordinate; however, that choice should always come from that person.

Once the person in charge has been contacted, give a short briefing describing the inspection purpose and procedure. Request that person or his designated representative accompany you on the inspection. This serves several purposes. First, it will assure you access to all parts of the occupancy. The representative must have keys necessary to access locked areas, and he will be more familiar with the general layout of the building than you will (Figure 2.10). Secondly, it will eliminate any later claims the occupant may make regarding damage, missing items, or improper procedure.

Figure 2.10 The building representative should have keys that provide access to all areas of the structure.

Fire inspectors must have a systematic approach to performing the inspection. You must establish a system of touring a building that assures all portions of the building are examined. It is common for the inspector to begin with an exterior examination of the structure. All things that could affect fire department response to the building must be inspected.

Once inside the structure, the fire inspector may choose to use one of three common methods for performing an inspection:

- Start from the roof and work downward.
- Start from the basement and work upward.
- Follow an establish manufacturing process within the structure from the raw materials' end to the finished product end, including all storage areas (Figure 2.11).

Regardless of the method used, the inspector should inspect each level of the structure in a

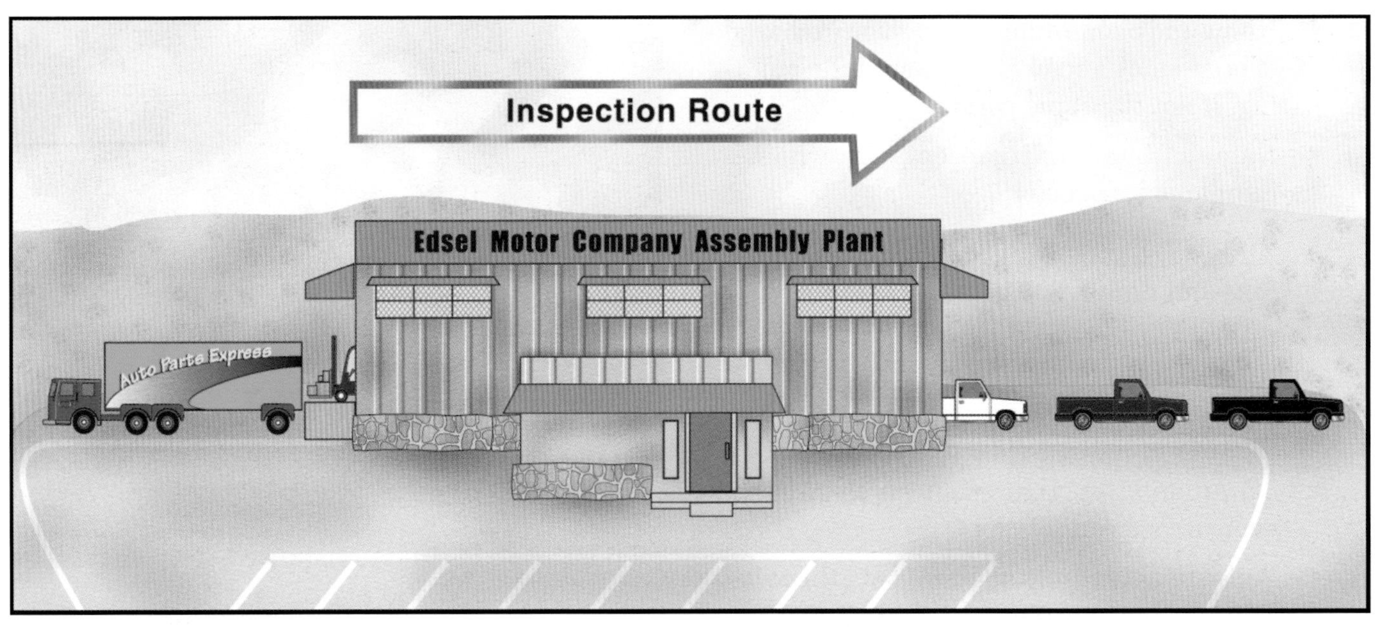

Figure 2.11 One method of choosing an inspection route is to follow a manufacturing process from start to finish.

systematic manner. Some inspectors like to use a clockwise or counterclockwise pattern to make sure all areas are covered. This is not unlike the right-hand or left-hand search pattern techniques that firefighters use when searching a structure for fire victims. If you find an area that is locked, you should _ask_ the host to unlock the area for you. If the host refuses to admit you to a particular area, you should note this and move on. Do not demand entry or stand there and argue with the host.

If the condition causing the host to prevent your entry into a certain area is based on concern for a secret business process, you may offer the host the opportunity to cover anything that you should not see. That way, you may look at the room without risking infringing on the occupant's business rights. If they still refuse, note this refusal and move on.

Photographs

Photographs are helpful when documenting serious problems or fire code violations found during the course of an inspection. Photographs, more so than sketches or drawings, are useful because they document the conditions at the time of the inspection. This is particularly helpful when dealing with violations that will be cited. Sketches and drawings establish the relative setting of the violation, but the photograph provides actual evidence of the violation's existence. This will be helpful if you need to defend the citation in a court proceeding.

Photographs may also be taken when an occupancy is found to be safe and in compliance with codes. These photographs may be kept on file. If code violations are noted in the future, the "before" and "after" photos could be very useful to inspection and code enforcement personnel.

Inspection Forms and Checklists

A well-constructed inspection form or checklist can be an invaluable tool for the fire inspector. The inspection checklist will serve as a reminder to the inspector of the common items that need to be checked during the inspection. However, the inspector must be cautious to watch for problems or code violations that are beyond those covered on the checklist. Fire departments may wish to purchase commercially produced inspection checklists or develop their own. If the fire department does not use checklists on a departmentwide basis, individual inspectors may wish to develop one for their own use. Information or blanks for responding to items that should be contained on the inspection checklist include:

- Inspection agency's name and address
- Inspector's name
- Date of inspection
- Name and address of occupancy being inspected

- Name and phone number of person in charge of occupancy and/or the owner of the building, if not the same person
- Name of building representative who accompanied inspector
- Occupancy class of building
- Size and number of stories of building
- Interior and exterior finishes
- Exit information
- Fire protection features
- All other areas that must be checked to ensure code compliance
- Space to note things that are not otherwise contained on checklist

Departments that choose to develop their own inspection forms or checklists should tailor them to the types of inspections and occupancies with which they commonly deal. This will make the inspector's job easier to follow. Appendix C contains several examples of good inspection forms and checklists.

For most routine inspections, a well-completed inspection form, along with appropriate drawings and sketches (which are discussed later in this chapter), will be the only documentation needed to record the inspection. Formal inspection reports (also discussed later in this chapter) are only required on major inspections or when serious code violations have been encountered.

Closing Interview

When the inspection is complete, the fire inspector should discuss the results with the person in charge of the property (Figure 2.12). The purpose of the closing interview is to note good conditions as well as those conditions that need correcting. You should discuss violations in general terms, indicating that specific details will be sent in a written report at a later time.

The reaction of the person you are talking to will vary. Some people view the inspection report as a positive thing, and they work immediately to correct hazards that have been noted. Most business owners accept the report as a part of the routine of doing business. Some individuals will display hostility toward your remarks and recommendations. In these instances, you should be polite but firm. Avoid arguments.

Figure 2.12 The inspector may review the preliminary results of the inspection with the occupant before leaving the premises.

In every instance, you should use the closing interview to express thanks for all courtesies extended by the occupant. If a reinspection will be necessary, inform the owner/occupant that it will be made. Inspectors must use their judgment, but for those who are reluctant to comply with the codes, it may be necessary to explain the enforcement procedure and appeals process.

Inspection Drawings

In order to make a thorough inspection, the fire inspector may be required to take the time necessary to prepare a good field sketch. The field sketch is a rough drawing of the building made during the inspection. Fire inspectors use sketches for documenting existing conditions at the time of inspection. These sketches are not to be confused with pre-incident planning. It is not important for the inspector to draw the field sketch to scale. Dimensions can be written on the sketch so that a scale drawing may be developed later. It is helpful to draw the field sketch in proportion, using a straight edge. The sketch should have all hazards and their locations specifically noted.

The field sketch is used to make the final inspection drawing. The final drawing(s) should include all the necessary building inspection information. The drawings should be done to scale and should feature standard mapping symbols (see Appendix D for standard mapping symbols). On or attached to the drawing(s) should be a legend explaining the mapping symbols. If an item is included for which there is no standard symbol, a circled numeral should be used. This numeral can be explained in an appendix attached to the drawing.

There are three general types of drawings used to show building information: the plot plan, the floor plan, and the elevation view. The *plot plan* is used to indicate how the building is situated with respect to other buildings and streets in the area (Figure 2.13). The *floor plan* shows the layout of individual floors and the roof (Figure 2.14). Most of the building detail can be shown on the floor plan. The *elevation drawing* is used to show both the number of floors in the building and the grade around the building (Figure 2.15).

On smaller occupancies, you can often put all the required information on one drawing. However, as building size and complexity increase, the

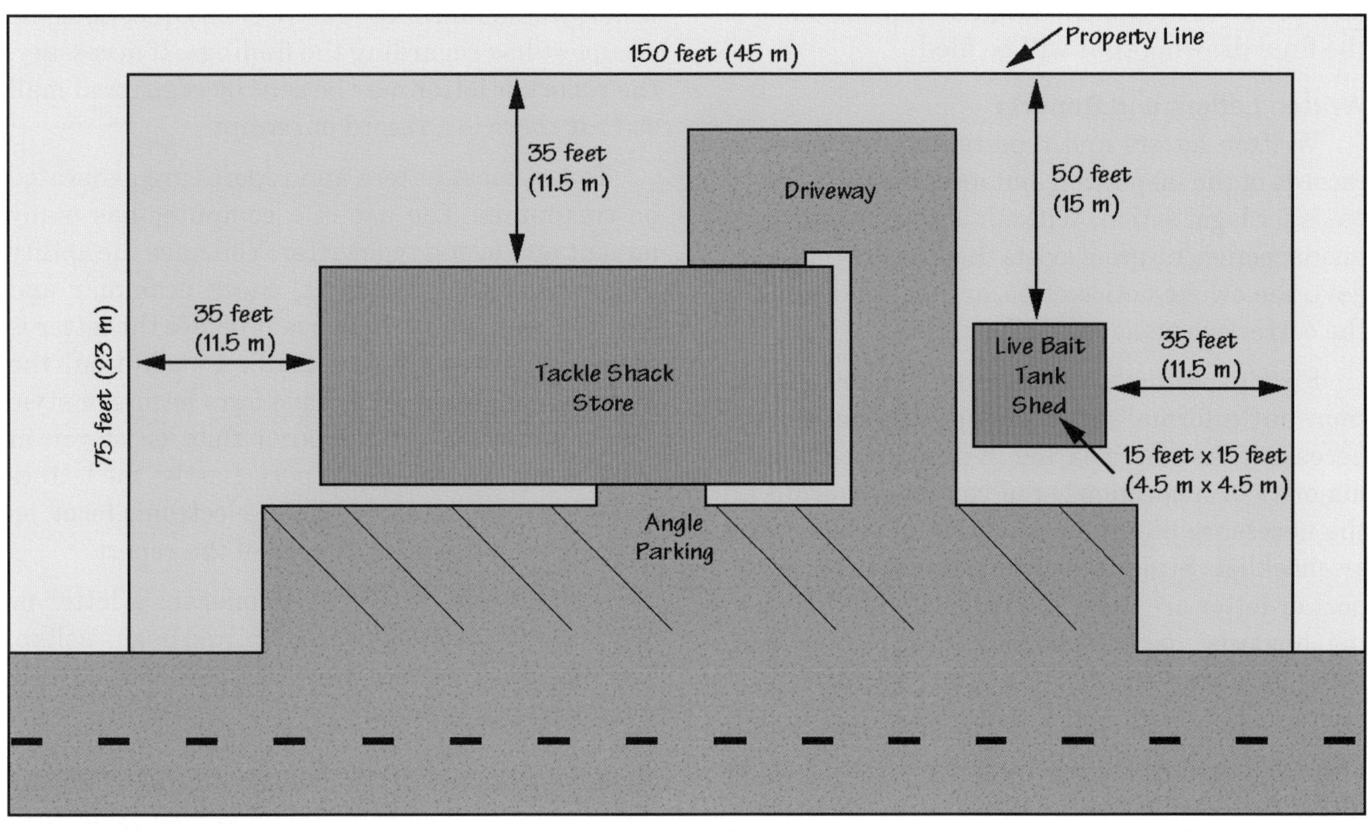

Figure 2.13 The inspector may draw a basic plot plan for the structure being inspected.

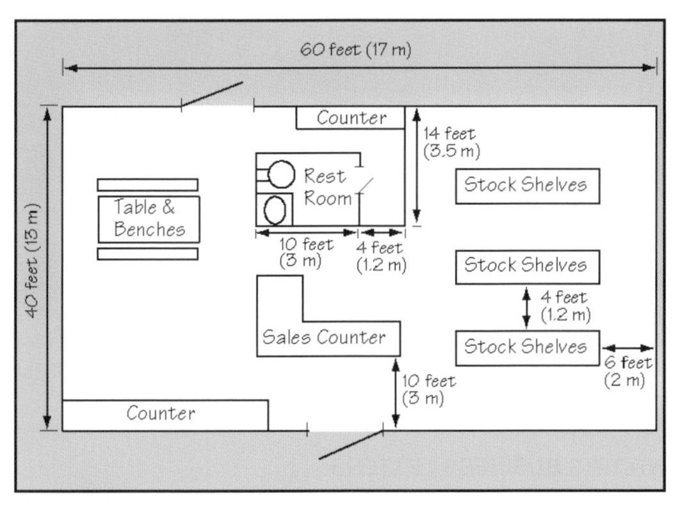

Figure 2.14 The floor plan shows the basic layout of the inside of the occupancy.

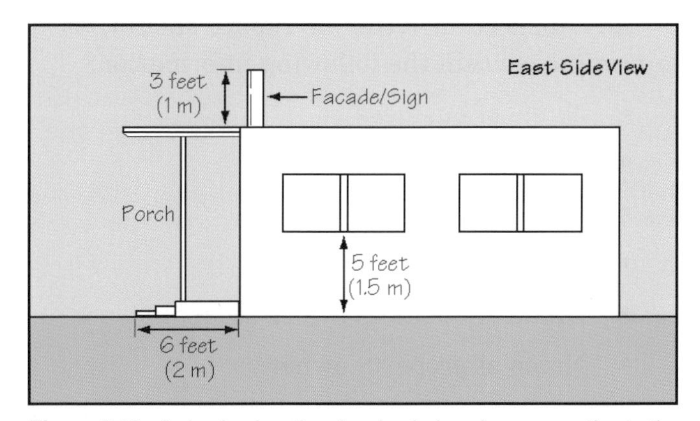

Figure 2.15 A simple elevation drawing helps give perspective to the inspection report.

number of drawings needed to show the necessary detail increase. Every drawing or page of a set of drawings should be labeled with a title that clearly indicates what type of information is included.

After you complete the final drawing, you should attach a set of clear, legible notes. In many cases, copies of this material must be given to other departments that need the information. They may also be included in pre-incident planning information for fire suppression companies. Some agencies also prefer to attach the original field sketches to the final drawing that will be filed.

Written Letters and Reports

Written letters and reports not only serve as records of the inspection, but also can be used as a basis for legal action. Without written evidence of an inspection, no proof exists that the fire inspector gave the owner notice of hazardous conditions or the corrective measures to be taken.

Fire inspectors must record every fire inspection, but a formal letter or report is not always necessary. As stated earlier in this chapter, on the majority of inspections, you can easily record all the necessary information on the inspection form or checklist. Situations that require a formal report or letter are those that involve life-threatening hazards, major renovations to comply with codes, or an extensive list of minor violations.

In an inspection letter or report, you are generally concerned with presenting facts and evidence to prove a point, draw a conclusion, or justify a recommendation. This information must be presented in a businesslike manner. The report should not be opinionated, biased, emotional, or unfair.

Any inspection letter or report should, as a minimum, contain the following information:

- Name of business
- Type of occupancy
- Date of inspection
- Name of inspector
- Name of business owner/occupant
- Name of property owner
- Name of person accompanying inspector on inspection
- Edition of applicable code, as a reference for future inspections
- List of violations and their locations stated in specific terms (code section numbers should be referenced)
- Specific recommendations for achieving compliance for each violation
- Date of follow-up inspection

In the final portion of any inspection report or letter, you should offer to answer any questions the occupant has regarding the findings. If necessary, the report or letter may be sent by registered mail so that there is a record of receipt.

Today, most letters and reports are generated on computers. The use of a computer has many advantages over a typewriter. You have the ability to complete the document, check grammar and spelling, and adjust the format before the letter is actually printed (Figure 2.16). In addition, the department may wish to have form letters or style sheets stored on the computer that assist you in completing the letter or report. Lastly, the letters or reports may be stored in an electronic form, as well as filing the printed copy of the report.

It is important that you generate a letter or report that is well written, clean, and businesslike. The following sections provide details on writing good reports and letters.

Figure 2.16 The inspector should be able to prepare a professional letter to the occupant detailing the results of the inspection.

WRITING BUSINESS LETTERS

Written communications are a reflection of the fire inspector and the inspector's agency. Letters that are misleading, rude, or poorly written will give the reader a negative impression of the fire inspector and the agency. All letters sent out must be grammatically correct, neatly typed, and reflect a positive professional attitude on behalf of the department.

Letters are an effective communication tool if they are written correctly and in a manner appropriate for the subject being discussed. An overly formal or pompous writing style makes the letter writer seem cold; a less formal, but grammatically correct style will accomplish the job in a more friendly manner. Compare the following examples:

> Change: It is requested that you send-us a copy....
>
> To: I would appreciate your send ing.....
>
> or Please send us a copy............

Both revisions send the same message but in a manner that is not confusing or difficult to understand. Letters should be concise and to the point without sounding rude. Excessive wordiness can cause the reader to lose interest or become confused as to the purpose of the letter. On the other hand, a letter that is too brief can read like a telegram. The letter should be checked after the first draft for wordiness or inadequate explanation. Several examples follow:

> Change: This is to acknowledge receipt of your letter...
>
> To: Thank you for your letter......

> Change: I would like to express my appreciation and thanks
>
> To: Thank you for......
>
> or I appreciate your assistance.....

The appearance of a letter is as important as the style because it is a critical part of the overall impression given to the reader. All letters should be neatly typed without obvious mistakes or corrections. It is assumed that a letter signed by an inspector has been either written or proofread by the inspector. The reader will interpret the letter's appearance as an example of the inspector's professionalism and attention to detail.

A helpful consideration for every letter writer is to realize that the reader will be reading the letter with interest. Let the letter be a substitute for an actual conversation; a natural and clear expression of thoughts. To accomplish this, the inspector needs to develop an attitude of "you" toward the reader. Maintain an awareness of the reader's impressions, thoughts, and feelings that may result from a letter. The combined use of natural expression of thoughts and a grammatically correct format will provide a good foundation for clear expression that will reflect positively on the inspector and the department.

It is important to consider how a letter is structured. To be consistent, all letters sent from the department should follow the same format. There are a few rules of accepted practice that should be followed (Figure 2.17). These are discussed in the following sections.

The Heading

The heading is usually the department's letterhead or logo. The heading should contain the return address, telephone number, FAX number, and electronic mail address of the writer so that a response can be returned if necessary.

Date Line

The date line is two lines below the heading, either on the left- or right-hand margin. Either location is acceptable. Again, write out all information such as the month and date. The proper way to express the date of the letter is, for example, July 6, 1998 or 6 July 1998. Do not use the method 7/6/98 or 7-6-98.

Inside Address

Allow four spaces below the date line for the inside address. The inside address must include the name, title, and address of the person receiving

GERYVILLE FIRE-RESCUE DIVISION

Headquarters:
112 North Quakertown Road
Geryville, PA 18073
Phone: (610) 555-5332
FAX: (610) 555-9346

Fire Marshal's Office:
2112 Geryville Pike
Geryville, PA 18074
Phone: (610) 555-0661
FAX: (610) 555-2841

March 12,1997

Mr. Dennis Parker
Cherrydale Farms Chocolate Co.
498 South Quakertown Road
Geryville, PA 18073

Dear Mr. Parker,

Thank you for the cooperation that was extended to me by you and your staff during my recent inspection of your facilities. Our goal of a fire-safe community relies on teamwork between our inspection staff and the citizens of the community.

In general, I found your facility to be compliant with the fire codes that are in effect in the City of Geryville. No serious fire or life safety hazards were noted during my inspection. However, there were several minor items noted that will require your attention and correction. These items are as follows:

1. There was an excessive amount of cocoa dust built up on all surfaces in the bean grinding area. This could be a potential source of a dust explosion or fire. These areas should be cleaned on a regular basis.

2. The exit sign over the door that leads from the employee cafeteria to the exterior was not illuminated. I believe it simply needs a new light bulb.

3. Tall grass and weeds obscured the view of the fire hydrant in the loading dock area. Grass and weeds should be moved to a level that allows fire personnel to easily locate the hydrant.

I will be in contact with you within the next two weeks to schedule a follow-up inspection to assure that these items are corrected to our satisfaction. If you have any questions on this matter, please do not hesitate to contact me in the Fire Marshal's office.

Sincerely,

Andrew C. Mount
Fire Inspector

ACM:kab

Figure 2.17 A sample inspection letter.

the letter. If the person holds an official title within the company, such as vice president, captain, or manager, the appropriate title follows the person's name.

Subject Line

Allow two lines below the inside address or two lines below the salutation for the subject line. The subject line is used to focus the reader's attention on the content of the letter. If the subject line is above the salutation, it is not necessary to use the word "subject." To avoid confusion for the reader, however, "subject" should be written if the subject line is below the salutation. The subject line is not necessary for all communications, but may be used in certain situations.

Salutation

This is the point where the writer addresses the reader (Dear Mr...). Avoid excessive salutations such as "My Dear Sir." Whatever salutation is used, it should contain the person's name and title as it appears in the inside address. If the name is unknown, the common practice is to use a subject line.

The Body

For appearance, there are some helpful spacing techniques that will make the letter more presentable. Average letters should be single-spaced between lines and double-spaced between paragraphs. Shorter letters should be double-spaced between lines and triple-spaced between paragraphs.

Do not split a date or a person's name between lines. Where there is a need to continue on to another page, the last paragraph on the first page should contain at least two lines. If two lines will not fit on the page, start the paragraph on the next page. Always carry at least two lines of the body text over to the second page; do not use a continuation page to type only the letter's closing. The second page should be typed on plain paper of quality equivalent to the letterhead. The second page should have a heading with the recipient's name, page number, and date. The heading may go in the upper left hand corner or across the page.

Closing

There are three standard parts to the closing of a letter. The following sections will discuss each of these.

Complimentary Closing

Space two lines after the last paragraph of the body for the complimentary close. Try to choose a simple closing such as "Sincerely." Again, the key is to be concise and appropriate.

Writer's Identificition

If appropriate, the writer's name and title should be typed on the fourth line below the complimentary close. If the letter is running short, you can leave up to six blank lines for the signature. If the letter is running long, you can reduce the signature space to two blank lines.

Reference Initials

Type the initials of the typist alone (or those of the writer and the typist) at the left margin, on the second line below the writer's name and title. If the writer's name is typed in the writer's identification block, the writer's initials are unnecessary here. However, if the writer wants his initials used, they should precede the intitials of the typist. If the writer and typist are the same, you may either type the initials on the left margin or omit them.

REPORT WRITING

The majority of report writing completed by the fire inspector consists of documenting serious or numerous fire code violations. Reports are designed to provide vital and useful information to the occupant and the fire department or inspection agency. Care should be taken to make each report neat in appearance and legible to the reader. In general, it is preferable that the report be typewritten or computer generated. The content of the report is even more important than its appearance. Information within the report should be delivered in a manner that is easily understood and not misleading.

To produce a well-written report requires the use of complete sentences, proper grammar, and appropriate use of words. These are basic writing techniques that apply to any type of report but plague many report writers. It is easy to write a report and have it appear to communicate the necessary information, only to find that the reader does not understand the report due to improper grammar or incomplete sentences. Two methods to overcome these problem areas of report writing are to practice writing skills and to have a second person proofread the report.

Misspelled words reflect poorly on the writer and will make the reader question the report's technical content. Any question about the spelling of a word should be settled by consulting a dictionary. Most computer word processing programs have built-in spelling correction features that allow the writer to make sure all words in the report are spelled correctly. Run-on or partial sentences make it difficult to understand the writer's meaning; it is always good writing practice to use short, clear sentences. Consult a good, easy-to-use style manual for correct word usage and punctuation.

Too many reports are plagued by misspelled words, poor punctuation, and inappropriate words. Writers sometimes fall into a habit of using unnecessary words or words with duplicate meanings. Always try to use words that identify exactly the meaning intended, and avoid words that may give the reader a different impression.

The following are some easy-to-see examples of unnecessary words. Word choice is an important consideration when writing effective, concise communications.

Example:	Should Be:
first of all	first
the actual truth	the truth
in the meantime	meanwhile
was definitely dangerous	was dangerous
for a period of two weeks	for two weeks
at this point in time	at this time; now

Taking time to prepare and adequately write a report will allow the opportunity to catch common writing mistakes. After the report is grammatically correct and proper use of words and appropriate punctuation have been assured, effort should be devoted to making the report neat in appearance.

Using these basic report writing guidelines will enable fire inspectors to present themselves and their agency in a positive and professional manner. Written communications are an integral part of a fire inspector's everyday activity, so it is crucial that clear and effective communications become second nature. It is beyond the scope of this manual to fully teach professional business communications; therefore, it is recommended that any individual in a fire inspector's position complete a business communication course.

FOLLOW-UP INSPECTIONS

Follow-up inspections are made to ensure that the recommendations made in the inspection report have been followed. Fire inspectors should confirm the time and date of the follow-up inspection with the occupant prior to arrival. When performing a follow-up inspection, it is not necessary to inspect the entire occupancy again. Only inspect the problem areas included in the inspection report to verify that the hazards have been corrected. This information should also be noted in the final inspection report.

If all hazards have been corrected, compliment the owner/occupant for taking the appropriate actions. Then, in order to close out the file, send a follow-up letter stating that the follow-up inspection found the violations to be corrected. This letter also gives you a second opportunity to thank the owner/occupant for cooperating.

If some hazards remain to be corrected, but the owner/occupant is making a conscientious effort to comply, they should be complimented for the progress that has been made to that point. Set a date and time for yet another follow-up inspection. Add a written update to the inspection files, with the original copy going to the owner/occupant.

If the hazards have not been corrected and it is apparent that the owner/occupant has made no effort to correct them, issue a final notice with a date for another inspection. The final notice should inform the owner/occupant exactly what legal action will be taken if full compliance is not attained by the date specified.

MAINTAINING FILES AND RECORDS

One of the most crucial functions carried out by any fire prevention bureau or division is that of maintaining accurate files and records of all occupancies they are responsible for inspecting. These records provide a historical perspective of fire prevention activities within that jurisdiction. They also provide the basis for all future code compliance and enforcement activities.

All documents and records pertaining to code enforcement activities should be kept within the agency's document management system. These include:

- Inspection reports, forms, and letters
- Violation notices
- Summonses
- Plans review comments, approvals, and drawings (Figure 2.18)
- Fire reports
- Investigations
- Permits and certificates issued

Figure 2.18 Storage of construction plans requires a lot of space.

It is most desirable to maintain files for all properties (excluding one- and two-family residences) within the jurisdiction. For many fire departments, this is not a realistic goal due to the size of their jurisdiction. As a minimum, files should be maintained on occupancies that:

- Have been issued a permit, certificate, or license of some type
- Contain automatic fire suppression or detection systems
- Contain hazardous materials or operations on a routine basis

It is recommended that records be maintained on a building for the entire life of that building. In other words, if the building is still standing, there should be records maintained on it by the fire prevention bureau. It is very common for a particular building to house many different owners and occupancies over the course of its lifetime. By maintaining a file on the structure throughout its life cycle, future inspectors will be able to note what changes have been made to the structure and how proposed changes may be affected by the previous uses of the structure. If a structure is demolished, the fire prevention bureau should maintain the files for a short period of time thereafter and then the files may be purged. Local policy and legal recommendations will dictate how long this period will be.

All files and records maintained by the fire prevention bureau are considered public domain documents. Obviously, there is some information contained on the reports that may be considered proprietary or confidential in nature. Officials who are responsible for overseeing the storage of inspection records should understand the open records laws and requirements for the jurisdiction they serve. This will provide guidance for them on what documents may be released to the public and those that must be withheld.

Written Records

Even in agencies that have been diligent and keeping up with technological advances, there tends to be a large number of written inspection records and documentation that need to be filed and maintained. This includes older inspection records that were generated before computerization and hard copy documents that are used in addition to computer files. Some agencies that have computerized their inspection documentation process have also chosen to print hard copies of the information and maintain them in a written file. This assures that new records are stored with older records to keep the file complete. It also serves as an additional backup should a serious failure of the computer system ever make the files inaccessible for any period of time.

Each inspected property should have a file that contains copies of all building and inspection records for that property (Figure 2.19). Each time an inspector has further contact with the occupant or the property, records of those actions should be added to the file. It is important that the file be kept as up to date as possible.

The methods for cataloging and storing the files will vary from jurisdiction to jurisdiction. Much of how this is done will be dependent on the size of the jurisdiction and the number of occupancies it is responsible for inspecting. Smaller jurisdictions may be able to store such documents in simple filing cabinets. Larger jurisdictions may have whole rooms dedicated to file storage (Figure 2.20). For agencies that must maintain large numbers of files and documents, older records may be reduced to microfilm to save storage space. This is particularly helpful on large documents such as building plans or architectural drawings.

Because the occupants in a particular building tend to change over time, it is generally not wise to catalog files alphabetically by occupant name. The most reliable method of cataloging inspection records is by the building's street address. This will allow the file to stay in the same location as occupants within the building change.

Figure 2.19 Proper files should be kept on all occupancies within the jurisdiction.

Figure 2.20 Large jurisdictions may have entire rooms dedicated to occupancy file storage.

Electronic Records

Many fire departments and other inspection agencies are taking advantage of advances in computer technology to assist them in efficiently maintaining and storing inspection records. Computerized inspection record systems can also assist in planning and scheduling future inspection needs. The level of sophistication of the computer system will be dependent on the fire department or inspection agency's needs. Most departments use a program that has been designed by an outside firm. Few departments have the resources to develop their own inspection data management system.

There are two primary methods by which data may be logged into the computer system:

- Inspectors use laptop computers or handheld electronic data recording equipment while making the inspection and then download the information into the system (Figure 2.21).
- Inspectors use written forms to record the information while performing the inspection and then manually enter the information into the computer system upon returning to the office/station.

The ability to electronically record information in the field and then download it into the computer system is the most efficient of the two methods. As further advances in technology make the use of small, portable computer equipment more accessible, this method will become a more common inspection practice.

Figure 2.21 Many inspectors use electronic devices, such as portable computers, in the field to assist with recording information during inspections.

There are many aspects of computer system management that must be given careful consideration when using the system to store inspection records and data. Several questions that must be answered are:

- How will the information be filed?
- How can the information be retrieved?
- What portion of the information will be stored in a read-only format so that records cannot be accidentally or purposely changed without authorization?
- What personnel will be given access to retrieve information from the system?

Fire inspectors who work for agencies that use computers for inspection records and data management must receive the appropriate amount of training on the system. Not even the most well-designed computer data management system will work unless the information put into it is done so in the proper manner.

3

Principles of Combustion and Fire Growth

This chapter contains information that will assist the reader in meeting the listed job performance requirements contained in NFPA 1031, ***Standard for Professional Qualifications for Fire Inspector and Plan Examiner*** (proposed 1998 edition).

Chapter 3 Fire Inspector I

3-3.14 Recognize a hazardous fire growth potential in a building or space, given field observations, so that the hazardous conditions are identified, documented, and reported in accordance with the policies of the jurisdiction.

Chapter 4 Fire Inspector II

4-3.11 Determine fire growth potential in a building or space, given field observations or plans, so that the contents, interior finish, and construction elements can be evaluated for compliance with applicable codes and standards and all deficiencies are identified, documented, and corrected in accordance with the policies of the jurisdiction.

Chapter 3
Principles of Combustion and Fire Growth

In order for the fire inspector to fully understand and recognize unsafe conditions or practices, he must have basic knowledge of the principles of combustion and the factors that affect fire growth and spread. The ability to recognize and correct these hazards is one of the most basic premises of the fire inspector's job. This chapter discusses the combustion process, the phases a fire goes through in its life cycle, the methods in which fire transfers heat, and the many factors within a structure that affect fire growth.

THE COMBUSTION PROCESS

Fire inspectors must always be alert for conditions or processes that might result in a fire. Fire is actually a by-product of a larger process called combustion. Fire and combustion are two words used interchangeably by most people; however, fire inspectors should understand the difference. *Combustion* is the self-sustaining process of rapid oxidation of a fuel, which produces heat and light. *Fire* is the result of a rapid combustion reaction.

Fuels are most commonly oxidized by the oxygen in air; although, in some cases, other oxidizers, such as those inherent in a specific material, are used. The normal oxygen content in air is 21 percent; nitrogen makes up 78 percent of air; and the remaining 1 percent is made up of other gases such as water vapor, neon, and carbon dioxide.

When air is not oxidizing a fire, chlorine or chemicals that release chlorine are oxidizing the fire. Some fuels, such as calcium hypochlorite and potassium chlorate, have their own oxygen tied up in their chemical formula. (This phenomenon can be compared to a firefighter with his own air supply functioning in an atmosphere that could not support breathing.) These self-oxidizing fuels can normally be recognized by the word oxide, oxalate, or peroxide in their name.

Combustion is defined as the process of rapid oxidation (resulting in fire). But oxidation is not always rapid. It may be very slow, or it may be instantaneous. Neither of these extremes produce fire (combustion) as we know it, but they are common occurrences in themselves. An example of very slow oxidation is *rusting*. (A light film of oil placed on metal prevents rusting because it keeps air off the metal so that it cannot react and oxidize the metal.) Instantaneous oxidation is an explosion, such as what occurs inside the casing of a bullet cartridge when the primer is ignited.

Rapid oxidation (combustion) can occur in two forms: smoldering fires and steady-state fires. Steady-state fires are sometimes called free-burning fires. The following sections look at the elements necessary to produce these two types of combustion.

The Fire Triangle

Historically, the fire triangle has been used to explain why fires are created and how they are extinguished. The triangle illustrates that oxygen, fuel, and heat in certain proportions create a fire; that if any one of the three elements is removed, a fire cannot exist (Figure 3.1). However, as research methods into the science of combustion improved over the years, it was realized that the fire triangle only accurately portrayed the smoldering mode of combustion. In order to support a flaming mode of combustion, an additional element is required. This is discussed in detail in the next section.

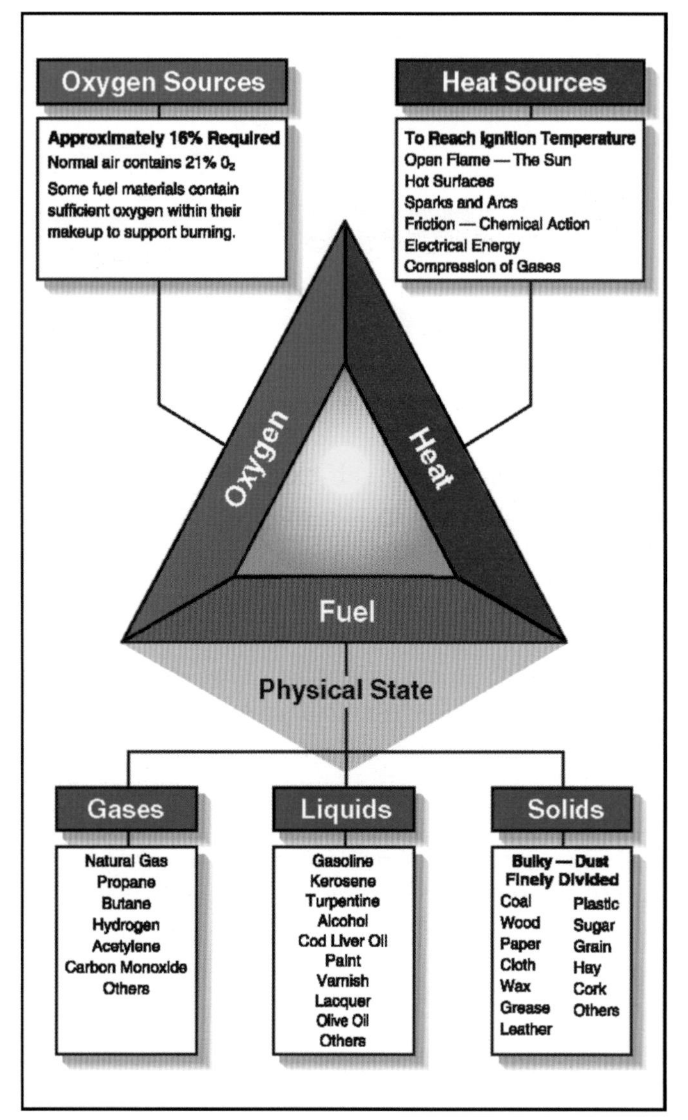

Figure 3.1 The fire triangle is used to portray the smoldering mode of combustion.

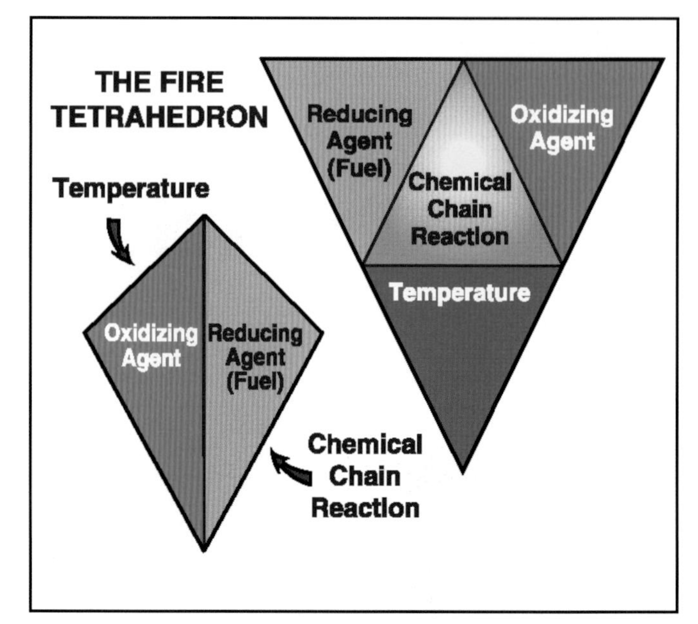

Figure 3.2 The fire tetrahedron reflects the flaming mode of combustion.

The Fire Tetrahedron

Modern fire science has recognized that in order to support flaming combustion a fourth element must be added to what was previously known as the fire triangle. In addition to fuel, heat, and oxygen, a chemical chain reaction is required to continue flaming combustion. The addition of this fourth element graphically creates a pyramid structure that is called the fire tetrahedron (Figure 3.2). As with the fire triangle, if any one of the four elements of the fire tetrahedron are removed, the fire is extinguished. Special extinguishing agents, such as halon, halon replacement agents, and some dry-chemical agents, interrupt the chemical chain reaction. This results in a dramatically fast extinguishment of the fire.

Sources of Heat Energy

The sources of heat energy are particularly important to fire inspectors. A clear understanding of these sources provides the fire inspector with much of the knowledge needed to recognize ignition hazards when performing inspections. Understanding the sources of heat energy also leads to the ability to make recommendations on how to minimize, or eliminate, these hazards.

Heat is a form of energy that may be described as a condition of "matter in motion" caused by the movement of molecules. All matter contains some heat, regardless of how low the temperature, because molecules are constantly moving. When a body of matter is heated, the speed of the molecules increases, and thus the temperature also increases. Anything that increases the motion of the molecules of a material produces heat in that material. The five general categories of heat energy are as follows:

- Chemical
- Electrical
- Mechanical
- Nuclear
- Solar

CHEMICAL HEAT ENERGY

Chemical heat energy is generated as the result of some type of chemical reaction. The four types of chemical reactions that result in heat production are heat of combustion, spontaneous heating, heat of decomposition, and heat of solution.

HEAT OF COMBUSTION

Heat of combustion is the amount of heat generated by the combustion (oxidation) reaction (Figure 3.3). The amount of heat generated by burning materials will vary depending on the material. This phenomenon is why some materials are said to burn "hotter" than others.

Figure 3.3 In some cases the heat of combustion is so great that firefighters are forced into an exterior fire attack.

SPONTANEOUS HEATING

Spontaneous heating is the heating of an organic substance without the addition of external heat. Spontaneous heating occurs most frequently where sufficient air is not present and insulation prevents dissipation of heat — heat that is produced by a low-grade chemical breakdown process. An example would be oil-soaked rags that are rolled into a ball and thrown into a corner. If ventilation is not adequate to let the heat drift off, eventually the heat will become sufficient to cause ignition of the rags. The speed of a heating reaction doubles with each 18°F (10°C) temperature increase. The inspector must recognize materials and conditions that might lead to fires caused by spontaneous heating. Bulk storage of materials such as hay, silage, combustible metal shavings, and manure are all examples of spontaneous heating hazards (Figure 3.4).

Figure 3.4 Spontaneous heating is a very common cause of hay barn fires.

HEAT OF DECOMPOSITION

Heat of decomposition is the release of heat from decomposing compounds, usually due to bacterial action. In some cases, these compounds may be unstable and release their heat very quickly; they may even detonate. In other cases, the reaction and resulting release of heat is much slower. This reaction can be commonly seen when viewing a compost pile. The decomposition of organic materials creates heat that can be seen on cold days when holes are poked into the pile. The heated vapors and steam can be seen rising from the holes in the pile.

HEAT OF SOLUTION

Heat of solution is the heat released by the solution of matter in a liquid. Some acids, when dissolved in water, can produce violent reactions, spewing hot water and acid with explosive force.

ELECTRICAL HEAT ENERGY

It is not uncommon to see "electrical" listed as the cause of fire that damaged a building or automobile. Electricity has the ability to generate high temperatures capable of igniting any combustible materials near the heated area. Electrical heating hazards are one of the more common things that fire inspectors will find themselves dealing with. Electrical heating can occur in a variety of ways. The following sections highlight some of the more common ways.

RESISTANCE HEATING

Resistance heating refers to the heat generated by passing an electrical current through a conductor such as a wire or an appliance. Resistance heating is increased if the wire is not large enough in diameter for the amount of current or there is a poor connection between conductors. Fires are caused when a simple extension cord is overloaded with too many appliances. Resistance heating can also be increased if the conductor is tightly wound. Fires are often started by cheap (small capacity) electrical extension cords that are wound around the leg of a table or chair to take up the slack or run under carpets to prevent people from tripping over them (Figure 3.5).

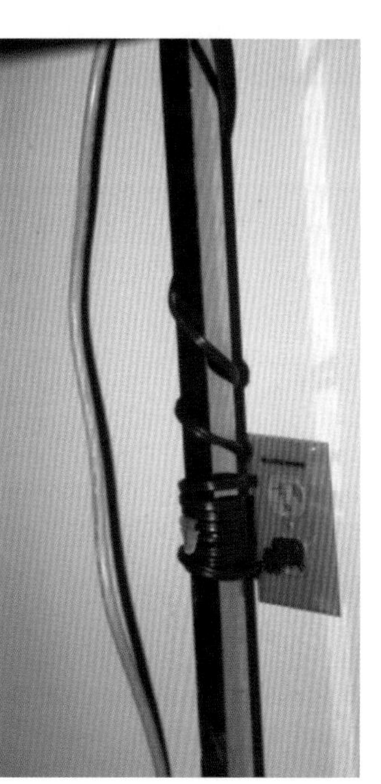

Figure 3.5 Resistance heating is created when an electrical cord is wound tightly around an object.

DIELECTRIC HEATING

Dielectric heating, used in microwave ovens, occurs as a result of either direct current (DC) or alternating current (AC) pulsating at high frequency on a nonconductive material. The nonconductive material is not heated by the dielectric heating; however, it is heated by being in constant contact with electricity. This is somewhat similar to bombarding an object with many little lightning bolts.

LEAKAGE CURRENT HEATING

Leakage current heating occurs when a wire is not insulated well enough to contain all the current. Some current leaks out into the surrounding material such as inside the wall of a structure. This current causes heat and can cause a fire.

HEAT FROM ARCING

Heat from arcing is a type of electrical heating that occurs when the current flow is interrupted. Interruption occurs when there is a loose connection. Arc temperatures are extremely high and may even melt the conductor. A common arc used in industrial applications is the arc welder. Here the welding rod (conductor) melts away as metals are joined together.

STATIC ELECTRICITY

Static electricity is the buildup of a positive charge on one surface and a negative charge on another surface. The charges are naturally attracted to each other and seek to become evenly charged again. This condition is shown when the two surfaces come close to each other, as in the case of a person's finger and a metal doorknob — an arc occurs producing the familiar spark and small shock.

Static electricity is often to blame when a fire occurs as flammable liquids are transferred between containers that are not properly electrically bonded together. This is why airplanes and fuel tankers have to be grounded and bonded by a wire during fuel transfer operations. Extra bonding is not necessary when filling the fuel tank of an automobile because gasoline has special additives that make the fuel act as a ground and the metal nozzle in contact with the filling pipe serves as a bonding source (Figure 3.6).

Figure 3.6 The metal nozzle in contact with the fill pipe creates an effective electrical bond.

Heat generated by lightning is static electricity on a very large scale. The heat generated by the discharge of billions of volts from either earth to cloud, cloud to cloud, or cloud to ground can be in excess of 60,000°F (33 300°C).

MECHANICAL HEAT ENERGY

Mechanical heat is generated two ways: by friction and compression. *Heat of friction* is created by the movement of two surfaces against each other. This movement results in heat and/or sparks being generated. A common example of heat generated by friction is that of a loose belt on a pulley. This is a common cause of automobile fires.

Heat of compression is generated when a gas is compressed. Diesel engines ignite fuel vapor without a spark plug by the use of this principle. It is also the reason that SCBA bottles feel warm to the touch after they have been filled.

NUCLEAR HEAT ENERGY

Nuclear heat energy is generated when atoms are either split apart (fission) or combined (fusion). In a controlled setting, fission is used to heat water to drive steam turbines and produce electricity. Currently, fusion cannot be controlled and has no commercial use.

SOLAR HEAT ENERGY

The energy transmitted from the sun in the form of electromagnetic radiation is called *solar heat energy*. Typically, solar energy is distributed fairly evenly over the face of the earth and in itself is not really capable of starting a fire. However, when solar energy is concentrated on a particular point, as through the use of a lens, it may ignite combustible materials.

The Burning Process

The first step in the burning process is for the fuel and oxygen (usually from air) to mix in proper proportions to support combustion. When the proper fuel vapor/air mixture has been achieved, it must then be raised to its ignition temperature or the point at which self-sustained combustion will continue. Fire burns in two basic modes: flaming or surface combustion. As previously described, the flaming mode of combustion, such as the burning of logs in a fireplace, is represented by the fire tetrahedron (fuel, temperature, oxygen, and the uninhibited chemical chain reaction). The surface, or smoldering mode of combustion, is represented by the fire triangle (fuel, temperature, and oxygen).

The fuel segment of both diagrams is any solid, liquid, or gas that can combine with oxygen in the chemical reaction known as oxidation. A fuel with a sufficiently high temperature will ignite if an oxidizing agent is present. Combustion will continue as long as enough energy or heat is present. Under most conditions, the oxidizing agent will be the oxygen in air. However, some materials, such as sodium nitrate and potassium chlorate, release their own oxygen during combustion and can cause fuels to burn in an oxygen-free atmosphere. While scientists only partially understand what happens in the combustion chemical chain reaction, they do know that heating a fuel can produce vapors that contain substances that will combine with oxygen and burn. These vapors must continue to be produced for combustion to continue.

A self-sustaining combustion reaction of solids and liquids depends on radiative feedback. *Radiative feedback* is radiant heat providing energy for continued vaporization. When sufficient heat is present to maintain or increase this feedback, the fire will remain constant or will grow, depending on the heat produced. When heat is fed back to the fuel, this is known as a *positive heat balance*. If heat is dissipated faster than it is generated, a *negative heat balance* is created. A positive heat balance is required to maintain combustion.

The amount of oxygen available to support combustion is important. Air contains about 21 percent oxygen under normal circumstances. However, fire inspectors may frequently encounter situations where less than 21 percent oxygen is present. The most common situations where this would be found are in empty storage tanks, oxygen-limiting silos, manholes, and valve vaults (Figure 3.7). Often, storage tanks are being purged or are filled with residual gases or vapors that result in an oxygen-deficient atmosphere. If a worker or inspector enters the tank without breathing equipment, he will soon become unconscious and may die.

Figure 3.7 Oxygen-deficient atmospheres are commonly encountered in oxygen-limiting silos.

Phases of Fire

Fires may start at any time of the day or night if a hazard exists. If the fire happens when the area is occupied and/or protected by automatic suppression and detection systems, chances are that it will be discovered and controlled in the beginning (incipient) phase. If the fire occurs when the building is closed, deserted, and without fixed protection systems, the fire may go undetected until it has gained major headway (Figure 3.8).

Fire in a confined room or building has two particularly important characteristics. The first characteristic is that there is a limited amount of oxygen. This differs from an outdoor fire where the oxygen supply is unlimited. The second characteristic is that the fire gases that are given off are trapped inside the structure and build up, unlike outdoors where they can dissipate. Fire confined to a building or room can be best understood by an investigation of its three main progressive phases: incipient, steady-state burning, and hot smoldering. Fire inspectors must also be aware of the variety of potentially hazardous conditions that may be intertwined within the three main phases. These hazards include rollover, flashover, and backdraft.

Figure 3.8 Buildings that lack fire suppression and detection systems are, when unoccupied, susceptible to large fires.

INCIPIENT PHASE

The *incipient phase* is the earliest phase of a fire, beginning with the actual ignition (Figure 3.9). The fire is limited to the original materials of ignition. In the incipient phase, the oxygen content in the air has not been significantly reduced, and the fire is producing water vapor (H_2O), carbon dioxide (CO_2), perhaps a small quantity of sulfur dioxide (SO_2), carbon monoxide (CO), and other gases. Some heat is being generated, and the amount will increase as the fire progresses. The fire may be producing a flame temperature well above 1,000°F (537°C), yet the temperature in the room at this stage may be only slightly increased.

Rollover

Rollover, sometimes referred to as flameover, takes place when unburned combustible gases released during the incipient or early steady-state phase accumulate at the ceiling level. These superheated gases are pushed, under pressure, away from the fire area and into uninvolved areas where they mix with oxygen. When their flammable range is reached, they ignite and a fire front develops, expanding very rapidly and rolling across the ceiling (Figure 3.10). Rollover differs from flashover in that only the gases are burning and not the con-

Figure 3.9 The earliest phase of a fire is called the *incipient phase*.

Figure 3.10 Rollover occurs when fire gases and flames move across the upper levels of the room.

tents of the room. The rollover will continue until its fuel is eliminated. This is done by extinguishing the main body of fire. The rollover will cease when the fire itself stops producing the flammable gases that are feeding it.

STEADY-STATE BURNING PHASE

For purposes of simplicity, the *steady-state burning phase* (sometimes referred to as the free-burning phase) can generally be considered the phase of the fire where sufficient oxygen and fuel are available for fire growth and open burning to a point where total involvement is possible (Figure 3.11). During the early portions of this phase, oxygen from the air is drawn into the flame, as *convection* (the rise of heated gases) carries the heat to the uppermost regions of the confined area. The heated gases spread out laterally from the top downward, forcing the cooler air to seek lower levels, and eventually igniting all the combustible material in the upper levels of the room. This early portion of the steady-state burning phase is often called the flame-spread

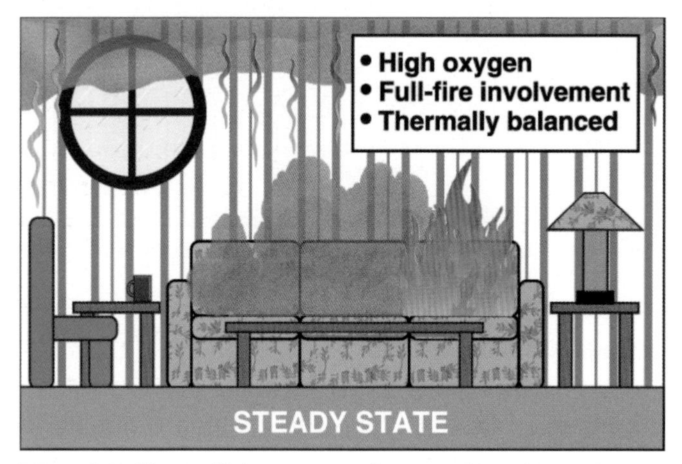

Figure 3.11 Given sufficient oxygen, a fire may achieve the steady-state phase.

phase. At this point, the temperature in the upper regions can exceed 1,300°F (700°C).

If conditions are perfect, and they rarely are, the fire may achieve what is commonly referred to as "clear burning." Clear burning is accompanied by high temperatures and complete combustion. Little or no smoke is given off. This fire is usually seen only when very clean fuels, such as methanol-based race car fuels, burn. Thermal columns will normally occur with rapid air movements upward from the base of the fire. In a confined space, the fire continues to consume the free oxygen until it reaches the point where there is insufficient oxygen to react with the fuel. The fire is then reduced to the smoldering phase, but this fire needs only a fresh supply of oxygen to burn rapidly.

Flashover

Flashover occurs when flames flash over the entire surface of a room or area. The actual cause of flashover is attributed to the buildup of heat from the fire itself. As the fire continues to burn, all the contents of the fire area are gradually heated to their ignition temperatures. When they reach their ignition point, simultaneous ignition occurs, and the area becomes fully involved in fire (Figure 3.12). This actual ignition is almost instantaneous and can be quite dramatic.

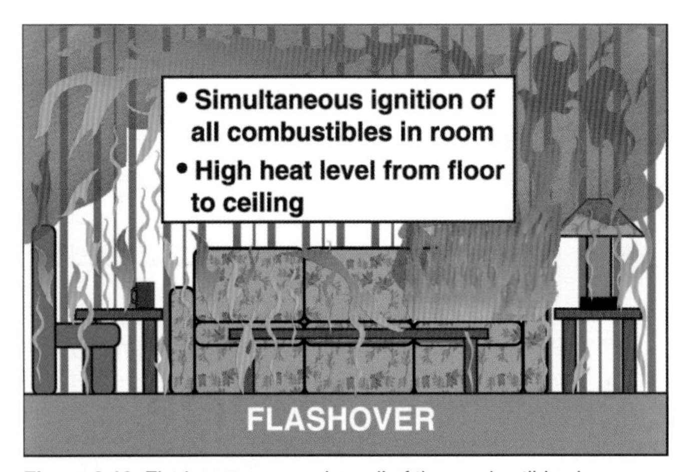

Figure 3.12 Flashover occurs when all of the combustibles in an area reach their ignition temperature at the same time.

HOT-SMOLDERING PHASE

After the steady-state burning phase, flames may cease to exist if the area of confinement is sufficiently airtight (Figure 3.13). In this instance, burning is reduced to glowing embers. As

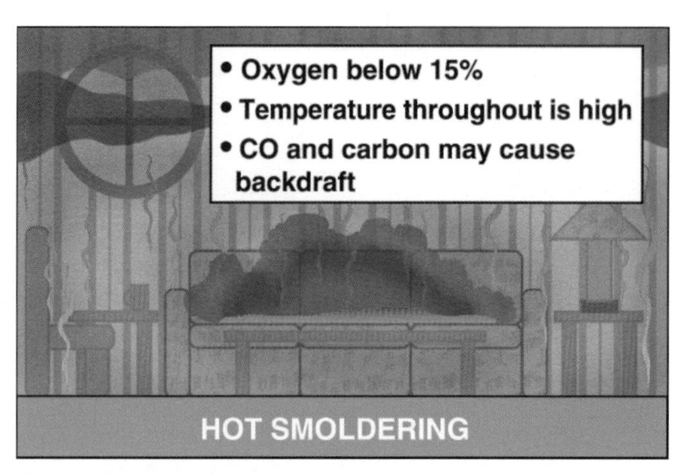

Figure 3.13 Fires that begin to deplete their oxygen supply may enter the hot-smoldering phase.

the flames die down, the room becomes completely filled with dense smoke and gases. Air pressure from gases being given off may build to the extent that smoke and gases are forced through small cracks. Room temperatures in excess of 1,000°F (537°C) are possible. The intense heat will have liberated the lighter fuel fractions, such as methane, from the combustible material in the room. These fuel gases will be added to those produced by the fire and will create the possibility of a backdraft if air is improperly introduced into the room. If air is not introduced into the room, the fire will eventually burn out, leaving totally incinerated contents.

Backdraft

Firefighters responding to a confined fire that is late in the steady-state burning phase or in the hot-smoldering phase risk causing a backdraft (also known as a smoke explosion) if the science of fire is not considered in opening the structure.

In the hot-smoldering phase of a fire, burning is incomplete because of insufficient oxygen to sustain the fire. However, the heat from the steady-state burning phase remains, and the unburned carbon particles and other flammable products of combustion are available for instantaneous combustion when more oxygen is supplied. Improper ventilation, such as opening a door or breaking a window, supplies the dangerous missing link — oxygen. As soon as the needed oxygen rushes in, the stalled combustion resumes; it can be devastating in its speed, truly qualifying as an explosion (Figure 3.14).

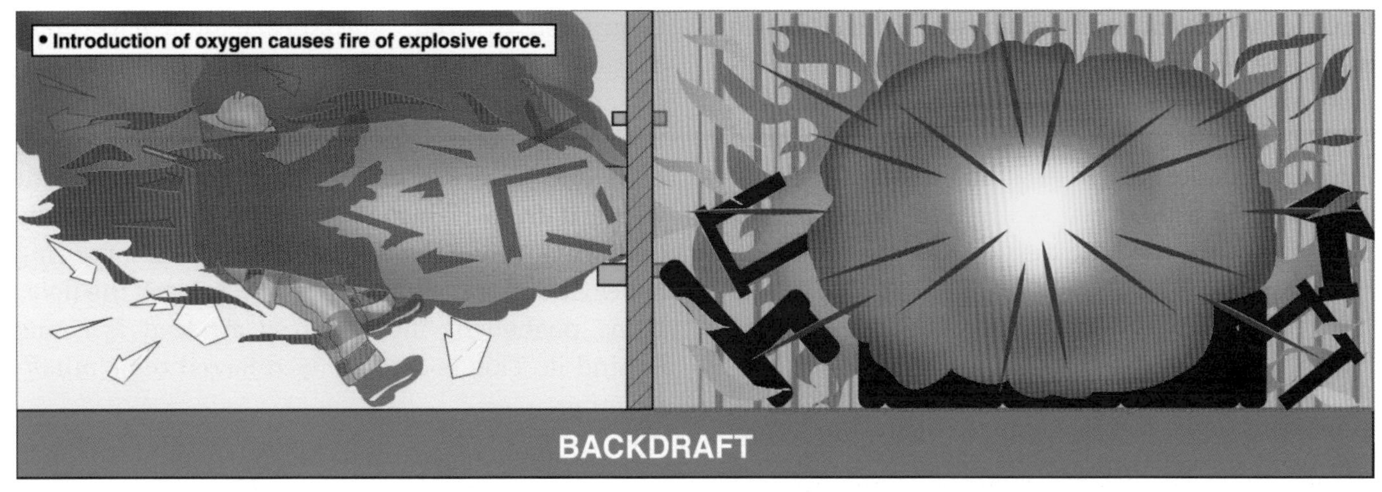

Figure 3.14 A backdraft may occur when oxygen is improperly introduced into a hot-smoldering phase fire.

Backdraft can be the most hazardous condition a firefighter will ever face.

Heat Transfer

Fire inspectors must be very concerned with the potential for fire to be transmitted through a structure. A number of the natural laws of physics are involved in the transmission of heat. One is called the *Law of Heat Flow*; it specifies that heat tends to flow from a hot substance to a cold substance. The colder of two bodies in contact will absorb heat until both objects are at the same temperature. Heat can travel throughout a burning building by one or more of three methods: conduction, convection, and radiation. The following sections describe how this transfer takes place.

CONDUCTION

Heat may be conducted from one body to another by direct contact of the two bodies or by an intervening heat-conducting medium. An example of this type of heat transfer is a basement fire that heats pipes enough to ignite the wood inside walls several rooms away (Figure 3.15). The amount of heat that is transferred and its rate of travel depend upon the conductivity of the material through which the heat is passing. Not all materials have the same heat conductivity. Aluminum, copper, and iron are good conductors; however, fibrous materials, such as felt, cloth, and paper, are poor conductors.

Liquids and gases are poor conductors of heat because of the movement of their molecules. Air is a relatively poor conductor. This is why double

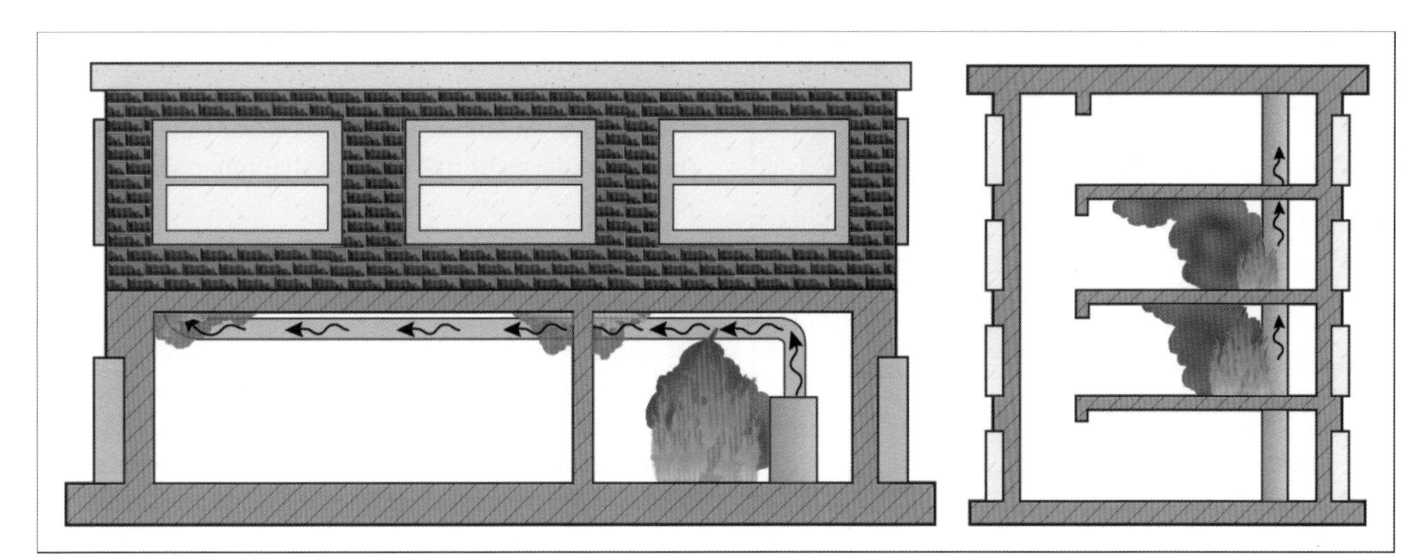

Figure 3.15 A heated pipe may conduct sufficient heat to start a fire in remote locations.

building walls and storm windows that contain an air space provide additional insulation from outside air temperatures. Certain solid materials, such as fiberglass shredded into fibers and packed into batts, make good insulation because the material itself is a poor conductor and there are air pockets within the batting.

CONVECTION

Convection is the transfer of heat by the movement of air or liquid. When water is heated in a glass container, the movement within the vessel can be observed through the glass. If sawdust is added to the water, the movement is more apparent. As the water is heated, it expands and grows lighter, hence, the upward movement. In the same manner, as air near a steam radiator becomes heated by conduction, it expands, becomes lighter, and moves upward. As the heated air moves upward, cooler air takes its place at the lower levels. When liquids and gases are heated, they begin to move within themselves. This movement is different from the molecular motion discussed in conduction of heat and is responsible for heat transfer by convection.

Heated air in a building will expand and rise. For this reason, fire spread by convection is mostly in an upward direction; however, air currents can carry heat in any direction (Figure 3.16). Convection currents are generally the cause of heat movement from floor to floor, from room to room, and from area to area. The spread of fire through corridors, up stairwells and elevator shafts, between walls, and through attics is caused mostly by the convection of heat currents. If the convecting heat encounters a ceiling or other barrier that keeps it from rising, it will spread out laterally (sideways) along the ceiling. If it runs out of ceiling space, it will travel down the wall toward the floor, being pushed by more heated air that is rising behind it. This is commonly referred to as *mushrooming*.

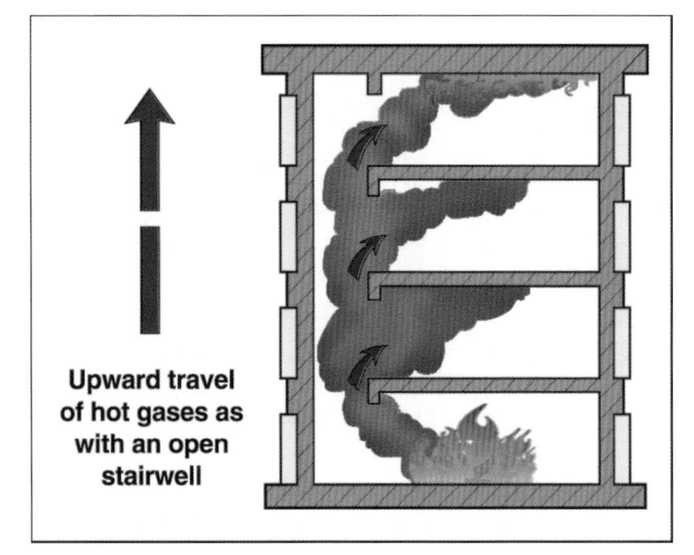

Figure 3.16 Convection is the most common method of heat and fire travel within structures.

Although often mistakenly thought to be a separate form of heat transfer, direct flame contact is actually a form of convective heat transfer. When a substance is heated to the point where flammable vapors are given off, these vapors may be ignited, creating a flame. As other flammable materials come in contact with the burning vapors, or flame, they may be heated to a temperature where they, too, will ignite and burn.

RADIATION

Although air is a poor conductor, it is obvious that heat can travel where matter does not exist. The warmth of the sun reaches us even though it is not in direct contact with us (conduction), nor is it heating gases that travel to us (convection). This method of heat transmission is known as radiation of heat waves. Heat and light waves are similar in nature, but they differ in length per cycle. Heat waves are longer than light waves, and they are sometimes called infrared rays. Radiated heat will travel through space until it reaches an opaque object. As the object is exposed to heat radiation, it will in return radiate heat from its surface. Radiated heat is one of the major sources of fire spread to exposures (Figure 3.17).

FACTORS AFFECTING FIRE GROWTH AND SPREAD

There are many factors within a structure that affect a fire's ability to grow and spread throughout the structure. Understanding each of these factors and their impact is considered basic knowledge for fire inspectors. The following sections examine some of the more important factors with which fire inspectors must be concerned.

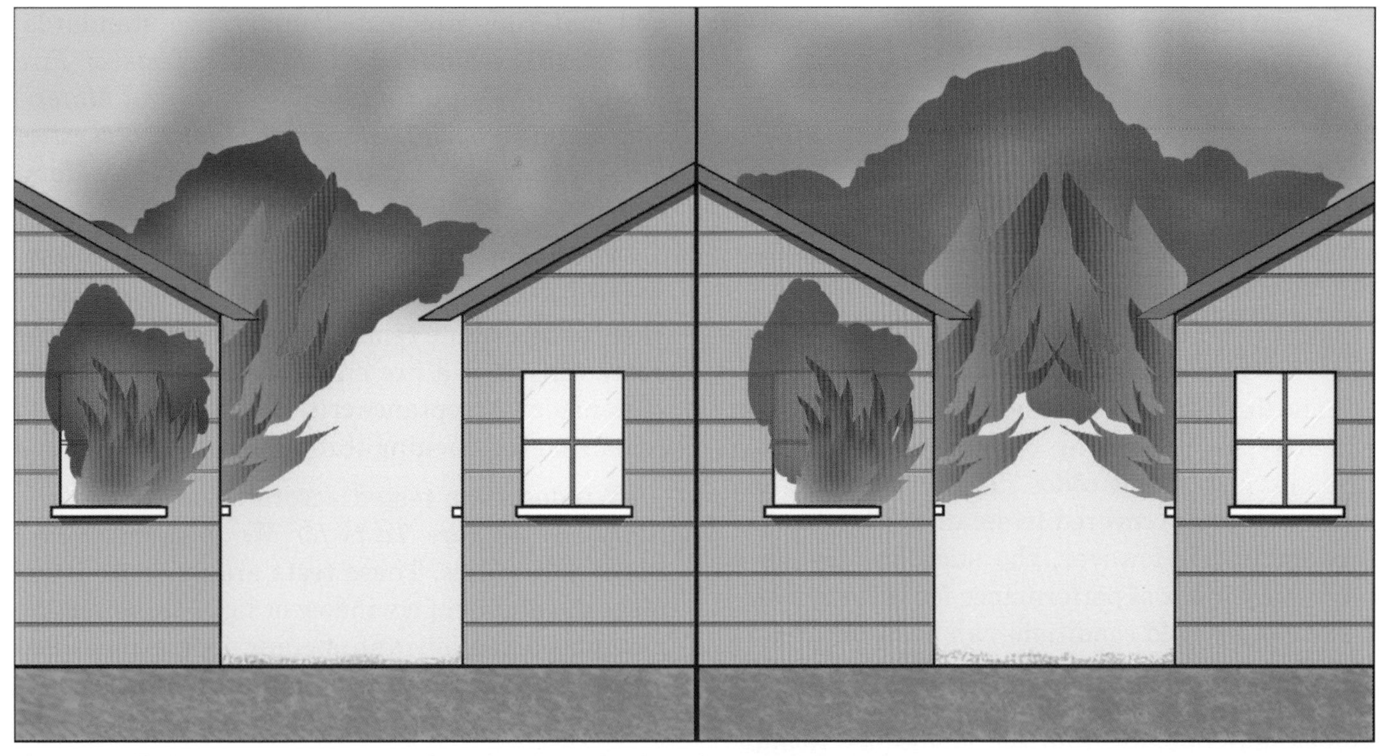

Figure 3.17 Radiation is the most common form of heat associated with fire spread to exposures.

Fire Resistance

Fire resistance is defined as the ability of a structural assembly to maintain its load-bearing ability under fire conditions. In the case of walls, partitions, and ceilings, it also means the ability of the assembly to act as a barrier to the fire. The fire resistance of a structural component is a function of various properties of the materials used. This includes their combustibility, thermal conductivity, chemical composition, and dimensions.

The fire-resistance ratings of materials are determined by fire test procedures simulating fire conditions using the standard time-temperature curve (Figure 3.18). Fire-resistance ratings are given for assemblies of structural elements such as floors, floor-ceiling assemblies, columns, walls, and partitions. It is necessary to test such assemblies as erected in the field so that values can be assigned that are meaningful to the fire protection engineer or building designer.

In the standard fire test, the furnace temperatures are regulated to conform to the time-temperature curve. The temperature rises to 1,000°F (538°C) in five minutes and then to 1,700°F (927°C) in one hour. The *fire-resistive rating* is the period of time that the assembly will perform satisfactorily when exposed to the standard test fire. Failure of an assembly is determined by one of several criteria. These include failure to support the load, passage of flame through the assembly, and excessive increase in temperature on the unexposed side of the assembly. NFPA 251, *Standard Methods of Tests of Fire Endurance of Building Construction and Materials* contains specifications of the test procedures.

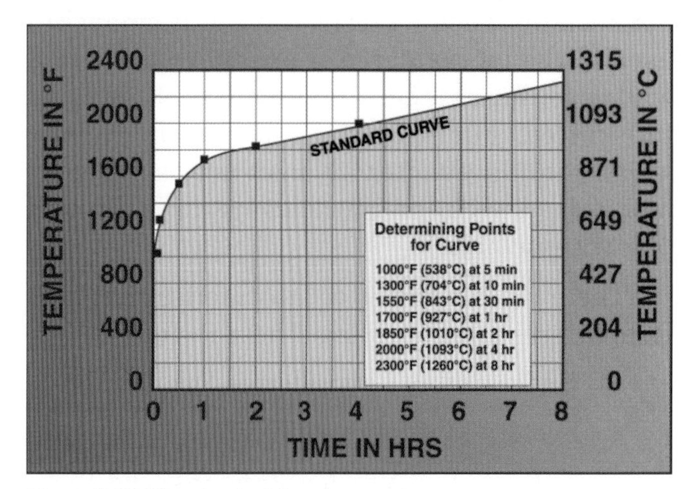

Figure 3.18 The standard time-temperature curve.

Although an assembly may fail at anytime, fire resistance is expressed in certain intervals, such as:

- 20 minutes
- 30 minutes
- 45 minutes
- 1 hour
- 1½ hours
- 2 hours
- 3 hours
- 4 hours

The fire resistance-ratings form the basis upon which types of building construction are recognized in the building codes. (The types of building construction are covered in detail in Chapter 5 of this manual.) However, the standard test is a laboratory index of performance for various materials. Not all field conditions can be duplicated in the laboratory. Obviously, size restrictions do not permit testing entire buildings. The field performance may deviate from the laboratory results. This is possible when building components have not been installed with the same craftsmanship used in the laboratory. For example, joints in a wall or ceiling assembly may not have been carefully fitted. Furthermore, some of the lighter weight building materials may not be properly maintained, such as when holes are made through plaster assemblies.

There are a number of laboratories that perform fire-resistance testing. These include:

- Underwriters Laboratories, Inc. (UL)
- Factory Mutual System (FM)
- National Institute for Standards and Technology (NIST)
- Forest Products Laboratory
- Portland Cement Association
- Southwest Research Institute

The largest and most well-known of these is Underwriters Laboratories. UL publishes test results annually in the *Fire Resistance Directory*.

Construction assemblies, such as roof assemblies, fire doors, and fire windows, are tested under a variety of fire conditions to determine the structural integrity of the assemblies during and after fire exposure. Building construction techniques and materials are tested under two standards: NFPA 251, *Standard Methods of Tests of Fire Endurance of Building Construction and Materials*; and NFPA 255, *Standard Method of Test of Surface Burning Characteristics of Building Materials*.

Fire doors are tested according to NFPA 252, *Standard Methods of Fire Tests of Door Assemblies*, to determine relative fire integrity. Tests for doors consist of a fire endurance test and a hose stream test. Acceptance criteria require the door to remain in the opening throughout both tests.

Windows are tested according to NFPA 257, *Standard on Fire Tests for Window and Glass Block Assemblies*. These tests are made to determine the ability of a window or light-transmitting assembly to remain in an opening during a specified level of fire exposure for 45 minutes. Like the test for door assemblies, tests for window assemblies involve both fire endurance and hose stream tests. Knowledge of all of these tests is extremely important for all inspection personnel.

Flame Spread Ratings

A critical factor in overall building fire safety is the combustibility of the materials used for the interior finish. Of particular interest is the speed at which flames may travel over the surface of the material. Some of the most tragic fires — in terms of loss of life — throughout history can have their severity traced to highly combustible interior finishes. Examples of these include the Coconut Grove Nightclub fire (Boston Massachusetts, 1942; 492 fatalities) and the Beverly Hills Supper Club fire (Southgate, Kentucky, 1977; 165 fatalities) (Figure 3.19).

Figure 3.19 Highly combustible interior finishes were a significant factor in the massive scope of the tragedy at the Beverly Hills Supper Club fire in Southgate, Kentucky in 1977.

Fire inspectors should be knowledgeable of flame testing and other fire testing methods. Inspectors who are familiar with fire test methods can further ensure that a building is reasonably fire safe. Interior finish contributes to fire impact in four ways:

- It affects the rate of fire buildup to a flashover condition.
- It may contribute to fire extension through flame spread over its surface.
- It may add to the intensity of a fire by contributing additional fuel.
- It may produce smoke and toxic gases that can contribute to life hazard and property damage.

Materials that exhibit high rates of flame spread, contribute substantial quantities of fuel to a fire, or produce hazardous quantities of smoke or toxic gases are considered to be undesirable from a fire protection standpoint. Naturally, the surfaces of some materials, such as wood veneers, pose more of a fire hazard than other materials such as plaster. Once materials have been evaluated, restrictions can be placed on the use of those materials that are most hazardous.

The most widely used and recognized test for determining the surface burning characteristics of interior finishes is the Steiner Tunnel Test. The test also has the designations ASTM E84 (American Society for Testing and Materials), NFPA 255, *Standard Method of Test of Surface Burning Characteristics of Building Materials* and UL 723, *Flammability Studies of Cellular Plastics and Other Building Materials Used For Interior Finish.* When materials are subjected to this test, their surface combustibility can be expressed in a numerical form called the *flame spread rating*. The flame spread rating provides a means of determining the relative hazard presented by interior surface finishes as compared to standard materials.

The Steiner Tunnel is a 25-foot (7.6 m) long horizontal furnace (Figure 3.20). The inside of the furnace is 17½ inches (445 mm) wide and 12 inches (305 mm) high. The top of the tunnel furnace is removable. The specimen to be tested is attached to the underside of the furnace top and the assembly

Figure 3.20 The Steiner Tunnel is used to test the flame spread characteristics of various materials. *Courtesy of Underwriters Laboratories, Inc.*

is lowered into place. A gas burner located at one end of the tunnel produces a flame that is projected against the specimen. The flame is adjusted to produce approximately 5,000 Btus (5 270 kJ) per minute. The extent of flame travel along the material is observed through view ports along the side of the tunnel.

Steiner Tunnels exist in several laboratories within North America. In addition, some manufacturers of building materials have tunnel furnaces for product development testing purposes. Because the flame spread rating cannot be accurately determined in the field, the inspector must research the ratings of materials noted during the inspection. These ratings can be obtained from the *Building Materials Directory* published by UL.

To derive the numerical flame spread rating, the flame travel along the test material is compared to two standard materials: asbestos cement board (given an 0 rating) and red oak wood (given a 100 rating). Obviously, the higher the flame spread rating, the more hazardous the material.

Building codes usually classify interior finish materials according to their flame spread rating. A building code will specify what class of material can be used for interior finish in various occupancies or portions of a building. For example, corridors in a hospital would be limited to a lower flame spread rating than corridors in an office building where all the occupants are capable of moving themselves.

In addition to the flame spread rating, the tunnel test provides two other measures of combustibility. These are the fuel contributed and the smoke developed. The measure of fuel contributed by the specimen in the test is rarely used in the building codes and is not even listed for many modern materials.

The smoke-developed rating is sometimes used in codes. The smoke-developed rating is not a measure of the toxicity of the products of combustion of a particular material. It is a measure of the visual obscurity created by the smoke generated by that material. This is determined by passing a beam of light through the exhaust end of the tunnel furnace. The light beam impinges on a photoelectric cell that is located at the opposite end.

A sample rating for a typical building material is as follows:

Acoustical Ceiling Tile

Flame Spread: 25

Smoke Developed: 10

Fuel Contributed: 5

Fire Load

The arrangement of materials in a building directly affects fire development and severity and must be considered when determining the possible duration and intensity of a fire. Because of this, one manner by which occupancies are classified is by their fire load. The concept of fire loading is used in most building/fire codes and by government agencies.

Fire load is defined as the maximum heat that can be produced if all the combustible materials in a given area burn. Maximum heat release is the product of the weight of each combustible multiplied by its heat of combustion. In a normal building, the fire load is calculated for the following:

- Structural components
- Interior finish
- Floor finish
- Combustible contents

If fire inspectors can estimate the fire load when conducting inspections of a building, they can subsequently estimate the loss potential. This information can be extremely useful when preparing pre-incident plans, calculating water flow requirements for automatic sprinklers, calculating fire flow demands, and determining fire extinguisher placement.

Historically, fire loading characteristics have been based on a system that was developed by the U.S. National Bureau of Standards (NBS) in the 1950s. (**NOTE:** Since that time, the NBS was renamed the National Institute of Standards and Technology [NIST]). The NBS study developed an approximate relationship between fire loading and an exposure to a fire severity equivalent to the standard time-temperature curve that is used in fire protection. Based on their evaluations, basic fire load classifications were derived for "typical" occupancies. They include:

Slight: Typical Fire Load 5 PSF (Pounds per Square Foot [24 kg/m²]) **(Figure 3.21)**

- Well-arranged office, metal furniture, noncombustible building
- Welding areas containing few combustibles
- Noncombustible power house
- Noncombustible building, small amount of combustible contents

Moderate: Typical Fire Load 10 PSF (49 kg/m²) **(Figure 3.22)**

- Cotton and waste paper storage (baled) in a well-arranged noncombustible building
- Paper manufacturing, noncombustible building
- Noncombustible institutional building with combustible occupancy

Moderately Severe: Typical Fire Load 10 to 15 PSF (49 kg/m² to 73 kg/m²) **(Figure 3.23)**

- Well-arranged combustible storage in a noncombustible building
- Machine shops with noncombustible floors

Severe: Typical Fire Load 15 to 20 PSF (73 kg/m² to 98 kg/m²) **(Figure 3.24)**

- Manufacturing areas with combustible products in a noncombustible building

- Congested combustible storage areas in a noncombustible building

***Very Severe: Typical Fire Load Greater Than 20 PSF (98 kg/m²)* (Figure 3.25)**

- Flammable liquids
- Woodworking areas
- Office with combustible furniture in a combustible building
- Paper working and printing operations
- Furniture manufacturing
- Machine shop with combustible floors

Figure 3.21 A slight fire load can be anticipated in an office building of noncombustible design.

Figure 3.22 This school building would be considered to have a moderate fire load.

Figure 3.23 A machine shop contains a moderately severe fire load.

Figure 3.24 A manufacturing operation in a noncombustible structure is considered a severe fire load.

Figure 3.25 Printing operations contain very severe fire loads.

Although the listed figures for fire loads are still used in many codes and regulations, their importance in the field of fire protection and fire protection engineering has decreased in recent years. At best, the numbers determined by the NBS were rough estimates and not totally accurate. In particular, the fire load ratings are not an accurate indicator of fire severity with combustibles that have a high heat of combustion and produce temperatures significantly higher than those used on the standard time-temperature curve.

Today, the effects of the design and contents of a building on fire growth and spread may be more accurately determined by using computerized fire modeling programs. These programs not only more accurately predict fire conditions, but they can also determine smoke and heat movement patterns that can be anticipated within the structure. As these programs become more prevalent, traditional fire load ratings will become obsolete. These programs are based on heat release rates for specific fuels as opposed to traditional fire load ratings (in PSF).

Roof Coverings

The roof covering is the final outside layer that is placed on top of a roof deck assembly. Common roof coverings include composite and wood shingles, tile, slate, tin, and asphaltic tar paper (Figures 3.26 a through f). The combustibility of the surface of a roof is a basic concern to the fire safety of the entire community. Some of the earliest fire regulations ever imposed in the United States, hundreds of years ago, related to combustible roof covering issues. These were a result of several conflagrations that were caused by flaming brands flying from roof to roof.

Figure 3.26a A composite shingle roof.

Figure 3.26b A wood shake roof.

Figure 3.26c A tile roof.

Figure 3.26d A slate roof.

Figure 3.26e A tin roof.

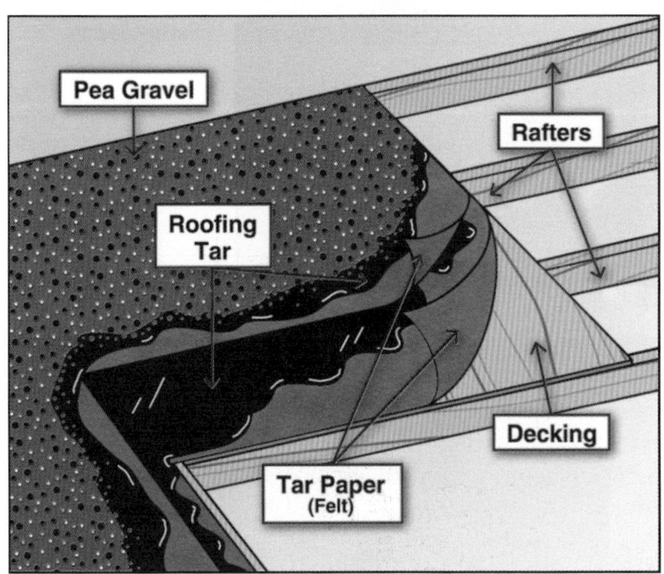

Figure 3.26f The elements of a tarpaper, or built-up, roof.

Test methods have been developed to evaluate the fire hazards of roof coverings. NFPA 256, *Standard Methods of Fire Tests of Roof Coverings* describes the appropriate procedures. The test evaluates the flammability of the roof covering, the protection it provides to a combustible roof deck, and the potential for producing flaming brands. To receive one of the classifications, the roof covering is given a series of fire tests of varying degrees of severity. In order to test roof coverings, several test decks of each type of roof are constructed. The following tests are conducted on each roof deck:

- Intermittent flame exposure test
- Flame spread test
- Burning brand test
- Flying brand test
- Rain test

The tests involving flame require a gas burner for the flame source, a wind tunnel, and an air blower.

INTERMITTENT FLAME EXPOSURE TEST

In this test, a gas flame of a specified temperature is applied to the test deck in an on-and-off fashion. The length of application varies for each class of roof. The roof covering is then observed to determine:

- If sustained flaming occurs on the underside of the roof deck
- Whether flaming or glowing brands are produced
- If portions of the test samples are displaced
- Whether portions of the roof deck are exposed and fall away

FLAME SPREAD TEST

In this test, the gas flame is applied continuously for 10 minutes to Class A and B test decks or 4 minutes to Class C test decks. During and after the test, observations are made to determine the distance to which flame spreads, whether flaming or glowing brands are produced, and if portions of the test sample are displaced.

BURNING BRAND TEST

The burning brand test is used to determine whether burning brands are likely to ignite roof coverings. Wood brands are constructed of varying sizes, depending upon the class of the roof test deck. The appropriate brands are then ignited and secured to the test decks. (For Class C test decks, the burning brands are applied at 1 to 2 minute intervals.) For all types of roofs, the tests are continued until the brands are totally consumed. Observations are made to determine if:

- Sustained flaming occurs on the underside of the test deck
- Flaming or glowing brands are produced
- Any part of the test sample is displaced
- Any part of the roof deck is exposed or falls away

FLYING BRAND TEST

In this test, a flame is applied to the roofing for a specified time with a 12 mph (19 km/h) air current. Observations are made to see if flying brands will develop.

RAIN TEST

The rain test is used to determine if the fire-retardant abilities of the roof are adversely affected by rain. In this test, spray nozzles are mounted above the test decks. These nozzles deliver an average of 0.7 inches (18 mm) of water per hour. All test decks are exposed to 12 one-week cycles; each cycle consists of 96 hours of water application followed by 72 hours of drying time. When the rain test is complete, the intermittent flame test, burning brand test, and flying brand test are repeated.

After all roof covering tests have been conducted, roof coverings are classified based upon test results:

- Class A — Roof coverings that are effective against severe fire exposures. They afford a high degree of fire protection to the roof deck, do not slip from position, and do not present a flying brand hazard.
- Class B — Roof coverings that are effective against moderate fire exposures. They afford a moderate degree of fire protection to the roof deck, do not slip from position, and do not present a flying brand hazard.
- Class C — Roof coverings that are effective against light fire exposures. These coverings provide a light degree of fire protection to the roof deck, do not slip from position, and do not present a flying brand hazard.

Class A materials possess the best fire-retardant properties. Class C materials possess the least. Remember that the rating is for the whole deck assembly, not just the material itself. For example, a roof deck constructed of asphalt-asbestos felt-assembled sheets of four-ply thickness has a Class A rating. However, the same material in a single thickness would have a Class C rating.

Fire Walls and Partitions

A *fire wall* is a wall with a specified degree of fire resistance that is designed to prevent the spread of fire within a structure or between two structures. (**NOTE:** Some model codes do not recognize the term "fire wall." They may be known as area or fire separation walls). The fire wall should extend from the foundation of a structure through and above combustible roofs so that it will stop flame spread on the roof covering and to adjacent structures (Figure 3.27). This protection is accomplished by topping the fire wall with a parapet.

Because fire walls are designed to protect against the spread of fire, no combustible construction is permitted to penetrate these walls. Fire walls are also self-supporting so that they can maintain their stability in the event of a structure collapse on either side of the wall.

Self-supporting fire walls are most often found in one- and two-story industrial occupancies, storage facilities that have unprotected steel frame construction, and garden apartment and townhouse residential occupancies (Figure 3.28). The minimum height will be specified by the respective building code, and the maximum height will be limited by economics and practicality. Free-standing fire walls are designed to be totally self-supportive under vertical loads; however, strong horizontal loads, such as seismic forces (earthquakes), wind, or failure of an attached structure, may cause them to collapse.

There are several important items that fire inspectors must check when performing an inspection in a building that contains fire walls. The walls must be of the proper fire resistivity. The code being enforced should specify the required rating of fire walls; typical requirements will range from two to four hours. Walls should be checked for

Figure 3.27 A proper fire wall should extend completely through the roof.

Figure 3.28 Fire walls are used to separate each occupancy in a townhouse or rowhouse development.

cracks, crumbling bricks, rotten mortar joints, and wooden lintels that carry masonry walls (in older buildings). Penetrations of the wall for purposes of utility extension should be limited where possible. All holes around utility penetrations should be sealed according to the applicable code. Ducts that penetrate the wall must be equipped with proper fire dampers to prevent the products of combustion from flowing through them.

Doors that penetrate fire walls must be examined to determine if they are defective, inoperative, or blocked open. Any of these conditions will compromise the effectiveness of the fire wall. The fire inspector should also evaluate manufacturing or operating processes in the structure to determine if there is a possibility of a flash fire or explosion that could destroy or damage the wall.

A *fire partition* differs from a fire wall in that the partition has a lesser degree of fire resistivity than a fire wall, and extends from one floor to the underside of the floor above or to the underside of a fire-rated ceiling assembly. Fire partitions are built with noncombustible or protected combustible materials. They are supported by structural members having a fire-resistance rating equal to or greater than that of the partition. The type of construction, the size of the area protected, and the severity of the fire hazard will govern the degree of fire resistance required. This generally ranges from ¾-hour to 2 hours.

Fire Doors and Door Assemblies

The number and size of openings in fire walls and fire partitions should be limited to the minimum number possible without interfering with building operations. Doorways and other openings in fire walls reduce the wall's effectiveness as a fire barrier; therefore, fire doors are installed to retain as much fire integrity as possible.

A *rated fire door* is a door that has complied with the requirements of the American Society for Testing and Materials (ASTM) Test E152, *Fire Test of Door Assemblies*. These tests are usually conducted at one of the research laboratories listed earlier in this chapter. NFPA 252, *Standard Methods of Fire Tests of Door Assemblies*, also describes the methods of fire tests applicable to various types of doors and door assemblies. Fire doors must be labeled as such. Labels or classification marks may be of metal, paper, or plastic, or may be stamped or diecast into the door (Figure 3.29). Labels are commonly found on the edge of the door. The hinged edge is the most common position; however, some doors also have them on the top edge.

Figure 3.29 Fire doors should be marked as such somewhere on the door.

A fire door assembly includes the fire door, the door frame, the door closing and latching hardware, and other accessories. When all of these elements are assembled, they are capable of providing a specified degree of fire resistivity. Standard operating designs for fire doors are:

- Horizontal sliding (Figure 3.30)
- Vertical sliding
- Single swinging (Figure 3.31)
- Double swinging (Figure 3.32)
- Overhead rolling (Figure 3.33)
- Counterbalanced (Figure 3.34)

Fire doors that are designed to serve as exit enclosures, such as into stairwells, must be of the hinged type.

Figure 3.30 A horizontal sliding fire door.

Figure 3.31 A typical single swinging fire door.

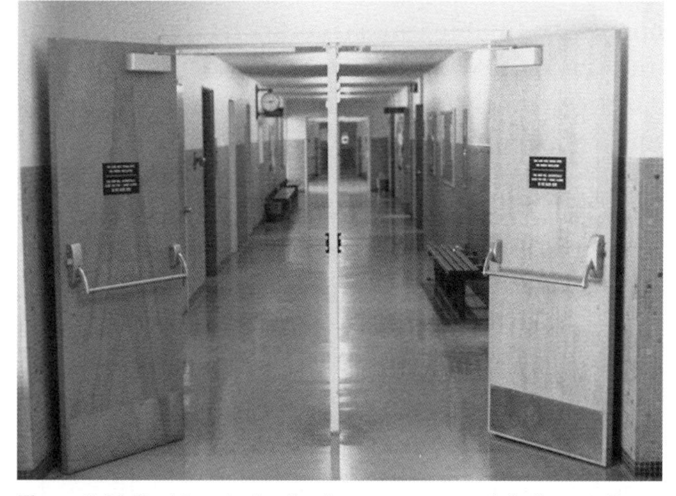

Figure 3.32 Double swinging fire doors are common in hallways of large structures.

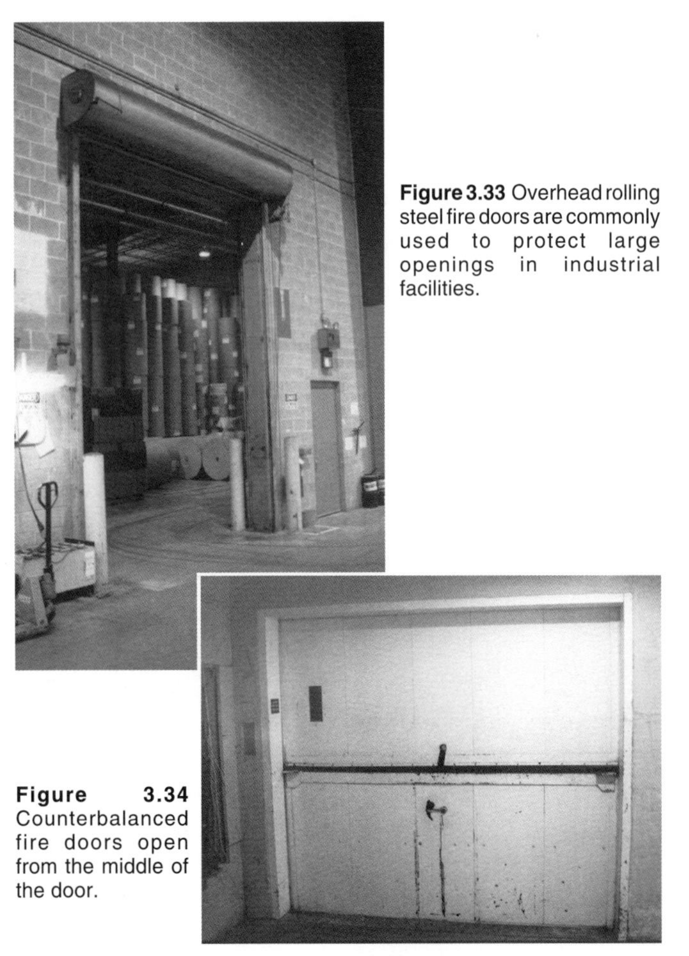

Figure 3.33 Overhead rolling steel fire doors are commonly used to protect large openings in industrial facilities.

Figure 3.34 Counterbalanced fire doors open from the middle of the door.

To provide an effective barrier against fire, fire doors must be closed during the fire. Self-closing doors automatically return to the closed and latched position after being opened. Automatic-closing fire doors include those that normally remain open but close when a fire situation occurs. There are several devices that are used to close the fire door when necessary. Doors may be held open with fusible links that are designed to melt and fall out when a specific temperature is reached, much in the same manner as an automatic fire sprinkler operates. Another method uses an electronically operated solenoid or magnet to hold the door open. This latch is triggered to allow the door to close when an automatic fire detection or suppression system is activated. All fire doors that are installed for *life safety* purposes must be of the electronic activation type.

When doors that serve as part of an exit are kept closed, they may be latched but not locked against the egress. Otherwise, people will not be able to use them unless they have a key. This is something for which fire inspectors must always be very alert. Fire inspectors must also make sure that rated fire doors are not wedged or otherwise blocked open in any manner other than using the approved opening hardware. This would prevent the door from closing under fire conditions and compromise the integrity of the entire fire wall.

Horizontal sliding fire doors are mounted close to the wall and require little floor space and room for operation. Materials stored against these doors may, however, render them inoperative. Horizontal sliding doors are generally used to close large openings in fire walls between sections in a large building. These doors are normally kept open. Swinging doors that fit into a rabbeted jamb generally fit tighter than sliding fire doors. However, an automatic closer is easier to employ on a sliding door. Vertical sliding doors are also normally kept open and arranged to close automatically. Overhead doors, such as rolling steel doors, may be installed where space limitations prevent the installation of other types of fire doors. Fire inspectors should make sure that no objects are in the door opening that would prevent the door from closing completely to the floor. Counterbalanced doors are generally used on openings to freight elevators. They are mounted on the face of the wall inside the elevator shaft.

Fire doors that are used in stairwells and corridors may have windows. These windows are acceptable if the door has been tested and rated with the window in place. Windows are not allowed in 3-hour rated doors or 1½-hour rated doors that are used in severe exterior fire exposure locations. The 1½-hour doors used at other locations may have a window that is no larger than 100 sq. in. (0.065 m^2). Doors rated for ¾-hour may have up to 1,296 sq. in. (0.84 m^2) of window, though no dimension of the window may exceed 54 inches (1.37 m). The windows used in fire doors must be of the wired-glass type (Figure 3.35). The wire in the glass distributes the heat, thereby lowering thermal stress and increasing the overall strength of the assembly. Even if the glass cracks from the heat, the wire will hold it in place and maintain much of the fire resistance of the assembly. Wire glass typically begins to weaken at about 1,500°F (816°C). Newer styles of heat-resistant glass that

Figure 3.35 Wired glass in a fire door.

do not contain wire mesh are now available. Some model codes require these new styles of glass instead of wire glass.

Inspection of fire doors is relatively simple. The inspector should check to ensure that the entire door assembly is of the appropriate rating and is listed and labeled for a specific use. Fire doors are given hourly fire-resistance ratings in much the same manner as walls and partitions. For example, fire doors that have a 3-hour rating are used to protect openings in fire walls that have a 3-hour resistance rating. Doors should be installed and maintained in accordance with local building codes, manufacturer's instructions, and NFPA 80, *Standard for Fire Doors and Fire Windows*. All doors should have a clear path for opening and closing and should move easily. Fusible links and rollers should not be painted. All window glass and fire-rated hardware should be in place and in good repair. Make sure that the doors close tightly and latch properly.

Draft Curtains (Curtain Boards)

Draft curtains, also called curtain boards, are designed to limit the mushrooming effect of heat and smoke as it rises in large areas of buildings that are not otherwise subdivided by partitions or walls. By allowing the heat and smoke to concentrate in the area between the draft curtains, venting systems will activate more quickly. This will speed the process to clear heat and smoke from the area.

Draft curtains may be constructed of any sturdy, noncombustible material that resists the passage of smoke through it. They must be fastened to the ceiling. The depth of the draft curtain should be at least 20 percent of the ceiling height in the protected area. Generally, they should not extend below 10 feet (3 m) from the floor. The maximum distance between draft curtains should not exceed eight times the ceiling height. They may be placed closer than this, but they should not be placed any closer than twice the ceiling height, unless the curtains extend down to a depth that is at least 40 percent of the ceiling height.

There is very little for fire inspectors to be concerned about when dealing with draft curtains during a routine inspection. Most of the important issues involving draft curtains must be addressed at the time of their construction. Once in place, the fire inspector just needs to make sure that no alterations have been made that would decrease their effectiveness under fire conditions.

Smoke and Heat Vents

Removal of smoke, heat, and toxic gases produced in a fire situation is frequently the determining factor in successful rescue and fire control operations. Studies have indicated that venting a building involved in a fire creates better visibility conditions for occupants and firefighters. However, ventilation also results in more rapid fuel consumption by the fire. Smoke and heat vents are most commonly used in the following types of buildings:

- Large, single-story structures
- Windowless structures (Figure 3.36)

Figure 3.36 Windowless buildings present a ventilation challenge.

- Underground structures or areas
- High-rise buildings
- Other buildings that have serious life safety concerns and cannot be vented by firefighters using traditional means.

When properly installed and appropriate in number, these vents will direct the flow of smoke and heated gases away from access points and escape routes, release them from the building, and restrict their spread within the structure (Figure 3.37). All vent installations should be in accordance with NFPA 204M, *Guide for Smoke and Heat Venting*.

Figure 3.37 Automatic smoke vents aid the vertical ventilation process during fire conditions. *Courtesy of The Bilco Co.*

In general, smoke and heat vents are most common in structures that are not protected by automatic sprinkler systems. The exceptions to this rule are high-rise and underground structures. There are varying opinions on the combined effectiveness of venting systems and automatic sprinkler systems. Research on the topic has drawn different results. Some research has indicated that the additional fresh air created by the venting may increase the intensity of the fire beyond the sprinkler system's ability to control it. The cooling effect of the sprinkler discharge may also work against the vent's ability to clear heat and smoke. Other research has shown the ill effects are minimal.

Older structures contained a variety of different vent systems. These included monitors, sawtooth skylights, continuous gravity vents, breakable glass, and stationary shutters (Figure 3.38). Many of these were not energy efficient under normal conditions and not always reliable under fire conditions. Today, there are three common types of smoke and fire vents in use:

- Manually operated vents
- Automatic unit vents
- Mechanical venting systems

Manual vents are typically metal roof hatches that are spring loaded and kept closed by a latch. Under fire conditions, a person must manually open them by pulling a wire or cable that is anchored somewhere near floor level. These are only effective if people are around to activate them. Inspectors looking at this type of vent should make sure that they operate as intended.

Automatic unit vents come in two designs: fusible link and drop-out panel. Both of these usually range in size from 16 to 100 sq. ft. (1.5 m^2 to 9.3 m^2). The fusible link type consists of spring-loaded, hinged metal doors or plastic domes set in a metal frame. The doors or dome are held closed by the fusible link (Figure 3.39). When the link melts, the doors or dome pop open. The fire inspector should make sure that the fusible link has not

Figure 3.38 Monitor vents may be used to vent heat and smoke.

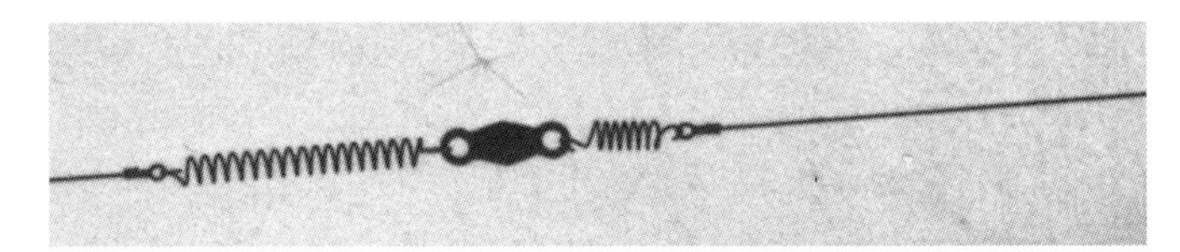

Figure 3.39 Fusible links may be used to hold automatic fire vents closed.

been painted or otherwise altered to adversely affect its operation. The fusible link should have a higher temperature rating than the sprinklers in the area to assure that sprinklers will operate before the vent opens. When performing a roof inspection, both manual and automatic vents should be checked to make sure that nothing is sitting on top of them to prevent them from opening. Inspectors should also make sure that security conscious building owners have not locked the vents in a closed position.

The drop-out panel type consists of a temperature-sensitive, transparent or translucent, thermoplastic dome or panel that deforms when exposed to heat and falls out of its mounting (Figure 3.40). Make sure that the mounting or panel has not been altered to the point that the panel will not readily drop out when exposed to heat. Sloppy painting can be one cause of this.

Mechanical venting systems are complex systems of ducts, shutters, and blowers designed to move smoke through and then out of a structure. These are most commonly used in high-rise and underground structures. Inspectors should make sure that all necessary information about these systems is obtained and passed on to fire companies who might respond to the structure. Fire companies will need to have a thorough understanding of these systems to make sure that they can operate them properly under emergency conditions.

Figure 3.40 Thermoplastic vents automatically drop out when they are subject to high levels of heat.

Fire and Smoke Dampers

A *fire damper* is a device that automatically interrupts air flow through all or part of an air handling system, thereby restricting the passage of heat. Most fire dampers have a fire-resistive rating of at at least 1½ hours; some are rated up to 3 hours. They may be of single-, multiple-, or interlocking-blade design. Fire dampers are usually activated by fusible links and most are designed for vertical installation. Fire dampers must be listed by an approved testing agency such as UL.

A *smoke damper* is a device that restricts the passage of smoke in an air handling system. It operates automatically upon activation of a smoke detector. Smoke dampers are required in air ducts that pass through smoke barrier partitions. They are supposed to stop the flow of smoke within the duct.

The use of fire and smoke dampers is governed in NFPA 90A, *Standard for the Installation of Air Conditioning and Ventilating Systems;* NFPA 90B, *Standard for the Installation of Warm Air Heating and Air Conditioning Systems*, and by local codes and ordinances. The model building codes tend to require dampers in more locations than do NFPA standards.

Fire Stops

Fire stopping is a means of preventing or limiting the spread of fire in areas such as:

- Hollow walls or floors
- Above suspended ceilings
- In penetrations for plumbing and electrical installations
- Cocklofts
- Crawl spaces

The most common form of fire stops that fire inspectors will be concerned with are those used in wood-frame construction. In this application, fire stops are pieces of 2-inch by 4-inch (50 mm by 100 mm) lumber placed between the studs in walls, partitions, and ceilings at each floor level and at the upper end of the stud channel in the attic (Figure 3.41). These wooden fire stops cut off the draft within the walls and prevent the spread of fire and smoke within the concealed space.

Other common fire stopping applications that fire inspectors will need to be familiar with are those involving closing penetrations through a fire

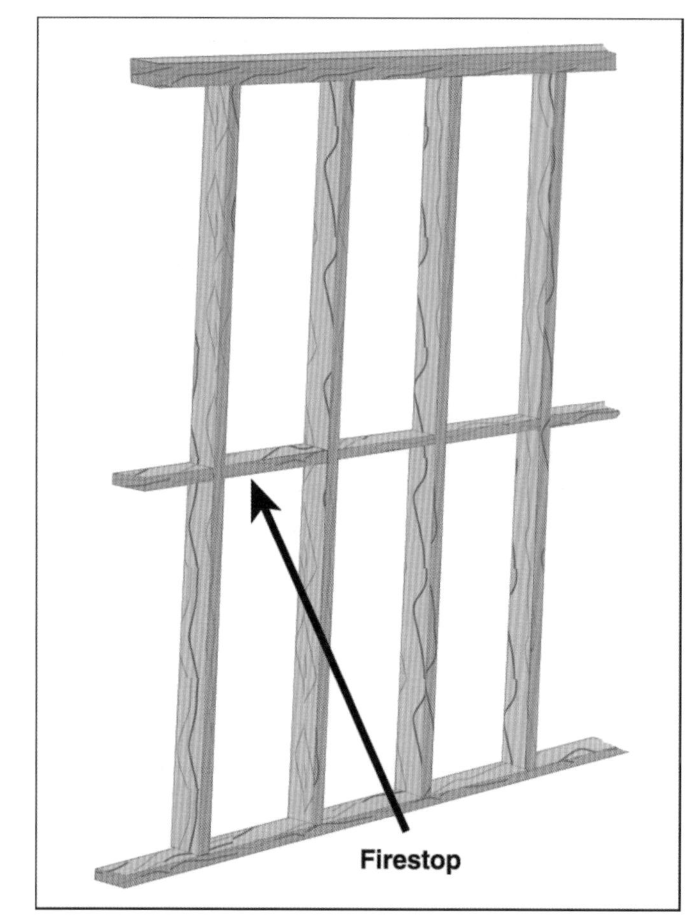

Figure 3.41 Fire stops within walls help limit vertical fire spread.

wall (Figure 3.42). These penetrations are typically created for building utilities such as electric, water, sewer, HVAC, and other services. There are two types of fire stops that are used in these situations: through-penetration fire stop devices and through-penetration fire stop systems. *Fire stop devices* are manufactured products used to protect small penetrations such as those involving wiring or cable installations. *Fire stop systems* are constructed on an as-needed basis to meet a specific application. They are commonly used to protect larger penetrations such as pipes, ducts, and cable trays. Recognized third-party testing agencies provide approved installation methods.

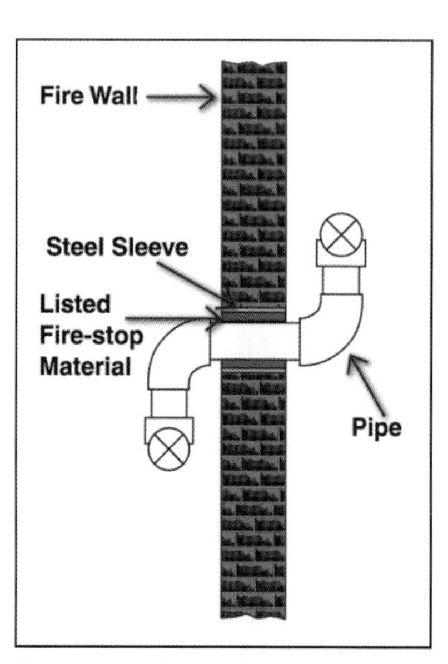

Figure 3.42 One example of a properly enclosed fire wall penetration.

Fire stop devices and systems must be constructed of materials that do not readily break down when exposed to heat or flame. Acceptable materials include:

- Sheet metal
- Gypsum board
- Brick
- Plaster
- Mineral fiber insulation (*not* fiberglass insulation)
- Cement grout
- Ceramic fiberboard
- Sand

The actual construction of fire stop systems must follow guidelines that are established by the American Society for Testing and Materials (ASTM) Standard E-814, *Methods for Fire Tests of Through-Penetration Fire Stops.* The standard provides for two types of hourly fire ratings for fire stops. These are called *F* and *T* ratings. The *F rating* is used to measure a fire stop's ability to maintain its physical integrity when exposed to fire and hose streams and to prevent the start or passage of flame on the unexposed surface of the stop. The *T rating* measures whether combustible materials on the unexposed side of the fire stop would be ignited by conduction of heat through the fire stop material or along the penetrating conduit.

Flammable Decorations

A common fire protection consideration encountered by fire inspectors is the issue of flammable decorations in a particular occupancy. These decorations may be of a permanent or temporary nature. Permanent decorations are typically found in places such as churches, nightclubs, restaurants, and theaters. Temporary decorations are commonly found in stores, schools, and college fraternity and sorority houses. As stated earlier in this chapter, rapid fire development and flame spread have been contributing factors in many large-loss-of-life fires, including the Beverly Hills Supper Club and Coconut Grove Nightclub fires.

When considering decorations with regard to fire safety, fire inspectors must realize that the degree of hazard involved depends upon the ease with which the decoration can be ignited by cigarettes, sparks, electrical defects, and other similar type heat or ignition sources. The inspector must be familiar with the requirements for decorative materials contained in whichever model code is used in that jurisdiction. Each of the model codes has varying requirements, depending on type of occupancy, for dealing with flammable decorative materials.

Occupants should be encouraged to use decorative materials that are nonflammable in nature. If the occupant wishes to use a decorative material that is inherently flammable, such as wood, cardboard, cloth, straw, etc., he should be required by the code used in that jurisdiction to use material that has been appropriately treated to be fire retardant. There are four basic methods for making a material fire retardant. They include:

- ***Chemical changes*** — This is done primarily with plastics and synthetic fibers. It involves altering the chemical compounds used to make the material so that it will be less susceptible to burning.
- ***Impregnation*** — This technique may be used on materials that are absorbent in nature. A fire-retardant agent, usually dissolved in water, is saturated into the material to make it fire resistive.
- ***Pressure Impregnation*** — This technique is used on dense, nonabsorbent materials such as wood. The fire-retardant agent is forced into the material using vacuum pressure. The fire retardant replaces the air cells inside the material to provide better fire resistance.
- ***Coating*** — While the three previous methods must be completed during the manufacture of the material, coating may be performed at anytime during the life of the material. It involves spraying any one of several recognized fire-retardant coatings over the material to decrease its susceptibility to fire.

The first three methods listed above will be carried out before the material is installed in an occupancy. The fire inspector simply needs to verify that the material meets the applicable code requirements. More commonly, the fire inspector will encounter situations where a nonfire material is in place, or desired to be put into place, and the occupant wishes to treat it with a fire-retardant coating. Questions about fire retardant coatings are commonly brought to the fire inspector's attention, particularly during Christmas holiday season. Members of the public frequently seek information on protecting Christmas trees and decorations. There are four basic types of fire-retardant coatings that may be used on wood or wood products. They include:

- ***Intumescent paints*** — Although a paintlike coating under normal conditions, when heated, these paints swell to a puffy form that resembles burnt marshmallows. This action provides protection by forming a heat insulative barrier, excluding oxygen from the fuel, producing diluent gases, and reducing the production of flammable gases.
- ***Mastics*** — These are thick substances that are sprayed or manually applied to the wood. They may form a hard finish or soft, tarlike finish.
- ***Gas-forming paints*** — These are paints that release a noncombustible gas when heated. This displaces oxygen from the fuel area and prevents ignition.
- ***Cementitious and mineral fiber coatings*** — These are coatings that in the past have been used primarily on steel structural members. In recent times, they have seen limited use on wood-based materials. Further research is being conducted on these applications.

These coatings are generally effective when properly applied and maintained. Fire inspectors should be aware of the common brands and proper application of fire retardants used in their jurisdiction. Make sure that only approved coatings are used. Many substances that are marketed as fire retardants (especially for cut Christmas trees), such as borax and boric-acid solutions, diammonium phosphate, and ammonium sulfate, are not truly effective and should be avoided. In reality, these susbstances deplete the moisture content in the tree. In the long run, this increases the fire risk posed by the tree.

If there is any question about the flammability of a material encountered by the fire inspector, a field test may be performed to determine if an immediate hazard exists. To perform a field flame test for hanging decorations or films, the fire inspector should remove a minimum of a ½-inch by 4-inch (13 mm by 100 mm) sample of the material for testing. The material should be suspended by one end from a spring clip or similar holder. Apply a small flame from a common kitchen match or similar material to the center of the bottom edge of the material, with the flame tip being ½-inch (13 mm) below the material (Figure 3.43). The flame should be held there for 12 seconds and then slowly moved away.

During exposure, flaming should not spread over the entire length of the sample. If the sample is longer than 4 inches (100 mm), flaming should not spread at least 4 inches (100 mm). The material should not support combustion or continue to flame for more than two seconds following the removal of the test flame. Materials that break or drip flaming particles should also be rejected.

WARNING

The test material may be highly flammable. Perform flame tests in such a way that they cannot result in a fire that can get out of control. The test material, or sample thereof, should be removed to a safe location for testing.

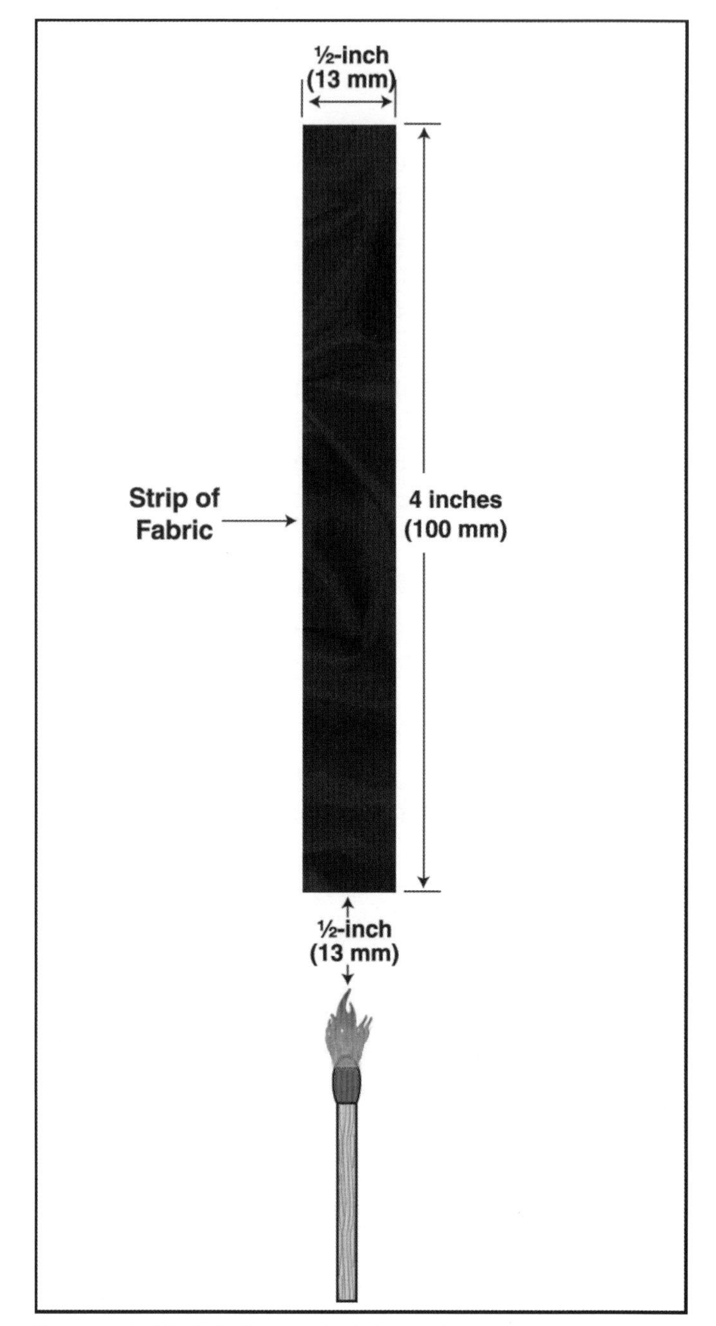

Figure 3.43 Hold the flame ½-inch (13 mm) from the test material.

For more information on field tests for flammable decorations, see NFPA 705, *Recommended Practice for a Field Flame Test for Textiles and Films.*

4

Fire Hazard Recognition

This chapter contains information that will assist the reader in meeting the listed job performance requirements contained in NFPA 1031, ***Standard for Professional Qualifications for Fire Inspector and Plan Examiner*** (proposed 1998 edition).

Chapter 3 Fire Inspector I

3-3.8 Recognize hazardous conditions involving equipment, processes, and operations, given field observations, so that the equipment, processes, or operations are conducted and maintained in accordance with applicable codes and standards, and all deficiencies are identified, documented, and reported in accordance with the policies of the jurisdiction.

Chapter 4 Fire Inspector II

4-3.6 Evaluate hazardous conditions involving equipment, processes, and operations, given field observations and appropriate documentation, so that the equipment, processes, or operations are installed in accordance with applicable codes and standards, and all deficiencies are identified, documented, and reported in accordance with the policies of the jurisdiction.

Chapter 4
Fire Hazard Recognition

Because the variety of occupancies that fire inspection personnel are responsible for checking is almost endless, there is a wide variety of hazards and processes with which the inspectors must be familiar. The inspector must be able to identify the process or equipment, know how and for what it is used, and understand the basic safety principles that apply to it.

It would be impossible in a manual of this size and scope to cover all the types of hazards and processes that fire personnel who perform inspections might encounter. However, through experience and statistics, we can account for those hazards and processes that historically have caused the greatest amount of concern and problems for the fire service. It is our intention to cover those hazards in this chapter.

ELECTRICITY

Historically, electricity has always been a common cause of fires. When principles of safety are not followed, electrical energy can produce unwanted and unexpected fires. Most electrical fires are caused by either arcs or overheating. For example, even in an ordinary household lamp, less than 10 percent of the electricity (current) is converted into light. Over 90 percent of the energy is converted into heat. Unless electrical energy is used in a very efficient manner, it almost always produces heat as an unwanted by-product.

Detailed electrical inspections are typically the responsibility of qualified electrical inspectors. However, personnel who perform fire inspections should have a basic knowledge of electrical theory so that they can understand what types of electrical energy cause fires and why they occur. In order to understand how electricity can cause a fire, it is important to know what electricity is and why it can be so dangerous if abused.

Basic Electrical Theory

Basically, *electricity* involves the transfer or movement of electrons between atoms. Materials that allow a free movement of a large number of electrons are referred to as *conductors*. Copper, steel, and aluminum are excellent conductors and are among the more common materials used in wiring. Electrical energy moves through these conductors as a result of electrons moving from atom to atom within the conductor. Each of the migrating electrons moves to a neighboring atom and pushes away one or more electrons that, in turn, move to the next atom. This same process occurs again and again. This phenomenon results in the flow of electrical energy. Poor conductors, such as glass, dry wood, or rubber, possess very few free electrons and thus impede the flow of electrical energy through the material. *Semiconductors* are materials that are neither good conductors nor good insulators; therefore, they may be used as either in some applications.

The flow of electricity through a wiring system is like the flow of water through a water distribution system. Water flow has three major components: quantity (gpm or L/min), pressure (psi or kPa), and resistance (friction loss). An electrical flow is also measured by quantity, pressure, and resistance. The quantity of electricity is expressed in terms of amperes (amps). Electrical pressure is termed *voltage*, and it is measured in terms of *volts*. Resistance in an electrical circuit is referred to as *ohms*. The amount of resistance will vary depending on the quality of the conductor or if a circuit or switch is opened. If a circuit or switch is opened, the resistance is infinite, and no current can flow, just as if a valve in a water system were closed. The relationship between water and electrons is shown in Table 4.1.

The relationship between these three variables may be expressed mathematically by Ohm's Law, which states:

$E = (I)(R)$ or $I = E/R$ or $R = E/I$

Where E = pressure in volts

I = current in amperes

R = resistance in ohms

By manipulating the variables in the equation, any value for a variable may be determined if the other two variables are known.

TABLE 4.1
Relationship Between Water And Electrons

Measurement	Water	Electrons
Quantity	gpm (L/min)	amperage
Pressure	psi (kPa)	voltage
Resistance	friction loss	ohms

Electrical Hazards

The four common causes of electrical fires are:

- Old or worn electrical equipment
- Improper use of electrical equipment
- Defective electrical installations
- Accidental causes

With these four causes understood, inspection personnel can take a more in-depth look at electrical equipment and installations. By doing this, they can correct these problems and prevent fires or injury.

In order to safely work around and inspect electrical installations and appliances, inspectors must first be familiar with electrical hazards. Perhaps the greatest threat posed involves the possibility of electrical shock. The seriousness of the shock depends largely on the amount of current (amperage) that passes through the body, the type of current (alternating [AC] or direct [DC]), the path the electricity takes through the body, and the length of time of contact. The damage produced by an electrical shock depends upon the number of vital organs affected by the electricity. Body resistance to the passage of electricity varies from approximately 1,000 to 500,000 ohms, if the person's skin is dry and unbroken. AC currents of 100 milliamperes or less may cause ventricular fibrillation of the heart (rapid, ineffective contractions of the heart). Currents of 100 to 200 milliamperes or more may be lethal.

Many electrical incidents that result in injury, fires, or damaged machinery occur during maintenance and repair operations. Thus, all such operations should be performed by qualified and authorized personnel.

The proper grounding of electrical equipment is extremely important because a poor or improperly connected ground is more dangerous than no ground. Improper grounding is extremely dangerous because the average appliance user has no practical way to discover a poor ground and develops a false sense of security. Incorrectly wired grounds may literally energize the metal shell of an appliance, thus providing a means for the user to be shocked. Three-wire cords with polarized plugs and ground pins or double-insulated tools are highly recommended (Figure 4.1).

Figure 4.1 Electrical equipment that uses three prong plugs is the safest type.

Another serious hazard is present during battery recharging. Hydrogen gas produced during this operation can ignite explosively. Fire inspectors must be sure that proper ventilation is in place, smoking and open flames are prohibited near these operations, explosion-proof electrical equipment is in place, and personnel understand proper procedures, particularly for shutting down chargers and unhooking cables.

CAUTION: Inspection personnel MUST NOT activate switches, or alter or switch electrical components.

Static Electricity

Static electricity refers to the presence of a nonflowing electrical charge that may develop on almost any surface. Normally, each atom in a given material has a balanced number of electrons and protons surrounding it. If any of the electrons is removed or lost due to friction or some other means, there will be more positively charged protons than negatively charged electrons. If this material is then brought against or close to a normally charged material, there will be a momentary electric current, known as a *static discharge*.

Sometimes materials take on an electrical charge when they undergo a process of physical contact and separation. This electrical charge is known as static electricity. Static electricity is hazardous if a charge is present on the surface of a nonconductive material and does not immediately dissipate. In the presence of a mechanism for generating a static charge, a static discharge or arc can occur when there is no good electrical path between the two materials. This static charge may become an ignition source in a hazardous atmosphere.

Many common processes generate static charges. Some examples include:

- Nonconductive fluids flowing through pipes
- Liquids breaking into drops and the drops hitting liquid or solid surfaces
- Air, gas, or steam flowing from an opening in a hose or pipe
- Pulverized materials traveling through chutes or pneumatic transfer devices (Figure 4.2)
- Belts in motion (Figure 4.3)
- Moving vehicles

Fire inspectors must answer four questions concerning static electricity as an ignition source:

- Is there a source that generates a dangerous amount of static electricity?

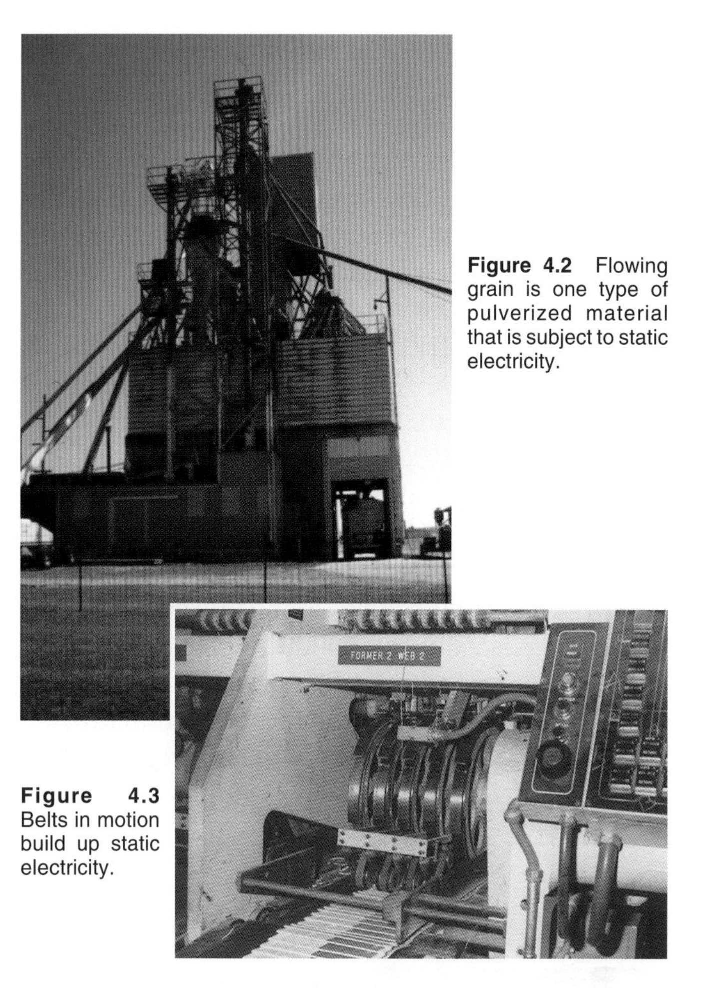

Figure 4.2 Flowing grain is one type of pulverized material that is subject to static electricity.

Figure 4.3 Belts in motion build up static electricity.

- Is there a conductor that will accumulate the charges generated and maintain a suitable difference of electrical potential between the materials involved?
- Will there be a spark discharge of sufficient energy to create a hazardous situation?
- Is there an ignitable mixture present?

Fire inspectors cannot prevent static electricity from being generated, but they can require control measures that will dissipate the generated charges. One method involves controlling the humidity of an area. Maintaining a relative humidity of 60 percent to 70 percent greatly reduces the static electricity problems associated with the manufacturing of paper, cloth, and fiber. Unfortunately, this method is not practical for all occupancies; furthermore, it is somewhat ineffective in controlling static electricity on heated materials and oils.

Bonding and grounding are two other means of dissipating static charges. Bonding involves connecting two objects that conduct electricity with something that is also a conductor (Figure 4.4). This procedure reduces the difference in electrical potential between the two objects that conduct electricity. *Grounding* involves connecting an object that conducts electricity to the ground with something that is a conductor (Figure 4.5). In this case, the procedure reduces the difference in electrical potential between the object and the ground. In either procedure, the bonding or grounding wires (the conductor) must be large enough to carry the greatest amount of current that may be produced in a given situation. Solid conductors may be used for connections that are permanent. However, if the connection will be made and broken frequently, flexible conductors should be used. They may be insulated or noninsulated. Temporary connections should use battery, magnetic, or other clamps that provide metal-to-metal contact, while permanent connections should be pressure-type brazed or welded clamps.

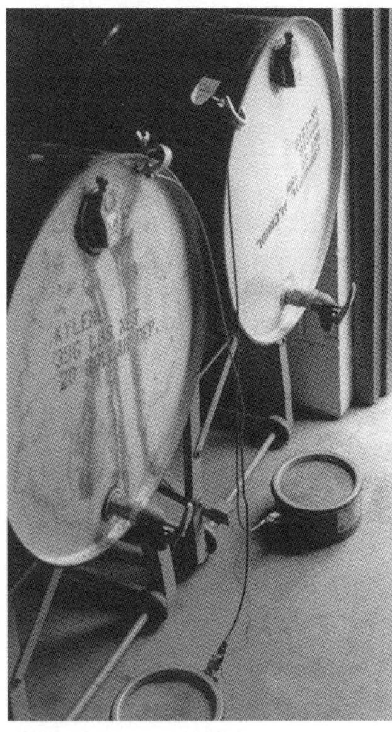

Figure 4.4 Flammable liquid containers should be bonded together before transferring liquid.

Figure 4.5 Flammable liquid containers should also be grounded.

Another acceptable means of dissipating static charges is to ionize the air in the immediate vicinity where the charges accumulate. Several methods can be used for ionization. In the first method, a grounded *static comb* (a metal bar equipped with a series of fine needle points) is brought close to the charged body. The air ionizes at the needle points, and the accumulated charge "leaks" away or dissipates. Static combs are suitable for processes involving paper, fabrics, or power belts.

In the second method, the electrical charges are neutralized with a high voltage device. This device produces a conducting, ionized atmosphere in the area of the charged surface and thereby provides a means by which the charges may dissipate. This method is applicable for processes involving cotton, wool, silk, paper, or printing.

In a third method, an open flame is used to dissipate the electrical charges. This method is frequently used in the printing industry to remove static charges from sheets of paper as they come off the press. However, the method is somewhat more hazardous than the others because the flame has the potential for igniting materials with which it comes into contact.

MATERIALS STORAGE

Materials storage in any operation presents several potential hazards that fire personnel should watch for when performing inspections. Problems, such as storing products too close to ceilings or sprinkler equipment and improper storage of flammable liquids, can be found in many different types of materials storage occupancies. In the following sections, we will examine some of the more common materials storage facilities found in most communities. These include:

- Lumberyards
- Recycling facilities
- Waste-handling operations
- Warehouse/High-Rack Storage operations

Lumberyards

Lumberyards present one of the largest fire hazards in small and medium-sized communities (Figure 4.6). History has shown that large undivided stacks, congested storage conditions, delayed fire detection, inadequate fire protection, and ineffective fire fighting tactics are the principal factors that allow lumberyard fires to reach serious proportions. Fire hazards can be effectively controlled by sound fire prevention practices.

A lumberyard presents an enormous challenge for inspection personnel. Obviously, the primary hazard associated with these occupancies is the

Figure 4.6 Lumberyards can pose a significant hazard to small communities. In this case the lumber is stored within an old mill building of heavy timber construction.

excessive fire load presented by the large amount of combustible material stored on site. Assuring compliance to all applicable codes will lessen the risk for the business owner and the hazard to firefighters in the event of an incident. There are several issues in particular that the inspector should look for while conducting an inspection in a lumberyard.

For open yard storage, lumber stacks should be on solid ground, preferably paved or surfaced with materials such as cinders, fine gravel, or stone (Figure 4.7). The heights of stacks should not exceed 20 feet (6 m). This will assure that they remain relatively stable. Driveways should be spaced so that a maximum grid system of not more than 50 feet by 150 feet (15 m by 45 m) is produced. Driveways between the stacks should be a minimum width of 15 feet (5 m) and have an all-weather surface capable of supporting fire department apparatus. The turning radius of all drives should meet the requirements of the largest fire apparatus that could respond to that location.

Figure 4.7 There should be adequate clearance between all of the buildings and piles of lumber. The lot should be paved or gravel.

For exposure protection, it is recommended that the least combustible materials (sand, stone, etc.) be stored or stacked on the perimeter of the yard to act as a barrier between the yard and adjacent properties or buildings. Unsprinklered buildings containing hazardous manufacturing operations should have at least 50 feet (15 m) of clear space to the nearest lumber stack, shed, or warehouse (Figure 4.8). Boundary markers should be used to assure separation of materials.

It is important to keep vegetation out of the storage area. In dry conditions, vegetation can easily ignite and communicate fire to the material it surrounds. Areas prone to vegetative growth should be sprayed with an effective weed control substance. It is also important for the inspector to make sure such waste materials as bark, sawdust, chips, and other debris, do not accumulate and create an unnecessary fire hazard. Smoking should be allowed only in designated areas. Signs should be posted so that workers know when and where they are allowed to smoke. Consider the need for fire apparatus to be able to travel around a lumber storage yard. Gates and driveways should be wide enough to accommodate your department's largest vehicle (Figure 4.9). Portable heating devices or open fires should not be allowed in the lumber storage area. Heating devices should be limited to approved-type equipment installed in an approved manner. Finally, it is important to consider lightning protection (grounding) in areas prone to lightning strikes.

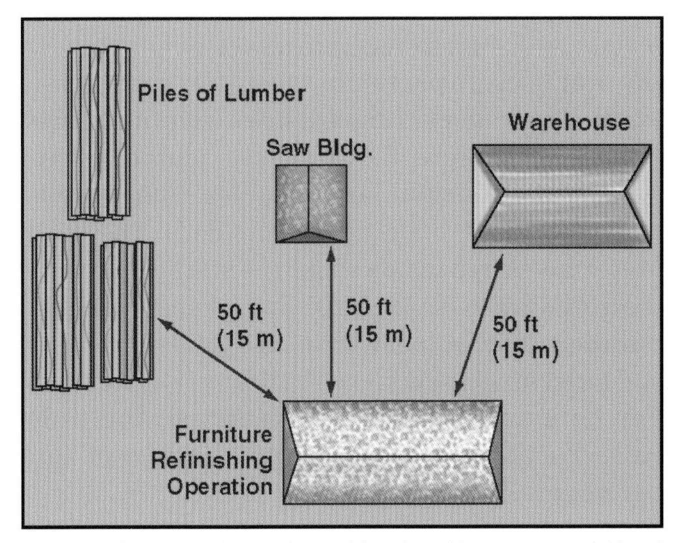

Figure 4.8 There should be at least 50 feet (15 m) between unsprinklered buildings and other hazards.

Figure 4.9 The gate into the lumberyard should be large enough to accommodate the largest fire vehicle that would be expected to pass through it.

Most of the model building codes discussed in the previous chapter have specific requirements for lumberyards. Consult the code used in your jurisdiction for more specific information.

Recycling Operations

Worldwide environmental consciousness has led to a dramatic increase in the field of waste recycling over the past decade. Landfills and dumps have served as large trash receptacles for years, and the effects of this practice are now being realized. With contamination of soil and water around these areas, recycling is a positive alternative. Measures are in place or being developed to be able to recycle almost all waste in our daily lives.

There are two main hazards associated with recycling plants: bulk storage of combustible materials and hazardous processes conducted on the premises. In particular, facilities that handle waste paper and cardboard have extremely high fire loads (Figure 4.10). These materials are usually contained in large bundles. They may be stored inside a building or outside in the yard. Materials stored inside should be stacked no closer than 2 feet (0.6 m) to the ceiling in unsprinklered buildings and at least 18 inches (450 mm) below sprinklers in a sprinklered building (Figure 4.11). For bulk storage outside a building, the same rules discussed above for outdoor storage of lumber generally apply.

Figure 4.10 Fires in bulk paper storage can present a significant challenge to firefighters.

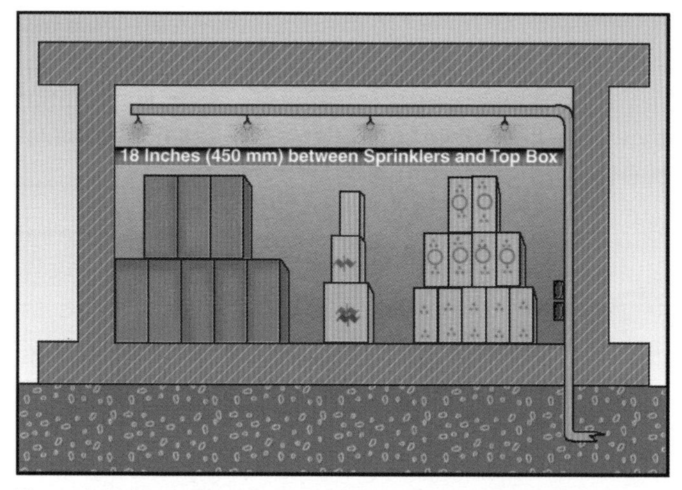

Figure 4.11 There should be a minimum of 18 inches (450 mm) between the sprinklers and the highest objects being protected.

There are many other types of materials that can be recycled in addition to paper and cardboard (Figure 4.12). These include most types of metals and plastics. It is extremely common to have recycling operations that contain different

Figure 4.12 Many communities have recycling centers that collect all sorts of materials.

Figure 4.13 Recycling centers frequently contain large compacting machines that make the material easier to handle for employees.

Figure 4.14 Many industrial facilities and communities have large-scale incinerators to handle the bulk of their trash.

types of materials on the same property. Many community-sponsored recycling programs have collection facilities that handle and process all of these materials.

Inspectors should check all materials handling equipment within the occupancy to make sure it is properly maintained and operated in a safe manner. This includes:

- Waste chutes and handling systems
- Shredders
- Extruders
- Compacting machines (Figure 4.13)

Common sense should prevail when dealing with issues involving the hazards of processing and the handling of different types of materials in the same area. For example, piles of combustible materials should not be stored in the same area where cutting torches are used to salvage scrap metal. Hot ashes, coals, cinders, or materials subject to spontaneous heating should be stored in approved metal containers, and these containers should not be located within 10 feet (3 m) of stored combustible materials.

Waste-Handling Operations

Waste-handling operations refer to those types of operations that involve the disposal of nonrecyclable materials. This can include anything from municipal trash to hazardous or biological materials. This issue may seem simple to some, but it is really a complex area.

A simple solution is to simply burn the waste (Figure 4.14). Depending on the type of waste, this is a viable option. Incineration of most domestic garbage, once metals and glass have been removed, is acceptable in many areas if the proper environmental controls are followed. Incineration can reduce the bulk of the waste by as much as 95 percent. It is also effective in destroying medical, chemical, and biological wastes that might otherwise pose a threat in a conventional landfill. Incineration is not without hazard. Overheating, structural failure, corrosion due to scrubber acids, and failure of the scrubber systems are all hazards which are common to incinerating operations.

When inspecting incinerators, look for some of the following signs that the unit is being operated in an appropriate manner:

- Fuel-fire incinerators are preheated for 30 minutes prior to use.
- Competent, trained operators operate the unit at all times.
- The feed door closes fully, after loading, for the entire combustion cycle.
- The waste material ash compartment is cleaned on a regular basis.
- No dense smoke or serious odor is emitted during operation.

The largest concern is with hazardous waste. Hazardous waste is waste that must have permits and must be disposed of in compliance with the proper regulations. Because of the hazards associated with hazardous waste in the United States, the Environmental Protection Agency (EPA) and the Occupational Safety and Health Administration (OSHA) have implemented strict guidelines to protect operators. Some of these guidelines call for written safety programs, hazardous site identifications, training of employees about hazards, medical surveillance of employees, personal protective equipment, written emergency response plans, and decontamination and sanitation facilities on site.

Hazard reduction can be accomplished in several ways. One way is to separate incompatible chemicals. Disposing of the waste immediately will also limit the exposure time for an accident to occur. Separating combustible waste from areas with buildings is another method of hazard reduction. A continuous quality control and inspections program, which marks containers and areas properly and ensures their integrity, is also important.

A fire department inspector should be aware of all the different wastes and their characteristics on the site. Make sure not to enter any hazardous areas without appropriate personal protection and a briefing on the hazards that might be encountered.

Warehouses

Warehouses present an extreme challenge for fire inspection and suppression personnel because of the magnitude of their fire potential. Warehouses have the potential for rapidly developing large fires because of the large fuel load that is typically contained within them (Figure 4.15). The configuration of the fuel within the structure also lends itself to rapid fire development.

Another challenge for the inspector is the fact that warehouses typically contain a wide variety of materials. In many cases, this inventory changes on a regular basis. No matter how often inspectors visit a warehouse, the next time they return — even if it is the next day — the stock has changed. Attempts to keep incompatible products away from each other are often futile battles in these situations.

There are three primary methods of storing materials within a warehouse. *Palletized storage* consists of materials that are stored on pallets (Figure 4.16). Most pallets are made of wood, although you may occasionally encounter some made of metal or plastic. Generally, materials are stacked 3 to 4 feet (1 m to 1.3 m) high on each pallet, and several pallets may be stacked on top of each other. Depending on the material on the pallets, stacks may approach 30 feet (10 m) in height. The air space between the top and bottom layers of the pallets (generally about 4 inches [100 mm]) creates

Figure 4.15 Large warehouses can contain massive fire loads.

Figure 4.16 Fire may easily travel through the channels created by pallets.

a significant fire hazard in that a natural conduit is created to promote the spread of fire. The fact that pallets may be stacked several layers high also prevents water from sprinklers reaching the bottom layers.

Rack storage consists of structural framework onto which pallets or other materials are placed (Figure 4.17). The height of these racks will vary depending on the inside ceiling height of the building and the equipment available to load and retrieve items from them. Rack storage of heights up to 25 feet (8 m) or so are common. Special automated rack storage systems may exceed 100 feet (30 m) in height. The fire hazards associated with rack storage are similar to those concerns expressed for palletized storage.

Figure 4.17 Rack storage is common in large-scale warehousing operations.

Solid piling involves stacking boxes, bags, bales, or the products themselves directly on top of each other (Figure 4.18). This may be done by hand or forklift. When compared to the other two methods of storage, solid piling gives a fire the least chance to develop because of the lack of air space between the fuel.

Figure 4.18 Solid piled storage, such as with these 3,000 pound (1 360 kg) rolls of paper, is used in some warehouse operations.

Each of the model codes will have specific information on material storage methods, aisle width, and related fire protection issues. You need to be familiar with the requirements of the code you use. One method for assuring a reasonable level of fire safety in warehousing operations is to require properly designed and maintained fire sprinkler systems. In addition to the typical sprinklers found at the ceiling level, warehouses with rack storage may have sprinklers located within the racks themselves (Figure 4.19). More information on these requirements may be found in NFPA 231C, *Standard for Rack Storage of Materials*.

Figure 4.19 Sprinklers within the racks will help to curtail the spread of fire.

When the inspector visits a warehouse facility, it is important to quiz the management or materials handling personnel about how inventory and conditions change. If it is determined that there is a pattern of when the warehouse is near capacity or when it is busier, then the inspector should return at these times and conduct an inspection. The most critical thing for the inspector to check is the automatic sprinkler system. The inspector should make sure that the system is functional and should ask to see records to make sure that the system is being serviced as required (more information on how to do this is contained in Chapter 7). Next, the inspector should examine the storage heights of the stock to see if it is too close to the sprinkler heads, rendering that head ineffective. Information gathered by the inspector during an inspection at a warehouse should include the following:

- Building construction characteristics
- Location of fire walls, fire doors, and barriers
- Types of commodities stored within the warehouse
- Type of packing material used
- Method of storage
- Commodities typical storage levels
- Type and design of automatic sprinkler system
- Minimum flow and pressure required for the system; adequate water system
- Location of water sources, hydrants, standpipes, and fire department connections
- Locations and types of smoke and fire detectors
- Ventilation locations
- Plant personnel's level of training
- Disposal method/storage of waste and packing materials

The inspector must make sure that the aisles are kept clear so that in the event of a fire, a group of boxes on the floor do not allow the fire to travel to the next rack of storage (Figure 4.20). Aisles and exterior doors must be kept clear so that firefighters have an adequate path of travel during fire fighting operations. For more information on warehousing operations consult any of the following:

- The code used in your jurisdiction
- NFPA 231 C, *Standard for Rack Storage of Materials*
- NFPA 231, *Standard for General Storage*

Figure 4.20 Boxes and other materials should be kept out of warehouse aisles.

WELDING AND THERMAL CUTTING

Cutting and thermal welding practices can present significant hazards to the area in which they are being performed. They inherently provide two of the three sides of the fire triangle: an ignition source and oxygen in air. All that is needed is a fuel source and you have the potential for a serious fire. Historically, welding and thermal cutting operations have caused a significant percentage of fires in commercial and industrial occupancies (Figure 4.21). It is important that the inspector be aware of the hazards that exist and aware of the necessary measures to eliminate or lessen the degree of these hazards.

Figure 4.21 Welding operations are a common cause of industrial and commercial fires. Note the welder in the foreground of this picture admiring his work.

Very simply, *welding* operations are those where two pieces of metal are heated and joined together. *Thermal cutting* operations involve using the heat source to melt the metal and cut it apart. There are two basic processes by which welding and thermal cutting operations are performed. The first is by an electrical arc (Figure 4.22). The electrical arc creates a heat source that can be used for melting or joining metals. There are many different types of electrical arc equipment, including:

- Shielded metal arc
- Gas metal arc
- Flux cored arc
- Gas tungsten arc
- Resistance welding
- Flash welding
- Electroslag welding

The second type of welding and thermal cutting equipment utilizes a combination of oxygen and fuel gas to produce a flame. The most common gas that is used for this purpose is acetylene. These are referred to as oxy-fuel gas welding (OFW) or oxy-fuel gas cutting operations (Figure 4.23).

Welding and thermal cutting operations may be conducted in a fixed location within a structure or on an as-needed basis. Examples of a fixed location basis would be a welding business or the welding department in a manufacturing facility. An example of an as-needed welding operation would be workers with portable equipment doing plumbing repairs to a structure.

The two primary fire safety concerns with these operations are making sure the equipment itself is well maintained and in proper working order and making sure combustible materials are kept well away from the areas where these operations are conducted. Electric welding and thermal cutting equipment should be maintained in accordance with manufacturer's instructions and applicable electrical codes. The primary fire hazard associated with oxy-fuel gas equipment is the storage of the oxygen and fuel gases in cylinders. For more specific information on compressed gas safety and inspection principles, see Chapter 10.

Cutting, as well as certain arc welding operations, produces literally thousands of ignition sources in the form of sparks and hot slag (Figure 4.24). One of the primary things that personnel inspecting these occupancies must assure is that there is no chance of the sparks or slag coming in contact with combustible materials. There should be no storage of such materials in the welding or cutting areas. Shop rags or towels should be kept in approved metal containers with lids on them at all times (Figure 4.25).

Figure 4.22 Arc welders use electricity to provide cutting or welding energy.

Figure 4.23 Acetylene torches use a gas flame for cutting or welding.

Figure 4.24 Tremendous amount of hot slag or sparks can be produced by cutting and welding operations.

Figure 4.25 Approved safety containers should be used to store dirty shop rags. *Courtesy of Justrite Mfg., Inc.*

Precautions should also be in place to assure that sparks or slag do not fly from the immediate work area. Openings in walls, floors, or ducts should be covered if within 35 feet (11 m) of the work. Fire-resistance tarps or curtains may be placed around the work area to prevent sparks or slag from flying very far.

When welding and thermal cutting work must be performed on an as-needed basis, the area should be made fire safe by removing combustibles or protecting combustibles from ignition sources. Welding and cutting should never be allowed:

- In sprinklered buildings while the system is not in operation
- In the presence of explosive atmospheres, and drums, tanks, or other containers that have previously held flammable liquids
- In the proximity of large quantities of exposed combustible materials

Where combustible materials, such as paper clippings, shavings of wood, or textile fibers, are on the floor, an area with a radius of at least 35 feet (11 m) must be cleared. Where impractical to produce this radius, the materials must be covered with approved fire-resistant guards or curtains.

A fire watch must be maintained for at least thirty minutes after operations cease. A fire watch is required in locations where fires may develop, when appreciable combustible material is used in the building construction, or combustible contents are closer than 35 feet (11 m) from the point of operation. The fire watch must also monitor adjacent areas when wall or floor openings are within a 35-foot (11 m) radius or exposed combustible materials are in adjacent areas. This includes concealed spaces, walls, and floors. In addition, a fire watch is needed anytime a fire could spread due to conduction or radiation. Fire watch personnel must be trained in the use of fire extinguishers and in procedures to warn occupants and summon the fire department in the event of a fire. Fire extinguishers shall be present at all times when an operation is in progress.

A hot work program should be developed by the proprietor and evaluated by the inspector. The majority of fires occur with portable equipment. Hot work programs can lessen the hazard of fire when proper compliance is followed. A hot work program involves the use of a form which requires the person issuing the permit to ensure that adherence to all safeguards for the prevention of fire have been achieved. An inspection of the area where the work is going to be performed must be done before the permit can be issued. Also, all hazards present in the area, as well as the emergency procedures, must be discussed with the person issuing the work permit.

The minimum information to be contained on this permit includes:

- Date
- Work to be performed
- Period for which the permit is valid
- Location of the job
- Type of suppression equipment available
- Notification if fire watch is assigned
- Inspection of area before work begins
- Authorization signature of individual in charge
- Final signature of the individual after the work has been completed

For more information on these types of operations, consult NFPA 51B, *Standard for Fire Prevention in Use of Cutting and Welding Processes* or the Factory Mutual Data Sheets.

HEATING, VENTILATING, AND AIR CONDITIONING (HVAC) SYSTEMS

All personnel who perform fire inspections must be knowledgeable about heating, ventilating, and air conditioning (HVAC) systems. To effectively inspect these systems from a fire safety standpoint, inspectors must become familiar with the various types and operating conditions of a variety of these systems. There are installation and operating requirements for HVAC systems contained in all of the model building/fire codes and several NFPA standards. Because of this fact, the information in this section is meant to be generic in nature. Consult the code used in your jurisdiction for specific information.

HVAC systems provide a variety of services, including:

- Heating or cooling of air
- Humidifying or dehumidifying of air
- Filtering or cleaning of air

The major components within an HVAC system are heaters, air conditioners, fans, ducts, heat exchangers, and thermostatic controls. Mechanical equipment, such as the heater itself, must be located in a room that is separated from the rest of the building and has a minimum of a one-hour fire-resistance rating (Figure 4.26). Access to this room should be restricted to authorized personnel only. The HVAC room should be free from any storage.

Air intakes for the system should be selectively located to prevent fire, fire gases, and smoke from being drawn into the system. The intakes must also have grates over them to prevent objects from being drawn into the system (Figure 4.27).

Figure 4.26 HVAC equipment should be located in a separate, limited combustible room.

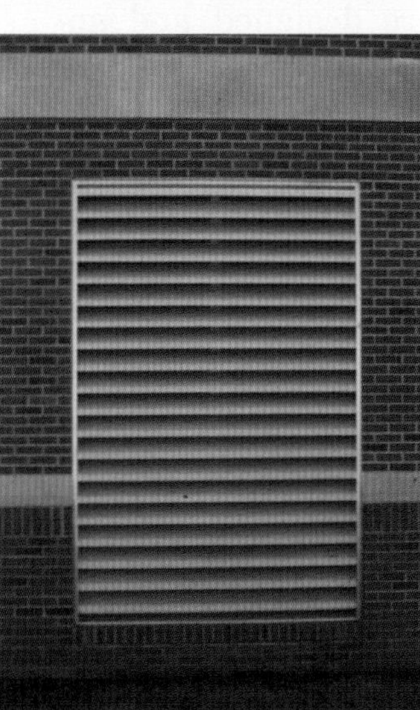

Figure 4.27 HVAC intakes should have grates to protect their openings.

Heating Appliances

Although there are a variety of fire scenarios that can be attributed to HVAC systems, the two primary hazards they pose are fire hazards caused by the heating appliance and the spread of fire and products of combustion through the air handling system. By far the biggest concern of these two are fire hazards posed by the heating appliance. Common causes of fires include improperly adjusted or worn fan drive belts, clogged filters, combustible lint or debris near the burners, and poorly maintained motors. Inspectors may encounter any of three primary types of heating appliances: central heating appliances, unit heaters, and room heaters.

CENTRAL HEATING APPLIANCES

There are four basic types of central heating appliances:

- Boilers
- Warm air furnaces
- Floor furnaces
- Wall furnaces

Boilers are used for a variety of industrial processes including generating steam to power machinery and providing heat required by industrial processes (Figure 4.28). There are two common types of boiler furnaces: fire-tube and water-tube. In *fire-tube boilers*, the combustion gases pass through tubes that are immersed in circulating water, which is converted to steam. A *water-tube boiler* is a steam-generating unit in which steam and water circulate through a series of small drums and tubes while the combustion gases pass over the outside of these steam- and water-containing elements. Almost all of the big boilers are of the water-tube type. The combustion process in a boiler-furnace results from a continual introduction of fuel and air in a flammable mixture. If any one of the inputs is interrupted, an explosive atmosphere can develop.

Boilers are typically fueled by natural gas, coal, or oil. In an oil-fueled boiler, oil is atomized at the burner which causes it to mix with air and burn most efficiently. Natural gas is either supplied to the burner premixed or externally mixed. In a premix, the air and fuel are mixed before they

Figure 4.28 Boilers are used to generate steam in industrial applications. *Courtesy of Ed Prendergast.*

reach the burner. Coal-fired units deliver air and pulverized coal to the burners for combustion.

Explosion is the principal hazard associated with boilers. When a mechanical failure occurs, gases can build up in the system and ignite. In an oil-fired boiler, the hazards come from the fuels and buildup of materials on the nozzles and in the pipes. In natural-gas-fired boilers, the main hazard is the leaking of gas and the buildup of fuel-rich mixtures if a burner fails. Coal-fired boilers present hazards when debris is mixed with the coal. This alters the rate at which the coal is being fed, and it damages the equipment.

Fire and explosion prevention for boilers should first be considered in the design stage. When these vessels are built, the manufacturer should strictly adhere to all applicable codes. The entire system should be composed of interlocks, detectors, alarms, and sensors to shut down fuels and so forth when a problem occurs. Operator training should be an important part of any hazard reduction plan. Workers who are educated in the process and function of the equipment around them will be better able to eliminate a potential hazardous situation. Finally, housekeeping and preventive maintenance are critical in fire prevention.

Warm-air furnaces are the most common type of central heating appliance in use today. They operate on a variety of principles (Figure 4.29). Gravity furnaces operate primarily by circulation of air due to gravity. Forced-air furnaces rely on a fan to move heated air through the system. Because these units contain plenums and ductwork that may become hot enough to ignite adjacent, unprotected woodwork, they require the use of appropriate clearances and insulation.

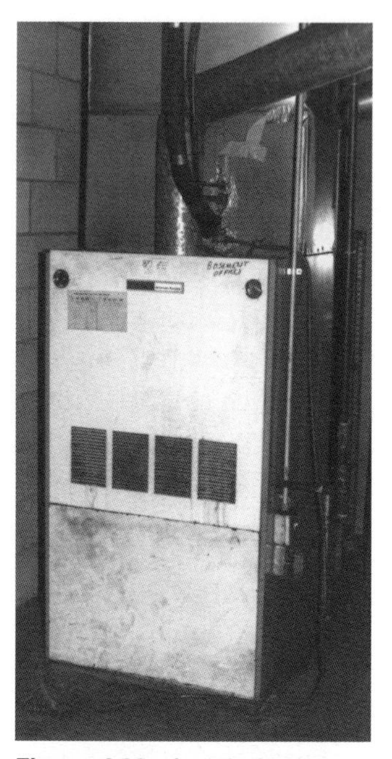

Figure 4.29 A typical warm air furnace. *Courtesy of Des Plaines (IL) Fire Department.*

Automatic controls are provided to shut off the heating device when the temperature within the ductwork or plenum reaches 250°F (121°C). Fire problems with warm-air furnaces are principally due to inadequate clearances, lack of proper limit controls, heat exchanger burnout, improper installation, or improper maintenance procedures.

Most *floor furnaces* are designed and approved to be installed underneath combustible floors (Figure 4.30). They also require temperature limit controls and proper clearances from miscellaneous combustibles. Inspection personnel should warn individuals in occupancies with these units not to cover the register or attempt to dry wet items, such as clothes or newspapers, over the top of them (Figure 4.31).

Figure 4.30 Floor furnaces are more common in older structures.

Wall furnaces are self-contained electric, indirect-fired gas, or oil heaters installed within or on a wall (Figure 4.32). Gas or oil heaters may have a direct vent to the exterior or be equipped with an indirect vent or chimney.

UNIT HEATERS

The second major type of heating appliance is the unit heater. Unit heaters are self-contained devices that are automatically controlled. They may be mounted on the floor or suspended from a wall or ceiling (Figure 4.33). The heating element and fan are enclosed in a common operating unit.

Figure 4.31 Floor furnaces are a frequent ignition source for house fires. This furnace had combustible materials placed over it while the family was in the process of moving.

Figure 4.32 A typical wall furnace.

Figure 4.33 Some unit heaters are suspended from the ceiling. *Courtesy of Des Plaines (IL) Fire Department.*

ROOM HEATERS

The third major type of heating appliance is the room heater. Room heaters use the circulation of radiant heat as the heating medium. They are self-contained units that are designed to heat the immediate surrounding area. Wood and coal stoves, electric and gas logs, and open front heaters are examples of room heaters (Figures 4.34 a and b). They usually incorporate manual or thermostatically controlled drafts and require that nearby floors and walls be properly protected. Solid-fuel room heaters have poor fire safety records, largely due to inadequate clearances from combustible materials and inadequate maintenance. There are numerous hazards associated with room heaters, including:

Figure 4.34a Wood and coal stoves are common in many colder regions.

Figure 4.34b Some occupancies are equipped with unvented, gas-fired space heaters.

- Overfiring the device
- Careless handling of fuel and ashes
- Inadequate clearances from fuel cans or combustible materials
- Accumulation of creosote in flues and chimneys
- Use of insufficient extension cords with electric heaters
- Refueling with the wrong fuel (pouring gasoline into a kerosene heater, for example)

HEATING APPLIANCE SAFETY CONSIDERATIONS

Primary safety controls are a basic requirement on all heating appliances. These controls stop fuel from flowing to the unit in the event of an ignition or flame failure. Other common controls that inspection personnel should look for include:

- Air-fuel interlocks
- Oil temperature interlocks
- Pressure regulation interlocks
- Atomizer-fuel interlocks
- Manual restart mechanisms
- Remote shutoffs
- High temperature limit switches
- Safety shutoff valves

Inspectors should also assure that other common causes of heating appliance related fires are avoided. No combustible materials should be stored within the same room as central heating units. Solid fuels for wood or coal stoves should be stored a safe distance from the stoves themselves. Provisions should be included for the safe removal and storage of ashes and cinders. Make sure that all chimneys and flues are kept clear of excessive creosote.

Air Conditioning and Ventilation Systems

As previously stated, the second primary fire safety concern associated with HVAC systems is the potential for transmission of fire, fire gases, and smoke through the air distribution system. The air distribution system is typically tied into both the heating appliance and the air conditioning equipment. We have covered the hazards associated with heating appliances; however, air conditioning equipment and ventilation systems present their own unique hazards as well. Air conditioning systems are a form of refrigeration system. Any one of a number of gases, such as freon, is compressed into a liquid within a coil and then allowed to expand back into a gas. During the expansion process, the gas (called a refrigerant) cools. Air is blown across the coils to cool the surrounding air.

Most commercial refrigerants have a classified level of toxicity and flammability, as determined by the American National Standards Institute (ANSI) Standard B79.1 and the American Society of Heating, Refrigeration, and Air Conditioning Engineers (ASHRAE) Standard 34, *Number Designation and Safety Classification of Refrigerants.* Refrigerants are assigned to one of two toxicity classes (A or B) based on what would be a permissible exposure. Class A refrigerants are not considered toxic when a person is exposed to concentrations of less than 400 parts per million (ppm) by volume. Class B refrigerants are toxic when a person is exposed to concentrations less than 400 ppm by volume. People performing inspections must keep in mind that they may be dealing with a refrigerant that is not very flammable but is very toxic.

To clean the air and remove particulate dust and pollens, filtering devices are used. There are three types of filters used in HVAC systems: fibrous media unit filters, renewable media filters, and electronic air cleaners. The hazard posed by the first two filters exists because if ignited when full of trapped particulates, they can generate large amounts of smoke. The primary fire hazard associated with electronic air filters is the potential for an electrical malfunction that leads to a fire in the unit. Underwriters Laboratories classify filters into two categories based on flame propagation and smoke development: Class 1 and Class 2. Class 1 filters, when clean, will not contribute fuel to a fire, and they emit very small quantities of smoke when attacked by flames. Class 2 filters, when clean, will burn moderately, and they emit moderate quantities of smoke.

As previously stated, ductwork for HVAC systems can provide a means for the products of combustion to travel throughout the entire area served by the system (Figure 4.35). In addition to traditional ducts, exit corridors and crawl spaces may be used as plenums for air movement as well. Provisions must be made to control the flow of smoke within the building through HVAC systems. Although some HVAC systems are actively used to exhaust smoke from a building during fire conditions, most are not. Most systems are intended to shut down immediately upon detection of fire conditions within the HVAC system. Instead of using the HVAC system, most buildings use a system of passive smoke control: the compartmentation concept. This concept involves shutting down the HVAC system to prevent the further spread of smoke or fire. Once fire conditions are detected, the HVAC system fan shuts down and smoke or fire dampers within the system close. These dampers are typically located at points where the HVAC ducts pass through smoke or fire barrier walls within the building.

Figure 4.35 HVAC ducts provide a route of travel for smoke, heat, and fire. The insulation on the ductwork may further aid the process.

Smoke detectors within the HVAC system itself are generally found in the main supply ducts, downstream of the air filters and cleaners (Figure 4.36). They are also located in the main return ducts prior to the point where the duct discharges exhaust from the building or joins the fresh air intake.

An active smoke control system may be separate from the building's HVAC system. Alternately, it may use the building's HVAC system to create and maintain differential pressure that prevents smoke from moving from one area to another and exhausts products of combustion from the structure.

When the HVAC system is used for smoke control, the system operates in a special mode that no longer delivers supply air to the fire area. Air drawn from the fire area is discharged to the

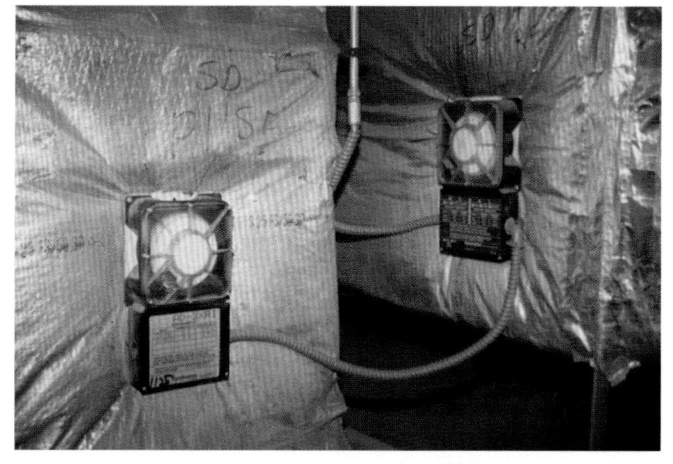

Figure 4.36 HVAC ducts should be equipped with smoke detectors.

outside without recirculating or contaminating the air intake. At the same time, the return air supply to all or part of the building shuts down while the system continues to supply air from the outside to this space. In this type of smoke control system, fire dampers are usually omitted. These types of systems are common in atriums, auditoriums, stadiums, and high-rise structures (Figure 4.37). Active smoke control systems are especially applicable in high-rise structures because a lengthy time is needed for evacuation, aerial ladders cannot reach upper floors, and there are strong air flow patterns in vertical shafts (known as the "stack effect"). The system must be capable of maintaining smoke- and heat-free exit routes for occupants. It must also allow for sufficient evacuating time to either leave the structure or to move to various refuge areas.

When an inspector begins to examine an HVAC system, the following questions should be answered:

Figure 4.37 Smoke control systems are common in high-rise structures.

- What is the type of building construction and its occupancy?
- Is the building compartmentalized?
- Are the utility spaces cut off by walls, floors, and ceilings of adequate fire resistance?
- Where are the fire suppression systems/drains?
- What type of fire detection system exists?
- What is the location of fresh air intakes?
- What type of air handling fans, power drive, and bearings are used?
- How is the ductwork, material, construction, and insulation in the system?
- What type of filter coatings/adhesives are used?
- How is the air heated?
- How is the air cooled?
- Does the system utilize a direct or indirect heat transfer system?
- How is the air humidified/dehumidified?
- Does the system recirculate?
- What is the type and location of smoke/fire dampers?
- Is the system designed to function as a smoke control system?
- Was a satisfactory acceptance test conducted?

TEMPORARY/PORTABLE HEATING EQUIPMENT

Historically, temporary or portable heating equipment has caused a significant number of serious fires involving major structural damage and loss of life. Portable heating equipment may be found in virtually every type of occupancy, including those that are still under construction. The common types of portable heating devices that inspection personnel will encounter include:

- Kerosene heaters — These are most commonly found in residential occupancies (Figure 4.38).
- Portable electric heaters — Common in all occupancies (Figure 4.39).
- Salamander/Bullet heaters — Small LPG-, propane-, or solid-fueled heaters that are common on construction sites or in industrial occupancies (Figure 4.40).

Figure 4.38 Kerosene heaters pose a significant hazard and are outlawed in many communities. *Courtesy of Springfield (IL) Fire Department.*

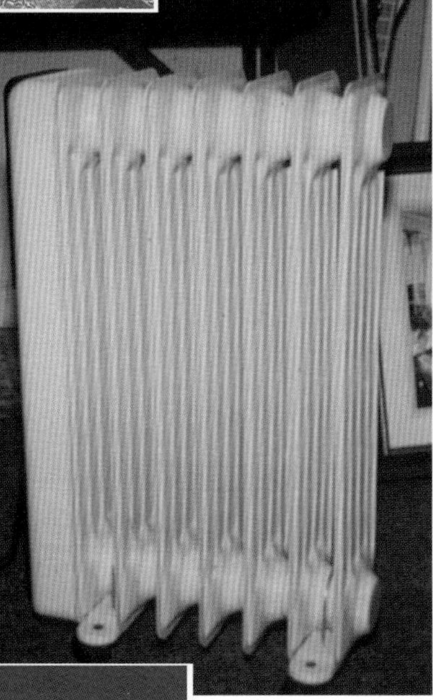

Figure 4.39 Portable electric heaters are perhaps the most common portable heating device.

Figure 4.40 Gas-fired salamanders are commonly found on construction sites or in industrial locations. *Courtesy of Ed Prendergast.*

Portable heaters present significant hazards because of their portability, misuse, and misplacement. Many fires could be avoided if people would be more aware of the surroundings when placing a portable heater. Often, heaters are placed too close to combustibles such as bedding, clothing, and furniture. Proper venting of kerosene heaters and salamanders is imperative so that the occupants will not be overcome by carbon monoxide gas.

There are a number of issues relative to temporary or portable heaters that inspectors should look for. The most basic of these issues is whether or not it is appropriate for the heater to be there in the first place. The inspector must be aware of local code requirements or ordinances regulating the use of portable heaters. Many communities have enacted laws prohibiting certain types of heaters in specific types or all types of occupancies. If this is the case, the only inspection issue is to make sure the heater is removed from the premises as soon as possible. If the heater is permissible, the inspector should check the following:

- The heater should not be placed in a route of egress from the structure (Figure 4.41).
- The heater should appear to be well maintained and in correct working condition.
- The heater should be equipped with safety switches that turn the heating element off if the heater is tipped over.

Figure 4.41 Portable heaters must not be placed in a means of egress.

- Electric heaters should not be plugged into otherwise already overloaded electrical circuits or extension cords.
- Heaters should not be operated in close proximity to flammable or combustible materials.
- Fuel for kerosene heaters should not be stored within the structure, nor should heater refilling be done within the structure.
- Make sure the appropriate fuel is used for the heater.
- The heater should be listed by some formal testing agency such as Underwriters Laboratories (UL) or Factory Mutual (FM).

COOKING EQUIPMENT

Cooking is another very common cause of fires. Occupancies with commerical cooking equipment represent a very large percentage of the total occupancies that inspection personnel will enter (Figure 4.42). The inspector must be aware of the hazards associated with cooking equipment and be able to recognize them. Common cooking equipment that inspectors encounter includes ranges, ovens, fryers, mixing equipment, cutting equipment, exhaust equipment, and warming equipment.

Cooking equipment must have a clearance of at least 18 inches (450 mm) to any combustible material, unless it is specially designed for a lesser clearance. All cooking equipment should be in proper working order. Excess grease and dirt should

Figure 4.42 Cooking equipment will be found in restaurants, such as this 250-year-old country inn.

not be allowed to accumulate on the equipment. Provisions should be made for the removal of excess fat and grease.

In commercial cooking establishments, much of the inspector's attention will be focused on hood, exhaust, and fire protection systems installed above the cooking area (Figure 4.43). These systems should meet the requirements set forth in NFPA 96, *Standard for Ventilation Control and Fire Protection of Commercial Cooking Operations.* This equipment will be protected by wet- or dry-chemical extinguishing systems. All dry-chemical systems must meet the requirements set forth in NFPA 17, *Standard for Dry Chemical Extinguishing Systems*. Wet-chemical systems must meet the requirements in NFPA 17A, *Standard for Wet Chemical Extinguishing Systems.* Requirements for both types of systems are also contained in Underwriters Laboratories (UL) Standard 300, *Fire Testing of Fire Extinguishing Systems for Protection of Restaurant Cooking Areas.*

Figure 4.43 The inspector should thoroughly check all components of the cooking equipment.

A dry-chemical system can be grouped into either of two broad design categories: engineered and pre-engineered. An *engineered system* is specifically calculated and constructed for a specific hazard. Engineered systems tend to be large, expensive systems. The most common system, the *pre-engineered* or package system, is calculated to protect areas of a given size and may be installed at any location. Most kitchen cooking area and hood/duct systems are of the pre-engineered variety. Basically, all dry-chemical systems have the following components:

- Storage tank for expellant gas and agent
- Piping to carry the gas and agent
- Nozzles to disperse the agent
- Actuating mechanism

There are no standard container sizes for agents. The storage container may contain both the agent and the pressurized expellant gas (stored pressure), or the agent and the gas may be stored separately. A pressure gauge attached to the container is an indication of a stored pressure. The expellant gas is either nitrogen or carbon dioxide. The containers are basically the same as those used for portable fire extinguishers except that the system containers are much larger (Figure 4.44). Most tanks are in the 30 to 100 pound (13.6 kg to 45.4 kg) range. The tanks must be located as close to the discharge point as possible. They must also be in an area that will fall within the -40°F to 120°F (-40°C to 49°C) temperature range.

Dry-chemical agent is delivered to the hazard through nozzles attached to a system of fixed piping. The piping is specially designed to account for the unique flow characteristics of the dry-chemical agent. There are no standard nozzles; however, they can be generalized into two broad types: one-position and two-position nozzles (Figure 4.45). One-position nozzles shoot a straight stream of agent at the hazard while two-position nozzles project a fan-shaped fine stream of agent. Each system manufacturer has its own nozzle models.

Figure 4.44 Wet and dry chemical agents are stored in tanks that are remote for the hazard being protected.

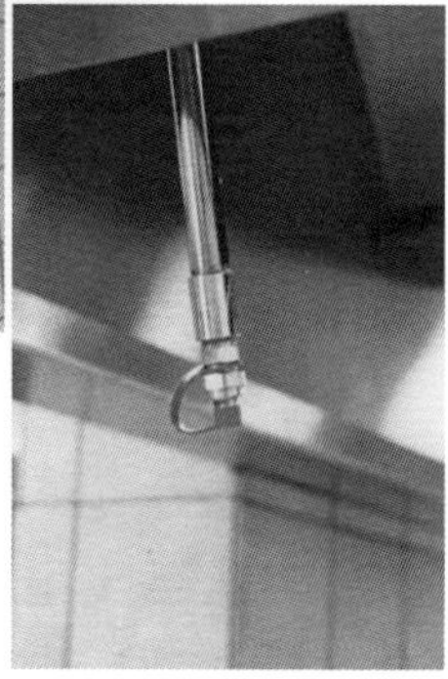

Figure 4.45 Discharge nozzles are located directly over the cooking area.

Dry-chemical is released into the piping system in response to detection devices. Historically, actuation has most often been mechanical in response to the melting of fusible links (Figure 4.46). The fusible links then trigger a mechanical or electrical release that in turn triggers the flow of expellant gas and agent. Systems that have automatic actuation should also be equipped with audible warning signals that will ensure prompt evacuation of the area. This will lessen the chance of occupants suffering from reduced visibility or breathing difficulties as a result of the agent discharge. The majority of fixed systems must also be capable of manual release and must be equipped with automatic fuel or power shutdown (Figure 4.47).

Figure 4.46 Fusible links are used to allow the system to discharge automatically.

Figure 4.47 Manual discharge controls are found somewhere near the cooking area.

There are a number of things that inspectors should look for when inspecting hood systems and their associated fire protection equipment. The following is a brief list of these items:

- Check for mechanical damage.
- Check aim of nozzles.
- Check all parts of the system to ensure that they are in their proper location and that they are connected.
- Inspect manual actuators for obstructions.
- Inspect the tamper indicators and seals to ensure that they are intact (Figure 4.48).
- Check the maintenance tag or certificate for proper placement and to assure that the system has been inspected by a qualified technician within the code-specified time parameters.

Figure 4.48 The tamper seal is intended to let the inspector know if the system has been discharged or worked on.

- Examine the system for any obvious damage.
- Check the pressure gauges to ensure that they read within their operable ranges.

If the system actuator is controlled by a fusible link device, the fusible link should be replaced at least annually. If it appears to be distorted from frequent exposure to heat, it may need to be replaced more often. Dry-chemical cylinders that are less than 150 pounds (68 kg) must be hydrostatically tested every 12 years. Larger tanks have no hydrostatic test requirements.

Wet-chemical systems are also used to protect commercial cooking hoods and associated cooking appliances. The wet-chemical system operates and is designed similarly to a dry-chemical system. The wet-chemical agent is typically a solution that is composed of water and either potassium carbonate or potassium acetate, and it is delivered to the hazard area in the form of a spray. It is an excellent extinguishing agent for fires involving flammable liquids, gas, grease, or ordinary combustibles such as paper, wood, or charcoal. It is not recommended for electrical fires because the spray may act as a conductor.

A wet-chemical system is most effective on fires caused by cooking hazards. The nature of the chemical is such that it reacts with animal or vegetable oils and forms a surfactant (soap). When using a wet-chemical agent, grease or oil fires are extinguished by fuel removal, cooling, smothering, and flame inhibition. Wet-chemical systems are messy, particularly when food greases are involved.

For the most part, the components and actuation of wet-chemical systems are the same as those for dry-chemical systems. The inspection and testing procedures are also the same.

Solid-fuel cooking equipment is becoming very common in many restaurants. This includes equipment such as wood-fired brick ovens and meat grilling or smoking equipment (Figure 4.49). This equipment must still be protected with a hood system to capture escaping smoke and gases. Spark arrestors must be located in these hoods ahead of the filter system. Grease removal devices should be incorporated into the system. Solid-fuel burning systems that have a fire box greater than 5 cubic feet (0.14 m^3) shall be protected by a fixed water system with attached hose providing at least 40 psi (280 kPa). Fire extinguishers are satisfactory for those with a fire box of less than 5 cubic feet (0.14 m^3). Once a week the interior surface must be cleaned and the flue or chimney inspected. Solid-fuel systems must be installed on noncombustible floors that extend 3 feet (0.91 m) from the outside of the appliance in every direction. Only a one-day supply of solid fuel is allowed in the same room as the appliance. Ashes must be removed from the firebox regularly so that the buildup of ashes does not interfere with the draft.

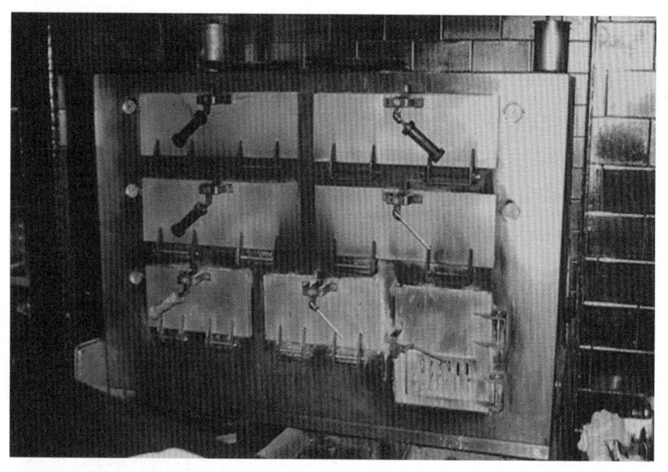

Figure 4.49 Wood-fired ovens are becoming increasingly popular in many restaurants. *Courtesy of Ed Prendergast.*

APPLICATION OF FLAMMABLE FINISHES

Facilities where flammable finishes are applied present many hazards for the fire inspector to check. It is obvious that any operation where flammable liquids or combustible powders are applied in an atomized form have the potential to easily create severe fire conditions. These finishing operations are commonly found in metal shops, auto body shops, and manufacturing facilities where metal and wood products are produced (Figure 4.50). The inspector should look at the finishing operations as a distinctly separate portion of the facility.

In general, the finishing area includes the following portions of an occupancy:

- The spray booth, or finishing room, itself
- The interior of ducts used for exhausting the spray area
- Any area in direct contact with the spray

Most of the model codes completely restrict certain types of occupancies from having finishing operations within them. The codes also require separation of the finishing area in other occupancies.

There are two basic processes for coating fabricated products: fluid coating and powder coating. Fluid coating most often uses an air spray gun which atomizes the fluid as it is sprayed onto the product. There are also two other methods for applying fluid coatings. One uses a hydraulic means instead of air. The second uses electrostatic forces and a disc that turns at high speeds creating a centrifugal force that delivers the paint in a 360-degree pattern.

Powder coating is a process that is gaining in popularity. In this process, powder is suspended in air and then electrostatically charged. The charged

Figure 4.50 Large auto body operations contain paint booths.

powder particles are then attracted to the part and, due to the electrostatic forces, remain on the part. The part is then passed through an oven where the powder coating is baked on.

There are several types of rooms or enclosures where fluid and powder coating are performed. One such area is a spray booth, which is a mechanically ventilated enclosure where spray applications are performed (Figure 4.51). This is the most common type of application room. Some special enclosures are continuous coaters and decorating machines. Continuous coaters utilize moving racks to hang parts from while passing them through an enclosure that is open on both ends. The parts are coated by spray from fixed spray guns. Mechanical ventilation is used to recover overspray. Decorating machines are used to apply paint stripes or other patterns on automobile parts.

Figure 4.51 A typical paint booth.

Several types of devices have been built to lessen the hazards associated with overspray and vapors. A baffle maze is made up of panels in which air from the ventilation process is directed. These panels accumulate particulate until their use is impaired. At this time, they are removed, mechanically cleaned, and returned to service. Dry filters are very similar to baffle mazes with one exception: At the end of a dry filter's life, it is discarded and replaced by a new filter. Waterfall and cascade scrubbers are used when spraying occurs extensively. The spray is captured by water spray from nozzles or by cascading water from a waterfall. The particulate is then removed mechanically from the water.

Hazards associated with fluid spraying processes are numerous. For this reason, these areas should be given special evaluation to limit the ignition sources and materials which could add to a potential fire. When brought into the facility, materials should be marked in accordance with NFPA 704, *Standard System for the Identification of the Hazards of Materials for Emergency Response*. Flammable liquids should be stored in accordance with NFPA 30, *Flammable and Combustible Liquids Code*. Mixing rooms should be built against an outside wall and explosion venting should be constructed into the room. Consult NFPA 68, *Guide for Venting of Deflagrations* for more detailed information on explosion venting. All electrical wiring should be done in conjunction with NFPA 70, *National Electrical Code®*.

The entire room should be enclosed by fire-resistive construction and protected by an approved fixed fire suppression system. Ventilation should be provided that moves an appropriate amount of air based on the room size and materials contained therein. Pumps in the mixing room should have fire alarm interlocks so that in the event of a fire, they do not continue to operate and feed fuel to the fire. Vessels should be grounded so that an electrostatic spark does not ignite any vapors which may be in the air. Only enough flammable liquids that may be needed for a process for one shift should be kept on hand. These flammable liquids should be kept in approved containers in compliance with NFPA 30. Housekeeping is extremely important in these rooms. The buildup of paint or other finishes on the walls and so forth are the same as a flammable solid. The buildup of materials can act as an insulator and render the grounding equipment useless.

The hazards that exist with powder coating are due to the use of combustible organic powder and the potential of explosion because of the accumulation of the dust in air. Static accumulations are the major cause of fires in powder coating processes. Fire protection measures for powder coating are flame detectors within the booth and interlocks for the ventilation and dust recovery units. Automatic pressure relief vents should also be used to vent anytime a pressure buildup occurs.

In either liquid or powder coating facilities, inspectors should make sure that the fire protection measures required by the code they use are in place. This includes both fixed fire protection systems and portable equipment. All of the model codes specify minimum sizes of portable fire extinguishers and the maximum travel distance they must be from the work area. Obviously, welding and cutting operations, as well as any other that involves open flames or potential ignition sources, must be kept clear of the finishing operation.

DIP TANKS/QUENCHING OPERATIONS

Dipping and quenching operations present another serious hazard in which personnel who perform inspections must be well versed (Figure 4.52). Dipping and quenching operations can cover a wide range of processes. In both of these operations, parts may be immersed into combustible liquids. The potential for hazards in quench or dipping operations not only comes from the combustible liquids but also from the ancillary equipment associated with these operations. Some of the equipment necessary for these operations includes flammable liquid storage tanks, conveyor pumps, heaters and heat exchangers, agitators, and ventilation and exhaust equipment. By examining this list, it is evident that the potential for fire is great.

Figure 4.52 A small-scale dip tank. *Courtesy of Justrite Mfg., Inc.*

Dip tanks most often utilize conveyor systems from which a part is hung, brought over the dip tank, immersed, and removed in a continuous operation. The tanks come in various sizes depending on the process being performed. Volume of the tank can vary from a small amount to hundreds of gallons (liters). The tanks should be made of noncombustible material of high integrity that can handle the operations and not react with the dipping medium.

The hazards associated with dip tanks include fires involving the liquids and explosions involving vapor and air mixtures. To lower the potential of one of these incidents occurring, the operations should be located away from other process areas. The dip tanks should be contained in one-hour fire-rated construction, and ignition sources should be identified and removed if possible. Splash boards, drains, and overflow protection devices are all components that should be included in dipping operations. These processes should never be located near an egress area. In some cases, it may be necessary to shield an operation, and this can be accomplished by hanging nonflammable curtains. In some instances where these processes are confined, explosion venting construction can be utilized. Ventilation and exhaust of the areas are important to remove hazardous vapors that could lead to an explosion and harm a worker.

In order to limit the accumulations of combustible materials, housekeeping is important around these operations. Maintenance is also important so that the buildup of combustible residues is kept to a minimum. Regular inspections of the process area, equipment, and storage of commodities associated with the process will limit the occurrence of incidents in a facility.

Quenching is the immersing of a metal part in a quench medium. Quenching is divided into batch operations and continuous operations. There are several mechanical components involved in a quenching process, including elevators, conveyors, hoists, cranes, and others. The mechanical means immerses a part, moves it through the quench medium, and removes it from the quench medium.

Mineral oils are commonly used as a quench medium. Regardless of the medium used, the properties of the medium must remain fluid and stable. There are two types of quenching: heated and unheated. Unheated quenching uses temperatures from 100°F to 200°F (38°C to 93°C). The quench medium used for unheated quenching normally has a flash point of 300°F (149°C). For heated quenching operations, temperatures range from

200°F to 400°F (93°C to 205°C). The normal flash point of the various mediums for heated quenching is 500°F (260°C). A quench oil should never be allowed to be within 50°F of its flash point.

There are several hazards associated with oil quenching. These hazards include, but are not limited to: vapors given off by the quench oil; oil that collects along walls and ceilings from vapors; the high temperatures and furnaces used in the process; and overflow, tank, and part configurations. Overflow of tanks can be caused by several factors. One may be that the part being dipped has displaced too much quench oil, and the tank overflows. To overcome this possibility, tanks should be built within dikes with drains capable of containing the contents of the tank and piping to move it into specialized holding tanks. Another way a tank could overflow is if the fire sprinkler system was to discharge and overflow the tank. To reduce the chance of this situation, drain boards can be installed and automatic-closing covers can be used.

Tanks should be located at grade level. This is important so that in the event of an overflow, flammable liquids will not be flowing freely to lower levels. All tanks should be designed with enough freeboard to keep the tank from overflowing. *Freeboard* is the distance from the liquid surface to the top of the tank when the tank is fully loaded. The freeboard should never be less than 6 inches (150 mm) (Figure 4.53). All tanks should have drains to protect from overflow, particularly those that contain 150 gallons (600 L) of quench oil or more. In emergency situations, it may be necessary to drain the tank so that the liquid medium does not add to the fuel of the fire. Any tank over 500 gallons (2 000 L) must have a bottom drain that will open automatically or manually in the event of a fire.

In the event of a fire, conveyor systems should stop automatically. All portions of the quench process should be interlocked so that they all cease operations. Hazard elimination around a quenching process includes the elimination of open flames, spark-producing equipment or processes, and equipment whose exposed surfaces exceed the autoignition temperature of the dipping or coating liquid. If these hazards cannot be moved, then they should be totally enclosed. It is important that all aspects of the process be electrically grounded in accordance with NFPA 77, *Recommended Practice on Static Electricity.*

Quenching operations should be housed in fire-resistive buildings away from the main production areas. The dip tanks should never be placed in

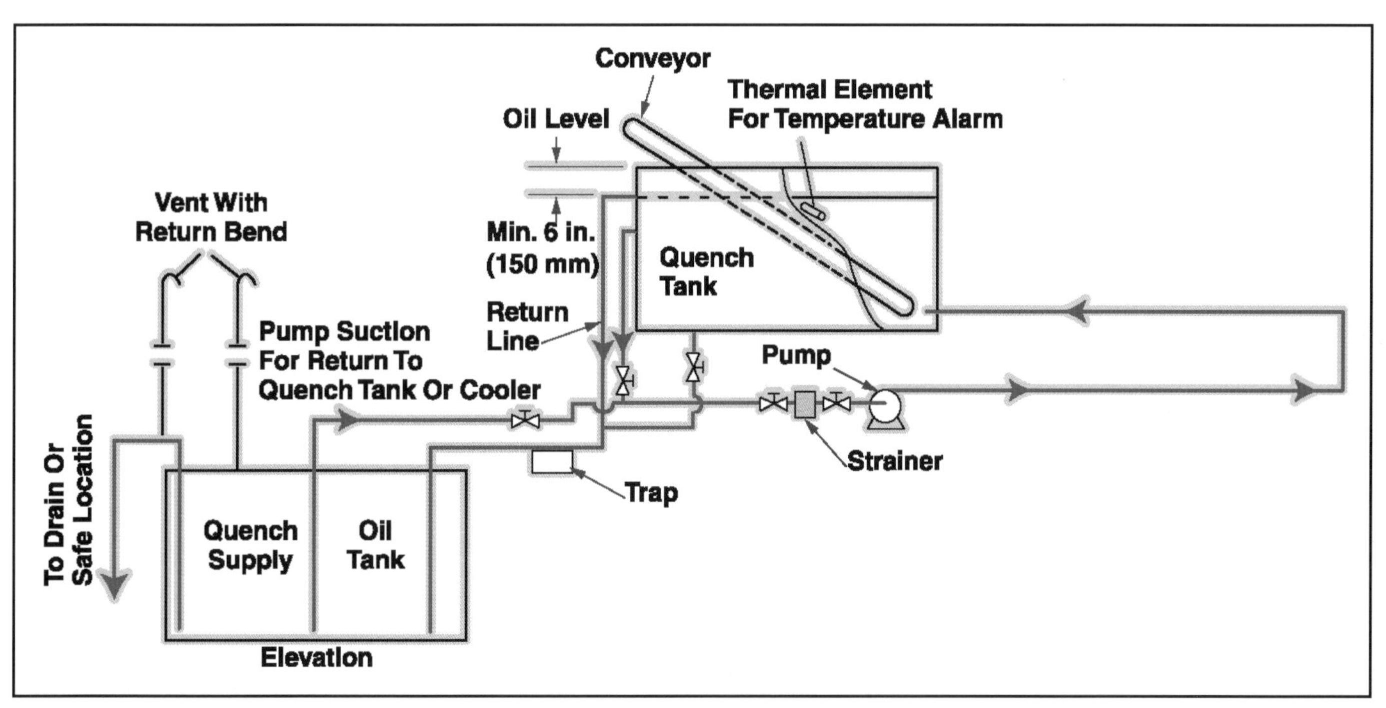

Figure 4.53 The layout of a typical dip tank or quenching operation. There should be a minimum of 6 inches (150 mm) of freeboard during operation.

basements because of the potential hazard from fire and the increased risk placed on the lives of those who would respond in an emergency. There should never be more than three flammable liquid storage cabinets in any single process area unless approved by the authority having jurisdiction. Areas in which quenching operations are performed should be protected by an automatic sprinkler system. It is important that workers be trained in the handling of flammable liquids so that hazards can be lessened. Fire extinguishers of several types are available for extinguishment of quench mediums.

DRY CLEANING

Dry cleaning is not normally thought of as a hazardous process because it seems to be the basic task of cleaning clothes. The truth is that the dry-cleaning process involves numerous hazardous chemicals. *Dry cleaning* is the process of removing dirt, grease, and stains from clothing or other textiles by the use of solvents (Figure 4.54). This may be done by any one of several methods, including:

- Immersion and agitation with the solvent in a closed machine
- Brushing or scouring with the cleaning solvent
- Dual-phase processing

Dual-phase processing involves dry-cleaning equipment that also performs standard laundering either before or after the dry cleaning.

Each of the model codes have specific requirements for occupancies containing dry-cleaning operations. Extensive information on these operations can also be found in NFPA 32, *Standard for Drycleaning Plants*. The following general information should be used when inspecting dry-cleaning plants.

Figure 4.54 Most communities have dry cleaning operations.

Dry-cleaning operations using combustible liquids should not be performed in the same building with other occupancies. Construction of buildings which contain dry-cleaning operations must have floors of fire-resistive construction and a wearing surface of noncombustible and solvent-resistant material. In addition, the floor or roof above a dry-cleaning room shall have a fire-resistance rating of not less than 1 hour. An emergency drainage system is needed in a dry-cleaning operation to direct solvent leakage and fire protection water to a safe location. To accomplish this, the use of curbs, scuppers, or a special drainage system to control the spread of fire may be utilized. A dry-cleaning room must have at least two doors as a means of egress located at opposite ends of the room, one of which must lead directly outside.

DUST HAZARDS

Dusts, when suspended in air in sufficient quantities, and in the presence of an ignition source, can cause powerful explosions capable of destroying an entire building. In order for a dust explosion to occur, four conditions must be met at the same time:

1. Combustible dust must be suspended in air.
2. The dust must be within its explosive range.
3. An ignition source must be present.
4. The dust must be in a confined space.

Evaluation of a combustible dust explosion hazard and the prevention techniques employed should be determined by means of actual test data. All combustible dusts that may produce a dust explosion should be tested so as to determine the following data:

- Particle size (surface area to mass ratio)
- Moisture content as received and dried
- Minimum dust concentration to ignite
- Minimum energy required for ignition
- Maximum rate of pressure rise at various concentrations

- Layer ignition temperature
- Maximum explosion pressure at optimum concentration

Dust explosions occur in series. The first explosion usually is not as severe as subsequent explosions. The first explosion will stir up dust that has settled on ledges, walls, or equipment. This second introduction of particles to the air generally results in larger and stronger explosions. There are many processes that produce the potential for a dust explosion. In the following sections, we will examine three of the most common facilities fire inspection personnel will encounter: grain, wood processing, and aluminum dust facilities.

Grain Facilities

Grain handling facilities have long been subject to deadly grain dust explosions. Grain handling facilities are comprised of storage structures (commonly called grain elevators), grain handling equipment, grain drying and cleaning components, and receiving and shipping areas (Figure 4.55). There are a variety of grains handled in these facilities such as wheat, rye, oats, and others.

Figure 4.55 Large-scale grain handling operations are significant dust hazards.

Figure 4.56 Grain elevators must have effective dust handling and collection systems.

Virtually every point in the grain handling process implements some type of dust control. Dust control methods and equipment employ passive and active means. Passive means of dust control include enclosures or reduced speeds at which grains are produced. Active dust controls are most commonly mechanical dust collection systems. Mechanical systems are capable of collecting 99.5 percent of the dust and storing it in bins outside the facility. These dusts are used in animal foods. It is important that dust collecting systems be located outside of the grain elevator (Figure 4.56). Buildings in grain handling operations should be built in accordance with the building code used in your jurisdiction, as well as the following NFPA standards:

- NFPA 61, *Standard for the Prevention of Fires and Dust Explosions in Agricultural and Food Products Facilities*
- NFPA 68, *Guide for Venting of Deflagrations*

Life safety issues for a grain handling facility should comply with the appropriate local code or NFPA 101. All subterranean tunnels and passageways 50 feet (15 m) or longer must have two means of egress as remote from each other as possible. Roofs and bin decks shall have two means of egress that are remote from each other so that a single fire or explosion event will not likely block both means of egress. Access doors or openings shall be provided to permit inspection, cleaning, and maintenance and to allow effective use of fire fighting techniques in the event of fire within the bin, tank, or silo.

In grain elevators, a method to prevent the escape of dust into surrounding areas shall be provided at leg boot sections, belt loaders, belt discharge or transfer points, trippers, turnheads, distributors, and on unfiltered vents from which dust could be emitted into interior areas with displaced air.

Fire protection for grain elevators is typical of other process industries. The building should have fire extinguishers in accordance with NFPA 10, *Standard on Portable Fire Extinguishers.* The NFPA recommends that wet or dry standpipes be provided to all operating levels over 75 feet (23 m) above grade. Wet or dry standpipes should also be installed in warehouses and packing areas with combustible contents. Each facility shall have a written emergency action plan that includes, but is not limited to, the following:

- Means of notification for occupants in the event of fire and explosion
- Preplanned evacuation assembly area
- Person designated to notify emergency responders, including the fire department
- Facility layout drawings showing egress routes, hazardous chemical locations, and fire protection equipment
- Location of material safety data sheets (MSDS) for hazardous chemicals
- Emergency telephone number(s)
- Emergency response duties for occupants

Training must be provided regarding the emergency action plan for all affected personnel. The emergency action plan and the fire department prefire plans must be coordinated with local emergency responders. Warning signs must be posted for areas containing inert protection systems, as well as explosion protection systems. Welding and cutting and the use of power tools are areas of concern because they present so many ignition sources. Spark producing portable power tools and propellant-actuated tools should not be used where combustible dust is present. When the use of spark-producing tools becomes necessary, all dust-producing machinery in the area must be shut down. All equipment, floors, and walls shall be carefully cleaned, and all dust accumulations in the area removed. An inspection must be conducted at the end of the job to assure that no spark-producing tools or debris, which could enter equipment, are left on the premises.

Housekeeping is also important in these operations. There should be no storage of sacks, nonessential uninstalled machinery or parts, or other supplies in areas where the only other combustible material is the agricultural commodity that is being stored. Miscellaneous storage must not impede facility housekeeping or fire fighting.

Woodworking and Processing Facilities

Woodworking in general, be it the production of lumber or making raw lumber into finished products, creates a substantial amount of waste (Figure 4.57). Most of this waste is in the form of small wood particles and dust. If not controlled properly, this dust can accumulate and ignite given the proper conditions. Although wood dust does present an *explosion* hazard similar to the other materials in this section, the primary hazards fire inspectors are concerned about relative to these operations are dust *fires*.

Figure 4.57 Large-scale woodworking and furniture manufacturing facilities may cover many acres under one roof.

The most common place for fires to occur in woodworking operations is at the dust hogger. This is a machine that turns scrap wood into splinters or chips. Small pieces of metal, called *tramp*, wear off the machine during the milling operation and accumulate in the hog trap. The tramp can become hot enough to ignite wood on contact. The fire can then spread into other portions of the machinery, including the dust collection system (Figure 4.58). These fires are often hard to control because they burrow through piles of sawdust. They also may be contained within an elaborate system of ducts, storage bins, and other associated equipment.

The keys to preventing fires in woodworking facilities are controlling the dust and ignition

Figure 4.58 Woodworking operations should have proper dust collection systems.

sources. In a woodworking facility, fine dust is removed by conveyors and pneumatic systems. The conveyors are magnetized to remove any trapped metal pieces that could spark and become an ignition source. The use of pneumatic means, such as cyclones, for dust collection must be carefully designed so that the air being returned to the plant does not contain any dust which may lead to a combustible mixture with air.

Housekeeping needs to be frequent to prevent the accumulation of dust. Dust must be continually cleaned up and placed in appropriate storage bins or containers. This may be done by manual sweeping, vacuuming, or other means. Metal must be collected separately so as not to get any metal chips in the wood-handling or processing equipment.

Electrical equipment in the plant needs to be explosion-proof. All equipment must be grounded to eliminate static electricity. Smoking is restricted to designated areas. Fire extinguishers must be provided throughout all buildings in accordance with the requirements of NFPA 10. Standpipe and hose cabinets must comply with NFPA 14, *Standard on Standpipes and Hose Systems.*

Metal Dust

Metal dusts, such as aluminum and magnesium, are commonly produced in machine shops and manufacturing operations. Fine metal dust has an enormous explosive potential. As the inspector visiting one of these facilities, there are several issues which must be addressed to assure the safety of those handling this material.

The inspector should view the company's policies on smoking and open flames. No smoking, open flames, electric or gas cutting or welding equipment, or spark-producing operations should be allowed where metal dusts exist. Employers need to have strict rules which emphasize that lighters, matches, and smoking materials may not be carried into processing or handling areas by employees. Flame-resistant clothing should be worn by employees working in metal dust processing or handling areas. It is imperative that workers be trained in emergency procedures. Workers need to be familiar with the equipment they can use in the event of a fire. The inspector should review the training program to make sure it emphasizes the different types of fires anticipated and the appropriate agents and techniques to be used. Inspectors should assure that the appropriate extinguishing agent for the type of metal being protected is in the area.

INDUSTRIAL FURNACES AND OVENS

Explosions and fires in fuel-fired and electric-heat-utilization equipment constitute a loss potential in life, property, and production. The cause of most failures can be traced to human error. The most significant failures are:

- Inadequate training of operators
- Lack of proper maintenance
- Improper application of equipment

Classes of Furnaces

NFPA 86, *Standard on Ovens and Furnaces,* classifies ovens and furnaces into one of four categories: A, B, C, or D. Class A ovens and furnaces are heat-utilization equipment that operate at approximately atmospheric pressure. There is a potential for explosion or fire hazard when flammable volatiles or combustible materials are processed or heated in the furnace.

Class B ovens and furnaces are heat-utilization equipment that operate at approximately atmospheric pressure, in which no flammable volatiles or combustible materials are being heated.

Class C ovens and furnaces are those in which there is a potential hazard due to a flammable material being used or a special atmosphere present.

This type of furnace can use any type of heating system, and it includes a special atmosphere supply system. Also included in the Class C classification are integral quench furnaces and molten salt bath furnaces.

Class D ovens and furnaces are those that operate at temperatures from above ambient to over 5,000°F (2 760°C). They also operate at pressures normally below atmospheric using any type of heating system. These furnaces can include the use of special processing atmospheres.

Hazards Associated with Ovens and Furnaces

Hazards associated with ovens and furnaces are related to the following:

- Combustibility of construction surrounding them
- Ventilation
- Location
- Heat, gas, and smoke removal
- Temperatures required
- Materials handling

Fire hazards can be lessened by proper housekeeping techniques that remove combustible items from the area of the oven or furnace. General dirt and debris should also be cleaned on a regular basis. Overheating and other malfunctions can be controlled by automated controls and alarms.

Safety controls are an important part of an oven or furnace operation. These controls must constantly evaluate the system and shut down if a problem occurs. The entire system should be interlocked so that in the event of a burnout or other malfunction, the equipment will shut down and vent off remaining fuel in the system. Switches, relays, detectors, valves, controllers, and interlocks are all components commonly found on the safety controls. All safety devices must be listed for the service intended. A shutdown of the heating system by any safety feature or safety device requires manual intervention of an operator for reestablishment of the normal operation of the system. It is important to check the schedule and the extent of the inspection, testing, and maintenance program. Ovens containing or processing sufficient combustion materials to sustain a fire must be equipped with automatic sprinklers or water spray.

CONSTRUCTION/DEMOLITION

The risk of fire rises sharply when construction or demolition is being performed on a structure (Figure 4.59). There are a number of reasons for this increased level of hazard. One contributing factor is the additional fire load and ignition sources brought in by the contractors and their associated equipment. Ignition sources, such as open flames from torches and sparks from grinding or cutting processes, are examples of hazards added by contractors.

Buildings under construction are subject to rapid fire spread because many of the protective

Figure 4.59 Fire inspectors should monitor demolition operations within their jurisdiction.

features, such as sheet rock, are not yet in place (Figure 4.60). The exposed wood framing can be likened to a vertical lumberyard. The lack of doors or other measures that would normally slow fire spread are also contributing factors to rapid fire growth.

Buildings that are being renovated or demolished are also subject to faster than normal fire growth. Breached walls, open stairwells, missing doors, and disabled fire protection systems are all potential problems. The potential for a more sudden building collapse during fire conditions is also a serious consideration. Arson is also a factor at construction or demolition sites because of the easy access into the building.

Measures to reduce the risk of fire include pre-incident planning, basic fire prevention methods, fixed suppression systems, and good communications. It is essential that a fire safety inspection program be established and adhered to while all aspects of construction and demolition are being completed. Fire inspectors must maintain close coordination with building inspection officials during this process. The owner needs to designate an individual to oversee and enforce the fire prevention program. This manager must develop pre-incident plans and share them with the local fire department. The fire protection manager will also be responsible for the purchase, weekly inspection, and maintenance of any fire protection equipment that is needed. Any alteration of fire protection equipment must be approved by the manager. The following are some other considerations at construction/demolition sites:

- There must be some means to call the department, and emergency numbers need to be posted near phones.
- A convenient location must be provided with plans, keys, emergency information, etc (Figure 4.61). A means must be provided to allow unobstructed fire department access.
- The building must meet the requirements of NFPA 101 or locally adopted codes.
- Fire department personnel must have access to fire protection systems and hydrants.
- Fire extinguishers must be provided in accordance with NFPA 10.

Figure 4.60 Massive, fast-moving fires have occurred within wood frame structures under construction. Firefighters often refer to these as "vertical lumberyards."

Figure 4.61 Keys, emergency information, and the like are generally kept in the on-site construction office.

When a demolition project is underway, the contractor should schedule the demolition systematically to keep the fixed fire protection systems operational as long as possible. If feasible, the system should be removed floor by floor and modified so that it can remain in service. The last piece of fire equipment to be removed from a building should be the standpipe. It is critical in a multistory building to leave the standpipe intact so that in the event of a fire, firefighters will have a means to get water to upper floors.

Due to the rising costs of new construction, renovating or rehabbing of old buildings is becoming more popular. This can be positive from a fire protection perspective because most jurisdictions require the owner to bring the building up to meet current code requirements. This means that these buildings must meet the most current codes on such things as means of egress, fire protection systems, ventilation, lighting, HVAC systems, and so forth. Special precautions must be taken when rehabilitation construction is ongoing. Since occupants and their belongings can remain in a building while construction is going on, it is critical to make sure that no means of egress is blocked that would violate building codes and prevent the egress of persons from the building.

In rehab situations, housekeeping is very important. With the accumulation of debris, new construction materials, and equipment, exits can easily be blocked. Care must be taken so that this problem does not arise. Fire detection or alarm systems should remain intact for the duration of construction and they may be covered to protect them, if necessary. In the event that a system will be down for any length of time, a fire watch should be posted for the duration. In any construction project, whether demolition or renovation, the fire department should be notified and pre-incident plans reviewed.

The following is a list of other concerns that fire inspection personnel should look for at construction and demolition sites:

- Make sure that all flammable and combustible liquids are stored in compliance with the appropriate code.
- Exhaust discharge from internal combustion engines should not be close to combustible materials.
- Asphalt and tar kettles should be equipped with lids that can be fully closed in the event of a fire. These are covered in more detail later in this chapter.
- Make sure appropriate regulations for temporary heating devices are followed during cold weather operations.
- Trash and rubbish should be cleaned up and disposed of on a daily basis.

For more information on requirements for construction and demolition sites, see your local codes or consult NFPA 241, *Standard for Safeguarding Construction, Alteration, and Demolition Operations.*

OPEN BURNING

Many local municipalities, as well as some county and state governments, enact regulations to control open burning within their jurisdictions. Open burning activities include:

- Trash in piles or barrels
- Piles of leaves
- Piles of brush or limbs that have been cleared off land
- Bonfires and campfires
- Prescribed burns of wildland or agricultural land (Figure 4.62)

Typically, fires used to cook food for human consumption (referred to as recreational fires in some codes) do not fall under the requirements for open burning.

Figure 4.62 Prescribed burns of agricultural land are common in many regions.

Most of the model fire codes have requirements that jurisdictions may adopt to control open burning. The requirements are fairly consistent in each of the model codes. All require that the party wishing to burn file an application for a burn permit prior to the date they wish to burn. During that period, a fire inspector must visit the site and make sure that conditions are safe for holding the burn. Once the site has been determined safe, a permit may be issued. The permit should contain the conditions under which the burn will be held.

This may include restrictions if the wind is blowing too hard on the day of the proposed burn.

In general, the following guidelines should be used to regulate open burning:

- Uncontained fires should not be allowed within 50 feet (15 m) of a structure.
- Fires in approved burn containers should not be within 25 feet (8 m) of a structure.
- Grounds around a burn pile or bonfire should be cleared of vegetation or cut very short to slow any spread of fire.
- A fire extinguisher, garden hose, or other similar equipment that can be used to extinguish spreading fire must be present during the burn.
- The burn operation must be attended by a person capable of preventing the spread of fire until the fire is extinguished.
- Fires should only be allowed when wind conditions are light and the danger of causing the fire to spread out of control is minimal.

TAR KETTLES

Asphalt or tar kettles are typically trailer-mounted devices that are used to heat and dispense tar or asphalt for use on roads and roofs (Figure 4.63). The tar or asphalt on the trailer is contained in a tank that is heated using an LPG burner or similar heat device. Because this equipment heats a flammable or combustible substance above its fire point, tar kettles have the potential to cause serious fires. Boilover of the hot contents can ignite other fuels and spread fire rapidly. The inspector must be aware of fire protection and safety practices associated with tar kettles so that the severity of the hazards are lessened.

Figure 4.63 Tar kettles and asphalt-holding equipment frequently catch on fire.

Kettles must not be operated inside a building or on the roof of a building. The area in which the kettle is operating should be identified by the use of traffic cones, barriers, and other suitable means as approved by local officials. The tar kettle must be attended by at least one employee knowledgeable of the operations and hazards. The employee must be within a reasonable distance of the kettle (this distance varies from code to code) and have the kettle within sight. Two approved 20B:C fire extinguishers shall be provided and maintained within 25 feet (8 m) of an operating kettle. Fire extinguishers must be mounted in an accessible and visible location. Roofing kettles must not block exits, means of egress, gates, roadways, or entrances. In no case should kettles be closer than 10 feet (3 m) from exits or means of egress.

LPG cylinders used to supply heating elements on the trailer must be secured to prevent turnover. Regulators are required on all cylinders. LPG containers for roofing kettles must not be used in any building. All kettles must have an approved working visible temperature gauge that indicates the temperature of the material being heated. They must also have a lid over the product tank that may be closed in the event of a fire. See NFPA 1, *Fire Prevention Code* for more information on tar kettle fire protection.

TIRE STORAGE

The bulk storage of tires presents a serious challenge to fire inspectors and an even more serious challenge to firefighters should they ignite. This is due to the flame intensity they produce, the massive amounts of smoke these fires generate, and the toxic oil that they produce as they are reduced by fire. This oil creates a serious pollution and ground water contamination threat.

There are two types of tire storage with which inspection personnel are likely to deal: inside and outside. Inside storage may be found in tire or automotive stores, warehouses, and tire manufacturing facilities (Figure 4.64). Inside storage of tires should be in accordance with local code requirements or NFPA 231D, *Standard for Storage of Rubber Tires*. The standard gives detailed instructions on the storage configuration for the tires and the types of racks that should be used. Because of the need to suppress a fire early and due to the fire load of tires, larger capacity sprinkler system flow capabilities are incorporated into the standard. High-expansion foam also can be used as an effective means of fire suppression where tires are stored inside structures. Outside storage is most commonly in conjunction with junkyards or scrap tire storage facilities (Figure 4.65). In recent years, the problems associated with these large-scale scrap tire storage facilities have increased and numerous serious fires have occurred. Fires involving tires burn long, hot, and smoky. Because of their configuration, tires allow for the spread of heat and flames readily between them.

The various fire codes and NFPA 231D provide guidance on outdoor storage of tires. The requirements vary between the standards; however, each generally specifies the maximum height, length, and width the piles may be. Also specified are the distances required between other piles, structures, and property lines. Weeds and vegetation must be cleared within 50 feet (15 m) of any tire pile.

Figure 4.64 Most communities have auto repair garages or tire stores that contain large numbers of tires within the structure.

Figure 4.65 Large-scale outdoor tire storage areas contain enormous fire fuel loads. *Courtesy of Ed Prendergast.*

POWERED INDUSTRIAL TRUCKS

Industrial trucks come in many types and sizes — too many to list in this brief section. For a list of truck types and for the hazards and types of atmospheres in which each type can be used, refer to NFPA 505, *Fire Safety Standard for Powered Industrial Trucks Including Type Designations, Areas of Use, Conversions, Maintenance, and Operation.*

In general, industrial trucks are fueled by diesel, gasoline, liquefied petroleum gas (LPG), compressed natural gas (CNG), electric batteries, or dual fuels (gasoline/LPG). Electric battery and LPG models are the most common types in use. The most common type of powered industrial truck is the forklift (Figure 4.66). However, there are many other kinds including personnel transportation vehicles, special materials handling vehicles, and emergency response vehicles (Figure 4.67).

In addition to NFPA 505, requirements for powered industrial trucks are contained in most of the model fire codes. In general, the fire inspector should look for the following conditions when inspecting facilities that contain powered industrial trucks:

Figure 4.66 Most warehouses have forklifts operating within them.

Figure 4.67 Facilities that handle special materials often have special equipment to move that material.

- Chargers for electric battery-operated units should be kept at least 5 feet (1.5 m) from combustible materials (Figure 4.68). Charging areas should not be accessible to the public. The charging area should be vented to prevent an accumulation of hydrogen gas.
- Liquid- or gas-fueled vehicles should not be fueled inside the building.
- Repairs should be performed in areas designated for that function.
- An industrial truck should only be used in the atmosphere for which it is specified. Hazardous atmospheres need special equipment that must be approved for those conditions. Signs should be posted to advise operators as to whether or not a certain type of truck is safe for that area.
- All powered industrial trucks should be marked with signs indicating the class of truck and any operational restrictions placed on that truck.
- Powered industrial trucks must be maintained in accordance with the instructional and training material provided by the manufacturer.
- Industrial trucks must be kept in a clean condition and reasonably free of lint, excess oil, and grease. Noncombustible agents are preferred for cleaning trucks. Flammable liquids must not be used. Combustible liquids can be used.

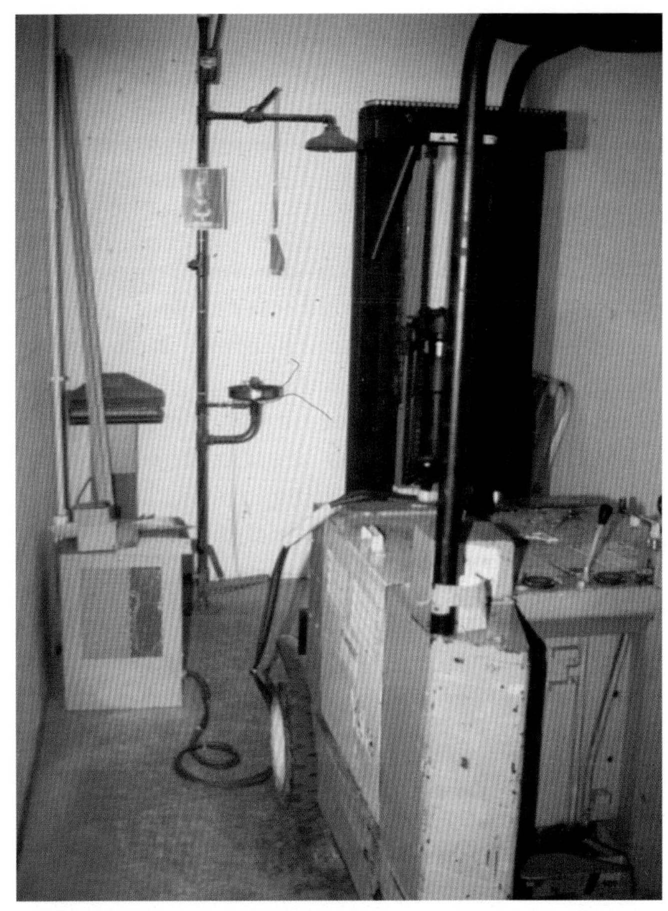

Figure 4.68 Electric-powered industrial trucks are charged somewhere within the occupancy.

- Fuel storage must be in compliance with NFPA 30.

SEMICONDUCTOR/ELECTRONICS MANUFACTURING

In today's modern world, people want things smaller and faster, especially when it comes to electronics. The components needed to make these electronic gadgets and modern conveniences are called semiconductors or chips.

The hazards with semiconductors do not lie in the chip itself but in the manufacturing process needed to develop the chip. Semiconductors are most usually made from silicon. The production of semiconductors includes the use of substances such as acids, alkalis, flammable liquids, flammable gases, pyrophoric gases, and toxic gases to name a few. Most of the model codes refer in general to these substances as *hazardous production materi-*

als (HPM). Because debris, such as dust or even flaking human skin, can ruin a chip, these chips must be manufactured in clean rooms. A *clean room* is a room that must maintain at least a particle level of less than 100,000 particles per cubic foot of air.

General office areas and nonhazardous production areas in semiconductor manufacturing facilities are inspected for the same fire hazards found in any general manufacturing facility. The areas of particular concern at semiconductor manufacturing facilities are the clean rooms, HPM storage and handling areas, and other areas where hazardous production processes are used.

Semiconductor clean rooms must be made of fire-resistive construction and contain a sprinkler system. The ventilation system in these rooms must be independent of those supplying the rest of the building. These ventilation systems need to have interlocks to shut down the system in the event of a chemical spill or fire. Minimum required portable fire extinguishers for clean rooms are 2A:40B:C. Dry-chemical extinguishers should not be used because the powder residue from the extinguishers will destroy all chips in the room. Rooms where HPMs are used must have special drainage systems capable of safely handling any of the liquids that are routinely disposed of or accidentally spilled.

Each of the model fire codes have extensive requirements on the storage and handling of HPMs. In general, HPMs may be stored in the following areas:

- *Within the work area.* Only small quantities that are expected to be used within the course of that work shift may be stored in approved containers and cabinets in this area.
- *Within the manufacturing building itself.* There may be specially designed rooms for storing HPMs within the manufacturing facility. Most of the codes require the storage area to be separated from the rest of the facility by at least a two-hour fire-resistive occupancy separation. At least one of the walls of the room must be an exterior wall and appropriate explosion protection features must be in place.
- *Outside storage.* The materials may be stored in a separate building from the manufacturing operation or in outside bulk storage tanks.

The fire inspector must ensure that these materials are all stored and handled in compliance with the code that is used within that particular jurisdiction.

TENTS

Prior to July 6, 1944, tents were not thought of as a major concern in most jurisdictions. However, on that day in Hartford, Connecticut, 168 people died when the circus tent under which they were seated caught fire. This event led to representatives of BOCA and the NFPA developing what eventually became NFPA 102, *Standard for Grandstands, Folding and Telescopic Seating, Tents, and Membrane Structures.* Most of the model building and fire codes also have extensive requirements for tents as well (Figure 4.69).

All of these standards and codes brought about great changes in the guidelines regulating the types of material from which tents could be made. Flame-resistant materials that are approved and tagged as such must be used for any tent (Figure 4.70). Materials that go inside the tent or on the

Figure 4.69 Large tents are often erected for special events, fairs, and carnivals.

Figure 4.70 Tent materials must be fire-resistant and should be marked as such.

tent, such as banners or decorations, must be treated with or made of flame-retardant material. All tent fabric must meet the requirements of the large-scale test contained in NFPA 701, *Standard Methods of Fire Tests for Flame-Resistant Textiles and Films*. The tents must be tagged or certified that they meet the requirements. The inspector should not attempt to test the flammability of the material on his own.

According to NFPA 102, tents should not cover more than 75 percent of the premises, unless otherwise approved by the authority having jurisidiction. Tents must meet all means of egress requirements contained in NFPA 101. There must be a minimum of 10 feet (3 m) between stake lines (Figure 4.71). Storage and handling of flammable or combustible liquids must be in accordance with NFPA 30. Flammable and combustible liquids must not be stored near means of egress. If a fuel has a flash point below 100°F (38°C), refueling is not allowed within the structure. Pyrotechnics and flames are prohibited in any tents or temporary membrane structures.

Figure 4.71 Tent stakes and ropes should be at least 10 feet (3 m) apart.

Railings or guards should not be less than 42 inches (1 070 mm) above the aisle surface or footrest, or 42 inches (1 070 mm) vertically above the center of the seat or seat board surface, whichever is adjacent. These guidelines must be followed for those portions of the backs and ends of all grandstands where the seats are more than 4 feet (1.3 m) above the ground.

Vegetation must be cleared both inside the tent and up to 10 feet (3 m) around the outside of the tent. It is up to the inspecting authority to agree if vegetation has been removed properly. No hay, straw, or shavings are allowed in the tent unless they have been treated with fire-retardant chemicals. Fire extinguishers are required as set forth in NFPA 10. Employees must be trained in fire extinguisher use. One or more methods to signal an emergency must be on hand. In many cases, a fire watch may be needed. Housekeeping is important to assure that egress areas are not blocked and that fire loads within the structure are not excessive.

5

Construction and Occupancy Classifications

This chapter contains information that will assist the reader in meeting the listed job performance requirements contained in NFPA 1031, ***Standard for Professional Qualifications for Fire Inspector and Plan Examiner*** (proposed 1998 edition).

Chapter 3 Fire Inspector I

3-3.1 Identify the occupancy classification of a single-use occupancy, given a description of the occupancy and its use, so that an accurate classification is made according to applicable codes and standards.

3-3.4 Verify the type of construction for an addition or remodeling project, given field observations or a description of the project and the materials being used, so that the construction type is classified and recorded in accordance with the applicable codes and standards and the policies of the jurisdiction.

Chapter 4 Fire Inspector II

4-3.2 Identify the occupancy classification of a mixed-use building, given a description of the uses, so that each area is properly classified in accordance with applicable codes and standards.

4-3.3 Determine the type of construction in a new building, given plan review and observations or a description of the building's height, area, occupancy, and construction features, so that the construction type is properly classified according to applicable codes and standards.

4-4.1 Classify the occupancy type, given a set of plans, specifications, and a description of a building, so that the classification is made according to applicable codes and standards.

4-4.5 Verify the construction type of a building or portion thereof, given a set of approved plans and specifications, so that the construction type complies with the approved plans and applicable codes and standards.

Chapter 5
Construction and Occupancy Classifications

Two of the most basic pieces of information that a fire inspector must know when preparing to inspect a building are how it is constructed and what it is used for. Technically, how the building is constructed is referred to as its *construction classification*. What the building is used for is referred to as the *occupancy classification*. Each of the model codes has its own construction and occupancy classifications. The fire inspector must have a thorough understanding of the construction and occupancy classifications so that he will apply the proper code requirements to the building being inspected.

There are four major model code organizations that produce codes which contain information concerning construction and occupancy classifications. Fire inspectors must be knowledgeable of the requirements within the code that has been adopted in their jurisdiction. The four code organizations and the codes they promulgate are:

National Fire Protection Association (NFPA)

NFPA 101, *Life Safety Code®* (Occupancy classifications)

NFPA 220, *Standard on Types of Building Construction*

Building Officials & Code Administrators International, Inc. (BOCA)

The BOCA Building Code® (BOCA)

International Conference Of Building Officials (ICBO)

The Uniform Building Code™ (UBC)

Southern Building Code Congress International, Inc. (SBCCI)

The Standard Building Code© (SBC)

Each of these code organizations also promulgates other codes, such as fire, plumbing, electrical, and mechanical codes. The fire inspector must be well versed in all portions of the codes that he will be responsible for enforcing. Appendix E contains more detailed information on each of these model code organizations.

This chapter examines construction and occupancy classifications that are relative to each of the four model code organizations previously listed. The information in this chapter is by no means a complete treatment of all of the requirements for each of the topics. For more detailed information, fire inspectors should consult a copy of the code used in their jurisdiction.

CONSTRUCTION CLASSIFICATIONS

Building construction is classified in different ways in different codes. In general, construction classifications are based upon materials used in construction and upon the hourly fire-resistance ratings of structural components. The following sections detail the construction classifications for each of the four major model codes.

NFPA 220, *Standard on Types of Building Construction*

NFPA 220 divides construction types into five basic classifications: Type I through Type V. Each classification is then further broken down into subtypes by use of a three-digit Arabic number code or several letters (for example, Type I-332 or Type IV-2HH). Each digit in the numerical chain has a specific meaning:

- The first digit refers to the fire-resistive rating (in hours) of the exterior bearing walls.

- The second digit refers to the fire-resistive rating (in hours) of structural frames or columns and girders that support loads of more than one floor.
- The third digit indicates the fire-resistive rating (in hours) of the floor construction.

A listing of construction types and the degree of fire resistance of each is shown in Table 5.1. For an explanation of the way fire-resistance ratings are determined, fire inspectors should refer to NFPA 251, *Standard Methods of Fire Tests of Building Construction and Materials*. Construction classifications from NFPA 220 will be further explained in the following sections.

TYPE I CONSTRUCTION (I-443 OR I-332)

Type I construction has structural members including walls, columns, beams, floors, and roofs of noncombustible or limited combustible materials. Type I construction is commonly referred to as *fire-resistive* construction (Figure 5.1). Table 5.1 shows the degree of fire resistance for the various components of Type I construction.

TYPE II CONSTRUCTION (II-222, II-111, OR II-000)

Type II construction is similar to Type I construction, except the degree of fire resistance is lower. Note that in II-000, construction materials with no fire-resistance rating are used. Type II construction is commonly referred to as *non-*

TABLE 5.1
Fire Resistance Ratings (In Hours) For Type I Through Type V Construction

	Type I		Type II			Type III		Type IV	Type V	
	443	332	222	111	000	211	200	2HH	111	000
Exterior Bearing Walls —										
Supporting more than one floor, columns, or other bearing walls	4	3	2	1	0	2	2	2	1	0
Supporting one floor only	4	3	2	1	0	2	2	2	1	0
Supporting a roof only	4	3	1	1	0	2	2	2	1	0
Interior Bearing Walls —										
Supporting more than one floor, columns, or other bearing walls	4	3	2	1	0	1	0		1	0
Supporting one floor only	3	2	2	1	0	1	0		1	0
Supporting roofs only	3	2	1	1	0	1	0		1	0
Columns —										
Supporting more than one floor, columns, or other bearing walls	4	3	2	1	0	1	0	H	1	0
Supporting one floor only	3	2	2	1	0	1	0	H	1	0
Supporting roofs only	3	2	1	1	0	1	0	H	1	0
Beams, Girders, Trusses & Arches —										
Supporting more than one floor, columns, or other bearing walls	4	3	2	1	0	1	0	H	1	0
Supporting one floor only	3	2	2	1	0	1	0	H	1	0
Supporting roofs only	3	2	1	1	0	1	0	H	1	0
Floor Construction	3	2	2	1	0	1	0	H	1	0
Roof Construction	2	1½	1	1	0	1	0	H	1	0
Exterior Nonbearing Walls	0	0	0	0	0	0	0	0	0	0

[white box] Those members that shall be permitted to be of approved combustible material.

"H" indicates heavy timber members; see text for requirements.

combustible or *limited combustible* construction (Figure 5.2).

TYPE III CONSTRUCTION (III-211 OR III-200)

Type III construction, also known as *ordinary construction*, has exterior walls and structural members that are of approved noncombustible or limited-combustible materials (Figure 5.3). Interior structural members including walls, columns, beams, floors, and roofs are wholly or partly constructed of wood. The wood used in these members is of smaller dimensions than are required for heavy timber construction. In addition, structural members must have fire-resistance ratings that are not less than those specified in Table 5.1.

Figure 5.1 A fire-resistive (Type I) construction office building.

Figure 5.2 A noncombustible (TypeII) construction office building.

Figure 5.3 An ordinary construction (Type III) educational facility.

TYPE IV CONSTRUCTION (2HH)

In Type IV construction, exterior walls, interior walls, and structural members that are portions of these walls are of approved noncombustible or limited-combustible materials. Type IV construction is commonly referred to as *heavy timber construction* (Figure 5.4). Other interior structural members including columns, beams, arches, floors, and roofs are of solid or laminated wood without concealed spaces. The requirements for these members are detailed in Chapter 3 of NFPA 220.

TYPE V CONSTRUCTION (V-111 OR V-000)

Type V construction, also called *wood-frame construction*, has exterior walls, bearing walls, floors, roofs, and supports made completely or partially of wood or other approved materials of smaller dimensions than those used for Type IV construction (Figure 5.5). In addition, structural members have fire-resistance ratings that are not less than those set forth in Table 5.1.

Figure 5.4 Notice the heavy timbers used to construct this old barn.

Figure 5.5 A wood-frame (Type V) construction garage.

The BOCA Building Code®

The *BOCA Building Code®* utilizes five construction classifications that are very similar to those described for NFPA 220. Each classification is given a numerical designation (Types 1 through 5). Some of the individual types are further broken down into subgroups that are given a letter designation (for example, Types 1A and 1B). These subgroups are more fully explained in the sections to follow.

TYPES 1 AND 2 CONSTRUCTION (1A, 1B, 2A, 2B, AND 2C)

Types 1 and 2 construction are considered *fire-resistive* or *noncombustible* types of construction. These types contain walls, partitions, structure elements, floors, ceilings, roofs, and exits that are constructed of approved noncombustible materials. Types 1 and 2 are similar in design with the primary differences being in the fire-resistance ratings of various components of the construction assembly. Types 1, 2A, and 2B require that structural elements be of protected construction. *Protected construction* consists of structural members that are constructed, chemically treated, covered, or protected so that the entire assembly has an appropriate fire-resistance rating. Type 2C may utilize unprotected structural members.

TYPE 3 CONSTRUCTION (3A AND 3B)

Type 3 construction, also known as *ordinary construction*, consists of those structures whose exterior walls are constructed of concrete, masonry, or other approved noncombustible materials. The interior structural elements, load bearing walls, partitions, floors, and roofs are constructed of any approved materials. The primary difference between Type 3A and 3B structures involves different fire-resistance ratings for certain elements of the construction assembly. As well, Type 3A requires protected structural members and Type 3B does not.

TYPE 4 CONSTRUCTION

Type 4 construction, also known as *heavy timber construction*, includes those buildings in which exterior walls are constructed of approved noncombustible materials and the interior structural members are of solid or laminated wood without concealed spaces. As well, the load bearing walls, partitions, floors, and roofs are constructed of any noncombustible materials permitted within this code.

TYPE 5 CONSTRUCTION

Type 5 construction, also called *wood-frame construction*, has exterior walls, load bearing walls, partitions, floors, and roofs constructed of any approved materials.

The Uniform Building Code™

The *Uniform Building Code™* also utilizes five construction classifications. Each classification is given a Roman numerical designation (Types I through V). Some of the individual types are further broken down into subgroups. The subgroups are named *fire-resistive (F.R.), one-hour*, or *no requirements (N)*. The basic difference in each of the subgroups is more fully explained in the sections to follow. For more detailed information on structural framework, walls, openings, stairway construction, and roofs for each construction type, consult the *Uniform Building Code™* directly.

TYPE I BUILDINGS

Type I buildings, also known as *Type I Fire-Resistive* or *Type I-F.R.* buildings, require that structural elements be of steel, iron, concrete, or masonry. Walls and permanent partitions must be of noncombustible fire-resistive construction, with the exception of permanent nonbearing partitions that have a one- or two-hour rating and are not part of a shaft enclosure. These walls or partitions may have fire-retardant-treated wood within the assembly.

TYPE II BUILDINGS (II-F.R., II-ONE-HOUR, AND II-N)

Type II, also known as *noncombustible construction*, is divided into three subgroups: II-F.R., II-One-Hour, and II-N. The primary difference between each of these subgroups involves hourly fire-resistance ratings and allowable materials in the construction assemblies. Type II-F.R. buildings are similar to Type I buildings in that structural elements must be of steel, iron, concrete, or masonry. Walls and permanent partitions must be of noncombustible fire-resistive construction, with the exception of permanent nonbearing partitions that have a one- or two-hour rating and are not part of a shaft enclosure. These walls or partitions may have fire-retardant-treated wood within the assembly. Types II-One-Hour and II-N structural elements must be of any approved noncombustible materials. For more detailed information on Type II requirements, see Table 5.2.

TYPE III BUILDINGS (III-ONE-HOUR AND III-N)

Type III buildings, also known as *ordinary construction*, are divided into two subgroups: III-One-Hour and III-N. The structural elements may be of any materials permitted in the code. The primary difference between Types III-One-Hour and III-N is that III-One-Hour requires one-hour fire-resistive construction throughout the entire structure. This is not required in III-N structures.

TYPE IV BUILDINGS (IV-H.T.)

Type IV buildings, also known as *heavy timber construction*, may have structural elements of any type permitted by the code. The wood used for interior structural members must be of the dimension that is required for heavy timber construction. Type IV buildings must have permanent partitions and members of the structural frame that have a minimum fire-resistance rating of at least one hour. See

TABLE 5.2
Types Of Construction — Fire-Resistive Requirements (In Hours)

	Type I	Type II			Type III		Type IV	Type V	
	Noncombustible				Combustible				
Building Element	Fire-resistive	Fire-resistive	1-Hr.	N	1-Hr.	N	H.T.	1-Hr.	N
1. Bearing walls — exterior	4 Sec. 602.3.1	4 Sec. 603.3.1	1	N	4 Sec. 604.3.1	4 Sec. 604.3.1	4 Sec. 605.3.1	1	N
2. Bearing walls — interior	3	2	1	N	1	N	1	1	N
3. Nonbearing walls — exterior	4 Sec. 602.3.1	4 Sec. 603.3.1	1 Sec. 603.3.1	N	4 Sec. 604.3.1	4 Sec. 604.3.1	4 Sec. 605.3.1	1	N
4. Structural frame[1]	3	2	1	N	1	N	1 or H.T.	1	N
5. Partitions — permanent	1[2]	1[2]	1[2]	N	1	N	1 or H.T.	1	N
6. Shaft enclosures[3]	2	2	1	1	1	1	1	1	1
7. Floors and floor-ceilings	2	2	1	N	1	N	H.T.	1	N
8. Roofs and roof-ceilings	2 Sec. 602.5	1 Sec. 603.5	1 Sec. 603.5	N	1	N	H.T.	1	N
9. Exterior doors and windows	Sec. 602.3.2	Sec. 603.3.2	Sec. 603.3.2	Sec. 603.3.2	Sec. 604.3.2	Sec. 604.3.2	Sec. 605.3.2	Sec. 606.3	Sec. 606.3
10. Stairway construction	Sec. 602.4	Sec. 603.4	Sec. 603.4	Sec. 603.4	Sec. 604.4	Sec. 604.4	Sec. 605.4	Sec. 606.4	Sec. 606.4

N — No general requirements for fire resistance. H.T. — Heavy timber.

[1]Structural frame elements in an exterior wall that is located where openings are not permitted or where protection of openings is required, shall be protected against external fire exposure as required for exterior bearing walls or the structural frame, whichever is greater.

[2]Fire-retardant-treated wood (see Section 207) may be used in the assembly, provided fire-resistance requirements are maintained. See Sections 602 and 603.

[3]For special provisions, see Sections 304.6, 306.6, and 711.

Table 5.2 for specific requirements related to Type IV buildings.

TYPE V BUILDINGS (V-ONE-HOUR AND V-N)

Type V buildings, also known as *wood-frame construction*, may be constructed in a similar manner to those described for Type V(5) construction in the other codes. Type V-One-Hour buildings must have a one-hour fire-resistive rating throughout the entire structure. See Table 5.2 for specific requirements related to Type V buildings.

The Standard Building Code©

The *Standard Building Code©* is the only one of the four discussed in this section that differs significantly from the others. The *Standard Building Code©* contains six construction types (Types I through VI). Several of the types are broken into two subgroups (1-Hour Protected and Unprotected). A listing of construction types and the degree of fire resistance of each are shown in Table 5.3.

TYPES I AND II CONSTRUCTION

As with the previously discussed model codes, Types I and II construction are considered to be *fire-resistive* construction. Both of these classifications feature exterior walls, interior bearing walls, columns, beams, girders, trusses, arches, floors, and roofs that are made of noncombustible materials. The primary difference between Types I and II is the degree of fire resistance that is required for each of the structural elements. These different ratings can be found in Table 5.3.

TYPE III CONSTRUCTION

In the *Standard Building Code©*, Type III construction is *heavy timber construction*. The code requires heavy timbers to be used that are of at least 8 x 8 inches (200 mm by 200 mm) when supporting roofs and at least 6 x 8 inches (150 mm by 200 mm) when supporting roof and ceiling loads only. Wood other than heavy timbers must have a fire-resistance rating of at least one hour. This can be achieved by avoiding concealed spaces under floors and roofs, using approved construction details and adhesives for structural members, and by providing the required degree of fire resistance in exterior and interior walls.

TYPE IV CONSTRUCTION (IV PROTECTED AND IV UNPROTECTED)

Type IV construction is also known as *noncombustible construction*. Both feature structural members including exterior walls, interior bearing walls, and columns that are constructed of noncombustible materials. The primary difference between Types I/II and Type IV buildings is the degree of fire resistance that is required for each of the structural elements. There are two subgroups within this classification: Type IV Protected and IV Unprotected. The different ratings can be found in Table 5.3.

TYPE V CONSTRUCTION (V PROTECTED AND V UNPROTECTED)

In the *Standard Building Code©*, Type V construction is *ordinary construction*. This type features exterior bearing and nonbearing walls that are constructed of noncombustible materials. Beams, girders, trusses, arches, floors, roofs, and interior framing are wholly or partly constructed of wood or other approved materials. There are two subgroups within this classification: Type V Protected and Type V Unprotected. See Table 5.3 for the fire-resistance ratings required of these two subgroups.

TYPE VI CONSTRUCTION (VI PROTECTED AND VI UNPROTECTED)

In the *Standard Building Code©*, Type VI construction is *wood-frame construction*. This type features exterior bearing and nonbearing walls and partitions, beams, girders, trusses, arches, floors, roofs, and interior framing that are wholly or partly constructed of wood or other approved materials. There are two subgroups within this classification: Type VI Protected and Type VI Unprotected. See Table 5.3 for the fire-resistance ratings required of these two subgroups.

OCCUPANCY CLASSIFICATIONS

The second manner by which buildings are grouped for fire inspections and code enforcement is by occupancy classifications. The *occupancy classification* can be defined as the use to which owners or tenants put all or a portion of a building. The premise upon which occupancy classifications are based is that certain occupancies, by their own

TABLE 5.3
Fire Resistance Ratings Required Fire Resistance In Hours

STRUCTURAL ELEMENT	Type I	Type II	Type III	Type IV 1-Hour Protected	Type IV Unprotected	Type V 1-Hour Protected	Type V Unprotected	Type VI 1-Hour Protected	Type VI Unprotected
PARTY AND FIRE WALLS (a)	4	4	4	4	4	4	4	4	4
INTERIOR BEARING WALLS	(l)								
Supporting columns, other bearing walls or more than one floor	4	3	2	1	NC	1(h)	0(h)	1	0
Supporting one floor only	3	2	1	1	NC	1	0	1	0
Supporting one roof only	3	2	1	1	NC	1	0	1	0
INTERIOR NONBEARING PARTITIONS		See 704.1, 704.2 and 705.2							
COLUMNS	(l)		See 605						
Supporting other columns or more than one floor	4	3	H(d)	1	NC	1	0	1	0
Supporting one floor only	3	2	H(d)	1	NC	1	0	1	0
Supporting one roof only	3	2	H(d)	1	NC	1	0	1	0
BEAMS, GIRDERS, TRUSSES & ARCHES	(l)		See 605						
Supporting columns or more than one floor	4	3	H(d)	1	NC	1	0	1	0
Supporting one floor only	3	2	H(d)	1	NC	1	0	1	0
Supporting one roof only	1½(e, p)	1(e, f, p)	H(d)	1(e,p)	NC(e)	1	0	1	0
FLOORS & FLOOR/CEILING CONSTRUCTIONS	(l) 3	2	See 605 H(o)	(n) 1	(n,o) NC	(n) 1	(m,n,o) 0	1	(o) 0
ROOFS & ROOF/CEILING CONSTRUCTIONS (g)	1½(e, p)	1(e, f, p)	See 605 H(d)	1(e,p)	NC(e)	1	0	1	0
EXTERIOR BEARING WALLS and gable ends of roof (g, i, j)	(% indicates percent of protected and unprotected wall openings permitted. See 705.1.1 for protection requirements.)								
Horizontal separation (distance from common property line or assumed property line).									
0 ft to 3 ft (c)	4(0%)	3(0%)	3(0%)(b)	2(0%)	1(0%)	3(0%)(b)	3(0%)(b)	1(0%)	1(0%)
over 3 ft to 10 ft (c)	4(10%)	3(10%)	2(10%)(b)	1(10%)	1(10%)	2(10%)(b)	2(10%)(b)	1(20%)	0(20%)
over 10 ft to 20 ft (c)	4(20%)	3(20%)	2(20%)(b)	1(20%)	NC(20%)	2(20%)(b)	2(20%)(b)	1(40%)	0(40%)
over 20 ft to 30 ft	4(40%)	3(40%)	1(40%)	1(40%)	NC(40%)	1(40%)	1(40%)	1(60%)	1(60%)
over 30 ft	4(NL)	3(NL)	1(NL)	1(NL)	NC(NL)	1(NL)	1(NL)	1(NL)	0(NL)
EXTERIOR NONBEARING WALLS and gable ends of roof (g, i, j)	(% indicates percent of protected and unprotected wall openings permitted. See 705.1.1 for protection requirements.)								
Horizontal separation (distance from common property line or assumed property line).									
0 ft to 3 ft (c)	3(0%)	4(0%)(b)	3(0%)(b)	2(0%)	1(0%)	3(0%)(b)	3(0%)(b)	1(0%)	1(0%)
over 3 ft to 10 ft (c)	2(10%)	2(10%)	2(10%)(b)	1(10%)	2(10%)	2(10%)(b)	2(10%)(b)	1(20%)	0(20%)
over 10 ft to 20 ft (c)	2(20%)	2(20%)	2(20%)(b)	1(20%)	NC(20%)	2(20%)(b)	2(20%)(b)	1(40%)	0(40%)
over 20 ft to 30 ft	1(40%)	1(40%)	1(40%)	NC(40%)	NC(40%)	1(40%)	1(40%)	0(60%)	0(60%)
over 30 ft (k)	NC(NL)	NC(NL)	NC(NL)	NC(NL)	NC(NL)	NC(NL)	NC(NL)	0(NL)	0(NL)

1 ft = 0.305 m NC=Noncombustible NL=No limits H=Heavy Timber Sizes

Notes:

a. See 704.5 for extension of party walls and fire walls. **b.** See 704.5 for parapets. **c.** See 705 for protection of wall openings. **d.** Where horizontal separation of 20 ft or more is provided, wood columns, arches, beams and roof deck conforming to heavy timber sizes may be used externally. **e.** In buildings not over two stories approved fire retardant treated wood may be used. **f.** In one story buildings, structural members of heavy timber sizes may be used as an alternate to unprotected structural roof members. Stadiums, field houses and arenas with heavy timber wood dome roofs are permitted. An approved automatic sprinkler system shall be installed in those areas where 20 ft clearance to the floor or balcony below is not provided. **g.** See 1503 for penthouses and roof structures. **h.** The use of combustible construction for interior bearing partitions shall be limited to the support of not more that two floors and a roof. **i.** Exterior walls shall be fire tested in accordance with 601.3. The fire resistance requirements for exterior walls with 5 ft or less horizontal separation shall be based upon both interior and exterior fire exposure. The fire resistance requirements for exterior walls with more than 5 ft horizontal separation shall be based upon interior fire exposure only. **j.** Where Appendix F is specifically included in the adopting ordinance, see F102.2.6 for fire resistance requirements for exterior walls of type IV buildings in Fire District. **k.** Walls or panels shall be of noncombustible material or fire retardant treated wood, except for Type VI construction. **l.** For Group A - Large Assembly, Group A - Small Assembly, Group B, Group E, Group F, Group R occupancies and Automobile Parking Structures, occupancies of Type I construction, partitions, columns, trusses, girders, beams, and floors may be reduced by 1 hour if the building is equipped with an automatic sprinkler system throughout, but no component or assembly may be less than 1 hour. **m.** Group A - Large Assembly (no stage requiring proscenium opening protection) and Group A - Small Assembly occupancies of Type V Unprotected construction shall have 1 hour fire resistant floors over any crawl space or basement. **n.** For Group B and Group M occupancies of Type IV or Type V construction, when five or more stories in height a 2 hour fire resistant floor shall be required over the basement. **o.** For unsprinklered Group E occupancies of Type III, Type IV Unprotected, Type V Unprotected or Type VI Unprotected, floors located immediately above usable space in basements shall have a fire resistant rating of not less than 1 hour. **p.** In buildings of Group A, B, E, and R occupancies, fire resistance may be omitted where structural members support a roof only and are 20 ft or more clear above any floor or balcony.

nature, will have higher fire loads and greater numbers of occupants within them than others. For example, the load of combustibles in an elementary school (educational classification) would not be expected to be as high as in a warehouse (storage classification) (Figures 5.6 a and b). By classifying structures using the construction classifications described in the first portion of this chapter and the occupancy classifications that we are about to discuss, fire officials and code enforcement personnel can gain a reasonable expectation of the level of hazard presented by a particular building.

Figure 5.6a Schools typically do not contain high fire loads.

Figure 5.6b Warehouses usually have high fire loads within them.

In earlier days, the four code organizations listed at the beginning of this chapter all contained significant differences in reference to occupancy classifications. However, representatives of each of the four organizations formed a consortium called the Board for Coordination of Model Codes (BCMC) to address these and other issues. This has resulted in three of the four organizations (UBC, BOCA, and SBCCI) adopting occupancy classifications that are essentially the same, with a few minor exceptions. NFPA 101, *Life Safety Code®* retained its old designations. This was because it was determined that the *Life Safety Code®* is not truly a building code and did not necessarily need to match the others. The duration of this chapter is dedicated to explaining these occupancy classifications.

NFPA 101, *Life Safety Code®*

NFPA 101 contains nine basic occupancy classifications that structures may fall under. These are:

- Places of Assembly
- Educational Occupancies
- Health Care Occupancies
- Detention and Correctional Occupancies
- Residential Occupancies
- Mercantile Occupancies
- Business Occupancies
- Industrial Occupancies
- Storage Occupancies

Places of Assembly include, but are not limited to, all buildings or portions of buildings used for gathering 50 or more persons for such purposes as deliberation, worship, entertainment, dining, amusement, or awaiting transportation (Figure 5.7). Assembly occupancies are further classified according to the number of persons that a structure can accommodate:

- Class A — Facilities capable of handling 1,000 persons or more
- Class B — Facilities capable of handling 300 to 1,000 persons
- Class C — Facilities capable of handling 50 to 300 persons

Figure 5.7 A restaurant is considered a place of assembly.

Examples of Places of Assembly include:

- Auditoriums
- Libraries
- Bowling alleys
- Passenger stations (train, bus, airports)
- College/University classrooms
- Restaurants
- Dance halls
- Theaters

Educational Occupancies include all buildings used for educational purposes up through the 12th grade by 6 or more persons, 4 hours per day or more than 12 hours per week. Examples of these occupancies include academies, kindergartens, nursery schools, day care facilities, and elementary and secondary schools (Figure 5.8). Other occupancies within educational facilities are governed by regulations for that specific occupancy. For example, the auditorium portion of a high school would be considered a Place of Assembly.

Health Care Occupancies are used for medical treatment or care of persons suffering from physical or mental illness, disease, or infirmity; or for the care of infants, convalescents, or infirmed, elderly persons. Included in this classification are hospitals, nursing homes, residential-custodial care facilities, and ambulatory health care facilities (Figure 5.9).

Figure 5.8 This high school is an educational occupancy.

Detention and Correctional Occupancies provide sleeping facilities for four or more residents and are occupied by persons who are generally prevented from protecting themselves because of security measures not under their control (Figure 5.10). Included in these occupancies are:

- Jails and prisons
- Detention centers
- Adult and juvenile correctional facilities
- Adult and juvenile substance abuse centers
- Juvenile detention and training facilities

Residential Occupancies are those in which sleeping accommodations are provided for normal residential purposes and include all buildings that are designed to provide sleeping accommodations.

Figure 5.9 Nursing homes are health care occupancies.

Figure 5.10 This multi-story county jail is an example of a detentional and correctional occupancy.

Examples of residential occupancies include:

- Hotels
- Apartments
- Dormitories
- Lodging and rooming facilities
- Board and care facilities
- Single- and two-family dwellings

Mercantile Occupancies include stores, markets, and other rooms, buildings, or structures used to display and sell merchandise. These include supermarkets, department stores, pharmacies, and shopping centers (Figure 5.11). Note that malls usually do not fall under this classification and are handled as a special occupancy. You should also be aware that small merchandising operations in buildings predominantly of another occupancy are subject to the requirements of the predominant occupancy. An example of this situation would be a gift shop in the lobby of a hotel.

Business Occupancies are those used for transacting business (other than those covered under Mercantile Occupancies), for keeping accounts and records, or other similar purposes (Figure 5.12). Examples of business occupancies include:

Figure 5.11 Strip shopping centers are one type of mercantile occupancy.

Figure 5.12 This county courthouse meets the definition of a business occupancy.

- Doctors' and dentists' offices
- General and government offices
- City halls
- Courthouses
- College and university classroom buildings with rooms for less than 50 people
- Instructional laboratories

Industrial Occupancies include factories making products of all kinds and properties devoted to such operations as processing, assembly, mixing, packaging, finishing, decorating, and repairing (Figure 5.13). Examples of industrial occupancies include:

- All factories
- Dry-cleaning plants
- Food-processing plants
- Aircraft servicing/maintenance hangars
- Power plants
- Refineries
- Saw mills
- Telephone exchanges

Storage Occupancies include warehouses, ministorage facilities, cold storage warehouses, freight terminals, truck and marine terminals, bulk petroleum storage facilities, parking garages, and barns (Figure 5.14).

Figure 5.13 Power plants are one type of industrial occupancy.

Figure 5.14 Many communities now have mini-storage facilities that fall within the definition of storage occupancies.

MIXED OCCUPANCIES

The determination of an occupancy classification for a structure that only has one type of operation within it is a relatively simple task. However, it is very common to be confronted by buildings that have two or more very different types of operations within them. These situations are often not so clear for personnel performing a fire inspection or code enforcement. Multiple-use occupancies usually require personnel to make a determination of which parts of the facility fall under a particular occupancy classification. Fire personnel making a routine inspection should have records available to them that indicate the occupancy classifications that have been used in the past. This information can be found on previous inspection reports or occupancy certificates. In most cases, inspection personnel will only need to verify that the structure still falls under the same classification. The fire inspector should consult with the building official if a change is found.

In general, the *Life Safety Code®* does not require the separation of occupancies in the same structure. Local code ordinances may choose to require separations, but this will vary from jurisdiction to jurisdiction. There are two situations concerning mixed occupancies that fire inspection and code enforcement personnel must be prepared to deal with:

1. Buildings that have different uses within them in distinctly different portions of the building.
2. Buildings that have different uses within them in an intermingled manner that makes it impractical to separate out portions of the building for different classifications.

An example of the first scenario would be the aforementioned high school with a large auditorium within the building. The majority of the building would fall under the auspices of an Educational Occupancy. However, the auditorium would be considered to be a Place of Assembly. Each portion of the building would be required to meet the code requirements that are applicable for that type occupancy.

Examples of occupancies fitting the second scenario are less likely to be encountered by most inspection personnel, yet they represent the most challenging cases. One example of this might be a large commercial printing operation (Industrial Occupancy) that has bulk paper storage (Storage Occupancy) throughout the facility (Figure 5.15). The *Life Safety Code®* recommends that the most restrictive life safety requirements of the occupancies involved be enforced in these situations. Depending on the particular situation, this could mean that all the requirements of one type of occupancy will take precedence or that various requirements from each of the two occupancies will be used.

Figure 5.15 Commercial printing operations are often considered mixed occupancies.

Model Building Code Occupancy Classifications

As discussed earlier, through the work of the Board for Coordination of Model Codes (BCMC), the *BOCA Building Code®*, the *Uniform Building Code™*, and the *Standard Building Code©* all generally use the same occupancy classifications. Some of these are similar to those in NFPA 101 and some are not. As well, there remain a few small differences between the three model building codes on this issue. These will be discussed in this section.

The occupancy classifications are found in Chapter 3 of each of the model building codes. The BCMC occupancy classifications are:

Group A — Assembly

Group B — Business

Group E — Educational

Group F — Factory/Industrial

Group H — Hazardous (called High Hazard by the SBC)

Group I — Institutional

Group M — Mercantile

Group R — Residential

Group S — Storage

Group U — Utility/Miscellaneous (not recognized by the SBC)

Each of the model codes tend to have different subgroups within each of the listed groups. The following sections will give a general description of each group and then will highlight the subgroups for each of the model standards. For more detailed information on these occupancies, consult the code that is used within your jurisdiction.

GROUP A — ASSEMBLY

In general, these occupancies are the same as those described for Places of Assembly occupancies in NFPA 101. The three model codes break this group into subgroups as described as follows:

The BOCA Building Code®

BOCA uses five subgroups to classify Group A occupancies:

- Group A-1, Theaters — Includes movie theaters, performing arts centers, and television or radio studios.
- Group A-2, Structures — Includes places without theatrical stages such as dance halls and nightclubs.
- Group A-3, Structures — Includes places without theatrical stages such as libraries, art galleries, lecture halls, museums, exhibition halls, and restaurants.
- Group A-4, Structures — Churches and other places of worship.
- Group A-5, Outdoor Assembly — Includes grandstands, bleachers, stadiums, amusement park structures, and fair or carnival structures (Figure 5.16).

Figure 5.16 BOCA would classify this football stadium as a Group A-5, outdoor assembly occupancy.

The Standard Building Code©

The SBC uses two subgroups to classify Group A occupancies:

- Group A-1 — Theaters and other large places of assembly without a stage and with an occupant load of at least 1,000 persons. Also, facilities with a stage and an occupant load in excess of 700 persons.
- Group A-2 — Theaters and other places of assembly with or without a stage and with an occupant load between 100 and 1,000 persons.

Assembly occupancies with an occupant load less than 100 persons are considered Group B occupancies in the SBC.

The Uniform Building Code™

The UBC divides Group A occupancies into five subgroups:

- Division 1 — A building or portion of a building with an occupant load of greater than 1,000 persons and a stage.
- Division 2 — A building or portion of a building with an occupant load of less than 1,000 persons and a stage.
- Division 2.1 — A building or portion of a building with an occupant load of greater than 300 persons and without a stage. Includes educational facilities that do not fall within Group B or E.
- Division 3 — A building or portion of a building with an occupant load of less than

300 persons and without a stage. Includes educational facilities that do not fall within Group B or E.

- Division 4 — Outdoor assembly structures.

GROUP B — BUSINESS

None of the three model codes have any specific subgroups for Group B occupancies.

GROUP E — EDUCATIONAL

In general, these occupancies are the same as those described for Educational Occupancies in NFPA 101. Neither BOCA nor the SBC have subgroups for Group E occupancies. However, the UBC contains the following subgroups:

- Group E, Division 1 — Any building used for educational purposes through the12th grade by 50 or more persons for more than 12 hours per week or 4 hours in any one day.
- Group E, Division 2 — Any building used for educational purposes through the12th grade by less than 50 persons for more than 12 hours per week or 4 hours in any one day.
- Group E, Division 3 — Any building or portion thereof used for day-care purposes for more than 6 persons (Figure 5.17).

Figure 5.17 The UBC would considered this daycare center a Group E, Division 3 occupancy.

GROUP F — FACTORY/INDUSTRIAL

This group covers the types of occupancies that were discussed in the Industrial Occupancies section for NFPA 101. The SBC does not have any subgroups for Group F occupancies. The UBC and BOCA break Group F occupancies into two subgroups. With the exception of slightly different names, these subgroups are the same and are described as follows:

- BOCA Group F-1 and UBC Group F, Division 1 structures are all those that are classified moderate-hazard factory and industrial occupancies (Figure 5.18). Examples of these occupancies include:
  - — Aircraft facilities
  - — Food processing
  - — Bakeries
  - — Machine shops
  - — Clothing manufacturing
  - — Printing or publishing
  - — Power generation facilities
  - — Textile mills
- BOCA Group F-2 and UBC Group F, Division 2 structures are all those that involve the manufacturing or fabrication of noncombustible materials that do not contribute to a significant fire hazard (Figure 5.19). Examples of these occupancies include:
  - — Nonalcoholic beverages
  - — Glass products
  - — Brick and masonry
  - — Ice
  - — Foundries
  - — Metal fabrication

Figure 5.18 This clothing manufacturing facility is considered to be a moderate hazard manufacturing facility.

Figure 5.19 Ice plants are considered lesser hazards than other manufacturing operations.

GROUP H — HAZARDOUS

This is an occupancy for which no comparable classification exists in NFPA 101. These are occupancies that are used for the manufacturing, processing, generation, or storage of hazardous materials. These materials generally constitute a high fire, explosion, or health hazard. Each of the model codes break Group H into multiple subgroups that are described in the following sections.

BOCA and SBC Group H Subgroups

BOCA and SBC divide Group H occupancies into four subgroups:

- Group H-1 — Buildings that contain materials that pose an explosion (detonation) hazard such as blasting agents, organic peroxides, oxidizers, and some pyrophoric materials.
- Group H-2 — Buildings that contain materials that present a hazard from accelerated burning (deflagration). These materials include, but are not limited to, combustible dusts and liquids, flammable gases and liquids, and oxidizing gases.
- Group H-3 — Buildings that contain materials that readily support combustion or present a physical hazard, including aerosols, combustible fibers, flammable solids, and tires.
- Group H-4 — Buildings that contain materials that present a health hazard. These include corrosives, highly toxic materials, radioactive materials, and irritants.

UBC Group H Subgroups

The UBC has its own method for breaking Group H occupancies into subgroups. It utilizes seven subgroups that are described as follows:

- Division 1 — Buildings that contain materials that pose a high explosion hazard such as explosives, detonatable organic peroxides, and oxidizers.
- Division 2 — Buildings that contain combustible dusts, materials with moderate explosion hazards, or a hazard due to accelerated burning (deflagration). These materials include pyrophoric, flammable, or oxidizing gases and flammable or combustible liquids stored under pressure.
- Division 3 — Occupancies where flammable solids, other than combustible dusts, are manufactured, stored, or generated. This category also includes a wide variety of other materials including flammable or oxidizing cryogenic fluids and flammable or combustible liquids NOT stored under pressure.
- Division 4 — Repair garages that do not fall under Group S requirements (discussed later in this chapter) (Figure 5.20).
- Division 5 — Aircraft hangars and heliports that do not fall under Group S requirements (Figure 5.21).
- Division 6 — Semiconductor fabrication facilities and comparable research and development facilities.

Figure 5.20 Auto repair garages with welding on site are considered Group H, Division 4 occupancies by the UBC.

Figure 5.21 Aircraft maintenance facilities with welding on site are considered Group H, Division 5 occupancies by the UBC.

- Division 7 — Buildings that contain materials that present a health hazard. These include corrosives, highly toxic materials, radioactive materials, and irritants.

GROUP I — INSTITUTIONAL

The Institutional Occupancy group in the model building codes combines the buildings that are covered in the Health Care and Detention and Correctional Occupancy groups in NFPA 101. Each of the three model codes has slightly different ways of dividing Group I into subgroups.

The BOCA Building Code®

BOCA separates Group I into three main subgroups. The third subgroup is further broken down into one of five conditions.

- Group I-1 — Buildings and structures that house at least six people, who because of age, mental disability, or other reasons, must live in a supervised environment. These people are capable of responding to an emergency environment without assistance. It includes buildings such as board and care facilities, group homes, drug and alcohol treatment centers, and convalescent facilities (Figure 5.22).
- Group I-2 — Buildings that house six or more people who are not capable of self-preservation in the event of an emergency. These include nursing homes, hospitals, and skilled nursing facilities.
- Group I-3 — Detention and correctional facilities for six or more people. Each Group I-3 is further classified as being under one of the following five conditions:

Figure 5.22 A substance-abuse treatment facility is considered a Group I-1 occupancy by BOCA.

1. Occupancy Condition I — Buildings in which free movement is allowed from sleeping areas and other spaces to the exterior without restraint.
2. Occupancy Condition II — Buildings in which free movement is allowed from sleeping areas and other occupied smoke compartments to any *other* occupied smoke compartment. Exits to the exterior are locked.
3. Occupancy Condition III — Buildings in which free movement is allowed within individual smoke compartments. Access from one smoke compartment to another is controlled by remote-controlled doors.
4. Occupancy Condition IV — Buildings in which free movement is restricted from an occupied space. Remote-controlled release is required to permit movement from all sleeping areas.
5. Occupancy Condition V — Buildings in which free movement is restricted from an occupied space. Staff-controlled manual release is required to permit movement from all sleeping areas.

The Standard Building Code©

The SBC divides Group I occupancies into two main subgroups.

- Group I Unrestrained Occupancies — Buildings and structures used for medical, surgical, psychiatric, nursing, or custodial care on a 24-hour basis for 6 or more people who are not capable of self-preservation. Includes detox facilities, hospitals, mental hospitals, and nursing homes.
- Group I Restrained Occupancies — Detention and correctional facilities for six or more people. Includes jails, prisons, and reformatories.

The Uniform Building Code™

The UBC separates Group I into four divisions:

- Division 1.1 — Buildings and structures used for medical, surgical, psychiatric, nursing, or custodial care on a 24-hour basis for

6 or more people who are not capable of self-preservation. This includes hospitals, nursing homes, mental hospitals, or detoxification facilities. Any child-care facility that accommodates full-time care of more than five children under the age of six years is also considered a Division 1.1 occupancy.

- Division 1.2 — Health-care centers for ambulatory patients receiving outpatient treatment that may render them incapable of unassisted self-preservation. This space must be for more than five patients.
- Division 2 — Nursing homes for five or more ambulatory patients and homes for five or more children under the age of six years.
- Division 3 — Mental hospitals and sanitariums, jails, prisons, and reformatories.

GROUP M — MERCANTILE

Group M occupancies in the model building codes generally follow the same requirements and examples listed in the Mercantile section of NFPA 101. None of the codes have any subgroups under Group M.

GROUP R — RESIDENTIAL

Group R occupancies encompass a wide variety of residential facilities. Each of the three model codes has different subgroups for Group R occupancies.

The BOCA Building Code®

BOCA divides Group R into four basic categories.

- Group R-1 — Buildings arranged for sleeping accommodations for more than five people who are basically transient in nature. This includes hotels, motels, and boarding houses.
- Group R-2 — Multiple-family dwellings that have more than two units and boarding houses that are not transient in nature. Includes apartment buildings, dormitories, and long-term boarding houses.
- Group R-3 — Buildings arranged for one- or two-family dwellings of less than five persons and multiple-family dwellings where each unit has separate means of egress and the units are separated by a two-hour fire wall (Figure 5.23). Child-care facilities that accommodate less than five children of any age are also Group R-3 occupancies.
- Group R-4 — Detached one- and two-family dwellings not more than three stories tall (Figure 5.24).

The Standard Building Code©

The SBC divides Group R occupancies into three main subgroups. Groups R1 and R2 are basically the same as BOCA Groups R-1 and R-2 respectively. SBC Group R3 is the same as BOCA Group 4.

Figure 5.23 BOCA considers these fire-wall-protected townhouses to be Group R-3 residential occupancies.

Figure 5.24 A single-family dwelling is a Group R-4 occupancy according to BOCA.

The Uniform Building Code™

The UBC uses three subgroups for Group R occupancies.

- Division 1 — Hotel, apartment houses, and congregate residences each accommodating more than 10 persons.
- Division 2 — Not used.

- Division 3 — Dwelling, lodging homes, and congregate residences each accommodating less than 10 persons.

GROUP S — STORAGE

Storage occupancies include all buildings or structures used primarily for storing or sheltering goods, merchandise, products, or vehicles. Each of the model codes has subgroups under the Group S occupancy classification.

The BOCA Building Code" and The Standard Building Code©

BOCA and the SBC divide Group S occupancies into two similar subgroups.

- Group S-1, Moderate-Hazard Storage Occupancies — Buildings for the storage of contents that are likely to burn with moderate rapidity but that do not produce poisonous gases or fumes or contain explosives (Figure 5.25).
  - — Barns
  - — Lumberyards
  - — Books and paper
  - — Motor vehicle repair garages
  - — Furniture
  - — Public garages and stables
  - — Grain silos
  - — Tobacco products

Figure 5.25 Motor vehicle garages that do not contain welding operations are considered to be Group S-1 occupancies by BOCA.

- Group S-2, Low-Hazard Storage Occupancies — Buildings occupied for the storage of noncombustible materials and low-hazard wares that do not burn rapidly (Figure 5.26). Examples of these occupancies, or the materials stored in them include:
  - — Beer or wine in metal, glass, or ceramic containers
  - — Meats
  - — Cement in bags
  - — Metal parts
  - — Electrical motors
  - — Open parking structures
  - — Fresh or frozen produce
  - — Washers and dryers

Figure 5.26 Beer distributorships are Group S-2 occupancies according to BOCA.

The Uniform Building Code™

The UBC divides Group S occupancies into five subgroups. The major difference between the UBC and the other two codes is that barns and agricultural buildings are not considered Storage Occupancies (they are Group U in UBC).

- Division 1 — Moderate-hazard storage occupancies.
- Division 2 — Low-hazard storage occupancies, used for the storage of noncombustible products on wood pallets or in paper cartons. Includes the examples listed for BOCA and the SBC, as well as ice plants, power plants, and pumping stations.
- Division 3 — Repair garages (no welding of flames required), gas stations, and enclosed parking garages.
- Division 4 — Open parking garages (Figure 5.27).
- Division 5 — Aircraft hangars where work does not involve open flames or welding (Figure 5.28).

Figure 5.27 Open parking garages are Group S, Division 4 occupancies according to the UBC.

Figure 5.28 Aircraft hangars that do not contain welding operations are Group S, Division 5 occupancies in the UBC.

GROUP U — UTILITY

Both BOCA and the UBC recognize Group U Occupancies. The SBC has no such designation. In general, Group U Occupancies include agricultural buildings, sheds, private garages, carports, and fences. BOCA does not divide Group U into subgroups. The UBC uses two subgroups:

- Division 1 — Private garages, carports, sheds, and agricultural buildings (Figures 5.29 a through c).
- Division 2 — Fences over 6 feet (1.8 m) high, tanks, and towers (Figure 5.30).

MIXED OCCUPANCIES

As described previously in the Mixed Occupancy section for NFPA 101, dealing with mixed occupancies within any of the three other model codes can be a challenge for fire inspection and code enforcement personnel. Most of the general principles discussed in that section regarding making occupancy determinations also apply when working with one of these three codes. Because the exact requirements for dealing with mixed occupancies vary significantly between the three codes, they cannot be covered in detail in a manual of this scope. Consult the code used in your jurisdiction for more specific information.

Figure 5.29a Private garages are Group U, Division 1 occupancies in the UBC.

Figure 5.29b The carport portion of this structure would be considered to be a Group U, Division 1 occupancy by the UBC.

Figure 5.29c Tool sheds are Group U, Division 1 occupancies in the UBC.

Figure 5.30 The UBC considers water towers to be Group U, Division 2 occupancies.

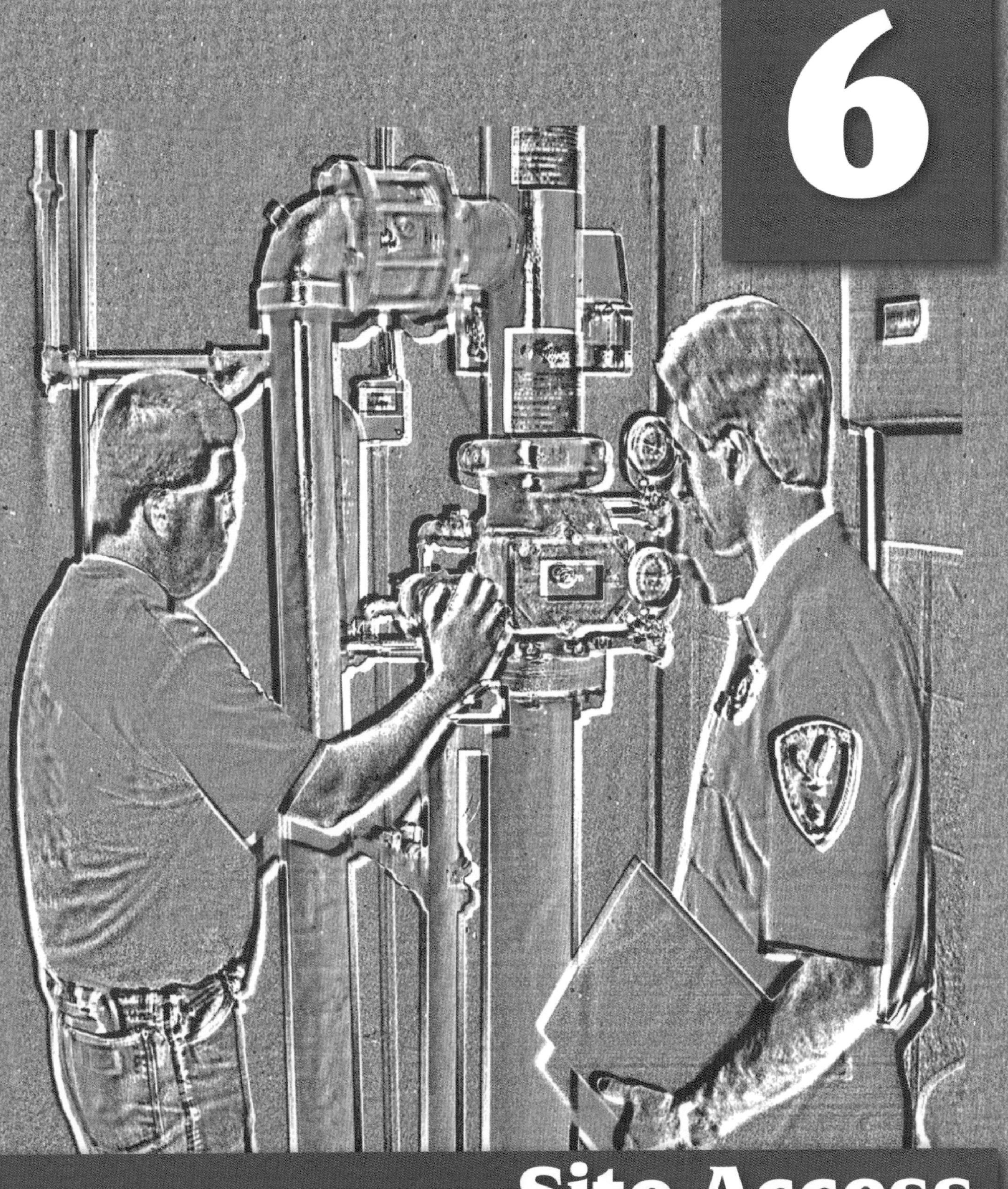

6

Site Access and Means of Egress

This chapter contains information that will assist the reader in meeting the listed job performance requirements contained in NFPA 1031, ***Standard for Professional Qualifications for Fire Inspector and Plan Examiner*** (proposed 1998 edition).

Chapter 3 Fire Inspector I

3-3.2 Compute the allowable occupant load of a single-use occupancy or portion thereof, given a detailed description of the occupancy, so that the calculated allowable occupant load is established in accordance with applicable codes and standards.

3-3.3 Inspect means of egress elements, given observations made during a field inspection of an existing building, so that means of egress elements are maintained in accordance with applicable codes and standards and all deficiencies are identified, documented, and reported in accordance with the policies of the jurisdiction.

3-3.10 Verify that emergency planning and preparedness measures are in place and have been practiced, given field observations, copies of emergency plans, and records of exercises, so that plans are prepared and exercises have been performed in accordance with applicable codes and standards, and all deficiencies are identified, documented, and reported in accordance with the policies of the jurisdiction.

3-3.11 Inspect emergency access for a site, given field observations, so that the required access for emergency responders is maintained or so that deficiencies are identified, documented, and corrected in accordance with the applicable codes, standards, and policies of the jurisdiction.

Chapter 4 Fire Inspector II

4-3.1 Compute the occupant load of a multi-use building, given field observations or a description of its uses, so that the maximum occupant load calculation is accurate and in accordance with codes and standards.

4-3.5 Analyze the egress elements of a building or portion of a building, given observations made during a field inspection, so that means of egress elements are provided and located in accordance with applicable codes and standards and all deficiencies are identified, documented, and reported in accordance with the policies of the jurisdiction.

4-3.7 Evaluate emergency planning and preparedness procedures, given copies of existing or proposed plans and procedures, to determine their applicability to the facility and their compliance with codes and standards.

4-3.12 Inspect emergency access for a site, given field observations, so that the required access for emergency responders is provided, approvals are issued, or deficiencies are identified, documented, and corrected in accordance with the applicable codes, standards, and policies of the jurisdiction.

4-4.2 Compute an occupant load, given a floor plan of a building or portion of a building, so that the calculated occupant load is in accordance with applicable codes and standards.

4-4.4 Verify that egress elements are provided, given a floor plan of a building or portion of a building, so that all elements are identified, checked against applicable codes and standards, and any deficiencies are discovered and communicated in accordance with the policies of the jurisdiction.

Chapter 6
Site Access and Means of Egress

The previous chapter discussed building construction and occupancy classifications with an emphasis on how these issues relate to life safety. However, there are some other aspects of site and building design that have a significant impact on life safety at a particular occupancy. The first is the ability to evacuate the structure rapidly and safely in the event of a fire or other emergency. Historically, many lives have been lost during emergencies because exits were blocked, locked, improperly marked, poorly designed, or otherwise inaccessible. Even properly designed and maintained means of egress cannot function effectively if the total occupant load has been exceeded and too many people are trying to exit at the same time. Some of the largest loss-of-life fires in history were glaring examples of exit inadequacies and exceeded occupant loads, including:

- Iroquois Theater fire, Chicago, IL, 1903; 602 deaths
- Cocoanut Grove Nightclub fire, Boston, MA, 1942; 492 deaths
- Beverly Hills Supper Club fire, Southgate, KY, 1977; 165 deaths

Although the previous examples are incidents that occurred many years ago, recent years have seen tragic fire incidents that show exit problems still exist. High fatality counts at the Happy Land Social Club fire in New York City (1990, 87 deaths) and the Imperial Food Products chicken plant in Hamlet, North Carolina (1991, 25 deaths) show that aggressive code enforcement of these issues is still necessary.

Another extremely important aspect of life safety is the accessibility to the site for fire apparatus and firefighters (Figure 6.1). If firefighters are unable to position their apparatus in functional positions close to the involved structure, the success of fire fighting operations can be severely affected.

Figure 6.1 Fire apparatus must be able to obtain functional positions adjacent to the structure. *Courtesy of Ron Jeffers.*

Fire personnel who perform inspections have a vital role in life safety. They must ensure that each occupancy is correctly designated and that the means of egress meet the specifications set forth in NFPA 101®, *Code for Safety to Life from Fire in Buildings and Structures* (herein referred to as the *Life Safety Code®*) or whichever of the model building codes is in effect in their jurisdiction.

Also important to the protection of life safety at any given occupancy is the adoption and practice of an emergency evacuation and disaster plan for that facility. All regular occupants of the structure should be familiar with the plan and know what is expected of them when a fire alarm is activated or other emergency condition exists.

Each of these three issues will be explored in this chapter. For the purposes of teaching the concepts of egress and occupant load calculation, NFPA 101 will be used as the basis of information for this chapter. Most of the egress requirements in the model building codes closely emulate the figures in NFPA 101. If your jurisdiction uses one of the other model codes and you require specific information on those requirements, consult that code for more information.

MAINTAINING ACCESS FOR FIRE APPARATUS AND PERSONNEL

During the course of a fire inspection, personnel should check for any condition that will affect fire department accessibility during emergency operations, either outside or inside the premises. In this section, we examine the major issues that affect fire department access to a structure.

Access to the Area Outside a Structure

A concern for fire fighting personnel at any given location is the ability to place fire apparatus in a position where it can operate effectively during emergency operations. Personnel performing inspections must be alert for conditions that will seriously hinder or completely block fire department accessibility to structures or endanger crews.

Local codes or ordinances should specify driveway and entrance requirements that facilitate easy access off the street for the largest fire apparatus that will be expected to respond to the occupancy. Driveway width and curves should be designed with the turning radius of fire apparatus in mind. Access close to the building must also be considered. This is particularly important for aerial apparatus that may be needed to ladder the structure. Fire lanes may need to be established to ensure that aerial apparatus or other fire equipment have the appropriate access to the building proper (Figure 6.2).

Figure 6.2 Fire lanes should be clearly marked and kept clear of parked vehicles.

In addition to the layout or dimensions of the driveways and parking lots, inspection personnel must also be concerned with the physical construction of those surfaces. While this is primarily an issue at the time of construction, inspectors may encounter parking lots and driveways at occupancies that were constructed prior to code enforcement in that community. In many cases, private driveways and parking lots are not constructed to the same specifications as public thoroughfares. Private driveways or parking lots may consist of either just a thin skin of asphalt over the top of gravel or a thin slab of concrete. These surfaces may not support the weight of fully loaded fire apparatus. While this may not necessarily prevent fire apparatus from accessing the structure, it could be important for aerial apparatus that need to set stabilizers and deploy an aerial device. These thin surfaces could provide an unstable surface for setting up the aerial device. Fire personnel should be made aware of these situations so that they can be compensated for by operating the aerial device over the front or rear of the apparatus.

The construction features and load capacities of the driveways and parking lots may be obtained from building construction documents or from information provided by the occupant. If neither of these is possible, personnel may get a general idea of the construction method by looking at the edges of the lot or driveway to determine the surface thickness (Figure 6.3).

Figure 6.3 Inspectors may sometimes be able to get an indication of the type and quality of parking lot construction by observing the edges of the lot.

Inspection personnel should be on the lookout for overhead obstructions that prevent the efficient operation of the aerial device.

These obstructions include such natural or man-made items as:

- Electrical lines or utility poles (Figure 6.4)
- Trees
- Parking lot lights
- Building canopies or overhangs (Figure 6.5)
- Flag poles
- Signs
- Satellite dishes

Figure 6.4 Overhead lines can pose operational limitations and safety hazards for firefighters.

Figure 6.5 Large overhangs on buildings may obstruct fire fighting operations.

Again, these things should be controlled by proper code enforcement during the plans review and construction process. However, this is not always the case, and it may be necessary for inspection personnel to note these items during routine inspections so that corrections may be required by the Authority Having Jurisdiction (AHJ).

Inspectors should also take note of occupancies that are subject to the habitual parking of vehicles within fire lane areas. It may be necessary to enlist the help of the property owner and/or local police to make sure that this situation is corrected. The inspector should be alert for fire lane areas that are free of obstructions during daytime inspections but are subject to illegal parking problems at other times. An example would be an apartment complex where many of the occupants are gone during the day, but they return during the evening. Police and property owners must be alert for these situations and take corrective measures.

In areas subject to snow fall, inspectors should make sure that adequate snow removal procedures are followed. Snow should not be piled in a manner that blocks access for fire apparatus or to important fire protection equipment.

Access to the Inside of the Structure

Personnel who perform inspections should pay attention to items that may hinder the access of firefighters to the inside of the structure. Ornamental walls, sun screens, burglar bars, hurricane shutters, and new or false fronts are all examples of items that can permanently block access to the inside of a building (Figures 6.6 a and b). In some cases, the fire inspector may be powerless to prevent the use of these items. However, the inspector should be familiar enough with the code used in that jurisdiction to know the things that he has the power to control. For example, the *Life Safety Code®* requires certain occupancies to have windows if they do not have sprinklers. The code further specifies the size of the window openings and the maximum distance that the bottom of the window may be from the floor. If all the windows in the structure are sealed (that is, bricked shut or covered), the building might be considered a windowless structure by the AHJ, and sprinklers may be required (Figure 6.7).

Figure 6.6a Barred windows present entry hazards for firefighters and exit hazards for occupants.

Figure 6.6b Some occupancies even have bars across the doors.

Figure 6.7 Bricked-in window openings limit firefighter access to a building.

Whenever inspection personnel find permanent changes in access to the inside of a structure, they should notify the appropriate people within the fire department. If inspection personnel find that required exits, exitways, or stairwells are blocked, they should immediately take steps to correct the situation. In many cases, if the inspector meets with the occupant of the structure and explains the hazards of the situation, the problem will be voluntarily corrected.

MEANS OF EGRESS

The "way out" or means of egress is one of the most important factors to be considered in determining whether the design and construction of an occupancy is safe. The *Life Safety Code®* and the three model building codes all define a *means of egress* as a "continuous and unobstructed way of exit travel from any point in a building or structure to a public way." A *public way* is a street, alley, or similar parcel of land essentially open to the outside and which is used by the public (Figure 6.8). All the codes specify that a public way should have a width and height of at least 10 feet (3 m). Because of the extreme importance for inspectors to understand the concepts associated with means of egress, the following sections will take a detailed look at the important aspects of this topic.

Figure 6.8 This alley behind a row of mercantile occupancies can be considered to be a public way.

Means of Egress Components

A means of egress consists of three distinct components:

- Exit access
- Exit
- Exit discharge

The *exit access* is that portion of a means of egress that leads to the exit. Hallways, corridors, and aisles commonly serve as exit access (Figure 6.9). The *exit* is the portion of a means of egress that is separated from the area of the building from which escape is to be made by walls, floors, doors, or other means that provide the protected path necessary for the occupants to proceed with reasonable safety to the exterior of the building. Examples of exits include:

- Door leading directly outside or through a protected passageway to the outside (Figure 6.10)

- Horizontal exit
- Smokeproof tower (Figure 6.11)
- Interior and exterior stairs or ramps (Figure 6.12)

The *exit discharge* is that portion of a means of egress that is between the end of an exit and a public way (Figure 6.13). An alley that joins a street or sidewalk is a typical exit discharge. In order to fulfill its intended function during an emergency, each component of a means of egress must be kept free from obstructions at all times. No furnishings or decorations may be allowed to obstruct or conceal an exit or exit access.

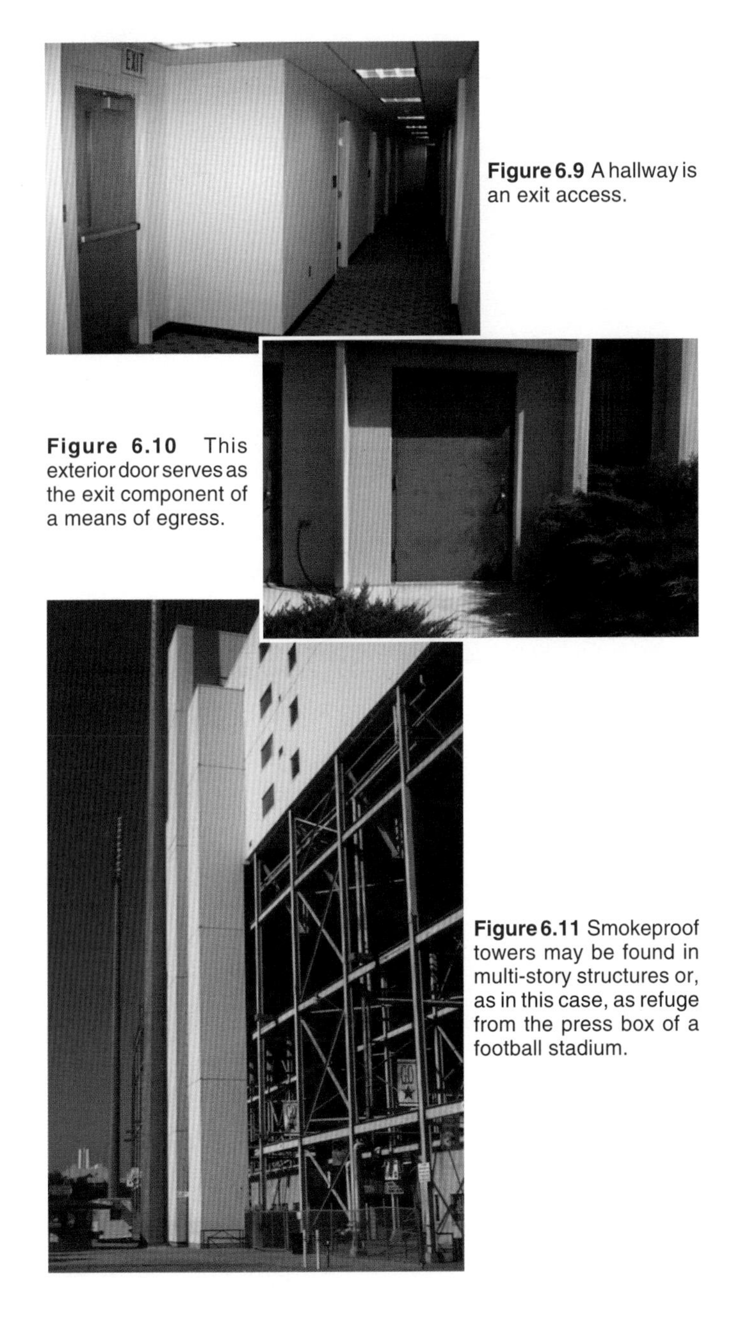

Figure 6.9 A hallway is an exit access.

Figure 6.10 This exterior door serves as the exit component of a means of egress.

Figure 6.11 Smokeproof towers may be found in multi-story structures or, as in this case, as refuge from the press box of a football stadium.

Figure 6.12 Exterior exit stairs.

Figure 6.13 The exit discharge leads to a public way.

Basic information on means of egress may be found in the means of egress chapters in the *Life Safety Code®* or any of the three major model building codes. In addition to that basic information, specific requirements for means of egress are contained in each chapter of the model codes specific to a particular occupancy.

There are many questions that must be answered when determining whether the means of egress meets the requirements of the code used in your jurisdiction:

- Are the components of a means of egress the type allowed for the specific occupancy classification?
- What is the total exit capacity in inches (mm), people, or both?
- Is the travel distance to the nearest exit within the specified maximums for that particular occupancy classification?
- Are all of the exits accessible and identifiable?
- Are the means of egress properly illuminated and marked?
- Do exit doors open easily, and are they equipped with panic hardware where required?

- Are the components of the means of egress free of obstructions?
- What is the maximum number of occupants allowed for the particular occupancy and building?
- Are interior finishes and decorations within the code-specified flame spread and smoke development limits for the particular occupancy?

Each of these questions will be addressed at various points within the information that is contained in the sections to follow.

DOORS

Two of the most important functions of doors, in terms of life safety, are to act as a barrier to fire and smoke and to serve as a component of a means of egress. When doors serve as a component of a means of egress, they must be constructed so that the way of exit travel is obvious. In general, doors should open in the direction of travel toward the exit (Figure 6.14). Each door opening must be wide enough to accommodate the number of people expected to travel through the door in an emergency. In new buildings, all the codes require doors serving as a component of a means of egress to be at least 32 inches clear (810 mm), but no more than 48 inches (1 220 mm) wide (Figure 6.15). There are exceptions to this for existing buildings and other special situations.

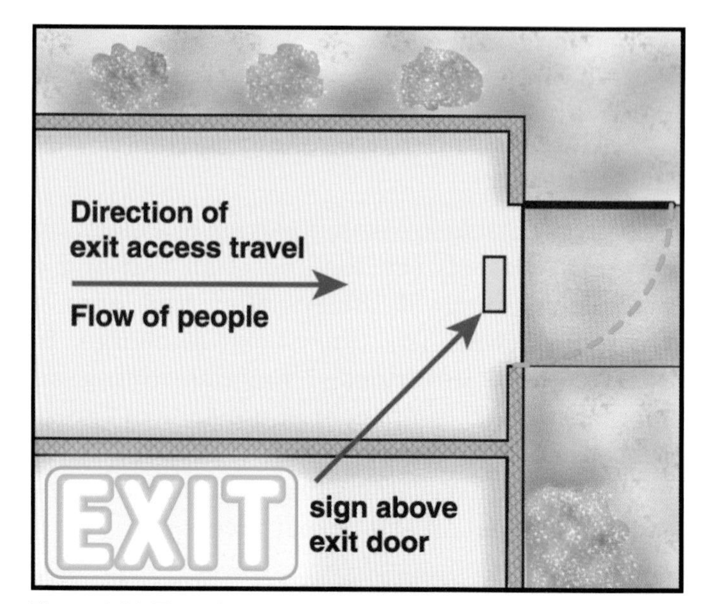

Figure 6.14 The exit door must open in the direction of exit travel.

Figure 6.15 The exit door opening must have 32 inches (810 mm) of clear width.

Figure 6.16 Exit doors must be equipped with some type of panic hardware.

The floor on each side of the door must be substantially level and must have the same elevation on both sides of the door. When a building is occupied, the exit doors must open easily. If 50 or more people are in a room or the occupancy is considered hazardous, the doors must open in the direction of travel. Panic hardware is not required for all exit doors (Figure 6.16). Situations where panic hardware is required are detailed under specific occupancy classifications. When panic hardware is required, occupants should be able to cause the latch to release by applying a force of not more than 30 pounds (134 N) and then to set the door in motion by applying a force of not more than 15 pounds (67 N) to the panic hardware.

For more information on fire and exit doors, refer to Chapter 3 of this manual; *Life Safety Code®*; NFPA 80, *Standard for Fire Doors and Fire Windows*; or any of the model building codes.

STAIRS

Exit stairs are a critical component of the means of egress in multistoried buildings. Stairways must be at least 44 inches (1 120 mm) wide, unless the total occupant load of all floors served by the stairway is less than 50 people (Figure 6.17). In this case, stairways must be at least 36 inches (910 mm) wide. Stair treads must give good footing, and landings must be provided to break up any excessively long individual flights. Rails are required for both sides of the stairs. Stairs that are exceptionally wide must have intermediate rails in the middle of the stairs.

Figure 6.17 An exit stairwell.

In order to provide a protected path of travel and to qualify as an exit, all interior stairs must be separated from other parts of the building by proper construction. The construction that encloses the exit must have at least a 1-hour minimum fire-resistance rating when the exit connects three stories or less. This minimum applies whether the stories connected are above or below the story at which the exit discharge begins. If the exit connects four or more stories, the separating construction must have at least a 2-hour fire-resistance rating. Any opening to the exit must be protected by a self-closing fire door (Figure 6.18). The fire door must have a 1-hour rating when used in a 1-hour rated enclosure. The fire door must have at least a 1½-hour rating when used in a 2-hour rated enclosure. The only permissible openings in an exit are those that allow people to enter the exit from the building and those that empty to the exit discharge.

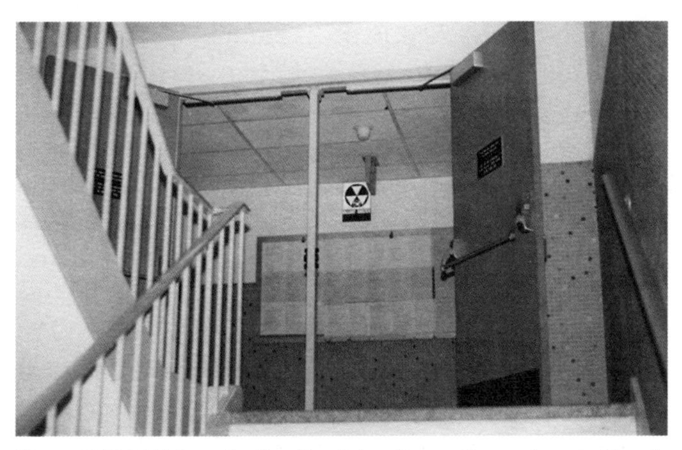

Figure 6.18 Exit doors that lead to stairwells must be equipped with self-closing hardware.

Exit stairs may be inside or outside a building. The codes generally require that exit stairs inside a building be enclosed. Outside stairs are permitted to serve as exit stairs if they meet the applicable code requirements for outside stairs. More information on exit stairs may be found in the *Life Safety Code®* or Chapter 10 in any of the model building codes. There may also be occupancy-specific requirements in the other chapters of the codes.

Inspectors should make sure that stairways as well as smokeproof enclosures — described immediately following this section — are not used for any other purposes than means of egress. These areas frequently tend to be depositories for trash, excess furniture, or product storage. All debris must be removed from the stair area so that egress is not potentially impaired (Figure 6.19).

Figure 6.19 Exit stairwells must be clear of debris.

SMOKEPROOF ENCLOSURES

Smokeproof enclosures are stairways that are designed to limit the penetration of smoke, heat, and toxic gases. Smokeproof enclosures provide the highest degree of fire protection of stair enclosures recommended by the model codes. All smokeproof enclosures must discharge into a public way or into an exit passageway. The stairway must be enclosed by fire barriers from its highest point to its lowest point. These barriers must have a 1-hour rating in buildings that are three stories or less and a 2-hour rating in buildings that are four stories or taller. Access to a smokeproof enclosure is made through a vestibule or outside balcony. This arrangement prevents smoke from entering the stairwell when corridor doors are opened. The stair enclosure may be made smokeproof by the use of natural ventilation, mechanical ventilation, or by pressurizing the stair enclosure.

HORIZONTAL EXITS

Horizontal exits are commonly used in high-rise and institutional-type buildings (Figure 6.20). There are two basic types of horizontal exits:

- A means of egress from one building to an area of refuge in another building on approximately the same level (Figure 6.21).

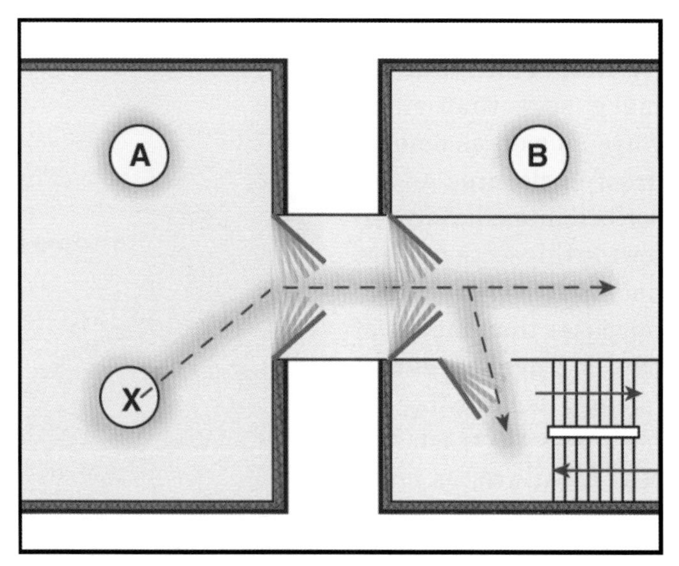

Figure 6.20 Horizontal exits are used to connect two distinctly different portions of a facility.

Figure 6.21 This horizontal exit provides a means of egress from one building to an area of refuge in another building on approximately the same level.

- A means of egress through a fire barrier or partition to an area of refuge at approximately the same level in the same building that provides protection from smoke and fire (Figure 6.22).

These types of exits require a separating wall or partition having at least a 2-hour fire-resistance rating. Horizontal exits may be substituted for other exits if they do not make up more than 50 percent of the total exit capacity of the building. If a space used as an area of refuge includes a horizontal exit, it must also have a standard exit, such as a stairway or a door, that leads outside. The area of refuge must be large enough to shelter occupants from the fire area as well as those in the area of refuge, allowing at least 3 square feet (0.28 m^2) per person.

Figure 6.22 An example of a horizontal exit.

RAMPS

A ramp may be used as a component of a means of egress (Figure 6.23). Ramps are more common in certain occupancies such as schools and institutions. In the United States, ramps are becoming more common in all types of occupancies because the Americans with Disabilities Act (ADA) requires public buildings to be handicapped accessible. Ramps are easier to traverse for the elderly, infirmed, or handicapped. The *Life Safety Code®* defines two main classes of ramps — Class A and Class B. Class A ramps must be at least 30 inches (760 mm) wide, with a maximum slope of 1 to 10 (that is 1 foot of rise for every 10 feet of horizontal distance). Class A ramps are not limited to a certain height between landings. This type of ramp is especially useful in certain health-care facilities

Figure 6.23 Some occupancies have ramps as a portion of the means of egress.

for moving patients in wheelchairs or on cots. Class B ramps must be at least 30 inches (760 mm) wide, with a maximum slope of 1 to 8. Several exceptions to this requirement are detailed in the specific occupancy chapters of the *Life Safety Code®*.

The other three model building codes also have requirements for ramps. The model codes do not break ramps into two class as does the *Life Safety Code®*. In general, these codes treat ramps in the same manner as corridors. Ramps serving occupant loads of 50 or more must be at least 44 inches (1 120 mm) wide. Ramps serving occupant loads less than 50 must be at least 36 inches (910 mm) wide. The slope requirements vary from code to code, depending on the situation.

Interior exit ramps must be enclosed and protected by construction in a manner similar to stairs. Exterior ramps must offer the same degree of protection as exterior stairs, with some exceptions for certain occupancies.

EXIT PASSAGEWAYS

An exit passageway serves as an access to or from an exit. It is not to be confused with an exit access corridor. Exit passageways have stricter construction protection requirements than do exit access corridors. Probably the most important use of an exit passageway is to satisfy the requirement that exit stairs discharge directly outside multistory buildings. Thus, if it is impractical to locate the stair on an exterior wall, an exit passageway can be connected to the bottom of the stair to convey the occupants safely to an outside exit door (Figure 6.24). In buildings of extremely large area, such as shopping malls and large factories, the exit passageway can be used to shorten the travel distance to an exit, which could otherwise be excessive. An exit passageway must be wide enough to accommodate the total capacity of all exits that discharge through it.

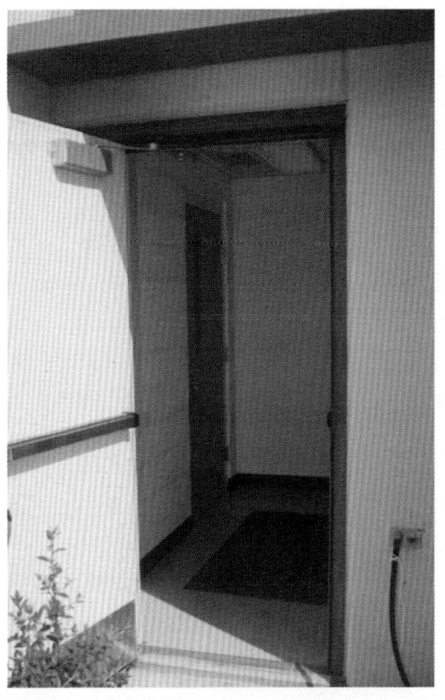

Figure 6.24 This exit passageway leads from the bottom of an exit stairwell to an exterior door.

ESCALATORS

None of the codes allow escalators and moving walkways to be a part of a means of egress in new occupancies (Figure 6.25). The *Life Safety Code®* does not allow them to be used as part of a means of egress in both new and existing occupancies. In limited cases, the model building codes allow escalators to serve as part of a means of egress in existing buildings. Inspectors should check the code used in their jurisdiction for specific information on this issue.

Figure 6.25 In some occupancies an escalator may lead to a means of egress.

FIRE ESCAPE STAIRS

Fire escape stairs may not be used as any part of a means of egress in new construction (Figure 6.26). Fire escape stairs may constitute only one-half of the required means of egress in existing buildings. There are a number of reasons that fire

Figure 6.26 Many older buildings are equipped with fire escape stairs.

escapes are not considered to be reliable exits. They are often poorly maintained and may be wet or icy, creating an additional hazard for those who must use them.

Many occupants are hesitant to use fire escape stairs in an emergency because they are not accustomed to using them on a regular basis. Furthermore, persons who have a fear of heights may find them uncomfortable to use, slowing the progress of other people behind them.

To avoid trapping occupants, fire escape stairs must be exposed to the fewest number of door or window openings as possible. The *Life Safety Code®* sets detailed guidelines for protection of openings that expose fire escape stairs. Consult the standard for specific requirements. Windows may be used as access to fire escapes in existing buildings if they meet certain criteria concerning the size of the opening. These windows must open with a minimum of effort. Access to fire escape stairs must be directly from a balcony, landing, or platform.

FIRE ESCAPE LADDERS AND SLIDES

These devices are not particularly common, but occasionally they may be encountered by some inspectors. Fire escape ladders are only allowed for limited purposes if approved by the AHJ. Fire escape slides, also called slidescapes, may be used as a means of egress where they are specifically authorized. They must be of an approved type and must be rated at one exit unit per slide with a rated capacity of 60 persons.

EXIT ILLUMINATION AND MARKING

Illumination and marking of exits vary with each type of occupancy classification and each model code. When exit illumination is required, it must be continuous during periods of occupancy. Floors have to be illuminated at not less than 1 footcandle (10.8 lumens [lx]) at floor level. A reduction of 1/5 footcandle (2.2 lx) is permitted in auditoriums, theaters, concert or opera halls, and other places of assembly during performances.

Any required exit illumination must be arranged so that the failure of any single lighting unit, such as a defective light bulb, will not leave any area in darkness. Battery operated units may not be used for primary exit illumination.

Emergency lighting may be required in certain occupancies (Figure 6.27). The emergency lighting system has to provide the proper amount of illumination (1 footcandle [10.8 lx] for 90 minutes) when normal lighting is interrupted.

Figure 6.27 Emergency lighting may be required in certain occupancies.

The requirements for marking exits may also vary according to the occupancy classification and the code being used. Generally, illuminated exit signs are required in most occupancies (Figure 6.28). Exit signs must be positioned so that no point in the exit access is more than 100 feet (30.5 m) from the nearest visible sign. The letters on exit signs must be at least 6 inches (150 mm) high and the principal strokes of the letters at least ¾-inch (19 mm) wide.

Floor level exit signs are being required in some jurisdictions. These signs are designed to allow occupants crawling low through smoke to identify when they have reached an exit (Figure 6.29). Floor-level signs are to be used *in addition* to standard ceiling-level signs; they are not a substitute. The requirements for illumination and character size for floor-level signs are the same as for those described earlier in this section. The *Life Safety Code®* specifies that the bottom of a floor-level exit sign must be between 6 and 8 inches (150 mm and 200 mm) above the floor surface.

Figure 6.28 Illuminated exit signs at the ceiling level are required by all the model codes.

Figure 6.29 Some jurisdictions require exit signs at floor level so they will be visible when occupants are crawling low in smoke.

Occupant Loads and Means of Egress Capacity

One of the functions of the model codes is to give code enforcement officials the ability to determine the number of people who may safely occupy a structure, room, or area and how many may safely exit that structure during an emergency condition. With regard to exits, the codes specifically provide the inspector the means to determine the following:

- Capacity of an exit
- Total exit capacity
- Number of exits required
- Maximum travel distance to an exit

The following sections will assist inspectors in learning how to determine each of these requirements.

OCCUPANT LOAD

The occupant load is the total number of persons who may occupy a building or portion thereof at any one time (Figure 6.30). For purposes of life safety, the inspector must assess occupant load in terms of the number of people who can safely occupy and exit a building at one time. The occupant load for a building or room should be established during the plans review process. In most cases, the inspector will simply be verifying that the occupant load for a given location is still correct. In some cases, a structure may change occupancy classifications and a new load must be calculated. In either case, all inspectors should be familiar with the methods for doing the calculation.

Figure 6.30 Some jurisdictions require building owners to post occupant loads in a visible location.

The primary method for determining occupant load is the same for all the model codes. The occupant load is determined by dividing the area of the building by the maximum floor area allowance per occupant or fixed seat (occupant load factor). The area may be net or gross based on the use (occupancy) of the space. The floor area may be taken from architectural plans or by physically measuring the building. The amount of floor area for each person varies with the type of occupancy. Each of the codes contain tables that list the required areas for each person. In general, these figures are the same for all four codes. Table 6.1 lists some examples of the required areas that are common to all of the model codes. Consult the code used in your jurisdiction for a complete list. The following examples show sample occupant load calculations.

TABLE 6.1
Maximum Floor Area Allowances Per Occupant

Occupancy	Floor Area Per Occupant In Ft² (m²)
Assembly, without fixed seats (tables and chairs)	15 (1.35) net
Dance floor	7 (0.6) net
Business areas	100 (9.2) gross
Industrial areas	100 (9.2) gross
Institutional—sleeping areas	120 (11.1) gross
Residential	200 (18.6) gross

Example 1: Your fire company has been assigned to perform an inspection in a building that was formerly a warehouse and has recently been turned into a nightclub. The public area of this facility is a room that measures 100 feet by 150 feet (30 m by 45 m). The fire marshal has asked your company to determine the legal occupant load for this business.

Step 1: Determine the square footage (floor area) of the public area of the nightclub by multiplying the room's length by its width:

100 feet x 150 feet = 15,000 ft^2

30 m x 45 m = 1 350 m^2

Step 2: Allow 15 ft^2 (1.35 m^2) per person as stated in Assembly Occupancy without fixed seats (Table 6.1).

$$\text{Occupant Load} = \frac{15{,}000 \text{ ft}^2}{15 \text{ ft}^2} = 1{,}000 \text{ persons}$$

$$\text{Occupant Load} = \frac{1\ 350 \text{ m}^2}{1.35 \text{ m}^2} = 1{,}000 \text{ persons}$$

NOTE: Keep in mind that the 1,000 person occupant load will also be dependent on an adequate number of exits being present.

Example 2: You have been given architectural plans for a new building being constructed within your jurisdiction. The building will be a manufacturing facility for charcoal grills and meat smokers. Given the following diagram of the building, determine the legal occupant load for the facility (Figure 6.31).

Step 1: Determine the floor area of the facility using the dimensions on the drawing. In the given example, the manufacturing and finishing areas are each 50 feet by 75 feet (15 m by 23 m). Thus:

(50 feet x 75 feet) x 2 = 7,500 ft^2

(15 m x 23 m) x 2 = 690 m^2

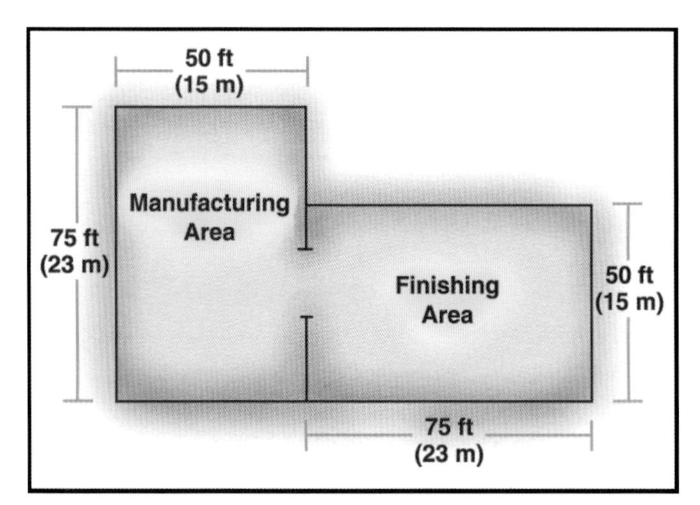

Figure 6.31 Example 2.

***Step 2*:** Consult the code used in your jurisdiction for the floor area per person that is required for this occupancy. In this example, we can use the figure for Industrial areas found in Table 6.1.

***Step 3*:** Use the following formula to determine occupant load:

$$\text{Occupant Load} = \frac{\text{Net Floor Area}}{\text{Area Per Person}}$$

$$\text{Occupant Load} = \frac{7{,}500 \text{ ft}^2}{100 \text{ ft}^2} = 75 \text{ persons}$$

$$\text{Occupant Load} = \frac{690 \text{ m}^2}{9.2 \text{ m}^2} = 75 \text{ persons}$$

The previous examples are pretty clear-cut cases where a building is being used for a single purpose. Inspection and code enforcement personnel will frequently encounter buildings that are used for more than one purpose or that contain more than one occupancy classification within the same structure. Inspection personnel must know how to handle these situations when they are encountered. In general, the procedure for handling these situations is the same for each of the model codes:

1. For a building or a portion of a building that has more than one use, the occupant load is determined by the use that allows for the largest number of persons to occupy the building.
2. For a building or a portion of a building that contains two or more distinct occupancies, the occupant load is determined by calculating the occupant load for each of the occupancies separately and then adding them together.

The following examples highlight each of these possibilities.

Example 3: Inspectors are assigned to determine the occupant load on a new youth-oriented entertainment facility being constructed within their jurisdiction. The building's principal room is 100 feet by 250 feet (30 m by 75 m). Depending on the day of operation, this room is used as either a skating rink or a dance hall.

***Step 1*:** Determine the floor area in the principal room.

100 feet x 250 feet = 25,000 ft^2

30 m x 75 m = 2 250 m^2

***Step 2*:** Consult the code used in your jurisdiction for the floor area per person that is required for this occupancy. In this example, all of the codes are consistent in their requirements. Both of the room uses (skating and dancing) fall within the Assembly Occupancy classification. However, there are different requirements for square feet per person for each of these uses. A dance hall is considered a *concentrated use* and each person is allowed 7 square feet (0.65 m^2) (see Table 6.1). A skating rink is considered an *unconcentrated use* and each person is allowed 15 square feet (1.35 m^2).

***Step 3*:** Use the following formula to determine occupant load.

$$\text{Occupant Load} = \frac{\text{Net Floor Area}}{\text{Area Per Person}}$$

Occupant Load when used as a skating rink =

$$\frac{25{,}000 \text{ ft}^2}{15 \text{ ft}^2} = 1{,}667 \text{ persons}$$

Occupant Load when used as a skating rink =

$$\frac{2\,250 \text{ m}^2}{1.35 \text{ m}^2} = 1{,}667 \text{ persons}$$

Occupant Load when used as a dance hall =

$$\frac{25{,}000 \text{ ft}^2}{7 \text{ ft}^2} = 3{,}571 \text{ persons}$$

Occupant Load when used as a dance hall =

$$\frac{2\,250 \text{ m}^2}{0.65 \text{ m}^2} = 3{,}571 \text{ persons}$$

Given these figures, and based on the provisions of statement #1, the legal occupant load for this building should be established at 3,571 persons.

Example 4: As a code enforcement official, you are given plans for a new restaurant with an attached clothing store in your jurisdiction. You are assigned to determine the occupant load for this structure. The restaurant portion of the building is 100 feet by 125 feet (30 m by 38 m) and contains tables and chairs that are not fixed. The clothing store portion is 75 feet by 75 feet (23 m by 23 m) (Figure 6.32).

***Step 1*:** Determine the floor area of the facility.

Floor area of the restaurant =
100 feet x 125 feet = 12, 500 ft^2

Floor area of the restaurant =
30 m x 38 m = 1 140 m^2

Floor area of the clothing store =
75 feet x 75 feet = 5,625 ft^2

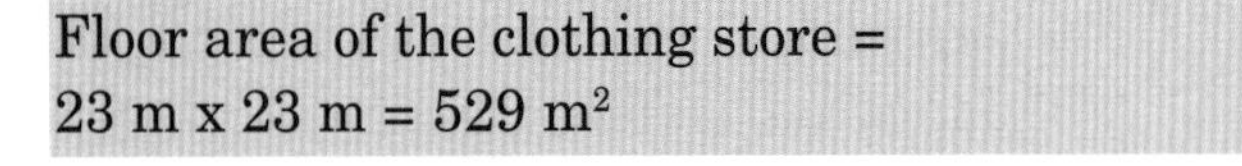

Floor area of the clothing store =
23 m x 23 m = 529 m^2

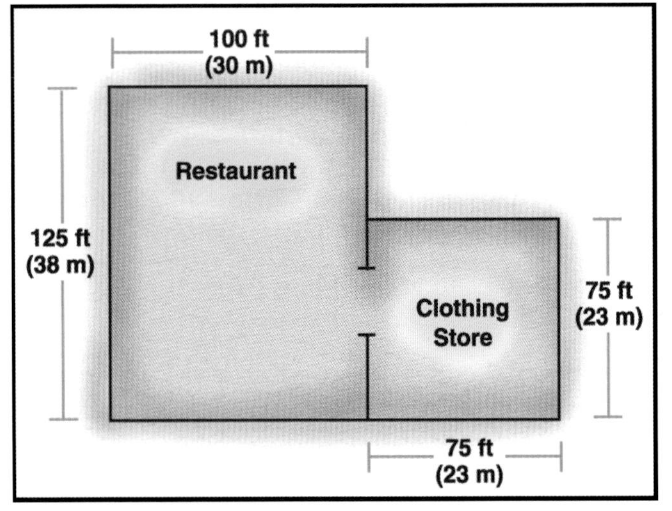

Figure 6.32 Example 4.

***Step 2*:** Consult the code used in your jurisdiction for the floor area per person that is required for these occupancies. In this example, all of the codes are consistent in their requirements. Restaurants without fixed seats are allowed 15 square feet (1.35 m^2) per person. The clothing store is considered a mercantile occupancy and is allowed 30 square feet (2.8 m^2) per person.

***Step 3*:** Use the following formula to determine occupant load.

$$\text{Occupant Load} = \frac{\text{Net Floor Area}}{\text{Area Per Person}}$$

Occupant Load for the restaurant =

$$\frac{12{,}500\ ft^2}{15\ ft^2} = 833 \text{ persons}$$

Occupant Load for the restaurant =

$$\frac{1\ 140\ m^2}{1.35\ m^2} = 833 \text{ persons}$$

Occupant Load for the clothing store =

$$\frac{5{,}625\ ft^2}{30\ ft^2} = 187 \text{ persons}$$

Occupant Load for the clothing store =

$$\frac{529\ m^2}{2.8\ m^2} = 187 \text{ persons}$$

***Step 4*:** Determine the occupant load for the whole building.

833 persons (restaurant) + 187 persons (store) = 1,020 person total occupant load

Note that Example 4 was performed using the requirements contained within NFPA 101. In some cases, the model building codes would consider the restaurant and clothing store *separate* occupancies and require tenent separation between them. Consult the code used in your jurisdiction for directions on how to handle these situations.

MEANS OF EGRESS CAPACITY

In addition to simply determining how many people may safely occupy a particular room or building, code enforcement officials must also determine that the room or building in question contains exits in sufficient quantity to safely evacuate the occupants under emergency conditions. The means of egress capacity must be at least equal to the occupant load determined by the floor area. If a building or room has a means of egress capacity that is lower than the floor area occupant load, additional exits must be constructed to handle the entire floor area occupant load. Reducing the official occupany load for the room or structure is another option; however, it is difficult to enforce.

Historically, means of egress capacity had been determined by a system that used units of exit width. A single unit of exit width was considered to be 22 inches (560 mm). This measurement was based on studies showing that the average shoulder width of a World War I soldier was 22 inches

(560 mm). The codes also considered 12 inches (300 mm) to be one-half unit of exit width. Any width less than 34 inches (860 mm) was considered to be one unit of exit. This system is no longer used by any of the model codes.

The modern method of determining the required width of means of egress components involves using numerical factors expressed in terms of inches or millimeters per person. The most common figures used by the codes are 0.3 inches (8 mm) per person for stairways and 0.2 inches (5 mm) per person for ramps or level exit components (Table 6.2). Each of the codes has factors other than these for selected occupancies. For example, buildings that are considered Hazardous Occupancies use the figures 0.7 inches (18 mm) per person for stairways and 0.4 inches (10 mm) per person for ramps or level exit components. Some codes also make adjustments based on whether or not the occupancy is sprinklered. Inspectors should consult the code used in their jurisdiction for specific information on means of egress width factors.

TABLE 6.2
Means Of Egress Width Factors

Use	Stairways In. Per Person (mm Per Person)	Level Components And Ramps In. Per Person (mm Per Person)
Board and Care	0.4 (10)	0.2 (5)
Health Care, Sprinklered	0.3 (8)	0.2 (5)
Health Care, Nonsprinklered	0.6 (15)	0.5 (13)
High Hazard Contents	0.7 (18)	0.4 (10)
All Others	0.3 (8)	0.2 (5)

Determining The Capacity Of An Exit

To determine the capacity of an exit, it is necessary to compute the capacity for each of the three components of a means of egress: the exit access, the exit, and the exit discharge. Each component must be measured in clear width at its narrowest point. The component that has the smallest capacity will determine the total capacity of the overall exit. The following example may be used to illustrate this point. Suppose we have a means of egress with the following dimensions:

- A corridor (the exit access) that is 42 inches (1 044 mm) wide
- A stairway (the exit) that is 44 inches (1 116 mm) wide
- An alley (the exit discharge) that is 12 feet (3.66 m) wide

The corridor is a level exit component; therefore, it is capable of handling one person for every 0.2 inches (5 mm). Thus:

$$\text{Corridor Capacity} = \frac{42\text{ in.}}{0.2} = 210\text{ persons}$$

$$\text{Corridor Capacity} = \frac{1\,044\text{ mm}}{5\text{ mm}} = 209\text{ persons}$$

As previously stated, the numerical factor for stairs is 0.3 inches (8 mm) per person. Thus:

$$\text{Stairway Capacity} = \frac{44\text{ in.}}{0.3} = 147\text{ persons}$$

$$\text{Stairway Capacity} = \frac{1\,116\text{ mm}}{8\text{ mm}} = 140\text{ persons}$$

The alley is considered a level component, so the 0.2 in. (5 mm) per person factor is used. In order for the calculation to be done correctly, the alley's dimensions must first be converted from feet (meters) to inches (mm).

12 feet x 12 inches/foot = 144 inches

3.66 m x 1 000 mm/m = 3 660 mm

$$\text{Alley Capacity} = \frac{144\text{ in.}}{0.2} = 720\text{ persons}$$

$$\text{Alley Capacity} = \frac{3\,660\text{ mm}}{5\text{ mm}} = 732\text{ persons}$$

The smallest number for the three computations is 147 persons (140 in metric), so the total capacity of that exit will be 147 (140).

Determining Total Exit Capacity

In order to compute the total exit capacity of an occupancy, it is first necessary to determine the capacity of each exit and then total those figures. The procedures for determining total exit capacity differ depending on whether the egress is level or nonlevel.

Example 5: You are reviewing plans for a new business occupancy that shows the building having three swinging exit doors that have a clear width of 36 inches (914 mm). These exits are all on ground level.

Step 1: Determine the numerical factor that should be used for a level egress in a business occupancy. From Table 6.2 we can determine that level components may be rated at 0.2 inches (5 mm) per person.

Step 2: Determine the capacity of each exit.

$$\text{Exit Capacity} = \frac{36 \text{ in.}}{0.2} = 180 \text{ persons}$$

$$\text{Exit Capacity} = \frac{914 \text{ mm}}{5 \text{ mm}} = 183 \text{ persons}$$

Step 3: Determine the total exit capacity. Because there are three doors of the same size, the calculation is relatively simple.

Total Exit Capacity =
180 persons x 3 = 540 persons

Total Exit Capacity =
183 persons x 3 = 549 persons

Example 6: You are assigned to determine the total exit capacity for a new four-story apartment building. There are two exits from each floor, and the stairways have a clear width of 44 inches (1 116 mm).

Step 1: Determine the numerical factor that should be used for a stairway in a residential occupancy. From Table 6.2 we can determine that stairways are rated at 0.3 inches (8 mm) per person.

Step 2: Determine the capacity of each exit stairway.

$$\text{Stairway Capacity} = \frac{44 \text{ in.}}{0.3} = 147 \text{ persons}$$

$$\text{Stairway Capacity} = \frac{1\,116 \text{ mm}}{8 \text{ mm}} = 140 \text{ persons}$$

Step 3: Determine the total exit capacity. Because there are two stairways of the same size, the calculation is relatively simple.

Total Exit Capacity =
147 persons x 2 = 294 persons

Total Exit Capacity =
140 persons x 3 = 280 persons

Stairways must be able to accommodate the largest number of people from any single floor above. Of course, everyone from all floors will not be in the same part of the stairwell at the same time. For example, by the time the people from the third floor reach the second floor, the people from the second floor should have exited the building.

Number Of Exits Required

Personnel must be familiar with the requirements for the minimum number of exits in the buildings they inspect. General information on minimum number of exits may be found in the *Means Of Egress* chapters of each of the model codes. In addition to the general requirements, there are special requirements and exceptions in the various chapters of the codes that pertain to specific occupancy types.

In most cases, all of the codes require that there be at least two exits from any balcony, mezzanine, story, or portion thereof that has an occupant load of 500 persons or less. For occupant loads greater than 500 persons, consult Table 6.3.

The codes also specify that the exits should be as remote from each other as possible. This helps minimize travel distance to an exit. It also increases the chance of finding an alternative exit should the closest exit be unreachable due to fire conditions.

TABLE 6.3
Required Number Of Exits

Occupant Load	Minimum Number Of Exits
501 to 1,000	3
Over 1,000	4

Each of the model codes has special exceptions that allow only one exit in certain situations. There is little consistency between the codes on this issue. Inspectors must consult the code used in their jurisdiction to determine the situations in which one exit will be acceptable. When inspectors encounter situations where it is apparent that the occupant load is consistently in an overload situation, all possible alternatives for correction should be explored within the framework of the code.

Maximum Travel Distance To An Exit

Each of the model codes establish the maximum allowable travel distance to the nearest exit. They also limit the length of dead-end corridors. The travel distance refers to the total length of travel necessary to reach the protection of an exit. The travel distance is measured on the floor or walking surface along the centerline of the natural path of travel (Figure 6.33). The measurement starts from the most remote portion of the occupancy, curving around any corners or obstructions with a 1-foot (0.3 m) clearance therefrom, and ending at the center of the exit doorway or other point at which the exit begins.

The maximum travel distance requirements to an exit are different for each of the model codes. Within each code, the distances will vary depending on the type of occupancy and whether or not the building is sprinklered. Generally, longer travel distances are allowed in buildings that are protected by automatic sprinkler systems. Inspection personnel must consult the code used in their jurisdiction to determine the figures they need to use.

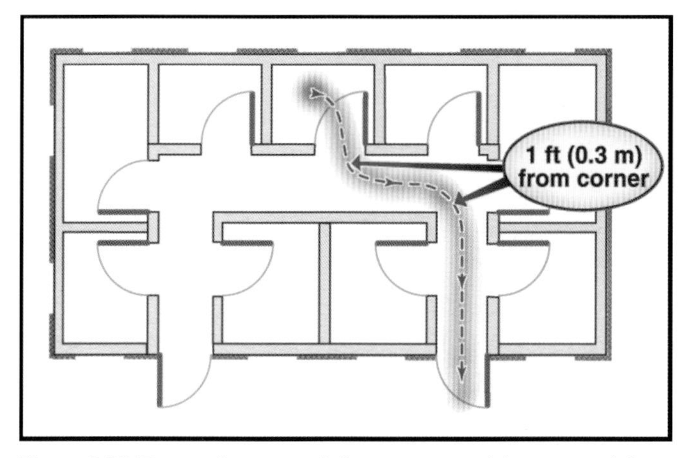

Figure 6.33 The maximum travel distance to an exit is measured down the center of the hall and one foot (0.3 m) from corners.

EMERGENCY PLANNING AND PREPAREDNESS PROCEDURES

Regardless of the type of occupancy, be it a school, factory, hospital, or any other, the key to successfully handling any emergency situation that may arise is having and executing an effective emergency plan. Fire department personnel must understand the importance of a good emergency plan for these facilities. Fire inspection personnel may be required to assist building occupants in the development, review, and testing of the emergency plan for a particular occupancy within their jurisdiction.

The complexity of the emergency plan will depend on the type of occupancy and the level of preparedness to which the occupants are organized. Obviously, the emergency plan for a petroleum refinery with a structural fire brigade will be significantly more complex than the plan for an elementary school (Figures 6.34 a and b). In a manual of this size and type, it is impossible to cover all of the information needed to develop detailed emergency plans for complex occupancies.

Figure 6.34a The emergency plan for this refinery may be very complex.

Figure 6.34b The emergency plan for this school would not be as complex as the plan for the refinery.

However, one element of an emergency plan that is common to all types of occupancies is the need to evacuate the structure in a timely and efficient manner. Fire inspection personnel should be very familiar with the proper components of an evacuation plan and the procedures for testing these plans using fire drills. There are three key elements to an evacuation plan:

1. Evacuation routes
2. Monitor duties
3. Employee duties

Each department or area of an occupancy should be provided with a primary and secondary route of exit from the structure. Maps showing these routes should be provided in conspicuous locations. The map should also note the location of the meeting point for occupants once they have left the structure (Figure 6.35). This will allow for a head count to be taken and information on the emergency to be passed along to the Incident Commander.

Figure 6.35 Employees should have a designated gathering place, such as the pavilion behind this structure.

Each area, department, or class should have a monitor assigned to oversee the evacuation process. Alternate monitors should be appointed to handle this duty in the event that the primary monitor is not available. The name(s) of the monitor(s) may be listed on the evacuation route maps; however, this information will have to be updated as personnel leave the organization. The monitor is responsible for making sure that all people under his charge know the emergency/evacuation plan. The monitor may also be responsible for conducting a headcount of those people who came from his area. This is also the role for teachers when evacuating their classes from a school.

All employees are responsible for making sure that their areas are completely evacuated. This includes making sure that any visitors are also led to safety. Employees may also be required to shut off equipment, close doors and windows, or perform other functions if conditions and time allow (Figure 6.36). Employees should provide information to the Incident Commander on the types of actions they have taken.

Figure 6.36 Employees should close their doors on the way out of the building.

As previously stated, evacuation plans should be tested on a regular basis using exit drills. The responsibility for planning and executing fire drills lies with the occupancy's fire loss prevention and control management staff. Plans for drills need to be discussed with both middle and line management so that they understand and cooperate in the effort. Following an exit drill, all members of management and the loss control staff should meet to critique the effectiveness of the evacuation plan and the drill.

Exit drills should be conducted in all types of occupancies. All occupants must be instructed on the actions they are to take during a drill or emergency. For business and educational occupancies, the instruction and drill may be very basic and involve nothing more than an efficient evacuation of the structure. Drills for occupancies such as health care, industrial, and correctional facilities will be more complex (Figure 3.37). These drills involve the movement of critical or bedridden pa-

Figure 6.37 Correctional facilities have unique evacuation procedures and problems.

tients, perimeter security of inmates, or process shutdown and industrial fire brigade response. All of these factors will have to be evaluated prior to and during the drill.

The frequency at which fire drills should be held will vary depending on the type of occupancy and local code requirements. Most occupancies should conduct at least two drills per year. Drills should be conducted at different times of the day. If the occupancy normally contains more than one shift of workers or occupants, drills should be conducted at the required intervals for each shift. In occupancies such as hospitals, hotels, assembly occupancies and stores, fire drills usually involve only staff personnel to avoid alarming patients, guests, and customers (Figure 6.38).

Figure 6.38 Generally only staff members will participate in emergency drills at hospitals.

Exit Drills in Educational Facilities

Exit drills are extremely important in educational facilities. Most states have laws that give certain individuals responsibility for conducting fire drills. These laws further dictate the frequency of drills and other details. Drills in educational occupancies should ensure that all persons in the building actually participate. Emphasis should be placed upon orderly evacuation under proper discipline rather than upon speed. If weather conditions may endanger the health of children during winter months, weekly drills may be held at the beginning of the school term to complete the required number of drills before the onset of cold weather.

Drills should be executed at different times of the day: during class changes, when the school is at an assembly, during recess or gymnastic periods, or during other special events. If a drill is called during the time classes are in session, the students should be instructed to form a line and immediately proceed to the nearest exit in an orderly manner. Fire officials should encourage local school officials to develop a plan that provides direction on where students should report if the fire alarm occurs during times other than normal class periods. Complete control of the class is necessary so that teachers can quickly and calmly control the students, form them into lines, and direct them as necessary. If there are students who are incapable of holding their places in a line moving at a reasonable speed, they should move independently of the regular line of march.

Monitors may be appointed from the more mature students to assist in the proper execution of all drills. The monitors should be instructed to hold the doors open in the line of march or to close doors when necessary to prevent the spread of smoke and fire. Teachers and other members of the staff should have responsibility for searching rest rooms or other rooms (Figure 6.39). Each class or group should proceed to a predetermined point outside

Figure 6.39 Teachers should search rest rooms for stray students.

the building and remain there while a check is made to see that all students are accounted for and that the building is safe to reenter. Assembly points should be far enough from the building to avoid danger from fire or smoke from the building, interference with fire department operations, or confusion among classes and groups. No one should be permitted to reenter until the drill is complete. Fire inspectors check on the frequency of exit drills and the time required to vacate the building.

Exit Drills in Health-Care Facilities

Hospitals and nursing homes require special evacuation procedures (Figure 6.40). A total evacuation is not warranted every time; the extent of the evacuation is determined by the severity of the emergency. Evacuation plans should proceed in a series of steps or phases.

Figure 6.40 Nursing homes require special evacuation procedures.

Phase I operations involve evacuating a single room. A common procedure used in many facilities involves following a system that uses the acronym "REACT":

R - Remove those in immediate danger.

E - Ensure that the room door is closed.

A - Activate the fire alarm if it has not done so on its own.

C - Call the fire department.

T - Try to extinguish or control the fire.

Phase II operations involve evacuating an entire zone of the building. Personnel should close all room and smoke-barrier doors. All rooms adjacent to the fire area should be evacuated first. Once the adjacent rooms are empty, concentrate on evacuating the remaining rooms within that building zone. Occupants should be moved to predetermined safe areas on that floor. Anytime patients are moved from their rooms, their medical charts/records should be taken with them.

Phase III operations involve evacuating an entire floor and zones on the floor above the fire floor. Evacuate the fire floor first. Once this is done, the necessary zones on the floor above the fire may be evacuated. Occupants may be moved down stairwells or elevators (if they are deemed safe and operable).

Phase IV operations require that the entire building be evacuated. The fire floor is evacuated first, followed by all floors above the fire. When these floors are clear, the floor(s) below the fire may be evacuated. Patients will need to be transported to other buildings or sites. These are complex incidents that work best when pre-incident plans are in place and practiced on a regular basis. There must be predetermined sites that are ready to accept the patients that need to be relocated. These locations must be able to provide adequate shelter and care for the special/medical needs of the patients.

Exit Drills in Correctional Facilities

Exit drills and evacuation procedures for correctional facilities provide a unique problem due to the inherent security features of the building. During an emergency, the lives of the inmates, visitors, and security personnel depend on the quick and effective actions of facility employees. Inspection personnel must make sure that staff are properly trained in appropriate evacuation routes and fire reporting procedures. Fire department access and coordination of fire department actions with those of security and police agencies must be an important part of emergency planning.

The vast majority of fires in correctional facilities are deliberately set by inmates. The most common reasons for setting fires are to cause malicious damage, increase chances of escape, commit suicide, or to show force during a riot. In a correctional facility, maintaining security is of primary importance. Thus, occupants are either protected where they are or evacuated to secure areas of refuge.

Correctional facilities should maintain an effective fire detection system, a key control system, and a written emergency plan. Copies of the plan should be posted for inmates to read. There should be at least two means of access to each main cell block. Inmates should be released by a reliable means, whether that be electrical, mechanical, or master keys. Personnel should ensure that prisoners remain segregated according to their detention classifications.

Exit Drills in Hotels and Motels

Hotels and motels provide a different problem for inspection personnel with regard to fire exit drills and evacuation procedures because of the temporary nature of their occupants. Often, the guests in these buildings have never been in the building before and will only remain there overnight or for a few days. Therefore, inspection personnel cannot rely on the occupants being familiar with the building if they should have to evacuate. For this reason, a reliable fire detection and alarm system must be installed in accordance with local, state, or provincial codes.

The fire detection and alarm system must include an occupant notification system so that hotel personnel can quickly relay evacuation instructions to the guests. Evacuation procedures should be posted in each room (usually on the door), at all fire alarm manual pull stations, and by exits (Figure 6.41). Inspection personnel should train employees in evacuation procedures and may use them as "fire wardens" to direct the flow of evacuees to ensure complete and efficient evacuation. The employees should attempt to take a head count of the guests and match that against the guest register to get a basic idea of how many guests are accounted for. It will be very difficult to be 100 percent accurate in these situations. Exit drills must be held frequently to keep all personnel familiar with the emergency evacuation plan and their assigned duties.

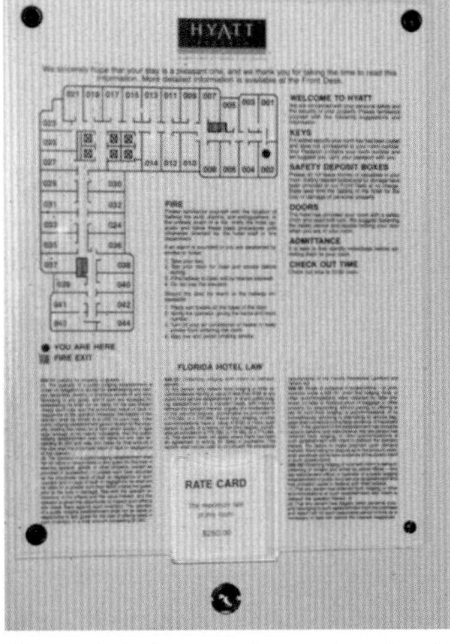

Figure 6.41 Evacuation routes should be posted on the back of the hotel room door.

Performing Fire Safety Standby in Public Assembly Occupancies

In some jurisdictions, local codes or policies dictate that fire inspection or suppression personnel stand by at major public events where large crowds will be present. There are a number of duties that standby personnel perform in these situations. However, before the event even takes place, fire personnel should check files to see when that occupancy was last subjected to a fire inspection, and review the recommendations made at that time. If the premises is past due for an inspection or there were major problems noted in the records, a complete inspection should be made before the event is allowed to take place.

Check with the person in charge of the event to determine if any hazardous event or display (in terms of fire safety) will take place. If so, the inspector should receive a detailed briefing describing exactly what is proposed. The inspector can then determine whether the display will be allowed, if a permit is required, and what precautions, if any, are necessary.

Other duties of the inspector or standby personnel are as follows:

- Ensure that all equipment brought in meets code requirements and is of an approved type.
- Check any temporary electrical wiring used for the event.
- Ensure that all exits are unlocked and exit lights are on.
- Ensure that emergency lights are working.
- Work with the building management to ensure that the occupancy limit is not exceeded.
- Announce the exit locations just before the event begins.
- Keep aisles open. Prevent people from sitting or standing in the aisles.
- Enforce the no smoking rule if applicable.
- Transmit an immediate alarm if there is a significant fire, and then direct an orderly evacuation. Make a special effort to control panic. Fight the fire if possible.

7

Water-Based Fire Protection and Water Supply Systems

This chapter contains information that will assist the reader in meeting the listed job performance requirements contained in NFPA 1031, ***Standard for Professional Qualifications for Fire Inspector and Plan Examiner*** (proposed 1998 edition).

Chapter 3 Fire Inspector I

3-3.5 Determine the operational readiness of fixed fire suppression systems, given test documentation and field observations, so that the system(s) is in an operational state, maintenance is documented, and all deficiencies are identified, documented, and reported in accordance with the policies of the jurisdiction.

Chapter 4 Fire Inspector II

4-3.4 Evaluate fire protection systems and equipment provided for the protection of a building or facility, given field observations of the facility and documentation, the hazards protected, and the system specifications, so that the fire protection systems provided are appropriate for the occupancy or hazard being protected and installed in compliance with applicable codes and standards, and all deficiencies are identified, documented, and reported in accordance with the policies of the jurisdiction.

4-3.8 Verify fire flows for a site, given fire flow test results and water supply data, so that required fire flows are in accordance with applicable codes and standards and all deficiencies are identified, documented, and reported in accordance with the policies of the jurisdiction.

4-4.3 Field verify the installation of a fire protection system, given shop drawings and system specifications for a process or operation, so that the system is reviewed for code compliance, installed in accordance with the approved drawings, and all deficiencies are identified, documented, and reported in accordance with the policies of the jurisdiction.

Chapter 7

Water-Based Fire Protection and Water Supply Systems

Personnel who perform inspections must be thoroughly familiar with all water-based fire protection systems they may encounter. These systems are designed to control or extinguish a fire in its incipient stage before the arrival of the fire department or fire brigade. There are a number of types of water-based fire protection systems and equipment with which inspection personnel need to be familiar, including the following:

- Automatic sprinkler systems
- Standpipe systems
- Fire pumps
- Water supply systems

Considered by many to be the first line of defense against fires, sprinkler systems, in their basic form, have been in use for over 100 years. Their origin dates back to the days of the large industrial mills that dotted the northeastern United States during the Industrial Revolution. Today, the automatic sprinkler system is an unsurpassed fire protection device. Fire loss data reveals that in buildings with automatic sprinklers, 96 percent of all fires were controlled or extinguished by these systems. Of the remaining fires that were not controlled in sprinkler-equipped buildings, failure was due to improper maintenance, an inadequate or shut off water supply, incorrect design, obstructions, or partial protection. As we will see later in this chapter, there are many types of systems, and they can be tailored to specific needs of an occupancy.

Standpipe systems are designed to provide a quick and convenient means for operating fire streams on all stories of buildings and in adjacent buildings. Depending on the type installed, the standpipe system may be used by firefighters, occupants, or both.

Both automatic sprinkler systems and standpipe systems are dependent on adequate water supplies to be effective. An adequate supply of water is one that is sufficient in both volume and pressure. In cases where the building is large or tall, the municipal water supply may not be adequate to meet the demands of the system(s). In these cases, it is necessary to provide on-site water supply sources and/or fire pumps to feed the system. The most common fire pumps are of the electric or diesel-driven variety. Because they may not be used in fire situations on a regular basis, these pumps must be regularly tested using a set procedure.

Fire inspectors are frequently involved in evaluating private or municipal water supply systems. This is done to determine whether the available amount of water is adequate for fire protection requirements. Inspection personnel must understand the proper method for determining the available water supply.

The installation, maintenance, and testing of fire protection equipment must be in compliance with local, state, and/or federal codes, ordinances, and standards. Many installations also follow applicable National Fire Protection Association, Underwriters Laboratories, or Factory Mutual guidelines. The material in this chapter is intended to be descriptive in nature and should not be used as the authority for examination of any legally constituted requirement. It is recommended that fire inspectors consult manufacturers' technical data sheets for answers to questions concerning design, installation, operation, and maintenance of system components.

AUTOMATIC SPRINKLER SYSTEMS

The value of automatic sprinkler systems to life safety has been proven again and again. According to the National Fire Sprinkler Association, there has never been a multiple loss of life (3 or more deaths) due to fire or smoke in a fully sprinklered building, except in case of intimate contact with the fire or an explosion.

Automatic sprinkler systems consist of a series of nozzlelike devices (called *sprinklers*) so arranged that the system will automatically distribute sufficient quantities of water to either extinguish a fire or prevent flashover until firefighters arrive (Figure 7.1). Water is supplied to the sprinklers through a system of piping. Most sprinklers (except deluge system sprinklers) are kept closed by fusible links or other heat-sensitive devices. Heat from a fire causes affected sprinklers above the fire to open automatically. Most fires are controlled with five or fewer sprinklers opening. Sprinklers can either extend upward from exposed pipe (upright sprinklers), protrude downward from exposed or hidden pipe (pendent sprinklers), or extend horizontally from the piping along a wall (sidewall sprinklers) (Figures 7.2 a through c).

A sprinkler system layout, as shown in Figure 7.3, consists of different sizes of pipe. The system starts with a feeder main that originates from a city or private water supply. The feeder main contains

Figure 7.1 The goal of automatic sprinklers is to control a fire before it gets to the stage where a major operation by the fire department is necessary. *Courtesy of the Phoenix (AZ) Fire Department.*

Figure 7.2a An upright sprinkler.

Figure 7.2b A pendent sprinkler.

Figure 7.2c A sidewall sprinkler.

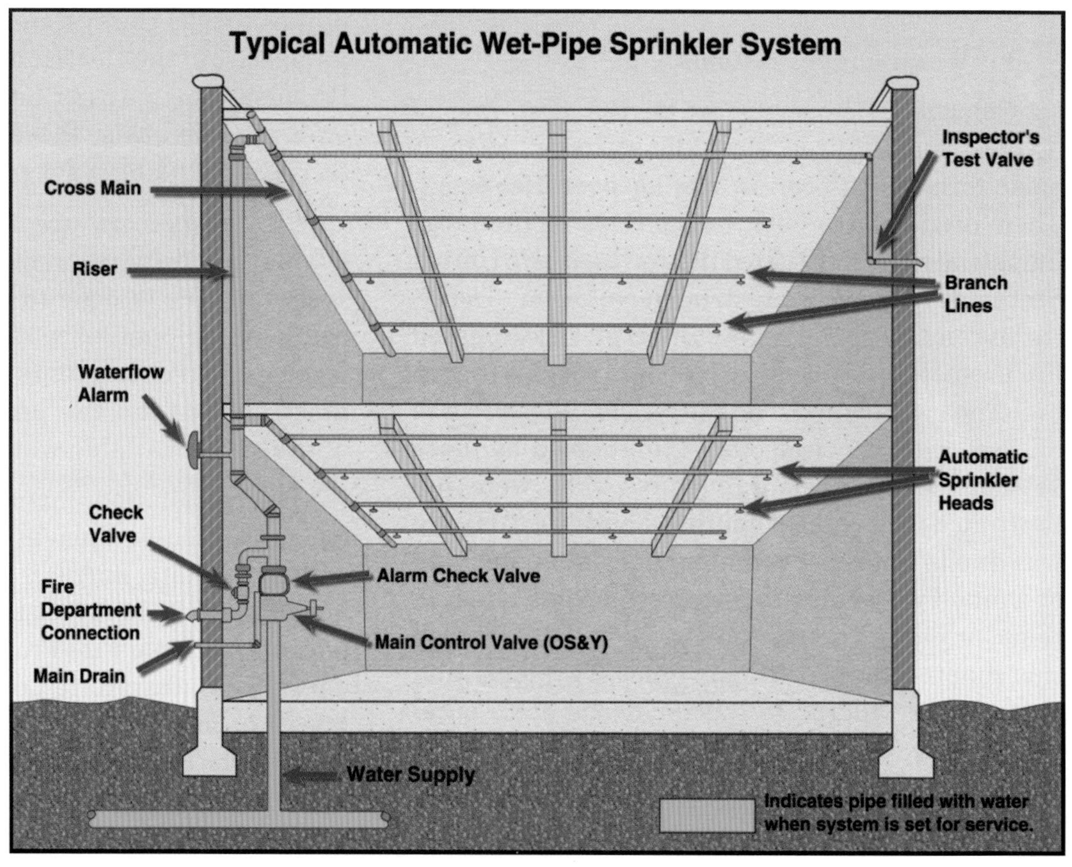

Figure 7.3 The components of a typical wet-pipe sprinkler system.

a check valve or backflow preventor to prevent sprinkler water from backflowing into the potable water supply. The feeder main also has a pipe to allow the fire department to augment the system through a fire department connection (Figure 7.4). *Risers* are vertical sections of pipe that connect to the feeder main. The riser has the system control valve and associated hardware that is used for testing, alarm activation, and maintenance. Risers supply the cross main. The *cross main* directly serves a number of branch lines on which sprinklers are installed. The entire system is supported by hangers and clamps and may be pitched to facilitate drainage.

Figure 7.4 The fire department connection allows the fire department to augment the permanent water supply system.

Sprinklers discharge water after the release of a cap or plug that is activated by a heat-responsive element (Figures 7.5 a through c). Sprinklers are kept closed by a number of devices. Some of the most commonly used release mechanisms are fusible links, glass bulbs, and chemical pellets, all of which fuse or open in response to heat (Figures 7.6 a through c).

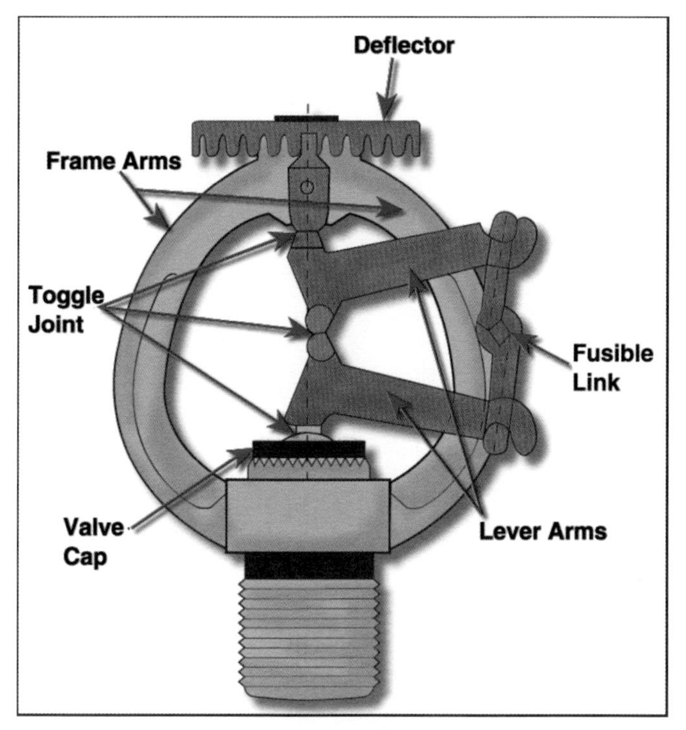

Figure 7.5a The major components of a fire sprinkler.

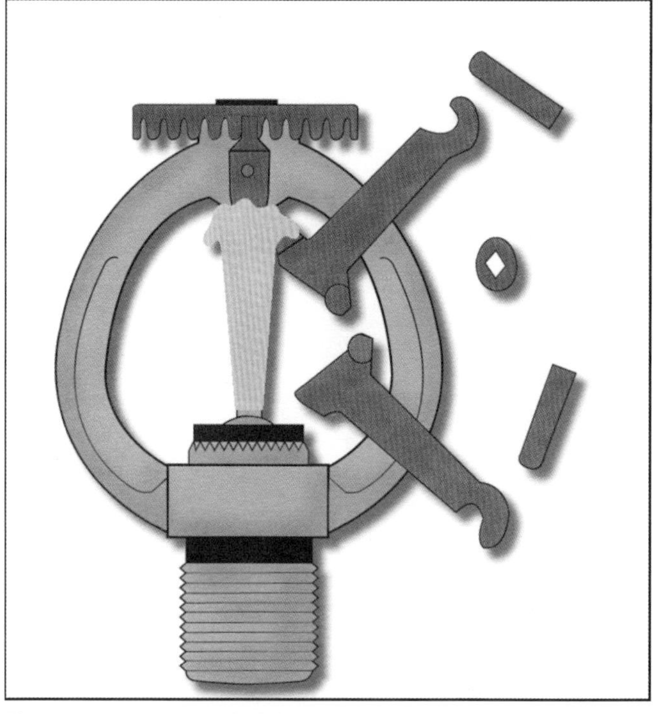

Figure 7.5b As the fusible link melts, the plug is thrown away from the discharge orifice by the water pressure in the piping system.

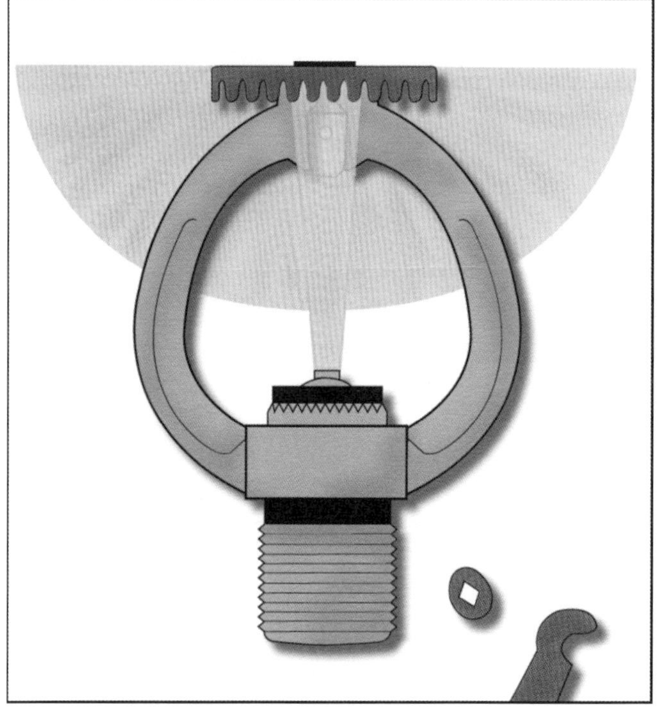

Figure 7.5c The water is given an effective discharge pattern by bouncing off the deflector.

The sprinkler used for a specific application should be based on the maximum temperature

Figure 7.6a Fusible link sprinklers.

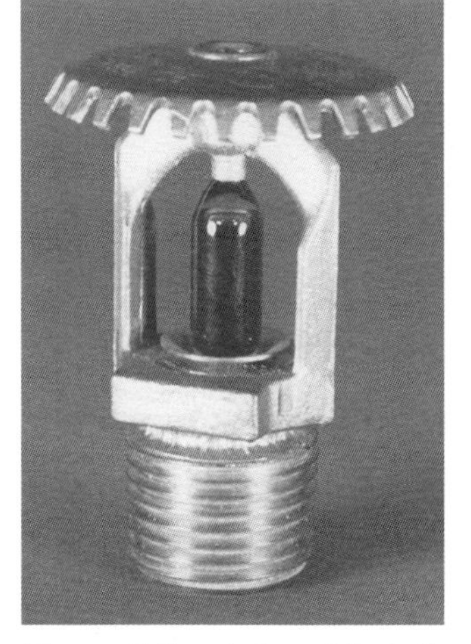

Figure 7.6b A frangible bulb sprinkler.

Figure 7.6c A chemical pellet sprinkler.

expected at the level of the sprinkler under normal conditions. Another important consideration is the anticipated rate of heat release that would be produced by a fire in the particular area. These temperature ratings are given in Table 7.1.

Certain other standards may require a head with a specified response time index (RTI). The *RTI* is a numerical value representing the speed and sensitivity with which a heat responsive device (like a fusible link) responds. For example, a residential head may have an RTI of 50 while a standard high temperature head is 450. The lower the RTI value, the faster the device will operate.

There are four basic types of sprinkler systems: wet-pipe, dry-pipe, deluge, and preaction. Wet-pipe systems contain water in the system at all times. Dry-pipe systems maintain air under pressure in the sprinkler piping. When sprinklers fuse, air escapes and water is automatically admitted into the system. A preaction system is a type of dry system that employs a deluge-type valve, fire detection devices, and closed sprinklers. This system only discharges water into the piping in response to a signal from the detection system. Water is then discharged onto the fire when individual sprinklers open. The deluge system is another type of dry-pipe system. It is equipped with open sprin-

TABLE 7.1
Sprinkler Temperature Ratings, Classifications, and Color Codings

Max. Ceiling Temp.		Temperature Rating		Temperature Classification	Color Code	Glass Bulb Colors
°F	°C	°F	°C			
100	38	135 to 170	57 to 77	Ordinary	Uncolored or black	Orange or red
150	66	175 to 225	79 to 107	Intermediate	White	Yellow or green
225	107	250 to 300	121 to 149	High	Blue	Blue
300	149	325 to 375	163 to 191	Extra high	Red	Purple
375	191	400 to 475	204 to 246	Very extra high	Green	Black
475	246	500 to 575	260 to 302	Ultra high	Orange	Black
625	329	650	343	Ultra high	Orange	Black

klers and a deluge valve. Upon fire detection, the deluge valve opens, permitting water to flow from all sprinklers at once.

Every sprinkler system is equipped with a main water control valve, or valves on either side of a backflow preventor, and various test and drain valves. *Control valves* are used to cut off the water supply to the system or zone when it is necessary to perform maintenance, change sprinklers, or interrupt operation. These valves are located between the water supply source and the sprinkler system or zone.

Control valves must be indicating valves; that is, they indicate whether the valves are open or closed. There are several common types of indicator control valves used in sprinkler systems: outside screw and yoke (OS&Y), post indicator valve (PIV), wall post indicator valve (WPIV), and the post indicator valve assembly (PIVA). The OS&Y valve has a yoke on the outside with a threaded stem or screw; the threaded portion of the stem is out of the yoke when the valve is open and inside the yoke when the valve is closed (Figures 7.7 a and b). The PIV is a hollow metal post that is attached to the valve housing. The valve stem inside the post has a target on which the words "OPEN" or "CLOSED or SHUT" appear (Figure 7.8). The WPIV is similar to a PIV except that it extends through the wall with the target and valve operating nut on the outside of the building (Figure 7.9). A PIVA is similar to the PIV except that is uses a butterfly valve, while the PIV uses a gate valve (Figure 7.10).

Figure 7.7a An open OS&Y valve will have the screw-thread stem visible outside the hand wheel.

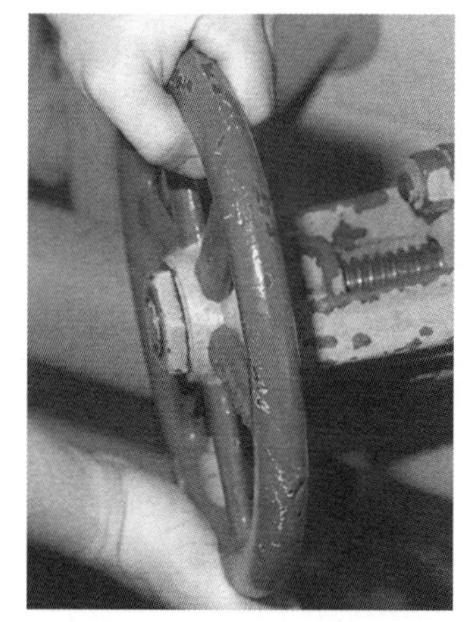

Figure 7.7b If the OS&Y valve is closed, the stem will not be visible.

Figure 7.8 A post indicator valve (PIV).

Figure 7.9 A wall post indicator valve (WPIV).

Figure 7.10 A post indicator valve assembly (PIVA).

In addition to the main water control valves, sprinkler systems employ various operating valves such as alarm test valves, globe valves, stop or cock valves, check valves, and automatic drain valves. The alarm test is located on a pipe that connects the supply side of the alarm check valve to the alarm line (Figure 7.11). This valve is provided to simulate actuation of the system by allowing water to flow into the alarm line and operate the water flow alarm devices. Some systems include a retard chamber in the alarm line (Figure 7.12). The retard chamber is used to eliminate false alarms caused by pressure surges within the water supply system.

Every sprinkler system must have an automatic water supply of adequate volume, pressure, and reliability. A water supply must be able to

Figure 7.11 The alarm test valve should be clearly marked.

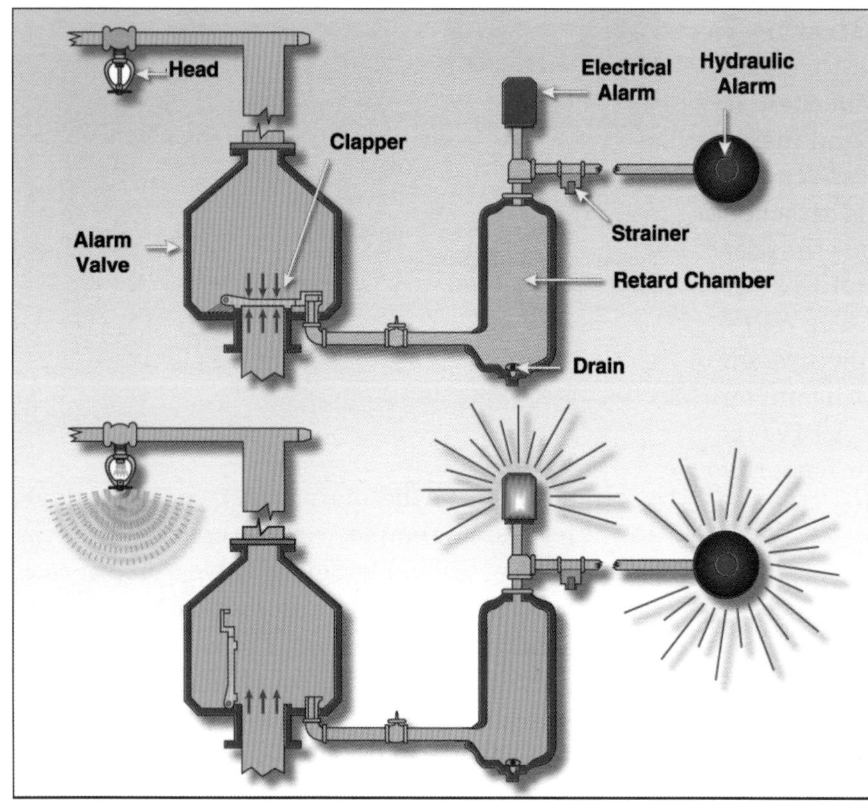

Figure 7.12 The retard chamber becomes active when the clapper valve is opened.

deliver the required volume of water to the highest sprinkler in a building at a minimum residual pressure of 15 psi (105 kPa) for systems designed by the pipe schedule method. Systems designed by the hydraulic calculation method require a minimum residual pressure of 7 psi (49 kPa) at the highest sprinkler. The minimum water flow required is determined by the hazard to be protected, the occupancy classification, and fire loading conditions.

A connection to a public water supply system that has adequate volume, pressure, and reliability is a good source of water for automatic sprinklers. This type of connection is often the only water supply. A gravity tank of the proper size also makes a reliable primary water supply. In order to give the minimum required pressure, the bottom of the tank should be at least 35 feet (11 m) above the highest sprinkler in the building (Figure 7.13).

In some instances, a second independent water supply is not only desirable but also required. Pressure tanks may be used as a secondary supply source, although in some cases (particularly with residential sprinkler systems) they may also be a primary source. This type of tank is filled two-thirds full with water and has an air pressure of at least 75 psi (525 kPa) (Figure 7.14). Fire pumps that take suction from large static water sources are also used as a secondary source of water supply (Figure 7.15). When properly powered and supervised, these pumps may be used as a primary water supply source.

General Inspection and Testing Principles for all Sprinkler Systems

Sprinkler systems require periodic inspections and maintenance in order to perform properly during a fire situation (Figure 7.16). Plant managers, maintenance personnel, and fire suppression and inspection personnel should be able to inspect systems and identify problems. Routine maintenance should be performed by competent plant personnel or a contracted sprinkler company. Fire department personnel should, however, be able to point out problems to assist building owners in maintaining system readiness. They should witness system tests for purposes of prefire familiarity and to verify system readiness.

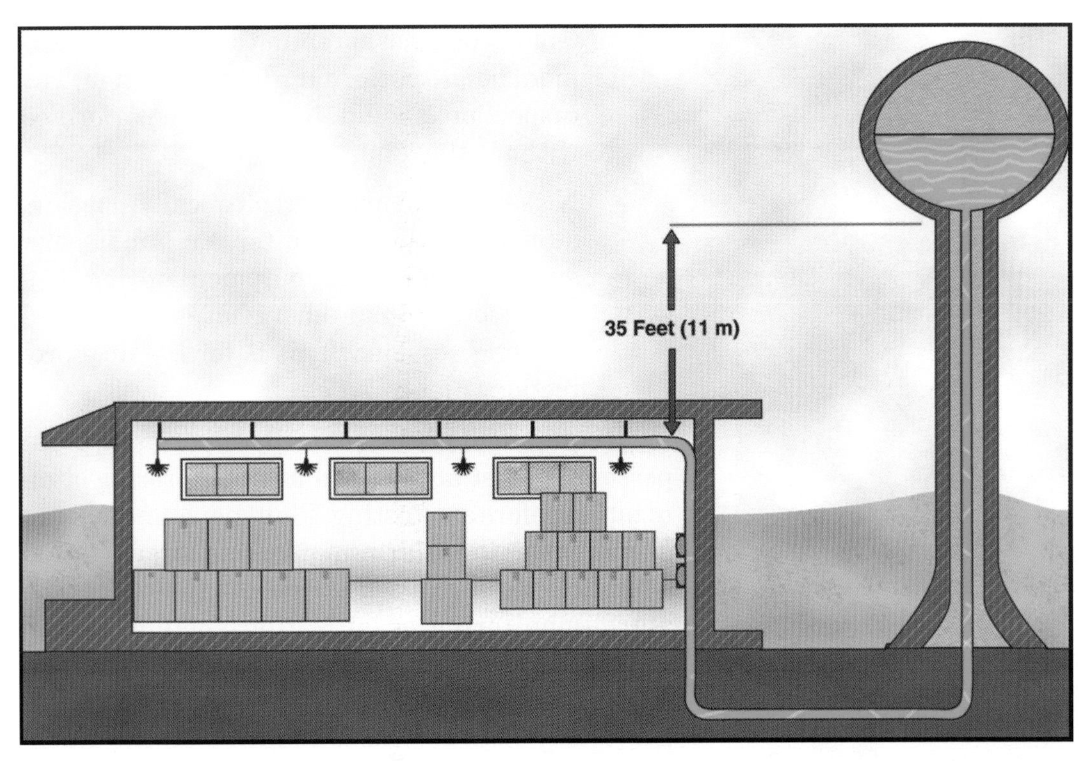

Figure 7.13 Systems supplied by elevated storage tanks must have the water at least 35 feet (11 m) above the highest sprinkler.

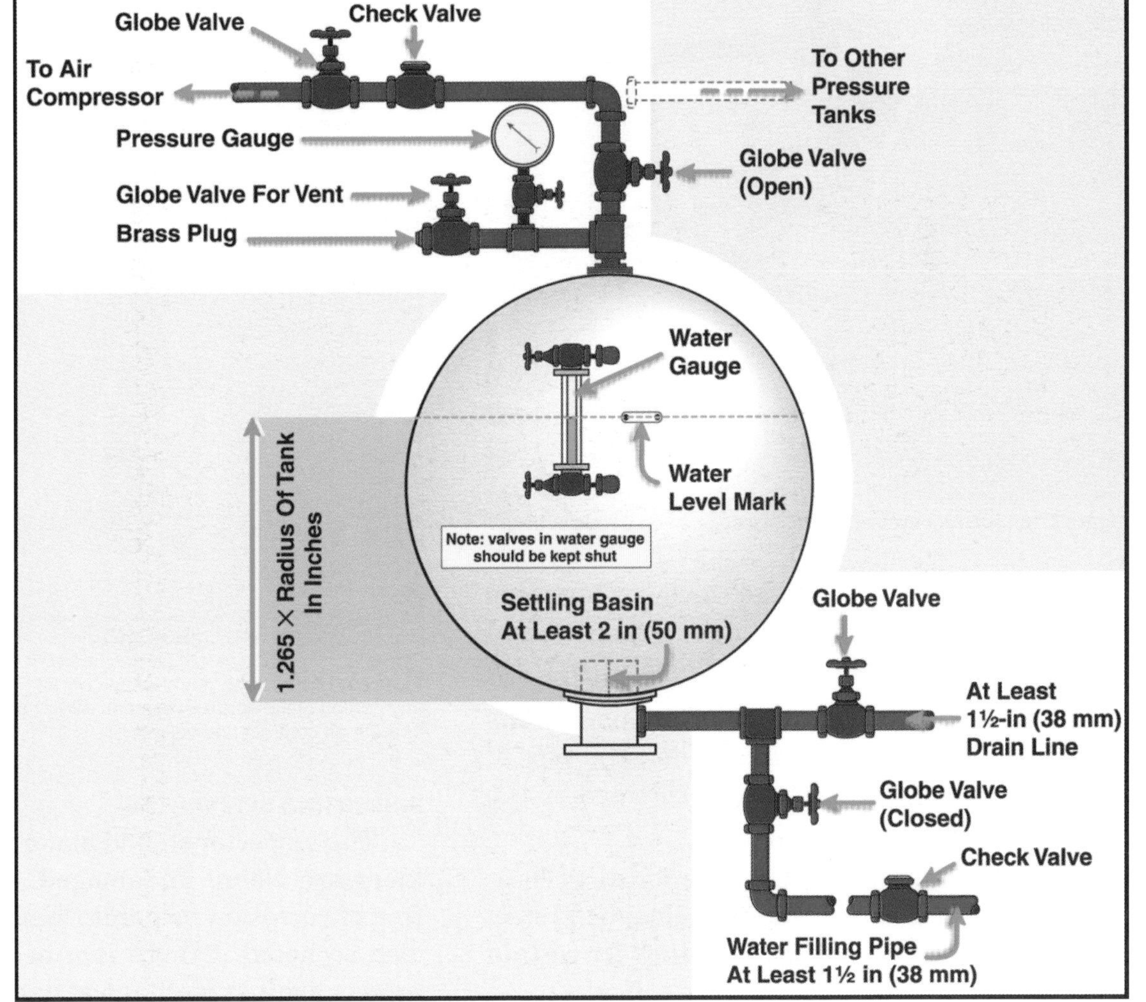

Figure 7.14 A diagram of a pressure tank assembly for a sprinkler system.

Figure 7.15 Fire pumps may be located in a building that is adjacent to the water supply.

Figure 7.16 Sprinkler systems must be inspected on a regular basis.

The inspector should take the following steps before performing any inspection or test on a sprinkler system.

- Review the records of prior inspections and identify the make, model, and type of equipment and the area protected by the system.
- Wear appropriate clothing for dirty locations such as attics and basements. Protective clothing may be necessary for certain manufacturing areas.
- Obtain permission from the plant management before performing any inspection. The inspection should NEVER be performed without this approval.
- NEVER personally or physically operate, adjust, manipulate, alter, or handle any sprinkler devices or equipment during situations other than emergencies or planned training sessions. This is for liability protection (Figure 7.17).

If equipment is electronically supervised, plant personnel must notify the alarm-monitoring organization before any testing. Plant personnel should also inform the alarm-monitoring organization when the testing is completed. At that time, the alarm-monitoring organization should confirm that the alarm equipment functions properly. If no alarms were received (water flow, valve supervision, etc.), service is required on the alarm equipment.

Figure 7.17 Fire department inspectors should never physically operate any portion of the sprinkler system. Allow the building representative to operate all controls and valves.

INSPECTING SPRINKLERS

The inspector should make sure that all sprinklers are clean, undamaged, unobstructed, and free of corrosion or paint (Figure 7.18). It should also be noted if there is a need for guards that protect against mechanical damage (Figure 7.19).

Figure 7.18 Sprinklers that have an excessive amount of paint or corrosion on them must be replaced immediately.

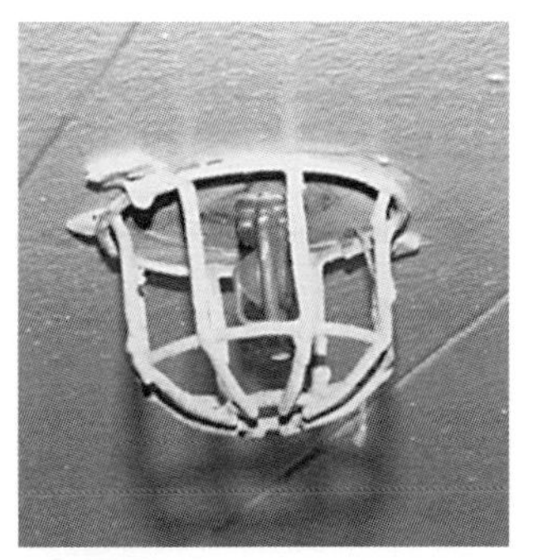

Figure 7.19 Some sprinklers are equipped with protective cages.

Sprinklers in buildings subject to high temperatures should be carefully examined. Carefully examine sprinklers in areas where changes of occupancy, fire hazards, heating, lighting, or mechanical equipment have occurred to determine if these changes may require the installation of different types or rated sprinklers.

Any sprinkler showing evidence of weakness or damage should be replaced with a sprinkler of the proper temperature rating. Weak sprinklers are indicated by a creeping or sliding apart of the fusible link (cold flow) or by leakage around the sprinkler orifice. *Cold flow* is caused by the repeated heating of a sprinkler to near its operating temperature. Cold flow problems can be eliminated either by using a higher temperature rated sprinkler or by using frangible bulb sprinklers. Sprinklers exposed to a corrosive atmosphere should have a special protective coating. Sprinklers that are corroded, painted, or loaded with foreign material should be replaced.

Partitions, stock, lights, or other objects should not obstruct the distribution of water discharge from sprinklers, and the discharge area should be free of hanging displays. A clearance of at least 18 inches (450 mm), which is measured from the deflector, should be maintained under sprinklers (Figure 7.20). For more detailed information on sprinkler obstructions, see NFPA 13, *Installation of Sprinkler Systems*.

A key aspect of the protection provided by an automatic sprinkler system is the continuity of that protection. To enhance system readiness, fire protection standards require that a supply of extra sprinklers be maintained at the protected premises (Figure 7.21). If sprinklers have been used to control a fire or if they are damaged, they can quickly be replaced and the system restored to service. NFPA 13 requires quantities of extra sprinklers as shown in Table 7.2. The supply of spares must include all types and ratings of sprinklers used in the building. In addition, a sprinkler wrench with which to change them must be provided.

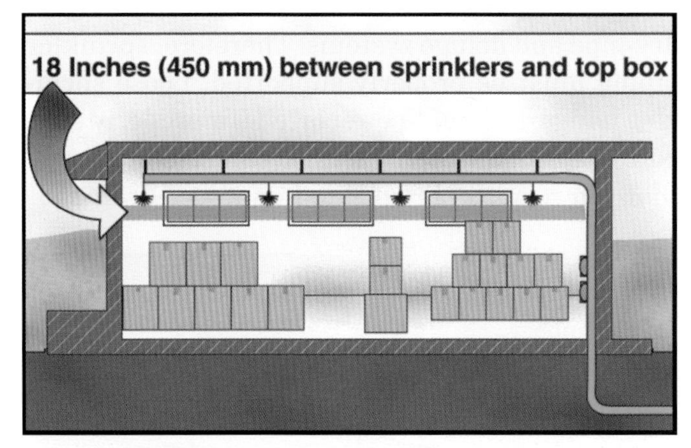

Figure 7.20 There should be at least 18 inches (450 mm) of clearance from all sprinklers.

Figure 7.21 NFPA 13 requires the occupants to maintain a supply of spare sprinklers.

TABLE 7.2
Required Number of Spare Sprinklers

Number of System Sprinklers	Spare Sprinklers (Minimum)
1-300	6
301-1,000	12
1,001 or more	24

INSPECTING SPRINKLER PIPING AND HANGERS

Inspect all sprinkler piping and hangers to determine that they are in good condition (Figure 7.22). Check for corrosion and physical damage to ensure that there are no leaks. Sprinkler piping is not to be used as a support for ladders, stock, or other material.

The flowing of water through sprinkler pipes can produce very significant forces, especially in dry-pipe and deluge systems. Therefore, sprinkler piping must be properly supported. Loose sprinkler hangers should be reported. Improperly supported sprinkler piping will be subjected to stress that can result in breaks and improper drainage.

Figure 7.22 Sprinkler hangers should be in good condition.

Figure 7.23 Check the fire pump control panel to make sure everything appears to be in working order.

Figure 7.24 The inspector should observe the incoming water pressure on the intake gauge.

INSPECTING SPRINKLER SYSTEM WATER SUPPLIES

All water supplies to sprinkler systems must be checked during an inspection. Particular attention should be paid to the electrical power for electrically driven fire pumps. The fire pump control panel must be visually checked to ensure that the circuit breaker or disconnect is closed and that the power indicating light is on (Figure 7.23). If the pump takes water under pressure (such as from a city water main), the incoming pressure gauge should be checked to verify that water is available (Figure 7.24). It is important to remember, however, that simply checking the incoming pressure will not disclose an obstruction, such as a partially closed valve, in the water supply. A flow test is necessary to check for obstructions.

Gravity tanks, pressure tanks, and ground-level reservoirs must be checked to ensure that they are full. Gravity tanks are equipped with various water-level devices. However, these devices have been known to fail. Unfortunately for the inspector, the only sure way to verify that a tank is full is to climb to the top and peer into it. Pressure tanks are equipped with sight gauges and pressure gauges to facilitate their inspection.

Detailed information on water supply and fire pump testing can be found later in this chapter.

CHANGES IN BUILDING OCCUPANCY

A frequent cause of a sprinkler system's inability to control a fire is a change in building occupancy or contents. Most modern sprinkler systems are hydraulically designed; that is, they are designed to supply a given amount of water based on the hazard to be protected. A change in occupancy may result in a hazard greater than the sprinkler system can protect. If, for example, a machine shop is converted to a tire warehouse, the increased fuel load will drastically reduce the ability of the sprinkler system to control a fire.

A change in occupancy can be the result of general remodeling as well as a change in contents. If walls are moved, partitions installed, or lighting fixtures relocated, the sprinkler discharge pattern can be obstructed. During inspections, the inspector should seek answers to the following questions:

- Has a change in the occupancy occurred?
- Have combustibles been added that will contribute to a greater fire load or a more rapid fire spread?
- Have alterations caused a need for the reconfiguration of sprinklers?

If the answer to any of these questions is "yes," the inspector will need to note this on the report and initiate corrective measures immediately. These measures might include immediate remodeling of the system or curtailment of activities in the occupancy until corrections can be made.

Inspecting and Testing Wet-Pipe Sprinkler Systems

A wet-pipe sprinkler system, as the name suggests, contains water within the piping at all times. Wet-pipe sprinkler systems are the most reliable of all sprinkler systems. Water under pressure is maintained throughout the system, with the exception of the piping to the water motor gong, the main drain piping, and piping from the fire department connection (Figure 7.25). The system is connected to a water source such as the municipal water system. When a sprinkler fuses, it immediately discharges a continuous flow of water.

Systems are designed so that the water flow actuates an alarm. This is accomplished by installing an alarm check valve or a waterflow indicator in the main riser. In systems with an alarm check valve, water flow lifts a clapper valve and allows water to flow into piping leading to an alarm device

Figure 7.25 A typical wet-pipe sprinkler riser.

(Figure 7.26). The alarm check valve is equipped with a false alarm prevention device called a *retard chamber*. The water chamber of the retard chamber must be filled with water before water will flow into the water motor gong. The retard chamber is equipped with a ball check valve leading to a drain. Water surges will partially fill the chamber but will subsequently drain, preventing false alarms.

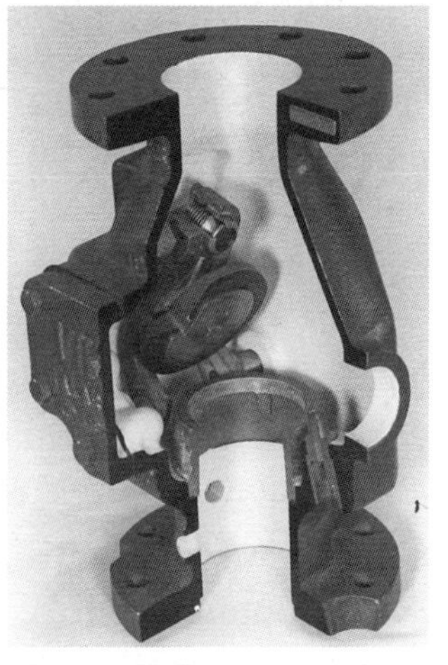

Figure 7.26 This cutaway shows the internal features of a wet-pipe sprinkler valve.

The *waterflow indicator* consists of a vane (or paddle) that protrudes through the riser into the waterway (Figure 7.27). The vane is connected to an alarm switch located on the outside of the riser. Movement of the vane caused by flowing water operates the switch that initiates an alarm. The vane must be sufficiently thin and pliable so that if many sprinklers operate, the water flow will flatten the vane against the inside of the riser, resulting in a clear waterway. Waterflow switches can be provided with a time-delay feature to prevent false alarms, just as the retard chamber does with an alarm check valve. Waterflow indicators are compact and are frequently used where only an electrical output from an alarm is desired.

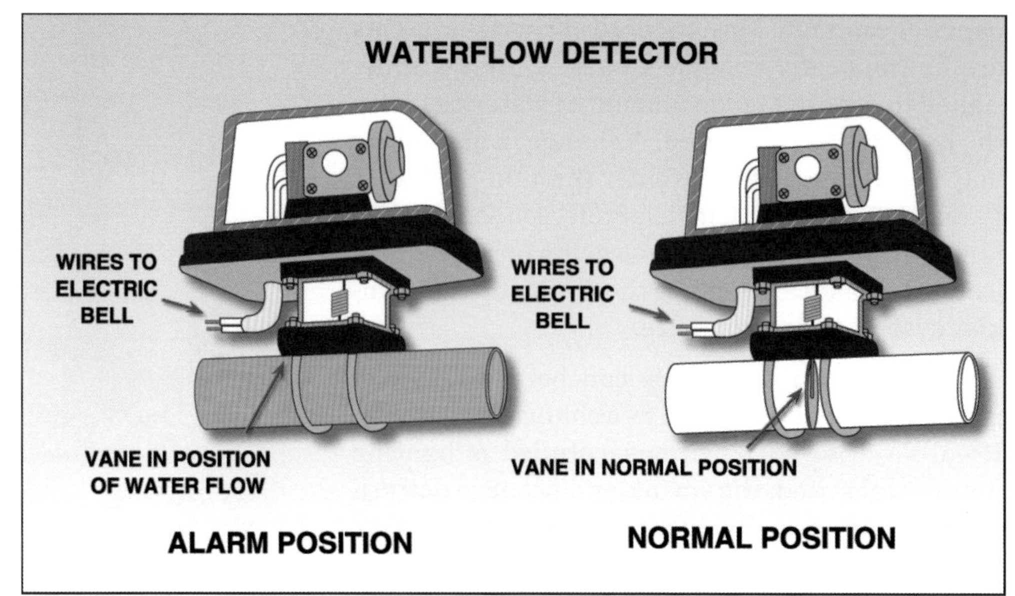

Figure 7.27 Waterflow indicators detect water movement within the piping system.

When a fire inspector is charged with the responsibility of performing inspections on wet-sprinkler systems, his primary concerns are four major areas: valves, sprinklers, piping, and water supply.

The inspector should ensure that all valves controlling water supplies to the sprinkler system and within the system (sectional valves) are open at all times. Anytime a valve is found closed, the inspector should report the condition to the responsible agency and the fire department. Examine each control valve for the following:

Figure 7.28 There must be some method of ensuring that the water supply valve stays open. In this case, the PIV is electronically supervised.

- Ensure that the valve is opened fully and secured or otherwise supervised in an approved manner (i.e., tamper switches, chained and padlocked in the open positions, etc.)(Figure 7.28).
- Check the valve operating wheel or crank to determine its condition (Figure 7.29).
- Make sure that the valve is accessible at all times. (**NOTE:** If a permanent ladder is provided to elevated valves, check to see that it is in good condition.)
- Check the valve operating stem to determine that it is not subjected to mechanical damage. Provide guards if necessary.
- Inspect post indicator valves (PIVs) to ensure that the operating wrench is in place (Figure 7.30). Try the wrench to feel the

Figure 7.29 Have the occupant check the OS&Y valve for proper operation.

Figure 7.30 The valve wrench should be securely stowed on the PIV.

spring of the rod when the valve is fully opened. The stem should be backed off about one-quarter turn from the fully open position to facilitate ease of operation and to prevent leaks caused by damage to the valve packing.

- Ensure that the target (open/shut sign) is properly adjusted and that the cover glass is in place and clean.
- Ensure that the PIV bolts are tight and the barrel casing is intact.

During normal conditions, the inspector should ensure the following:

- Alarm line shutoff cock is completely open.
- Valves to the pressure gauges are open.
- Static pressure above the clapper is equal to or greater than the static pressure below the clapper. (**NOTE:** Systems without alarm check valves will have only one pressure gauge on the riser.)
- Main drain valve, auxiliary drains, and inspector's test valves are closed.
- Automatic ball drip valve in the fire department connection moves freely and allows trapped water to drain out.
- Retard chamber automatic drip valve should move freely and allow water to drain out of the retard chamber (Figure 7.31).
- Fire department connection threads are unobstructed, in good condition, and the caps are in place.

Figure 7.31 Check the drip valve to make sure it operates properly.

Piping in wet systems must be protected against freezing. A frequent cause of frozen piping can be from windows left open at night or over a weekend during winter months. Branch lines near these windows can then freeze. Piping in or near loading docks can freeze when the loading dock doors are left open for extended periods. Piping over ceilings in the top floor of a building or in an attic may not receive enough heat during prolonged cold spells.

Freezing of sprinkler piping can stop the flow of water to sprinklers or cause the failure of control and alarm devices. The greatest danger from frozen piping is the rupture of a pipe or fittings, resulting in severe water damage or expensive repairs and interruption of protection. The inspector should be aware of the potential for this problem and make recommendations to the occupant to prevent pipe freezing from occurring.

ACCEPTANCE TESTS

In many jurisdictions, fire inspectors witness sprinkler system acceptance tests. These tests are conducted by a representative of the installation firm. This releases the fire department from liability resulting from damaged equipment because of improper operation or installation. The tests performed and the procedures for each are as follows:

- *Flushing of Underground Connections.* Underground mains and lead-in connections should be flushed **BEFORE** connection is made to the sprinkler piping. Flushing is continued until the water is clear.
- *Hydrostatic Tests.* All piping, including underground piping, is required to be hydrostatically tested at not less than 200 psi (1 400 kPa) for two hours. If the normal static pressure exceeds 150 psi (1 050 kPa), the system should be tested at 50 psi (350 kPa) above the normal static pressure. There should be no visible leakage while the system is pressurized.

Appendix F contains a variety of forms that may be used for sprinkler system inspections and testing.

WET-PIPE SYSTEM TESTING

Although in-plant personnel perform and record functional sprinkler system tests at the required frequencies, inspection personnel must be familiar with each test. This allows the inspector to better verify records that indicate that the test was performed and the system was found to be operational. On occasion, inspectors may be required to witness these tests (Figure 7.32).

Testing frequencies for various system components are specified in the applicable NFPA codes or local fire codes. Note that in some cases the occupant's insurance carrier may have more stringent testing requirements than the codes. If the system is equipped with a fire pump, it must be flow tested annually. Diesel engines that power fire pumps should be started weekly and electric motors monthly. The procedures for testing fire pumps are covered later in this chapter. The water level in gravity tanks should be inspected monthly. During cold weather, the heating system for the gravity tank and piping should be inspected daily. A written record for each system should be kept on file by the occupant after each test. When performing functional tests, make sure to notify the central station alarm company that the activation will be a test. The company must be notified again after the system has been restored to a point of readiness. The wet-pipe sprinkler system should be subjected to three main types of tests: alarm test, water flow alarm test, and main drain test.

Figure 7.32 On occasion, the inspector may be required to witness wet-pipe sprinkler testing.

Alarm Test

Alarms should be tested quarterly. The inspector's test connection should only be used during nonfreezing weather to avoid ice formation on sidewalks and roadways. Also, the alarm piping can be damaged or obstructed by ice formation. During tests, all components should also be visually checked.

On a wet-pipe system with an alarm valve, the inspector should witness the opening of the alarm bypass valve to test the alarm without unseating the valve (Figure 7.33). The pressure gauge readings should not change significantly, but water should flow to the retard chamber (if so equipped) and then to the alarm line. The water motor gong or electric alarm should sound. The retard chamber drain should empty the chamber after the alarm bypass valve is closed. If there is no retard chamber, the alarm line should be drained at the conclusion of the test.

Figure 7.33 Open the alarm bypass valve to check for proper operation.

Water Flow Alarm Test

On all types of wet systems, a water flow alarm test should be conducted by using the inspector's test connection. The *inspector's test connection* simulates the operation of a single sprinkler and ensures that the alarm will operate even if only one sprinkler is fused in a fire (Figure 7.34). In older systems, the inspector's test connection was usually located at the end of the most remote branch line. In newer systems, the inspector's test connection may be located near the sprin-

Figure 7.34 The inspector's test connection valve may be used to check the operation of waterflow indication equipment.

kler riser so that one person is capable of performing the test. It consists of a 1-inch (25 mm) pipe equipped with a shutoff valve and a discharge orifice equal in size to the smallest sprinkler in the system.

With an observer at the riser, another individual opens the inspector's test valve. The alarm should sound, and only a slight variation in pressure should be observed at the riser. After the alarm operates, the inspector's valve should be closed.

Main Drain Test

According to NFPA 25, *Standard for the Inspection, Testing, and Maintenance of Water-Based Fire Protection Systems*, the main drain test should be conducted quarterly. The inspector should visually inspect the system and witness the alarm test in conjunction with the main drain test.

Each sprinkler system riser has a main drain. Risers that are 4 inches (100 mm) or larger in diameter will be equipped with a 2-inch (50 mm) main drain (Figure 7.35). The primary purpose of the *main drain* is to simply drain water from the system for maintenance purposes. However, because a large volume of water flows when the main drain is opened, it can also be used to check the system water supply.

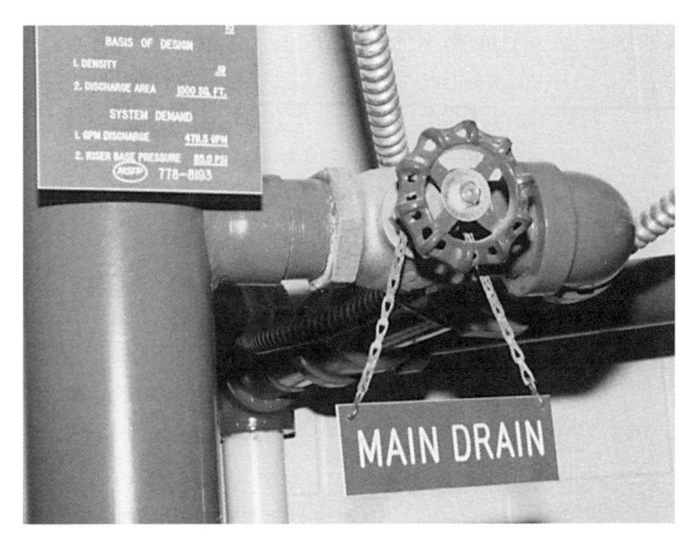

Figure 7.35 Most larger wet-pipe systems are equipped with a main drain valve.

The main drain test is useful for detecting impairments such as closed valves, obstructions, or gradual deterioration in the water supply. It can also be used to a limited degree as an indicator of the overall water supply in the area. To perform a main drain test, the inspector should use the following steps:

Step 1: Observe and record the pressure on the gauge at the system riser (Figure 7.36).

Step 2: A building representative fully opens the main drain. (**NOTE:** The main drain will usually discharge outside the building, and the area should be checked to make sure it is clear [Figure 7.37].)

Step 3: Observe and record the pressure drop.

Step 4: The building representative closes the 2-inch (50 mm) main drain slowly (Figure 7.38).

Figure 7.36 Record the system pressure prior to the test.

Figure 7.37 Make sure that the area near the discharge outlet is clear.

Figure 7.38 Have the building representative close the main drain valve.

These readings should be compared to previously recorded readings. If significant differences are noted, a supply valve may be partially closed or there may be an obstruction in the supply line. On a system using an alarm check valve, the pressure readings should be taken from the lower gauge because erroneously high static pressures can exist above the valve.

Inspecting and Testing Dry-Pipe Sprinkler Systems

Dry-pipe sprinkler systems are necessary and should only be installed where piping is subject to freezing. In a dry-pipe system, air or nitrogen under pressure replaces the water in the sprinkler piping above a device called the dry-pipe valve. The dry-pipe valve is located near the base of the riser in a heated area at the point where the supply main enters the premises. Dry-pipe valves are designed so that a small amount of air pressure in the sprinkler system piping holds back a much greater amount of water pressure (Figure 7.39). The *dry-pipe valve* balances the air pressure in the system against the incoming water pressure. This keeps the valve closed and prevents water from entering the sprinkler system piping. When sprinkler(s) are actuated, air escapes, which lowers the air pressure. Water pressure on the incoming side of the valve then causes the valve to open, allowing water to enter the system piping.

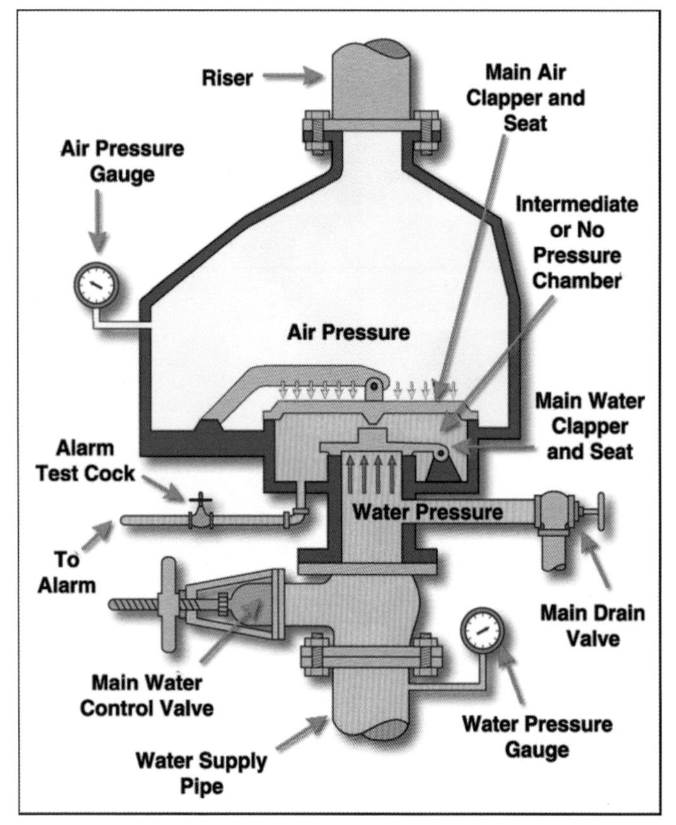

Figure 7.39 A typical dry-pipe sprinkler valve assembly.

During an inspection of a dry-pipe sprinkler system, inspectors should ensure that:

- All indicating control valves are open and properly supervised in the open position.
- Air pressure readings correspond to previously recorded readings.
- The ball drip valve moves freely and allows trapped water to seep out of the fire department connection.
- The velocity drip valve located beneath the intermediate chamber is free to move and allow trapped water to seep out. Inspectors can check this valve by instructing someone to lift a push rod that extends through the drip valve opening. Where an automatic drip valve is installed, the velocity drip valve can be checked by moving the push rod located in the valve opening.
- The fire department connection threads are unobstructed, in good condition, and the caps are in place.
- Any drum drips are drained to eliminate the moisture trapped in the low areas of the system (Figure 7.40). This should be done on a frequent enough basis to prevent clogging. This frequency will vary depending on the environmental conditions in that particular area.
- During freezing weather, the dry-pipe valve enclosure heating device keeps the temperature of the dry-pipe valve at or above 40°F (4°C).
- The priming water is at the correct level (Figure 7.41). If necessary, personnel can drain water by opening the priming water test level valve until air begins to escape.

(**NOTE:** If the system is equipped with a quick-opening device, opening the priming water test line could trip the system.)

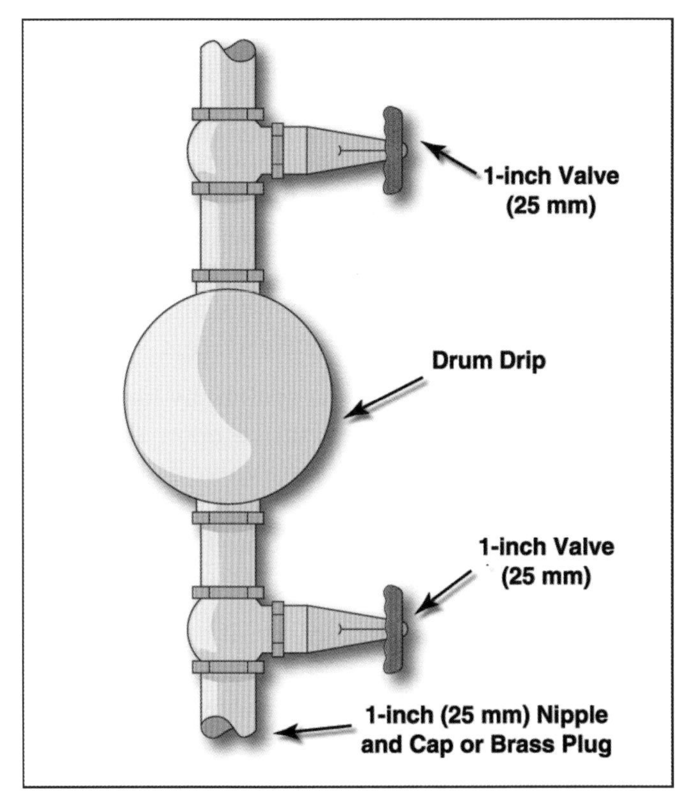

Figure 7.40 A drum drip valve.

Figure 7.41 It may be necessary to open the water test level valve to ensure that the water is at the correct level.

- The system's air pressure is maintained at 15 to 20 psi (105 kPa to 140 kPa) above the trip point and no air leaks are indicated by a rapid or steady air loss. If inspectors note excessive air pressure, they should have the system drained down.
- The system air compressor is approved for sprinkler system use, well-maintained, operable, and of sufficient size.

ACCEPTANCE TESTS

As with wet-pipe sprinkler systems, acceptance tests for dry-pipe systems are conducted by a representative of the installation firm but should be witnessed by fire inspection personnel. The tests performed and the procedures for each are as follows:

- *Flushing Underground Connections.* Underground connections should be flushed following the same procedures described for wet-pipe sprinkler systems.
- *Hydrostatic Testing.* Hydrostatic testing of dry-pipe systems is performed in the same manner as described for wet-pipe sprinkler systems. In freezing weather, however, dry-pipe systems are tested for 24 hours with not less than 40 psi (280 kPa) air pressure. If there is a loss of more than 1½ psi (10 kPa), the leaks should be located and corrected. When the weather warms, the system should be subjected to a hydrostatic test using water.

DRY-PIPE SYSTEM TESTING

Fire inspectors should encourage plant personnel to visually inspect dry-pipe sprinkler systems weekly. A weekly inspection is recommended because, as with wet systems, it provides plant personnel with system familiarity and early discovery of system deficiencies. In the case of dry systems, it is particularly important to check the system air pressure on a weekly basis. According to NFPA 25, heat for the dry-valve enclosure should be checked daily during cold weather. The main drain and alarm tests should be performed quarterly but not during freezing weather to avoid piping damage and ice accumulations.

Low-point drains should be tested each fall. A trip test of the dry-pipe valve should be made annually with the main control valve partially open. A full flow trip test is recommended every three years. According to NFPA 25, quick-opening devices should be tested semiannually.

Two types of tests, the main drain test and the trip test, are recommended for dry-pipe systems. The following sections detail the procedures for each of these tests.

Main Drain Test

The main drain test is conducted on dry-pipe systems for the same reasons as those discussed for wet-pipe systems. The same procedure is also used.

Trip Test

The operational test of a dry-pipe valve is known as a "trip test." A dry-pipe valve should be trip tested every three years to ensure that the air seat has not become stuck and that the clappers will move freely. The trip test of a dry-pipe valve permits water to flow into the system. Dry pipe valves with quick opening devices should be tested annually.

When planning a trip test, the inspector should be aware that this procedure can take from two to four hours to complete. The amount of time needed depends on the amount and size of the piping in the system and the capacity of the air compressor to pressurize the system. Also, older valves may prove more difficult to reset because of leaking seats or worn parts, and more than one attempt may be required. The inspector should use the following steps when performing a trip test:

Step 1: Have a building representative open the main drain fully. This action flushes any sediment or scale that may be in the water supply and prevents it from entering the sprinkler piping. Slowly close the drain valve.

Step 2: Check the system water control valve for freedom of movement by having the building representative slightly open the main drain and turn the water control handwheel or crank (Figure 7.42). The purpose of opening the main drain at this point is to prevent accidental tripping of the valve. Closing the water control valve without opening the drain may "squeeze" the water trapped between the control valve and dry-pipe clapper. For all practical purposes, water is incompressible; therefore, something has to give. In some cases, the clapper will be pushed off its seat.

Step 3: Leave the water control valve partially open and have the building representative close the main drain. Having the valve only partially open permits more rapid closing when the dry valve trips, thus minimizing the total amount of water that flows into the system.

Step 4: Bleed air from the system by having the building representative open the priming water valve or the valve body drain if the valve is so equipped (Figure 7.43). This should be performed while observing the reading on the air-pressure gauge.

Step 5: Record the tripping point of the dry-pipe valve as indicated by the air-pressure gauge, and close the control valve.

Figure 7.42 Have the OS&Y valve partially closed.

Figure 7.43 Air may be bled from the system using the priming valve.

Step 6: Check water and air-pressure gauges to ensure that pressure equalization has occurred after tripping (Figure 7.44).

Step 7: Verify that the local alarm and control panel, central station alarm, or fire department alarm has operated.

Step 8: Have the building representative open the main drain valve (Figure 7.45). Be sure that the system is completely drained before proceeding. A considerable amount of water will accumulate in the system riser, and it will rush out when the valve cover is opened if not properly drained.

Step 9: Have the building representative open the dry-pipe valve, and check to ensure that the clapper is latched in the open position (Figure 7.46).

Step 10: The building representative should clean the air and water clapper seats, and remove any debris from the valve housing (Figure 7.47). Check the condition of the air seat (this part is usually made of rubber). Check for any signs of impurity, such as oil, in the water.

Step 11: The building representative releases the clapper latch and reseats the valve.

Step 12: The building represetative closes the valve cover (Figure 7.48).

Step 13: The building representative adds priming water through the priming water fill pipe.

Figure 7.46 Remove the valve cover and check to see that the dry-pipe clapper is latched in the open position.

Figure 7.47 Clean any debris from the valve seat.

Figure 7.44 After the dry pipe has been tripped, the air and water gauges should read the same pressure.

Figure 7.45 Open the main drain valve to remove the water from the system piping.

Figure 7.48 Close the valve cover after the clapper has been reset.

Step 14: The building representative closes all drain valves.

Step 15: Once all drain valves are closed, the building representative pressurizes the system with air to the proper pressure (Figure 7.49).

Step 16: Check for a flow of water at the intermediate chamber drain. It is not unusual for water to drip out initially, but a steady stream of water usually indicates that the air seat has not properly seated.

Step 17: The building representative opens the main drain partially and then slowly opens the water control valve fully.

Step 18: The building representative slowly closes the main drain.

Step 19: Check the air- and water-pressure gauges. The air pressure should be lower than the water pressure. If the two gauges read the same, it is an indication that the valve has tripped, and Steps 9 through 18 must be repeated.

Step 20: The building representative notifies the alarm service that work has been completed.

Step 21: Attach to the valve a tag that shows the date the valve was tested, the air pressure at which the valve tripped, and the name of the person who performed the test (Figure 7.50).

Figure 7.49 Recharge the system with the appropriate amount of air pressure.

Figure 7.50 Tag the dry-pipe valve after it is back in service.

Inspecting and Testing Deluge and Preaction Sprinkler Systems

A deluge sprinkler system contains piping with *open* sprinklers. No water enters the piping until a device called a deluge valve opens. The deluge valve is tripped automatically by activation of fire detection devices that are installed in the same areas as the sprinklers (Figure 7.51). When the deluge valve operates, water flows from ALL the sprinklers.

A preaction sprinkler system contains piping with *closed* sprinklers. Air is maintained under pressure within the system piping in the same way as described previously for dry-pipe systems. Water is held back by a deluge valve, which is tripped when fire detection devices activate. This type of system is used when it is especially important that water damage be prevented. If a sprinkler is broken or piping is broken, no water will flow. If a detector malfunctions, water will enter the system but none will be discharged unless a sprinkler activates.

A combined dry-pipe and preaction sprinkler system may be used where it is important that failure of the fire detection system not prevent operation of the sprinkler system or vice versa. A combined system features closed sprinklers, air under pressure within the piping, and a dry-pipe

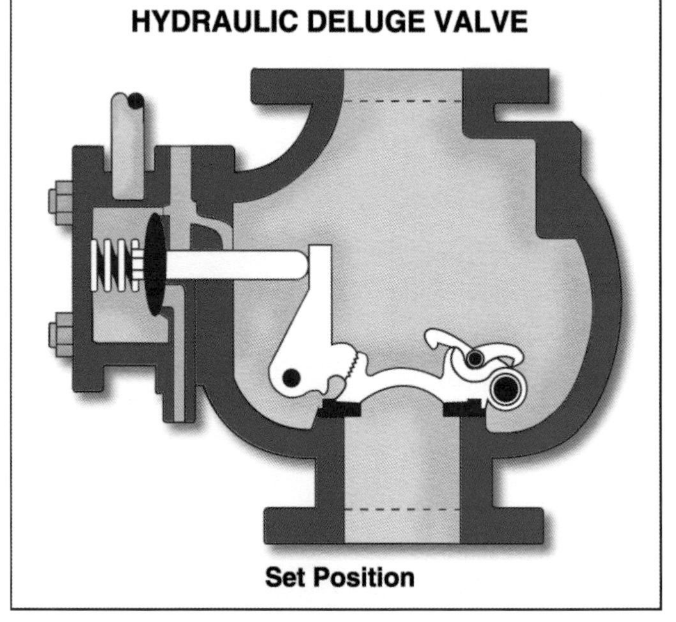

Figure 7.51 A typical deluge valve assembly.

valve. Operation of the fire detection system activates tripping devices that open the dry-pipe valve WITHOUT causing loss of air pressure in the system. Air exhaust valves at the end of the feed main act to speed the filling of the system and usually precede opening of sprinklers.

Deluge and preaction systems should be visually inspected monthly. However, because visual inspections provide continued familiarization and early problem discovery, a weekly inspection is recommended. The supervisory air pressure in a preaction system should be checked weekly. The detection system should be tested semiannually. The system itself should be trip tested annually. A written record of all tests and inspections should be made, compared to earlier tests, and filed.

Acceptance testing for deluge and preaction sprinkler systems should follow the same procedures as listed for dry-pipe systems. Deluge and preaction systems should have main drain, alarm, and trip tests performed. The following sections highlight the procedures for each of these tests.

MAIN DRAIN AND ALARM TESTS

The main drain test for deluge and preaction sprinkler systems serves the same purpose as it does on wet- and dry-pipe systems. It is conducted as previously described. Alarms on deluge and preaction systems may be somewhat more complicated than on wet or dry systems because they are usually designed to protect special hazards. However, some deluge systems make use of ordinary alarm lines that can be tested by opening an alarm bypass valve (Figure 7.52).

Figure 7.52 Some deluge systems are equipped with an ordinary alarm bypass valve.

TRIP TEST

The following procedure details the process for trip-testing deluge and preaction sprinkler systems. As with dry-pipe systems, the inspector should take into account that this test is time-consuming.

Step 1: Notify the supervising agency of the sprinkler system, such as a central alarm service or fire department, that a test is to be performed.

Step 2: The building representative replaces the open heads on a small deluge system with standard sprinkler heads or plugs (Figure 7.53). On large systems where the heads cannot be replaced or plugged, the main water control valve may be closed to within two turns from closed. This procedure permits rapid shutoff of the water when the valve trips. (Remember, in a deluge system all the heads are OPEN.)

Step 3: The building representative activates the system by using a heating or smoke device (as the case may be) on a detector or by using the manual trip (Figure 7.54).

Step 4: After the valve trips, the building representative immediately closes the main water control valve and drains the system using the main drain.

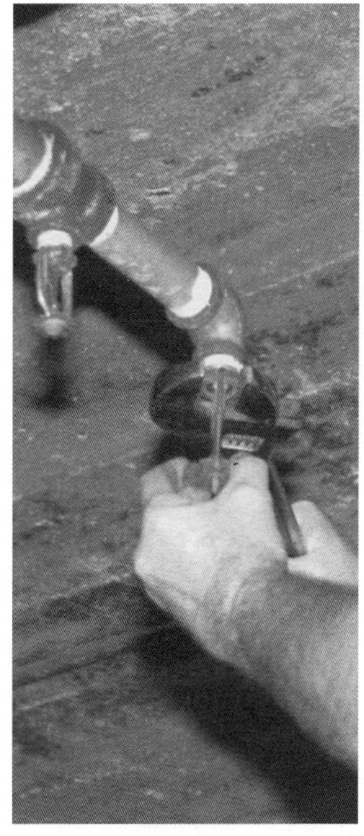
Figure 7.53 Replace the open sprinklers with closed sprinklers or pipe plugs/caps.

Figure 7.54 Use the manual control valve to trip the system.

Step 5: The building representative opens the valve cover, cleans the valve seat, and removes any debris from the valve housing.

Step 6: The building representative unlatches the clapper and resets the valve (Figure 7.55). (**NOTE:** Some deluge and preaction valves do not require resetting by removal of the cover. Refer to manufacturer's instructions for proper resetting procedure.)

Step 7: The building representative reinstalls the cover (Figure 7.56). (**NOTE:** Some deluge and preaction valves do not require resetting by removal of the faceplate. Refer to manufacturer's instructions for proper resetting procedure.)

Step 8: The building representative pressurizes preaction systems, if the pipe integrity is supervised, with air to the proper pressure.

Step 9: The building representative restores water pressure by opening the supply control valve with the main drain open.

Figure 7.55 Clean the valve seat before reseating the clapper valve.

Figure 7.56 Once the valve is reseated, replace the valve cover.

When the supply valve is completely open, the building representative slowly closes the main drain and checks the gauges.

Step 10: The building representative removes plugs or caps from sprinkler heads on deluge systems or reinstalls open heads.

Residential Sprinkler Systems

According to records kept by the NFPA, approximately 80 percent of fire fatalities occur in residential occupancies. It is logical, therefore, to extend the protection provided by automatic sprinklers into the residential environment. Within the last ten years, a very direct effort has been made to provide automatic sprinkler protection in residential buildings. The United States Fire Administration (USFA) funded research on the subject, and two residential sprinkler standards have been developed by the NFPA. These standards are as follows:

- NFPA 13D, *Standard for the Installation of Sprinkler Systems in One- and Two-Family Dwellings and Manufactured Homes*
- NFPA 13R, *Standard for the Installation of Sprinkler Systems in Residential Occupancies up to and Including Four Stories in Height*

These standards were developed so that systems could be installed economically in residential dwellings with the goal of preventing flashover to improve the chances of evacuation of the structure during an incipient fire. Fire inspection personnel should be familiar with residential sprinkler systems in case they are involved in a home inspection program.

There are several barriers to the application of conventional sprinkler technology to residential buildings, especially single-family dwellings:

- The hardware of standard sprinkler systems, such as fire department connections and alarm valves, is large and obtrusive if applied to residential dwellings under the same rules as used in commercial and industrial applications.
- Conventional industrial systems would be objectionably expensive if applied to ordinary dwellings and would be beyond the economic means of a large segment of the population.
- Water supply requirements are substantially reduced for residential systems. A minimum 10-minute supply is required for systems designed in accordance with NFPA 13D and a minimum 30-minute supply is required for systems designed in accordance with NFPA 13R.

To make sprinklers useful in residential applications, changes in design were made. These changes were needed not only to decrease the cost of the systems but also to enhance their effectiveness in protection of life. The following changes were made:

- Sprinkler design was modified, and fast-response residential sprinklers were developed.
- Water supply requirements were modified to accomplish realistic levels for residential protection.
- Areas of coverage for sprinklers were adjusted based on sprinkler design and typical residential fire loads.
- Alarms were made simpler and more realistic for residential applications.
- Valve arrangements were made so that they would be unobtrusive in a common residence.

Residential sprinklers differ from those used in conventional systems and are tested under a different standard by Underwriters Laboratories (UL). A major difference between residential sprinklers and standard sprinklers is their sensitivity, or speed, of operation (Figure 7.57). Residential sprinklers operate more quickly than standard sprinklers. Although a standard sprinkler may have an operating temperature of 165°F (74°C), the thermal lag of the fusible link may delay the operation of the sprinkler until the surrounding air temperature is considerably higher. By redesigning the fusible link, the sprinkler can be made to operate before conditions in the room become unsurvivable for the occupants.

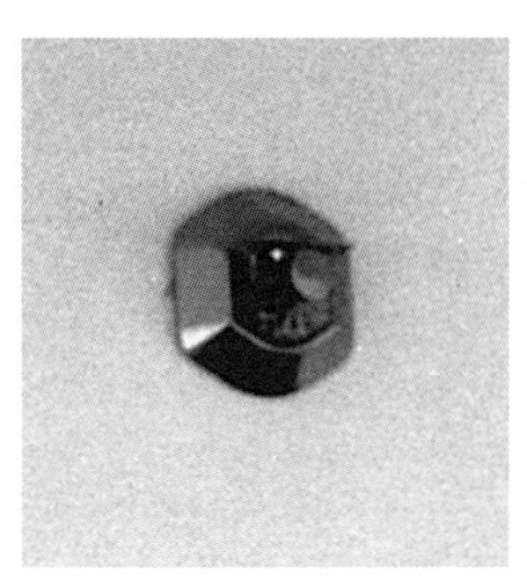

Figure 7.57 A typical sidewall residential sprinkler installation.

Residential sprinklers also have distribution patterns that are different from conventional sprinklers. Residential sprinklers are designed to discharge water higher on the enclosing walls of a room. This is to prevent a fire from traveling above the spray as might occur with burning drapes or in preflashover conditions.

Any type of piping, such as steel, copper, and plastic, that is listed by an approved testing agency may be used for residential sprinkler systems. The minimum pipe size that may be used in a residential system is ½-inch (13 mm).

WATER SUPPLY AND FLOW RATE REQUIREMENTS

The water supply requirements for residential sprinklers are less than those for commercial systems. NFPA 13D requires only 18 gpm (68 L/min) for any single sprinkler. If there are two or more sprinklers, each requires 13 gpm (49 L/min) as a maximum required water supply. In addition, the water supply needs only to supply this flow rate for 10 minutes.

For larger multiple dwellings, the designed flow rates are greater. NFPA 13R requires 18 gpm (68 L/min) for a single sprinkler and not less than 13 gpm (49 L/min) to a maximum of *four* sprinklers in a compartment. The water supply for these larger buildings is required to supply the sprinklers for 30 minutes.

The reduction in water supply is prompted not only by economics but also by the recognition that many residential buildings are served only by small domestic supplies. Many single-family dwellings, for example, are supplied by wells. Although a 10-minute supply may not completely control all fires, it will prevent room flashover and provide for occupant safety.

The water supply for residential sprinklers may be taken from several sources. These sources may include a connection to the public water system, an on-site pressure tank, or a tank with an automatic pump. A connection to a public water system is reliable and will usually provide adequate volume (Figure 7.58). However, rural homes may not be serviced by public water systems, thus requiring a pressure tank or automatic tank and pump (Figure 7.59).

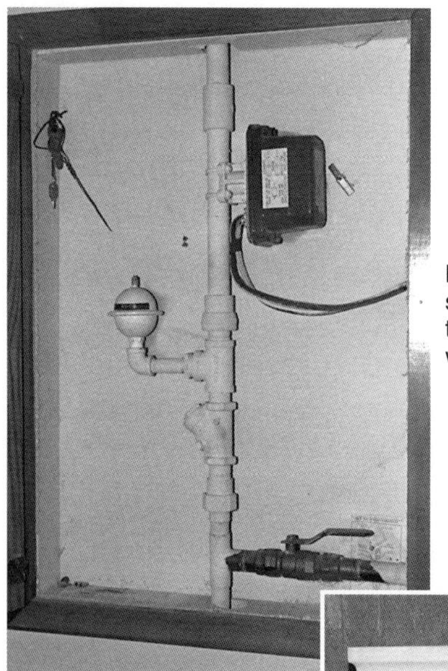

Figure 7.58 The sprinkler system should be connected to the intake for the domestic water supply system.

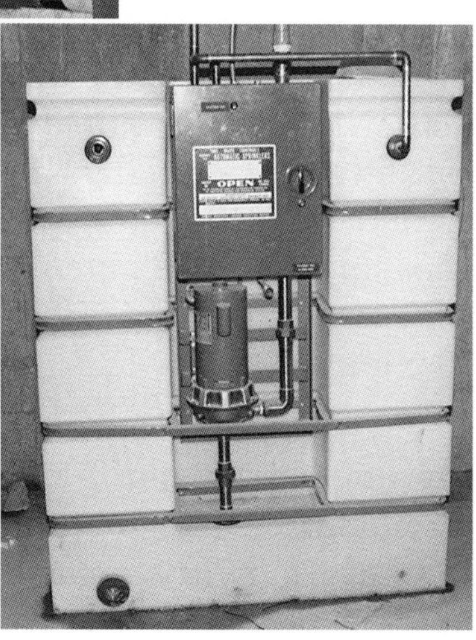

Figure 7.59 Some residential systems are equipped with supply tanks and automatic pumps.

To be of value, a residential sprinkler system, like any other system, must be continually in service. As in a standard system, inadvertent or deliberate closing of valves renders the system useless. When a residential sprinkler system is supplied from a public water system, the possibility of the supply valve being closed can be virtually eliminated by using one valve to control both the sprinklers and the domestic service (sinks and toilets). With this arrangement, the sprinklers cannot be shut off without shutting off the domestic supply. Even if the sprinkler system is viewed as unneeded, it is unlikely that a homeowner would be willing to go very long without the household water supply. Where plumbing or water department requirements do not permit this type of uninterrupted connection, NFPA 13D permits the sprinkler valve to be supervised or simply locked in the open position.

SPRINKLER SPACING

Sprinkler coverage in residential systems is not as extensive as in standard systems. In single-family dwellings, sprinklers can be omitted from many areas, including bathrooms not over 55 square feet (5.1 m^2), small closets, garages, porches, carports, uninhabited attics, and entrance hallways.

The basic spacing for sprinklers in residential systems is a maximum of 144 square feet (13.4 m^2) per sprinkler. The maximum spacing between sprinklers is 12 feet (3.7 m) with the maximum allowable distance of a sprinkler from a wall being 6 feet (1.8 m). However, sprinkler manufacturers have produced a variety of sprinkler designs, and the spacing of heads may be based on the spacing for the particular sprinklers that have been tested and listed. Some residential sprinklers, therefore, can be spaced to protect an area as large as 20 x 20 feet (6 m by 6 m). However, with this sprinkler spacing, the minimum discharge with one sprinkler operating is increased to 32.5 gpm (123 L/min) and 22.5 gpm (85 L/min) per head with two sprinklers operating.

Water Mist Systems

Limited use of water mist systems for fire protection can be traced back as far as the 1940s. Their use in these early days was limited primarily to passenger ferries. However, the phasing out of halon has resulted in a renewed interest in water mist systems. Research has indicated that water mist systems may be very suitable in many situations where halon was previously used. Continuing research also indicates that water mist systems may have successful applications in such places as residential occupancies and flammable and combustible storage facilities, which are normally protected by traditional sprinkler systems. Requirements for these systems are contained in NFPA 750, *Standard on Water Mist Fire Protection Systems*.

Water mist systems are currently used to protect the following types of target hazards:

- Gas jet fires
- Computer rooms
- Flammable and combustible liquids
- Hazardous solids, including plastic foam furnishings
- Ordinary Class A combustibles

Water mist systems produce a very fine spray of water that controls or extinguishes fire by absorbing heat, displacing oxygen, and blocking radiant heat production. In theory, the water mist raises the humidity of the room to 100 percent to halt the combustion process. An operating water mist fire protection system has a similar appearance to the water mist systems used to cool outdoor patios and eating areas in desert climates.

In general, the water mist system is composed of small-diameter, pressure-rated copper or stainless steel tubing. Very small-diameter spray nozzles are spaced evenly on the tubing (Figures 7.60 a and b). Depending on the design of the system, the spray nozzles may be of the open or closed head variety. These systems are designed to be operated at considerably higher pressures than standard sprinkler systems. There are three basic ranges in which these systems may be designed to operate:

- Low pressure systems — 175 psi (1 225 kPa) or less
- Intermediate pressure systems – 175 to 500 psi (1 225 kPa to 3 500 kPa)
- High pressure systems — 500 psi (3 500 kPa) or greater

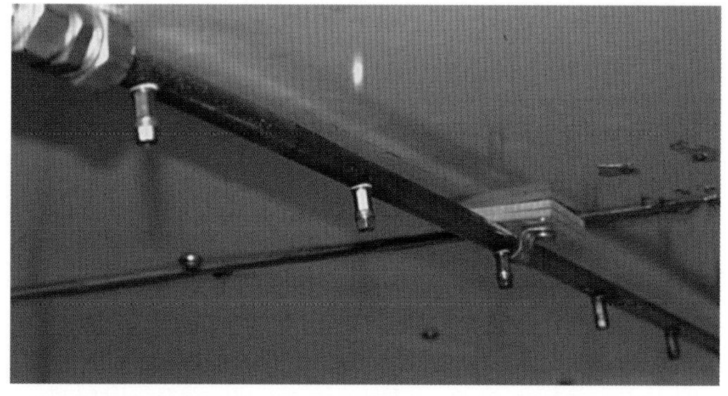

Figure 7.60a Water mist nozzles are located closer together than are traditional fire sprinklers. *Courtesy of Dan Gross, Maryland Fire & Rescue Institute.*

Figure 7.60b The water mist system emits a very fine spray of water. *Courtesy of Dan Gross, Maryland Fire & Rescue Institute.*

These high pressures are placed on the system by high-pressure water pumps or compressed air. Compressed air may be supplied to the system from storage cylinders or air pumps. Pressure may be imparted on the water through the water tubing itself (single-fluid system) or through a second tube to each spray nozzle (twin-fluid system).

The most common type of water mist fire protection system works similar to a traditional deluge sprinkler system. All of the spray nozzles in a particular room or zone are of the open head type and the system is activated by a product of combustion detection system. Generally, at least two detection devices in the affected zone or room must be activated for the deluge valve to be opened and water to be discharged from the spray nozzles. Other variations of the water mist system are also in service. These include closed head systems of the wet-pipe, dry-pipe, and preaction designs.

ACCEPTANCE TESTS FOR WATER MIST SYSTEMS

NFPA 750 lists detailed requirements for acceptance tests on new water mist systems. As with other types of sprinkler systems, the installing contractor or a representative of the building owner will be responsible for performing these tests. However, fire inspection personnel should witness these tests and should be familiar with what is required.

All underground water supply piping and lead-in connections to the system should be thoroughly flushed before they are connected. All mechanical and electrical components of the system should be visually and operationally inspected for correct installation and operation. This includes any product of combustion detection system that may be used to activate the water mist system. All piping and tubing systems should be hydrostatically tested. Low pressure systems should be able to maintain a pressure of 200 psi (1 400 kPa) for two hours. Intermediate and high pressure systems should be able to maintain 150 percent of their normal working pressure for this two-hour period. Dry-pipe and preaction systems should also be subjected to an air leakage test. These systems should be able to maintain 40 psi (280 kPa) of air pressure with no more than 1½ psi (10 kPa) of leakage for a 24-hour period. When feasible, full-scale operational tests of the system should be conducted.

INSPECTING AND SERVICE TESTING WATER MIST SYSTEMS

NFPA 750 also lists requirements for inspecting, maintaining, and service testing water mist systems. The types of tests and checks that should be performed and the frequency in which they should occur are contained in Tables 7.3 and 7.4. Systems that utilize air cylinders all require that the cylinders be hydrostatically tested on a regular basis. Empty cylinders must be tested before recharging if it has been more than five years since their last test. Cylinders that have not been discharged should be emptied and tested every 12 years.

TABLE 7.3
Water Mist System Inspection Frequencies

Item	Activity	Frequency
Water tank (unsupervised)	Check water level	Weekly
Air receiver (unsupervised)	Check air pressure	Weekly
Dedicated air compressor (unsupervised)	Check air pressure	Weekly
Water tank (supervised)	Check water level	Monthly
Air receiver (supervised)	Check air pressure	Monthly
Dedicated air compressor (supervised)	Check air pressure	Monthly
Air pressure cylinders (unsupervised)	Check pressure and indicator disk	Monthly
System operating components, including control valves (locked/unsupervised)	Inspect	Monthly
Air pressure cylinders (supervised)	Check pressure and indicator disk	Quarterly
System operating componenets, including control valves	Inspect	Quarterly
Waterflow alarm and supervisory devices	Inspect	Quarterly
Initiating devices and detectors	Inspect	Semiannually
Batteries, control panel, interface equipment	Inspect	Semiannually
System strainers and filters	Inspect	Annually
Control equipment, fiber optic cable connections	Inspect	Annually
Piping, fittings, hangers, nozzles, flexible tubing	Inspect	Annually

TABLE 7.4
Water Mist System Testing Frequencies

Item	Activity	Frequency
Pumps	Operation test (no flow)	Weekly
Compressor (dedicated)	Start	Monthly
Control equipment (functions, fuses, interfaces, primary power, remote alarm) (unsupervised)	Test	Quarterly
System main drain	Drain test	Quarterly
Remote alarm annunciation	Test	Annually
Pumps	Function test (full flow)	Annually
Batteries	Test	Semiannually
Pressure relief valve	Manually operate	Semiannually
Control equipment (functions, fuses, interfaces, primary power, remote alarm) (supervised)	Test	Annually
Water level switch	Test	Annually
Detectors (other than single use or self-testing)	Test	Annually
Release mechanisms (manual and automatic)	Test	Annually
Control unit/programmable logic control	Test	Annually
Section valve	Function Test	Annually
Water	Analysis of contents	Annually
Pressure cylinders (normally at atmospheric pressure)	Pressurize cylinder (discharge if possible)	Annually
System	Flow test	Annually
Pressure cylinders	Hydrostatic test	5-12 years
Automatic nozzles	Test (random sample)	20 years

Inspection personnel should also try to verify that the occupant or fire protection equipment company is servicing the system on a regular basis. Typical servicing functions include:

- Lubricating control valve stems
- Adjusting packing glands on valves and pumps
- Bleeding moisture and condensation from air compressors and air lines
- Cleaning strainers
- Replacing corroded or painted nozzles
- Replacing damaged or missing pipe hangers
- Replacing damaged valve seats or gaskets

Inspectors should also make sure that the required replacement components, such as extra spray nozzles, are present in sufficient quantity.

For more detailed information on sprinkler systems in general, consult the NFPA standards referenced in this section or IFSTA's **Private Fire Protection and Detection** manual.

STANDPIPE AND HOSE SYSTEMS

Standpipe and hose systems provide a means for the manual application of water on fires in large, one-story buildings or in multistory buildings. Horizontal standpipes are provided in large warehouses, factories, and shopping malls. Standpipes are required by building codes in buildings that are over four stories in height. Although a standpipe system is a necessity in a high-rise building, it cannot take the place of automatic sprinklers. Automatic sprinklers are still the most effective method of fire protection in high-rise structures.

Classification of Standpipe Systems

NFPA 14, *Standard for the Installation of Standpipe and Hose Systems*, is frequently used for the design and installation of standpipes. The standard recognizes three classes of standpipe systems: Class I, Class II, and Class III.

CLASS I STANDPIPE SYSTEMS

Class I standpipe systems are primarily for use by fire fighting personnel trained in handling large handlines (2½-inch [65 mm] hose). Class I systems must be capable of supplying effective fire streams during the more advanced stages of fire within a building or for fighting a fire in an adjacent building. Class I systems have 2½-inch (65 mm) hose connections or hose stations attached to the standpipe riser (Figure 7.61). The 2½-inch (65 mm) hose connections may be equipped with a reducer on the cap that allows for the connection of a 1½-inch (38 mm) coupling as well.

Figure 7.61 A Class I standpipe connection.

CLASS II STANDPIPE SYSTEMS

The Class II system is primarily designed for use by building occupants who presumably have no specialized fire training. These systems are limited to 1½-inch (38 mm) hose (Figure 7.62). This hose is typical of the single-jacket linen variety and is equipped with a lightweight, twist-type shutoff nozzle. There is some disagreement over the value of Class II standpipe and hose systems. The presence of the small hose may give a false sense of security to building occupants and create the impression that they should attempt to fight a fire when their safer course would be to flee. In any case, Class II hose on a standpipe cannot be depended on for fire control operations. (**CAUTION:** Fire department personnel should not use the supplied hose and nozzle for fire attacks. Fire personnel should carry their own hose packs into the structure, disconnect the house line, and connect their own equipment.)

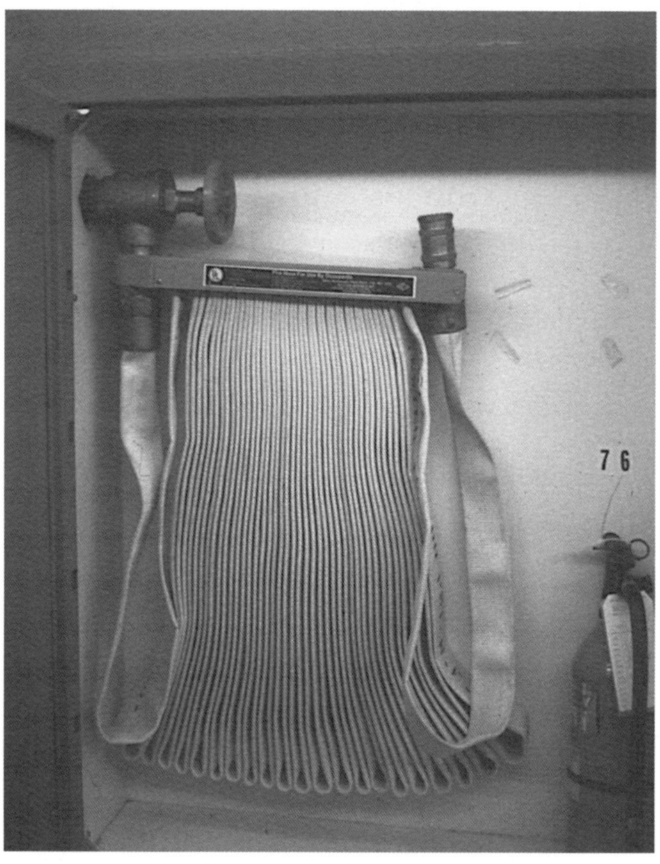

Figure 7.62 A Class II standpipe features only 1½-inch (38 mm) hose intended for the occupant's use.

CLASS III STANDPIPE SYSTEMS

Class III standpipes combine the features of Class I and Class II systems. Class III systems have both 2½-inch (65 mm) hose connections for fire department personnel and 1½-inch (38 mm) hose and connections for use by the building occupants (Figure 7.63). The design of the system must

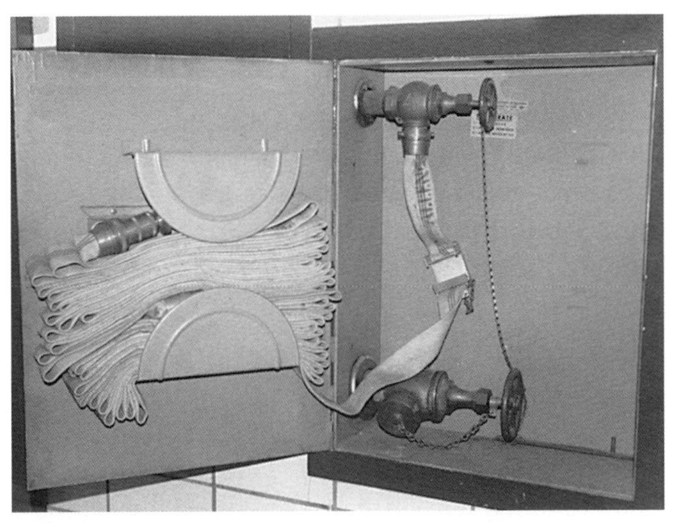

Figure 7.63 Class III standpipes have hose for the occupants, plus connections for fire department hoselines.

allow both the Class I and Class II services to be used simultaneously.

Water Supply for Standpipe Systems

The amount of water required for standpipe systems depends on the size and number of fire streams that are needed and the probable length of time the standpipe will be used. These factors are influenced by the size and occupancy of the building.

The water supply for Class I and Class III standpipe systems must provide 500 gpm (1 893 L/min) for at least 30 minutes, with a residual pressure of 100 psi (690 kPa) at the most hydraulically remote 2½-inch (65 mm) outlet. A minimum of 65 psi (448 kPa) is required for the most remote 1½-inch (38 mm) outlet. If more than one standpipe riser is needed to protect a building, the water supply must provide 250 gpm (946 L/min) for each additional riser to a maximum of 2,500 gpm (9 463 L/min). For a Class II standpipe, 100 gpm (378 L/min) must be provided for at least 30 minutes, with a residual pressure of at least 65 psi (448 kPa) at the highest outlet.

The current NFPA 14 minimum requirement for residual pressure is 65 psi (448 kPa). However, this is a minimum and may not be adequate to supply a fog nozzle on the end of a 100-foot (30 m) hose connected to the topmost hose outlet. Because of this, some other building codes and fire codes require higher minimum residual pressures. For example, the Uniform Building Code requires 100 psi (640 kPa). Consult the code used in your jurisdiction for minimum requirements in your occupancies.

In addition to supplying water for hose streams, many standpipe risers are also used to supply water for the sprinklers in high-rise buildings. In a fully sprinklered building, it is assumed that the sprinklers will act to extinguish or control an incipient fire and thereby reduce the water required for hoselines. Therefore, when determining the required water supply for a standpipe system in a fully sprinklered building, it is not necessary to add the sprinkler water demand to the water supply requirements for the standpipe. If, however, the sprinkler water demand is greater than the standpipe demand, the water supply must be adequate to meet the greater requirements of the sprinkler system.

The water supply for standpipes may be taken from several different sources such as public water supplies, automatic fire pumps, manual fire pumps, pressure tanks, and gravity tanks. Not all of these water sources are practical in every situation. In high-rise buildings, the water is usually supplied from automatic fire pumps that take suction from a municipal water main. Water supplies can be used in combination; for example, it is possible to incorporate both a tank supply and an automatic fire pump in the supply for a high-rise building.

Types of Standpipe Systems

In addition to the classes of standpipes, there are four types of standpipe systems. They are as follows:

- Wet standpipe system that has the water supply valve open and pressure maintained in the system at all times. When the hose valve is opened, water is immediately available.
- Dry standpipe in which water is admitted to the system by the operation of a valve controlled by an electrical switch or other device located at each hose station. Water is not available at the hose stations until the supply valve is opened from the control device.

- Dry standpipe that has air under pressure and admits water to the system automatically when a hose valve is opened through the use of a dry-pipe valve.
- Dry standpipe that has no permanent water supply and must be supplied totally from the fire department connection.

Some wet standpipe systems have a limited water supply that keeps the system riser full; these systems are known as "primed systems." Adequate water for fire fighting is not available until the fire department connects to the fire department connection or until a manual fire pump is started. The advantages of a primed system are as follows:

- Reduction of the time for water to reach the hose valves
- Reduction in water hammer
- Reduction in corrosive effects on the internal surfaces of the standpipe

A wet standpipe with an automatic water supply is the most desirable type of standpipe. With this type of system, water is constantly available at the hose station. However, wet standpipe systems cannot be used in cold environments, and a dry system may have to be used. Automatic dry systems have the disadvantages of greater cost and maintenance requirements. The main advantage of a dry standpipe with no permanent water supply is the reduction in cost.

Standpipes in High-Rise Buildings

The size of the standpipe riser is determined by the height of the building and the class of service. For Class I and Class III service, the minimum riser is 4 inches (100 mm) for building heights less than 100 feet (30 m) and 6 inches (150 mm) for heights over 100 feet (30 m). When a Class I or Class III standpipe exceeds 100 feet (30 m) in height, the top 100 feet (30 m) is allowed to be 4-inch (100 mm) pipe. Standpipes can also be sized hydraulically to provide the minimum required pressure at the topmost outlet.

For Class II service, a riser could be 2 inches (50 mm) for a building height less than 50 feet (15 m). For a building over 50 feet (15 m) in height, the minimum size riser is 2½ inches (65 mm). Class II systems in buildings over 275 feet (84 m) in height must be divided into sections.

In buildings with combined standpipe and sprinkler systems, the minimum riser size is 6 inches (150 mm). However, this requirement may be disregarded if the building is completely sprinklered and the system is hydraulically calculated to ensure that all water supply requirements can be met.

Current practice is to locate standpipes so that any part of a floor is within 130 feet (40 m) of the standpipe hose connection. This allows any fire to be reached with 100 feet of hose (30 m), plus a 30-foot (10 m) fire stream. Standpipes and their connections are most commonly located within noncombustible fire-rated stair enclosures so that firefighters have a protected point from which to begin an attack (Figure 7.64). If the building is so large that the standpipes located in the stairwells cannot provide coverage to the entire floor, additional stations or risers must be provided.

Figures 7.64 Fire department connections in high-rise structures are generally found in the stairwells.

The actual hose connections can be located no more than 6 feet (1.8 m) from floor level. These connections must be plainly visible and must not be obstructed. Any caps over the connections must be easy to remove.

Buildings equipped with Class I or III systems may be required to have a 2½-inch (65 mm) outlet on the roof. This is required when any of the following three situations are present:

- The building has a combustible roof.
- The building has a combustible structure or equipment on the roof.

- The building has exposures that present a fire hazard.

Pressure-Regulating Devices

Where the discharge pressure at a hose outlet exceeds 100 psi (682.5 kPa), NFPA 14 requires a pressure-regulating, or restricting, device to limit the pressure to 100 psi (682.5 kPa), unless otherwise approved by the fire department. The use of a pressure-regulating device prevents pressures that make hose difficult or dangerous to handle. This device also enhances system reliability because it extends individual zones to greater heights. In some instances, it may improve system economy because its use may eliminate some pumps. However, pressure-regulating devices make the system design more complex.

There are several different types of pressure-reducing devices. One type consists of a simple restricting orifice inserted in the waterway. The pressure drop through the orifice plate depends on the orifice diameter and the flow. The individual restricting orifice must be sized for different applications and will not be the same for each floor of a given building.

Another type of pressure-regulating device may consist of vanes in the waterway that can be rotated to change the cross-sectional area through which the water flows. A pressure-regulating device may also take the form of a pressure-reducing valve (Figure 7.65). There are several different pressure-reducing valves available from various manufacturers. Some of the valves are field adjustable, and others are set at the factory.

A pressure-regulating device must be specified and/or adjusted to meet the pressure and flow requirements of the individual installation. For factory-set devices, the pressure-regulating device must be installed on the proper hose outlet to ensure proper installation. When field-adjustable devices are installed, the manufacturer's instructions on making adjustments must be followed carefully. If a pressure-regulating device is not properly installed or is not properly adjusted for the required inlet pressure, outlet pressure, and flow, the available flow may be greatly reduced and fire fighting capabilities seriously impaired.

Fire Department Connections

Each Class I or Class III standpipe system requires one or more fire department connections through which a fire department pumper can supply water into the system (Figure 7.66). High-rise buildings having two or more zones require a fire department connection for each zone.

In high-rise buildings with multiple zones, the upper zones may be beyond the height to which a fire department pumper can effectively supply water. This height would be around 450 feet (137 m), depending on available hydrant pressure and other factors. For standpipe system zones beyond that height, a fire department connection is of no value, unless the fire department is equipped with special high-pressure pumpers and the system has high-pressure piping.

Standard requirements specify that there shall be no shutoff valve between the fire department connection and the standpipe riser. In multiple-

Figure 7.65 Some standpipe connections are equipped with pressure-reducing valves.

Figure 7.66 Standpipe fire department connections should be clearly marked.

riser systems, however, gate valves are provided at the base of the individual risers.

The hose connections to the fire department connection must be female and equipped with standard caps (Figure 7.67). Some jurisdictions require Storz-type (sexless) couplings that allow large diameter hose to be used to supply standpipes. It is important that the hose coupling threads conform to those used by the local fire department. The fire department connection must be designated by a raised-letter sign on a plate or fitting reading "STANDPIPE." If the fire department connection does not service the entire building, the sign must indicate which floors are serviced (Figure 7.68).

Figure 7.67 Some type of cap or cover should be provided for the fire department connection inlets.

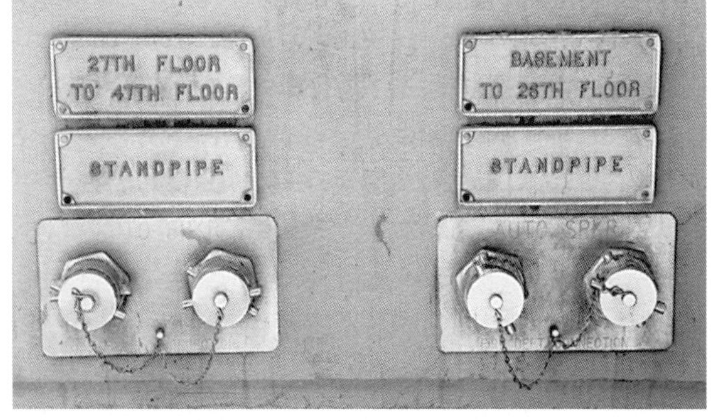

Figure 7.68 The appropriate fire department connections should be clearly marked for zoned standpipe systems.

Inspecting and Testing Standpipes

In order to ensure compliance with codes and operability, standpipes should be inspected when they are first installed and then periodically thereafter. The following sections highlight the inspection and test procedures. Appendix F contains sample standpipe inspections and testing forms.

INITIAL INSTALLATION INSPECTION AND TESTS

A standpipe system is a significant component in a building's design. Before system installation, detailed design plans should be submitted to the local fire department or building department in accordance with provisions of the local codes. As construction proceeds, the installation should be checked for conformity with the plans. In a high-rise building, it will be necessary to have the standpipe in partial operation as construction proceeds. It will provide protection to the structure should fire occur on the upper levels during construction.

When the installation is complete, the following tests and inspections should be performed:

- The system should be hydrostatically tested at a pressure of at least 200 psi (1 379 kPa) for two hours to ensure the tightness and integrity of fittings. If the normal operating pressure is greater than 150 psi (1 034 kPa), the system should be tested at 50 psi (345 kPa) greater than its normal pressure.
- The system should be flushed and flow tested to remove any construction debris and to ensure that there are no obstructions. This will also assure that the system is capable of flowing the required amount of water at the required minimum pressure.
- On systems equipped with an automatic fire pump, a flow test should be performed at the highest outlet to ensure that the fire pump will start when the hose valve is opened.
- The fire pump should be tested to ensure that it will deliver its rated flow and pressure.
- All devices used should be inspected to ensure that they are listed by a nationally recognized testing laboratory.
- Hose stations and connections should be checked to ensure that they are in cabinets within 6 feet (1.8 m) from the floor and are positioned so that the hose can be attached to the valve without kinking (Figure 7.69).
- Each hose cabinet or closet should be inspected for a conspicuous sign that reads

Figure 7.69 Standpipe connections should be within 6 feet (1.8 m) of floor level.

"FIRE HOSE" and/or "FIRE HOSE FOR USE BY OCCUPANTS OF BUILDING."

- Fire department connections should be checked for the proper fire department thread and for a sign indicating "STANDPIPE" with a list of the floors served by that connection.
- When a dry standpipe is used, check for a sign indicating "DRY STANDPIPE FOR FIRE DEPARTMENT USE ONLY."

IN-SERVICE INSPECTIONS

As with all fire protection systems, standpipe systems need to be inspected and tested at regular intervals. A visual inspection should be made at least monthly by the building management. Because interior fire fighting is dependent on the standpipe system, the fire department must also inspect standpipes at regular intervals. The actual testing of standpipes should be carried out by a fire protection contractor or the building operating staff if they are sufficiently knowledgeable. To avoid potential liability, it is prudent that fire department personnel not perform actual tests. However, fire department personnel should witness tests.

Fire department personnel should reinspect standpipe systems for the following:

- All water supply valves are sealed in the open position.
- Power is available to the fire pump (in other words, the pump is in running condition).
- Individual hose valves are free of paint, corrosion, and other impediments.
- Hose valve threads are not damaged (Figure 7.70).
- Hose valve wheels are present and not damaged.
- Hose cabinets are accessible.
- Hose is in good condition, is dry, and is properly positioned on the rack or reel (Figure 7.71).

Figure 7.70 Check the standpipe hose threads for damage.

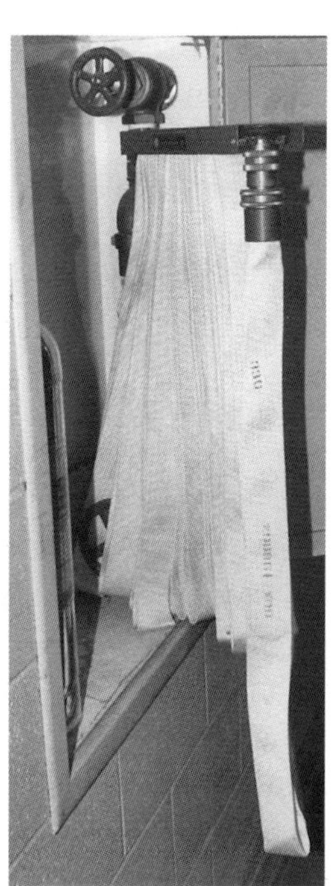

Figure 7.71 Make sure the hose is present and in good condition.

- Discharge outlets in dry systems are closed.
- Dry standpipe is drained of moisture.
- Access to the fire department connection is not blocked.
- The fire department connection is free of obstruction, the swivels rotate freely, and the caps are in place.
- Water supply tanks are filled to the proper level.
- If the system is equipped with pressure-regulating devices, those devices are tested as required by the manufacturer.
- Dry systems are hydrostatically tested every five years.

For more detailed information on standpipe and hose systems, consult NFPA 14 or IFSTA's **Private Fire Protection and Detection** manual.

FIRE PUMPS

Fire inspection personnel must be familiar with the fixed fire pumps that will be encountered in many commercial, institutional, and industrial facilities (Figure 7.72). The main function of a fire pump is to increase the pressure of the water that flows through it. Usually, a fire pump is needed to supply a sprinkler or standpipe system because the available water supply source, such as an elevated tank or ground storage tank, does not have adequate pressure to meet the fire suppression system demand. Water is available to a fire pump from sources such as municipal water mains, wells, storage tanks, and reservoirs. All fire pumps and their installations must meet the requirements set forth in NFPA 20, *Standard for the Installation of Centrifugal Fire Pumps.*

Types of Fire Pumps

There are three major types of stationary fire pumps used in fire protection systems. These pumps include the horizontal shaft, vertical shaft, and vertical turbine.

HORIZONTAL SHAFT PUMP

The horizontal shaft pump is the most common type of fire pump found in fixed fire suppression systems. This pump (sometimes referred to as a "horizontal split-case pump") has its shaft in a horizontal plane, with the pump on one end of the shaft and the driver (motor) on the other (Figure 7.73). This pump is used to boost the pressure from an incoming, pressurized water source such as a municipal water main. Because it is not a self-priming pump (it will not draft water from a static supply source into the pump on its own), a horizontal shaft pump cannot be used to supply water from a static supply source.

The most common gallonage rating for horizontal shaft pumps in use today are those in the 500 to 1,500 gpm (2 000 L/min to 6 000 L/min) range. However, pumps are available for flows as low as 150 gpm (600 L/min) to flows as high as 4,500 gpm (18 000 L/min). Unlike pumps on fire apparatus, fixed fire pumps are not rated at a particular pressure. Single-stage, horizontal shaft pumps are available with pressure ratings from as low as 40 psi (280 kPa) to as high as 290 psi (2 030 kPa).

Figure 7.72 A typical diesel motor driven fixed fire pump installation. *Courtesy of Conoco Oil Co., Inc.*

Figure 7.73 A horizontal shaft fire pump.

Firefighters and inspectors also need to be aware that in many cases fire pump pressure ratings or capabilities are often expressed in terms of feet of head rather than psi or kPa. A column of water 2.304 feet (0.7 m) high creates 1 psi (exactly 6.895 kPa) of pressure. Therefore, a pump rated at 231 feet (70.4 m) of head will be equal to one rated at 100 psi (689.5 kPa) (Figure 7.74). (**NOTE:** For ease of instruction, IFSTA generally uses a 1 psi = 7 kPa ratio when talking about pressure.)

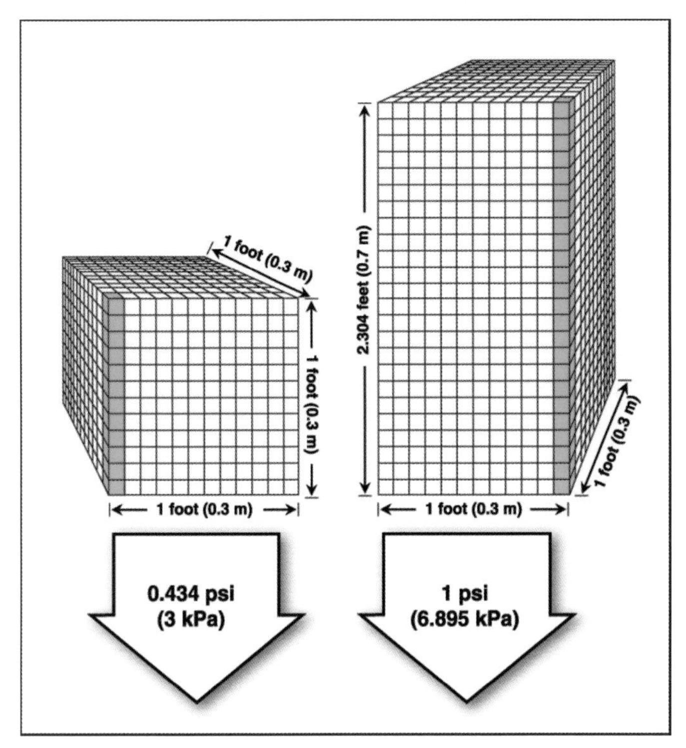

Figure 7.74 A one-foot (0.3 m) column of water will exert 0.434 psi (3 kPa) of pressure. Thus, it takes a 2.304 foot (0.7 m) column of water to develop 1.0 psi (6.895 kPa) of pressure.

Figure 7.75 A vertical shaft fire pump.

VERTICAL SHAFT PUMP

The vertical shaft pump is very similar to the horizontal shaft pump except that the impeller shaft runs vertically. This pump is always driven by an electric motor that sits on top of the pump (Figure 7.75). The main advantage of this pump is its compactness. Shaft bearings and the assembly frame are designed to support the weight of the unit in a vertical direction. The capabilities of these pumps are the same as those described for the horizontal shaft pump.

VERTICAL TURBINE PUMP

The vertical turbine pump is a very useful design when it is necessary to lift water from a source below the pump. This is sometimes the case when a below-grade reservoir is used as the water supply. In fact, vertical turbine pumps are commonly used as well pumps in nonfire protection applications.

The vertical turbine pump impellers are actually located within the water supply source (Figure 7.76). Water is drawn into the impeller and then discharged up through the impeller casing. Most of these pumps are multistage pumps. As the water exits one impeller, it enters the next, and so on until it is discharged into the fire suppression system piping.

The volume capabilities of vertical turbine pumps are consistent with horizontal and vertical shaft pumps. However, vertical turbine pumps are available with discharge pressure ratings of up to 500 psi (3 500 kPa).

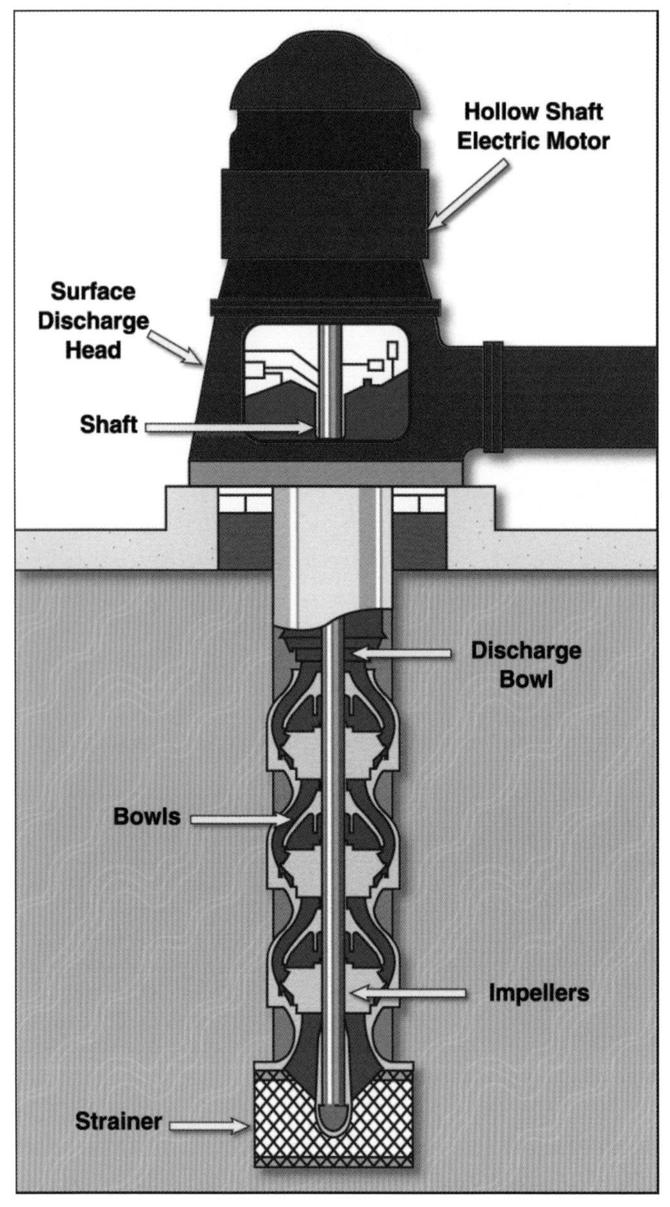

Figure 7.76 This diagram highlights the principal parts of a vertical turbine fire pump assembly.

Fire Pump Drivers

The source of power that operates the fire pump is called the *driver*. Fire pumps are commonly powered by one of three types of drivers: electric motor, diesel engine, or steam turbine. Other types of engine drivers, such as gasoline, natural gas, and liquefied petroleum, have been used in the past but are not currently recognized in NFPA codes.

ELECTRIC MOTOR DRIVER

An electric motor is the most common method for driving a fire pump (Figure 7.77). It is simple, reliable, and easily maintained. Electric motors used on fire pumps are not designed specifically for

Figure 7.77 Electric motors are used to drive some fire pumps.

that purpose, or for any other specific purpose for that matter. However, all electric motors must meet the requirements of the National Electrical Manufacturers Association (NEMA). The motor must have adequate horsepower to drive the fire pump. The required pump horsepower is determined by the pump capacity (gpm), the net pressure (discharge pressure minus the incoming pressure), and the pump efficiency. For a 1,000 gpm (4 000 L/min) pump rated at 100 psi (700 kPa), a motor of about 80 hp would be needed. Electric motors powerful enough to power fire pumps use a large amount of electricity and may require a larger electrical service to the building than would be needed otherwise.

DIESEL ENGINE DRIVER

A diesel engine is generally more expensive and requires more maintenance than an electric motor (Figure 7.78). A diesel engine is used in situations where a driver independent of the local electrical power supply is needed. However, this engine is more complex and requires an on-site fuel supply. Batteries are also required for starting the engine. A diesel engine is required to be tested weekly, and during testing it must run for at least 30 minutes.

Unlike an electric motor, the diesel engine used for fire pumps is tested and listed by testing laboratories. The diesel engine is required by testing agencies to be equipped with overspeed shutdown devices, tachometers, oil pressure gauges, and temperature gauges. The engine is required to have a closed-circuit-type cooling system. A fuel supply is required to provide at least 1 gallon (3.78 L) per

Figure 7.78 A diesel-driven fire pump.

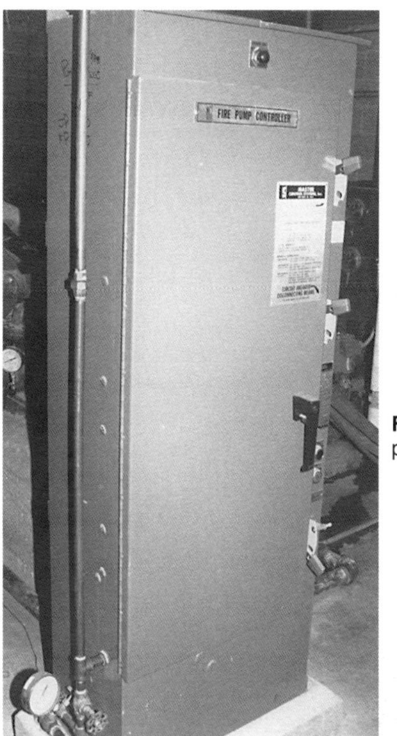

Figure 7.79a An electric fire pump controller.

horsepower. For example, a 185 hp motor requires a 185-gallon (740 L) fuel tank. Diesel engines that are contained in a room or other enclosure must also have an adequate flow of air through the room to ensure proper combustion and removal of exhaust fumes.

STEAM TURBINE DRIVER

Although permissible by NFPA 20, a steam-turbine-driven fire pump is not common. Its only feasible application is at such locations as electric generating stations, where large quantities of steam are already present.

Fire Pump Controllers

To a certain extent, the stationary fire pump used in fire suppression systems is more simple than those used on fire apparatus. However, a stationary fire pump differs in at least one fundamental respect: It is automatic in operation. A fire pump starts automatically whenever the fire suppression system it supplies operates, and it is frequently designed to stop automatically. This action is accomplished through the fire pump controller.

ELECTRIC MOTOR CONTROLLER

Most fire pumps are designed to start when a drop in pressure occurs in the fire suppression system (Figures 7.79 a and b). A pressure-sensing switch within the electric motor controller detects the drop in pressure resulting from the flow of water (Figure 7.80). The pressure switch then energizes a circuit that closes the contacts

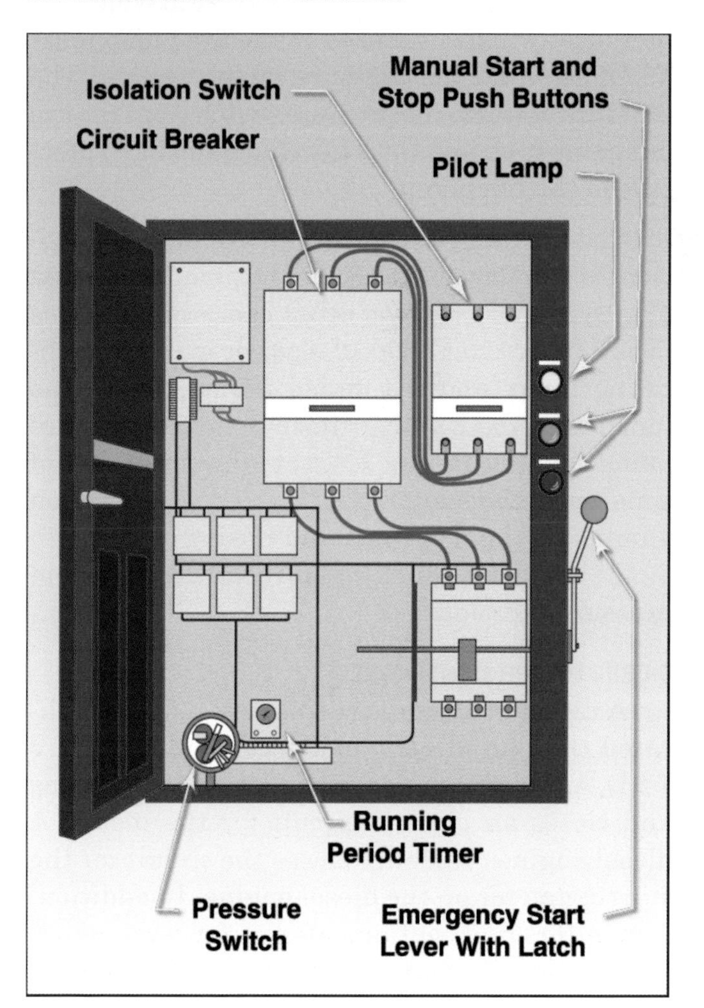

Figure 7.79b All electric fire pump controllers have the same basic features.

for the motor circuit and starts the fire pump motor. When the water stops flowing in the system, the resulting increase in pressure is detected by the pressure switch and the motor circuit is interrupted, thus shutting down the pump.

Figure 7.80 The pressure sensing switch determines when it is necessary to start the pump.

A pressure switch is adjustable; it must be properly adjusted for the individual fire suppression system. It is possible for a pressure switch to be set improperly. This may result in the pump not starting when needed, or it may not shut off when the system is shut down. The pressure at which the pressure switch is set to start the fire pump must be *higher* than the pressure in the system. The pressure at which the pressure switch stops the fire pump must be less than the churn (no flow) pressure of the fire pump.

The fire pump controller also contains a provision for starting and stopping the pump manually. The controller contains other operating features including a circuit breaker, a power-available indicating lamp, and a running period timer. The function of the running timer is to keep the fire pump motor running for a minimum period of time once the motor has started. This action eliminates rapid opening and closing of the main motor contacts, which could result from system pressure fluctuations.

DIESEL MOTOR CONTROLLER

A diesel fire pump controller is more complicated than an electric motor controller (Figure 7.81). An electric motor controller basically opens and closes an electric circuit for the motor. A diesel engine controller closes the circuit for the starting motor on the diesel engine. In addition, it monitors and contains alarms for low engine-oil pressure, high engine-coolant temperature, failure to start, engine overspeed shutdown, and battery failure.

Figure 7.81 A diesel fire pump controller.

Testing Fire Pump Installations

The following sections look at the testing of new and existing fire pump installations. It is critical that fire pumps be operated often and regularly. Fire protection literature is filled with accounts of fire pumps failing to operate when needed.

The results of pump failure are most often catastrophic. The way to prevent this from happening is to continually make sure that pump installations are in good operating condition. The recommended frequency for pump operation is weekly. Pumps should be operated from automatic starts, if so equipped, and brought up to full speed while pumping a substantial stream. Pump tests are most commonly performed by building maintenance personnel or occupant-employed fire protection specialists. Depending on local code requirements, fire department personnel may be required to witness the test on a periodic basis. Fire department personnel should never actually operate the fire pump.

PUMP TEST MANIFOLDS

Every pump is required to be provided with components that allow testing of the installation. The most prominent of these components is the test manifold (Figure 7.82). The test manifold piping should be connected to the pump discharge line between the check valve and the indicating control valve (Figure 7.83). There should also be an indicating control valve in the test manifold piping. This piping should terminate in a hose valve header located outside the pump room. The hose valve header should be equipped with 2½-inch (65 mm) hose connections with a shutoff valve for each connection (Figure 7.84). This manifold and test header permits flowing water from the pump installation through hoselines and nozzles for test purposes. The flow from the nozzles is measured using pitot tubes.

Figure 7.82 A common pump test manifold.

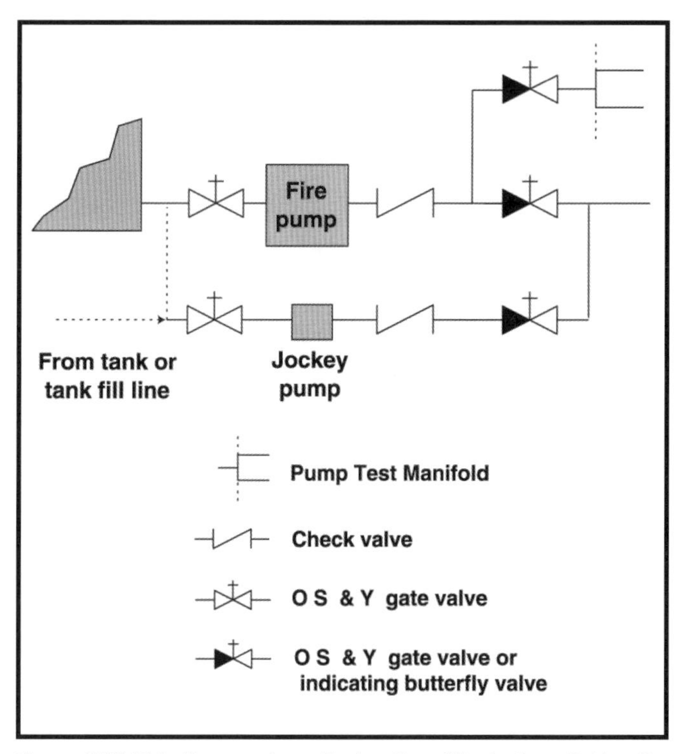

Figure 7.83 This diagram shows the location of the test manifold on the fire pump system.

Figure 7.84 Each discharge is equipped with its own valve.

The required size of the hose header supply and the number of hose valves required depend upon the rating of the pump. It is often possible to estimate the rating of the pump by counting the hose connections. For the 500, 750, and 1,000 gpm (2 000 L/min, 3 000 L/min, 4 000 L/min) pumps, there will be one 2½-inch hose connection for each 250 gpm (1 000 L/min) of pump rating. For example, a 500 gpm (2 000 L/min) pump usually has two hose connections, and a 750 gpm (3 000 L/min) pump usually has three hose connections. This method is not foolproof, however, as all pumps 1,250 gpm (5 000 L/min) and larger have six hose connections.

In some jurisdictions, more recent pump installations will not have the test headers and hose valves. Instead, they are equipped with a flowmeter that can be used to measure the gpm delivered by the pump. This is acceptable, but the meter line should discharge to a drain or back to the tank if a tank is used for the supply source.

TESTING PUMP INSTALLATIONS

Both the underground and aboveground piping must be hydrostatically tested in accordance with NFPA 24, *Standard for the Installation of Private Fire Service Mains and Their Appurtenances*, and NFPA 13 respectively. Both overhead piping and underground piping are required to be hydrostatically pressurized for 2 hours to 200 psi (1 379 kPa) or 50 psi (345 kPa) above the maximum static pressure, whichever is larger. For the overhead piping, any leakage at all constitutes failure. The underground pipe is permitted to leak a little (just a few quarts per hour) depending on the length of pipe and the number and type of valves and gaskets.

The underground pipe is also required to be flushed out before connection to the fire protection system piping. The reason for this is to blow out any accumulated debris in the piping. As might be expected, the required flow rate for flushing depends on the diameter of the underground pipe. If foreign materials are not flushed from the piping before connection to the fire protection system, these materials will end up inside the system piping. This can have a serious impact on the effectiveness of the fire protection system.

Before a pump installation is accepted from the installing contractor, the installation should be tested under the specifications of NFPA 20, *Standard for the Installation of Centrifugal Fire Pumps*. The installation should not be accepted if it cannot meet the standard specifications. Before a pump is shipped by the manufacturer, it is tested in the shop. The results of this test are plotted on graph paper. These plotted curves are called the *certified shop test curves* for the pump (Figure 7.85). The features of performance plotted are net pressure versus gpm delivered, and horsepower delivered versus gpm delivered. These characteristic curves are the manufacturer's guarantee of the new pump's capabilities. An important requirement of the acceptance test is that the pump operate at least as well as the pump characteristic curves.

The pump being tested must also meet the following three standard performance points:

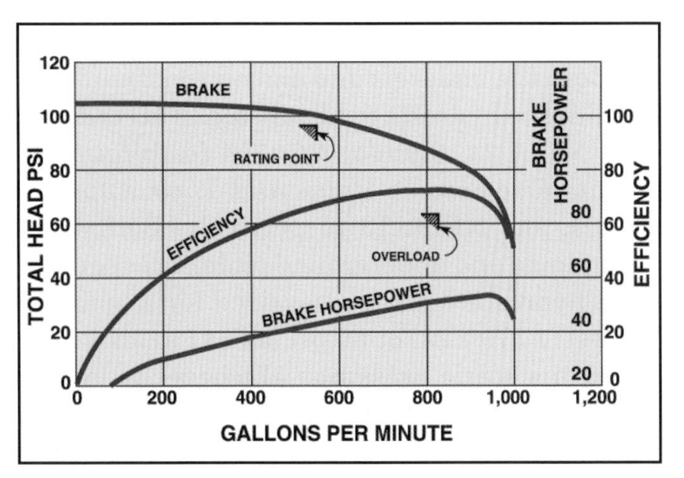

Figure 7.85 A typical fire pump shop curve.

- At shutoff, not more than 140 percent of the rated net pressure may be developed.
- It must develop at least the rated net pressure while delivering the rated flow.
- It must develop at least 65 percent of the rated net pressure while delivering 150 percent of the rated flow.

EQUIPMENT NEEDED FOR PUMP TESTS

In addition to the pump installation itself, the following is a list of basic equipment needed to conduct the test:

- One section of 2½-inch (65 mm) or larger hose for each hose connection on the test header, if it is not possible to connect the nozzles directly to the test header
- One Underwriters playpipe for each hoseline (Figure 7.86)
- Method for safely securing playpipes (Figure 7.87)
- Pitot tube and gauge (Figure 7.88)
- Method for measuring pump speed
- Voltmeter
- Ammeter

Figure 7.87 Some type of secure holder must be used to anchor the playpipes.

Figure 7.88 A pitot tube and gauge is used to measure flow (velocity) pressure.

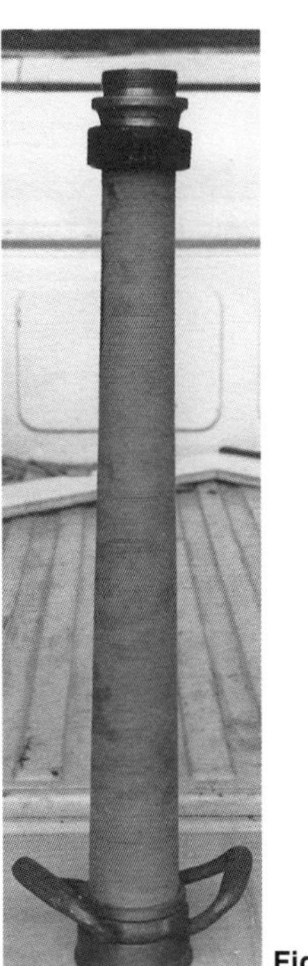

Figure 7.86 Underwriter's playpipes are used for the fire pump test.

If the system is equipped with a flow metering device, the first four items listed are not needed. Otherwise, at least one 50-foot (15 m) section of hose for each hose connection on the test header is needed. This hose must have 2½-inch (65 mm) couplings with threads compatible with the hose connections in the header and be at least 2½ inches (65 mm) in diameter.

There should be an Underwriters playpipe for each hoseline (Figure 7.89). It is possible to use a deluge gun where several hoselines supply a single nozzle. However, the pressure lost in such devices may limit the flow capacity to less than could be obtained with separate playpipes. If a deluge gun is used, it must be equipped with a straight stream nozzle.

Figure 7.89 Each hoseline is attached to a playpipe.

It is essential that the playpipes be firmly secured in place (Figure 7.90). The nozzle reaction on a playpipe connected to a 2½-inch (65 mm) line is tremendous. This reaction can actually lift a person off his feet. A 2½-inch (65 mm) hoseline cannot be safely handled by inexperienced and untrained personnel. It is therefore necessary to provide a means of securing playpipes and hoselines. The playpipes can be tied to some substantial structure, or special racks can be constructed.

Figure 7.90 In this case the playpipe holder is anchored by having a pickup truck parked on it.

WARNING

Playpipes MUST be firmly secured in place. Playpipes and hose that break loose during testing could strike people standing in the area and cause serious injury or death.

The purpose of the hoselines and playpipes is to allow measurement of the flow in gpm by means of a pitot tube and gauge. Pitot tubes come in all shapes and sizes. A model with an air chamber for a handle or with a liquid-filled gauge reduces needle vibration and gives more accurate readings. For pump testing, gauges calibrated to at least 100 psi (700 kPa) are generally needed. The blade of the pitot tube is inserted into the flowing stream at a point one-half of the nozzle diameter away from the nozzle. The reading registered on the gauge is the velocity pressure of the stream (Figure 7.91). This velocity pressure is then converted to gpm by use of the following formula:

$$Q = (29.83)(C_d)(D^2)(\sqrt{P})$$

Where: Q = Flow in gpm

C_d = Coefficient of discharge

D = Discharge orifice diameter

P = Nozzle Pressure

Figure 7.92 shows the common coefficient of discharges that might be used.

The rating of a pump is always for a specific pump speed. For example, a pump may be rated at 1,000 gpm and 60 psi at 1,770 revolutions per minute (rpm). The performance of such a pump will

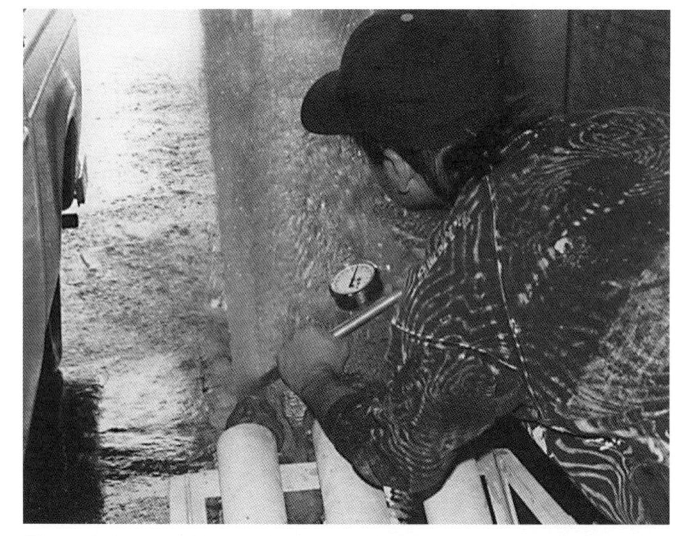

Figure 7.91 Insert the pitot tube into the stream to record the pressure.

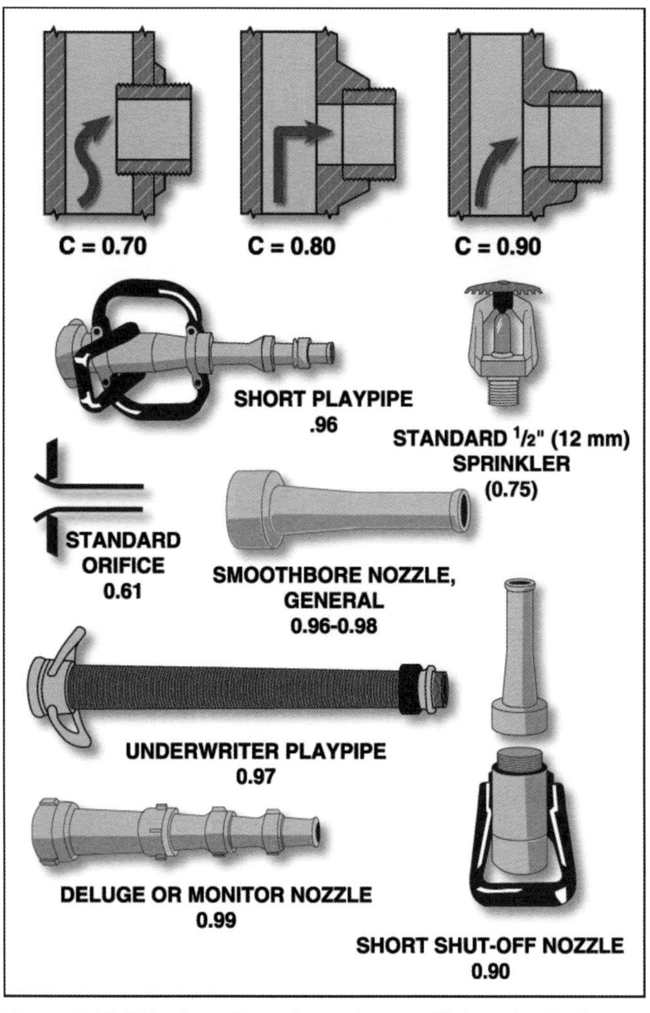

Figure 7.92 This chart shows the various coefficients for discharges that might be used for fire pump or hydrant testing.

be certified at the speed of 1,770 rpm. If the electric motor or diesel engine cannot turn the pump this fast, the pump cannot be expected to meet the performance specifications. For this reason, some means of measuring the rpm must be available. This can be accomplished by using either a hand-held revolution counter, a strobe-type tachometer, or the more modern digital tachometer. A voltmeter and an ammeter are needed for testing an electric-motor-driven pump.

ACCEPTANCE TEST ON AN ELECTRIC HORIZONTAL SPLIT-CASE PUMP

The following are the 10 steps required to perform an acceptance test on an electric horizontal split-case pump.

Step 1: Calculate the expected pitot pressure for 100 percent and 150 percent of the rated flow by algebraically manipulating the previous equation as follows:

$$P = \{Q/([29.83][C_d][D^2])\}^2$$

Step 2: Connect all the hoselines and nozzles. Make sure that all nozzles are securely fastened in place. All hose valves should be closed, and the control valve in the pipe to the test header should be closed (Figure 7.93).

Step 3: Close off the indicating control valve that separates the pump from the fire system (Figure 7.94). This will allow testing of the pump without subjecting the system piping to possible water hammers.

Step 4: Connect the ammeter and voltmeter to the test leads in the controller or at any other appropriate location. (**CAUTION:** Only experienced electricians should work with the wiring of an electrically driven pump.)

Figure 7.93 Connect all the hoses and nozzles.

Figure 7.94 Close the valve that leads to the sprinkler/standpipe piping system.

Step 5: Remove the end plate of the motor for access to the shaft if a handheld revolution counter is being used to measure pump speed. Because both the motor and the pump will be rotating at the same speed, measuring the speed of the motor also gives the speed of the pump. (**CAUTION:** Proper attention to safety is critical with respect to any rotating equipment. Do not wear loose-fitting clothing or ties.)

Step 6: Use a a strobe-type tachometer if the end of the shaft is not accessible. To establish the speed of the pump using the strobe-type tachometer, mark the shaft with a piece of chalk, and adjust the strobe impulse until the rotating chalk mark appears to be standing still. The pump speed can then be read from the tachometer dial. Several modern styles of digital tachometers are available. One style uses a small strip of reflective tape, which is placed on the shaft. The handheld sensor counts the rate of reflections as the shaft rotates.

Step 7: Start the pump when everything is ready. Initially, the pump should be operating against a closed system with no valves open and with no water flowing. This is the churn or shutoff phase. The pump can be started manually, or it can be started automatically (if equipped to do so) by bleeding off the water pressure. Once the pump is operating, both the suction and discharge pressures should be read and recorded, the rpm measured, and voltage and current readings taken. While the pump is operating at churn, the circulation relief valve should have opened automatically and be flowing a solid stream of water. If water is not flowing, the relief valve can be adjusted with a wrench until water begins to flow.

Step 8: Open the control valve in the line leading to the test header, and open the hose valves for the first gpm measurement.

Step 9: Open and adjust sufficient lines so that the exact required pitot pressures for 100 percent of the rated flow are read on the pitot gauge (Figure 7.95). While the velocity pressures are being measured outside, the rpm, voltage, current, discharge, and suction pressures are to be measured inside the pump room. When the first line is opened, the circulation relief valve should again be checked. When the first line was opened, the relief valve should have closed. If it continues to flow a solid stream, further adjustment is needed.

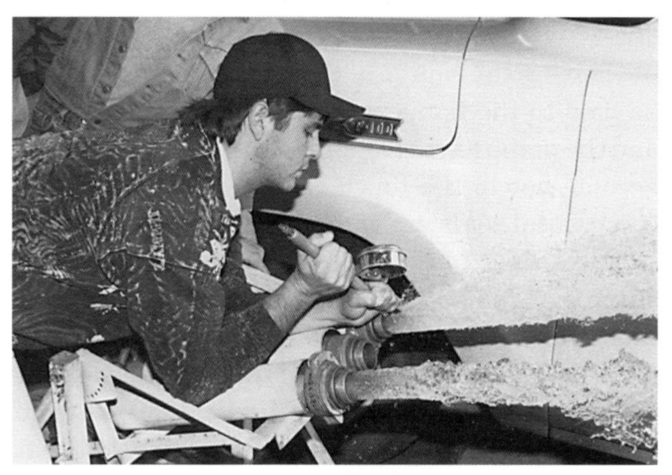

Figure 7.95 Record the pressure from each hoseline.

Step 10: Open additional hoselines when all readings are complete and recorded. Adjust these hoselines to the exact required pitot pressures for 150 percent of the rated flow. It will usually require flowing all of the hoselines to achieve 150 percent of the rated flow. It is important to calculate ahead of time what pitot pressure is needed at each nozzle to give both the 100 percent and the 150 percent points. The total flows required are divided by the number of hoselines to be used to determine the flow required from each line. This way the pitot reading required at each nozzle will be identical.

In addition to the first phase, manually controlled pumps shall be manually started and stopped at least ten times with the pump running at least five minutes each time. An automatically controlled pump shall be put through at least ten automatic operations plus ten manual operations with the pump running at least five minutes in each cycle. If the automatic controller is to start the pump in response to a fire protection system operation, such as a fire detection system, this feature also shall be tested.

For the electric-driven pump, one start-up is to be conducted with all hoselines open to see whether the pump will come up to rated speed under full load without pulling excess current and throwing the circuit breaker. All of these multiple operation tests are for determining whether the starting mechanism is operating properly. During all phases of the testing procedure, the pump is required to be in operation no less than one hour.

During the course of the test, attention should be paid to the temperature of the pump bearings and the pump itself. None of the components should become hot to the touch (Figure 7.96). Although mechanical seals are available, it is more common to find a fiber packing on both sides of the pump shaft that seals the shaft. This packing is water-cooled and lubricated, and some water will drip from the shaft at both ends. If the packing gland is adjusted too tightly, it will prevent water from cooling the fiber packing, and the packing will heat up. It can become hot enough to burn and smoke, causing the pump to seize. During the initial operations of the pump, gradual adjustment of the packing gland with a small wrench is often necessary. Ultimately, an adjustment should be achieved where about one drop every second is passing through the packing. More leakage than this requires tightening the packing, and less leakage requires loosening the packing.

Figure 7.96 Feel the pump motor housing for signs of overheating.

When the test has been completed, the data collected is used to construct performance curves that are compared with the manufacturer's certified curves. In constructing the pressure versus flow curve, it is the "net pressure" (discussed previously) that is used. If the performance curve falls very close to the characteristic curve, the velocity pressures should be considered. An increase or decrease in pipe size will cause pressure changes because of the change in water velocity. But for most practical applications, these pressure changes can be ignored.

Pump speed significantly affects the performance of the pump; this is the reason pump speed measurements must be taken. If the pump is turning at a speed other than the rated speed, any comparison with the certified performance may not be legitimate. However, by use of mathematical relationships called the "affinity laws," pump performance at any pump speed can be correlated to the rated pump speed, thus enabling the inspector to determine if the fire protection requirements are actually being met. The affinity relationships are as follows:

$$\frac{Q_1}{Q_2} = \frac{rpm_1}{rpm_2}$$

$$\frac{P_1}{P_2} = (rpm_1/rpm_2)^2$$

$$\frac{hp_1}{hp_2} = (rpm_1/rpm_2)^2$$

Where: Q = Flow in gpm or L/min

P = Pressure (net)

hp = Horsepower

rpm = Revolutions per minute

Increasing the speed by 10 percent increases the number of gpm delivered by 10 percent. The pressure, however, is related to the square of the pump speed. Thus, doubling the pump speed quadruples the pressure (twice the pump speed = four times the pressure). The horsepower developed is proportional to the cube or third power of the pump speed. Doubling the speed increases the horsepower by eight times.

Example

A pump rated at 1,770 rpm delivers 800 gpm while developing a net pressure of 72 psi and 33.6 horsepower. However, the pump speed is measured to be only 1,740 rpm instead of 1,770. These test results can be corrected to 1,770 rpm by use of the affinity relationships as follows:

$$\frac{Q_1}{Q_2} = \frac{rpm_1}{rpm_2}$$

$$Q_1 = (Q_2)\frac{rpm_1}{rpm_2}$$

$$Q_1 = (800 \text{ gpm})\frac{1{,}770 \text{ rpm}}{1{,}740 \text{ rpm}}$$

$$Q_1 = 814 \text{ gpm @ } 1{,}740 \text{ rpm}$$

and

$$\frac{P_1}{P_2} = (rpm_1/rpm_2)^2$$

$$P_1 = (P_2)(rpm_1/rpm_2)^2$$

$$P_1 = (72 \text{ psi})(1{,}770 \text{ rpm}/1{,}740 \text{ rpm})^2$$

$$P_1 = 73.5 \text{ psi @ } 1{,}770 \text{ rpm}$$

In this example, the corrections are relatively small. However, if the pump is operating near the certified performance, even these small differences can be important.

The voltage and current measured for the electrically driven pump are used to evaluate other acceptance criteria. An electric motor should have a nameplate on which various information is stamped, including the service factor, full-load current rating, and rated voltage. During the acceptance test, the full-load current rating shall not be exceeded except as allowed by the service factor. The ratio of the measured current in amperes to the full-load current rating shall not exceed the service factor at any time during the test.

Finally, the measured voltage should never be more than 5 percent below or more than 10 percent above the rated voltage.

For the vertical-shaft, electrically driven pumps, the test procedure is essentially the same. The primary difference is that there is no suction gauge, and the pressure developed is calculated by the procedure outlined in the first part of this chapter.

The test of diesel-driven pumps is the same as for electrically driven pumps. Of course, voltage and current readings are not necessary.

Regardless of what type of pump or driver is used, the actual performance of the pump is compared to the certified shop test curves provided by the manufacturer. If the pump does not meet or exceed the characteristic curves or malfunctions in any way, the pump installation should not be accepted. The installing contractor, in conjunction with the equipment manufacturer, must be required to bring the installation up to standard.

Keep in mind that a large volume of water will be discharged during a test. A 1,000 gpm (4 000 L/min) pump operated for 1 hour will discharge at least 60,000 gallons (240 000 L) of water (Figure 7.97). Attention must be given to where the water is flowing or draining. Care should be taken to avoid erosion or other property damage. If possible, it is better if the hoselines can discharge back into the water source. This will conserve water that otherwise runs onto the ground and is wasted. In areas where water

Figure 7.97 A large amount of water will be flowing away from the test site.

conservation is a serious concern, fire pump testing may have to be curtailed until a suitable amount of water is available.

For more detailed information on fire pumps or fire pump testing, see IFSTA's **Private Fire Protection and Detection** manual or Fire Protection Publications' **Fire Protection Hydraulics and Water Supply Analysis** book by Pat D. Brock.

WATER SUPPLY SYSTEMS AND WATER SUPPLY ANALYSIS

Water is one of the most important tools firefighters use to control and extinguish fire. Inspectors must be familiar with the types of water distribution systems in their local communities in order to ensure that the systems are adequate to handle emergency situations. Inspectors can usually obtain information on virtually any aspect of the local water supply network from the water department. The local water department is usually a separate city utility whose main function is to provide sanitary water that is safe for human use. As with all other city organizations, it is important that the fire department maintain a good working relationship with the water department.

Water Distribution Systems

Water can be obtained through surface supplies (rivers or lakes) or ground supplies (wells or water-producing springs). There are four components of an effective water distribution system (Figure 7.98):

- Water supply source
- Processing or treatment facilities
- Means or methods of moving the water
- Delivery system, including storage

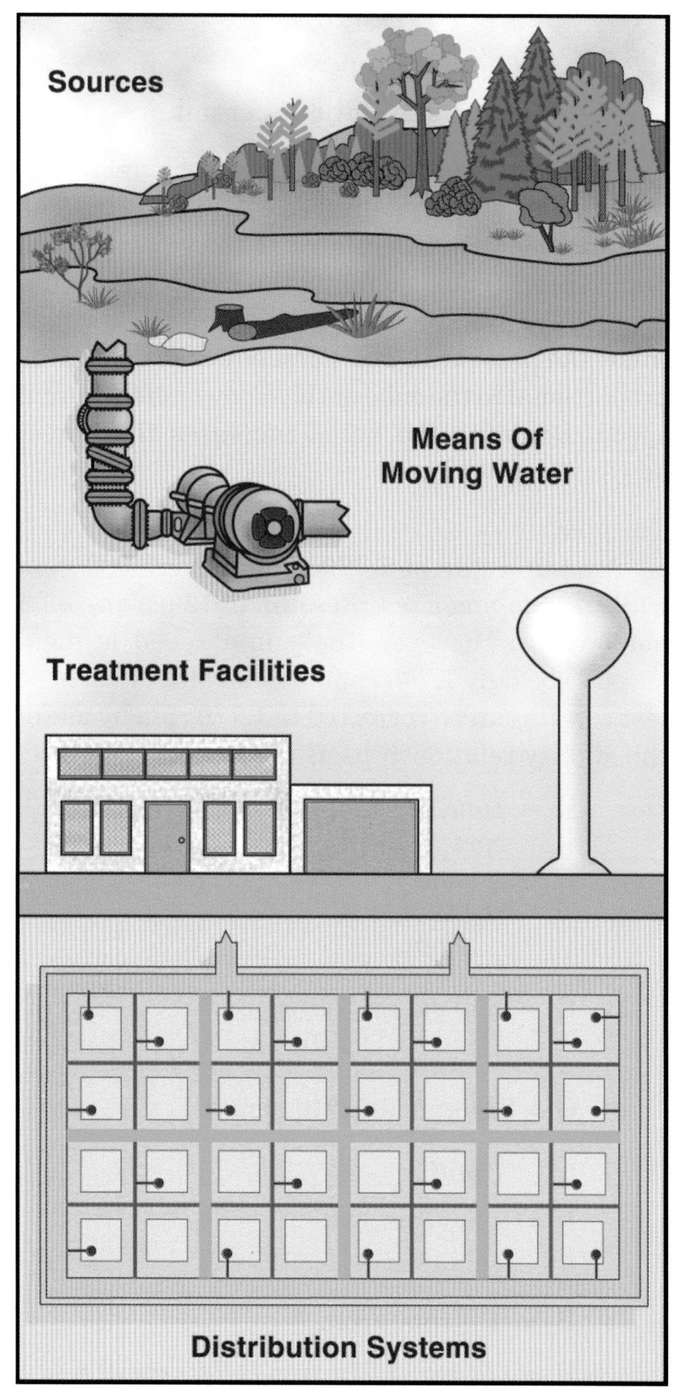

Figure 7.98 There are four basic components of any water supply system.

There are three types of distribution systems: gravity, direct pumping, and combination. A true *gravity system* delivers water from the source to the distribution system without pumping equipment (Figure 7.99). The natural pressure created by a difference in elevation provides the pressure within the distribution system. When elevation pressure cannot provide sufficient pressure for the community, a pump is placed relatively close to the water source to create the pressure within the distribution system. This is called a *direct pumping system* (Figure 7.100).

Most communities use *combination systems* — both gravity and pumping — to provide adequate

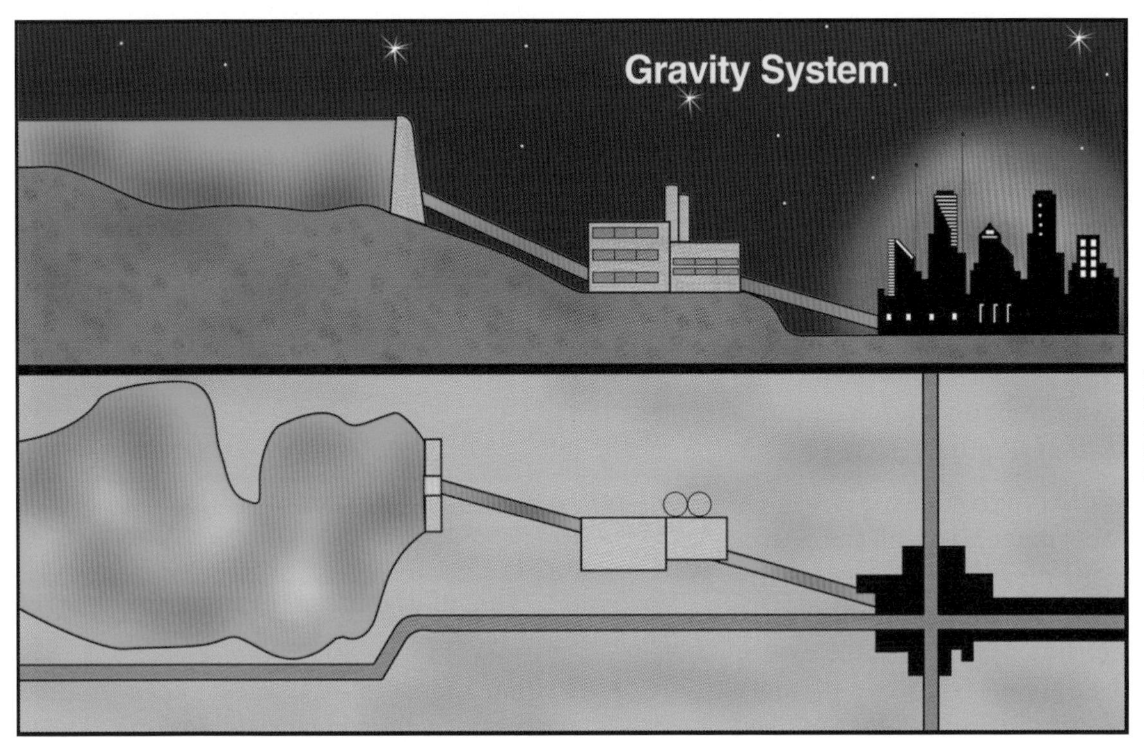

Figure 7.99 With a gravity system, the supply source must be located well above the area being served.

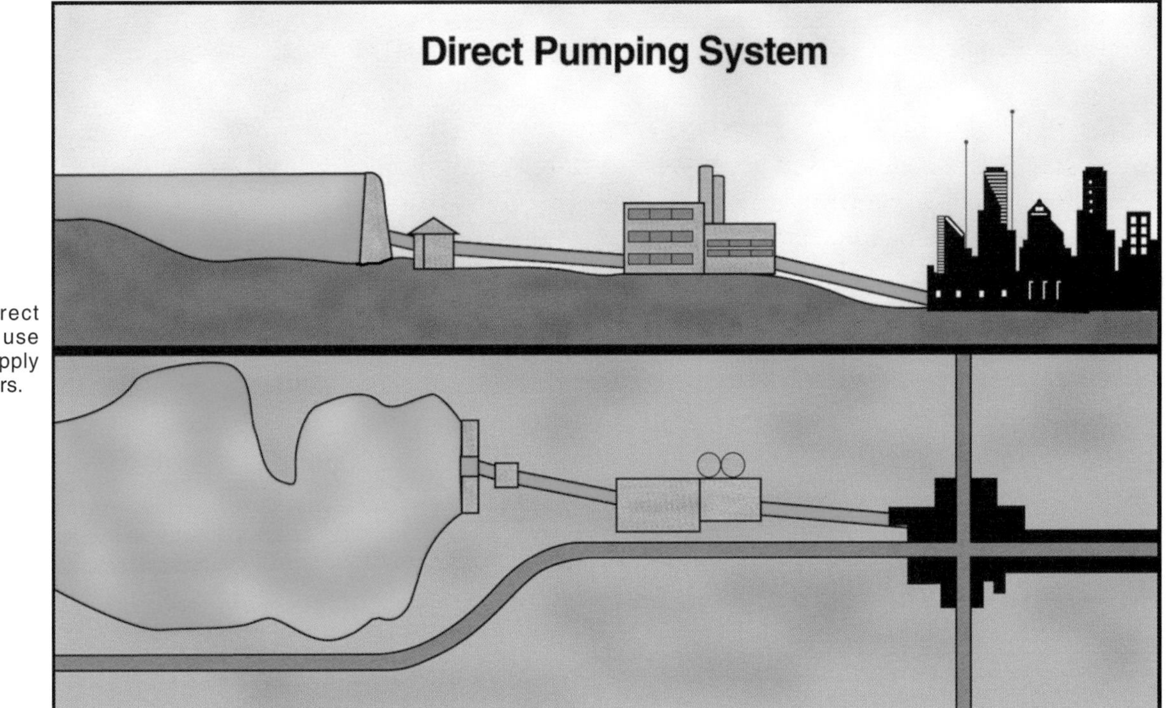

Figure 7.100 Direct pumping systems use large pumps to supply water to the customers.

pressure (Figure 7.101). Water is pumped into the distribution system and into elevated storage tanks (gravitational pressure). When the consumption demand is greater than the rate at which the water is pumped, the water flows from the storage tanks into the distribution system. Conversely, when demand is less, the water flows back into the storage tanks. Elevated storage reservoirs are usually constructed of steel or concrete. They vary in height and can hold as much as 2 million gallons (7 570 000 L), depending on pressure desired (Figure 7.102).

The distribution system receives the water from the pumping station/treatment facility and delivers it throughout the area to be served. Fire hydrants, gate valves, elevated storage, and reservoirs are supplementary parts of the distribution system. The term "grid" is sometimes used to describe the network of water mains that makes up a water distribution system (Figure 7.103).

When water flows through pipes, a pressure loss occurs due to the movement of the water against the inside of the pipe (friction loss). A fire hydrant that receives water from one side (called a *dead-end hydrant*) has less available water than a fire hydrant supplied from two or more directions (called a *circulating* or *looped feed hydrant*). The distribution system consists of three main feeders:

Figure 7.102 Elevated storage tanks are commonly used in combination water supply systems.

- Primary feeders — Largest pipes (mains) with relatively widespread spacing. These

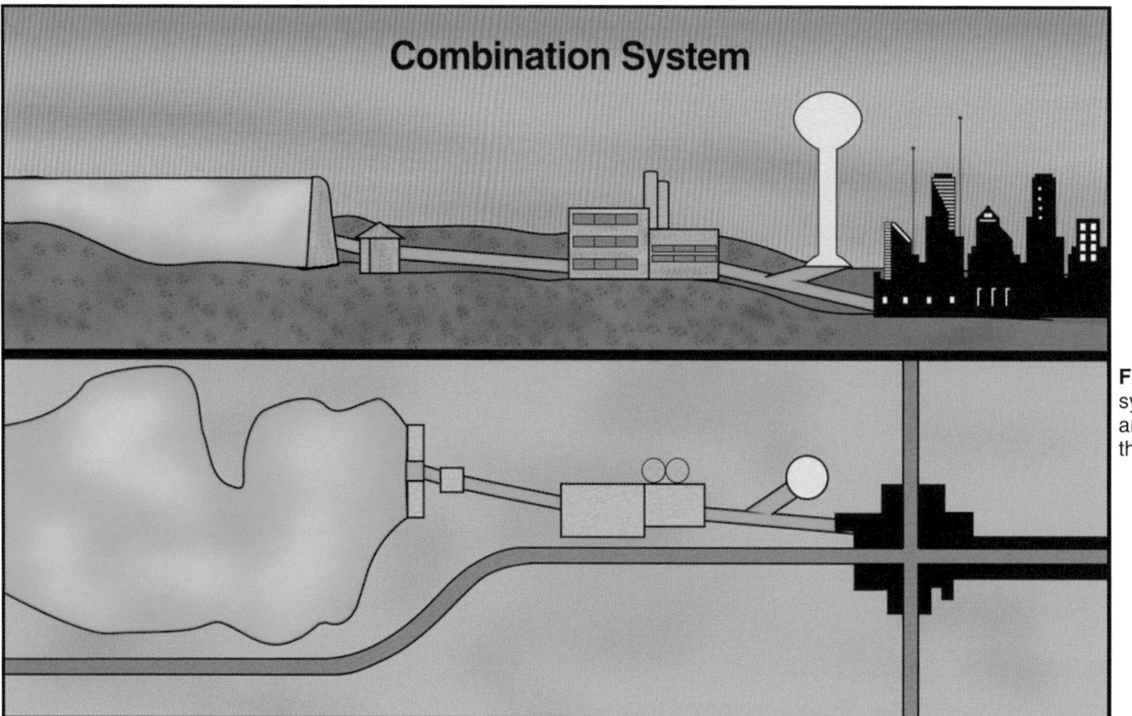

Figure 7.101 Combination systems use both gravity and pumps to move water through the system.

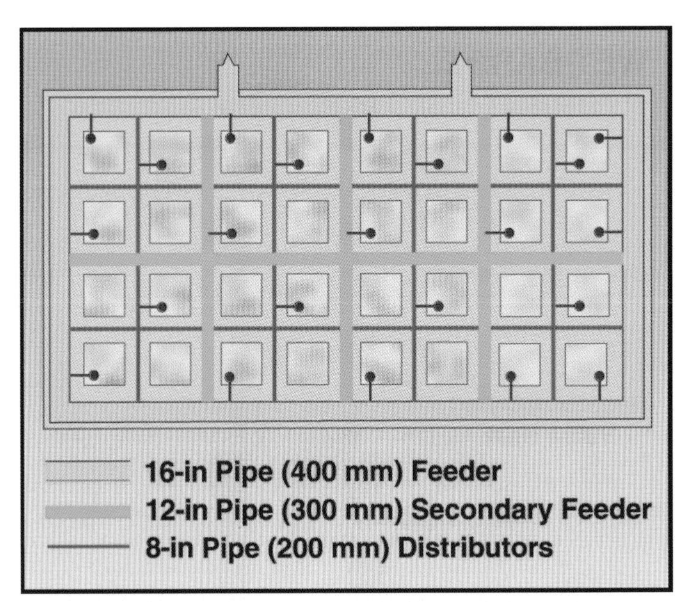

Figure 7.103 Water distribution grids are made of a variety of different sizes of pipe.

feeders convey large quantities of water to various points in the system for local distribution to the smaller mains.

- Secondary feeders — Intermediate pipes within the primary feeder network that reinforce the grid with a concentrated supply of water.
- Distributors — Smallest of the mains that serve individual fire hydrants and blocks of consumers.

The ability to deliver adequate water depends on the capacity of the system's network of pipes. Today, 8-inch (200 mm) pipe is becoming the *minimum* size used because of its increased flow capability over 4- and 6-inch (100 mm and 150 mm) pipes. Pipes that are 30 inches (750 mm) or larger are commonly found in modern municipal water supply systems.

FIRE HYDRANTS

Access to the underground water distribution network is made through the hydrant. The two main types of modern fire hydrants are dry-barrel hydrants and wet-barrel hydrants. Regardless of the design or type, the hydrant outlets are considered to be standard if there is at least one large outlet (4 or 4½ inches [100 mm or 115 mm]) for pumper supply and two outlets for 2½-inch (65 mm) couplings (Figure 7.104). Hydrant specifications require a 5-inch (125 mm) valve opening for standard 3-way hydrants and a 6-inch (150 mm) connection to the water main. The threads on all hydrant outlets must conform to those used by the local fire department. The principal items covered by the standard are the number of threads per inch and the outside diameter of the male thread. For exact details, refer to NFPA 1963, *Standard on Screw Threads and Gaskets for Fire Hose Connections*. Hydrant location is usually determined by the type, size, location of the protected occupancy(ies), and local code requirements.

Figure 7.104 A standard fire hydrant.

Dry-barrel hydrants are used in areas that have freezing temperatures (Figure 7.105). The dry-barrel hydrant has a base valve located below the frost line; the stem nut to open and close the base valve is located on the top of the hydrant. Any water remaining in a closed dry-barrel hydrant drains through a small valve that opens at the bottom of the hydrant when the main valve approaches a closed position.

The wet-barrel hydrant usually has a compression valve at each outlet, but may have another valve in the bonnet that controls the water flow to all outlets (Figure 7.106). This type of hydrant features the valve at the hose outlet and is used in mild climates where typical weather conditions are above freezing.

One of the most important periodic hydrant maintenance considerations is to check for leaks in the following areas:

- The main valve when it is closed
- The drain valve when the main valve is open but the outlets are capped
- The water mains near the hydrant

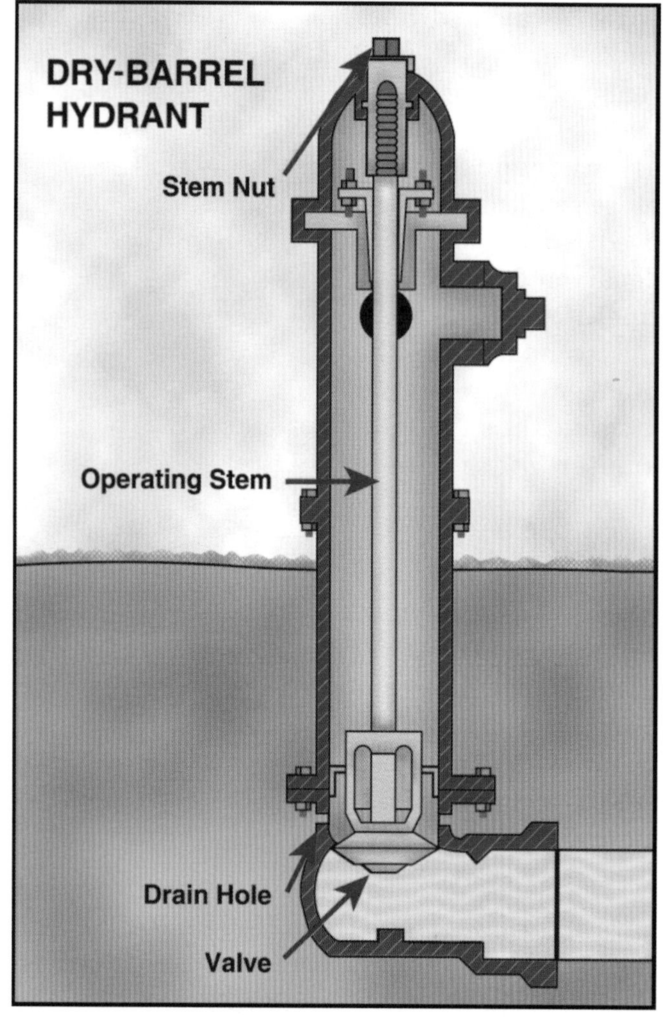

Figure 7.105 Dry-barrel hydrants utilize a long operating stem to keep the water well below ground when the hydrant is not in use.

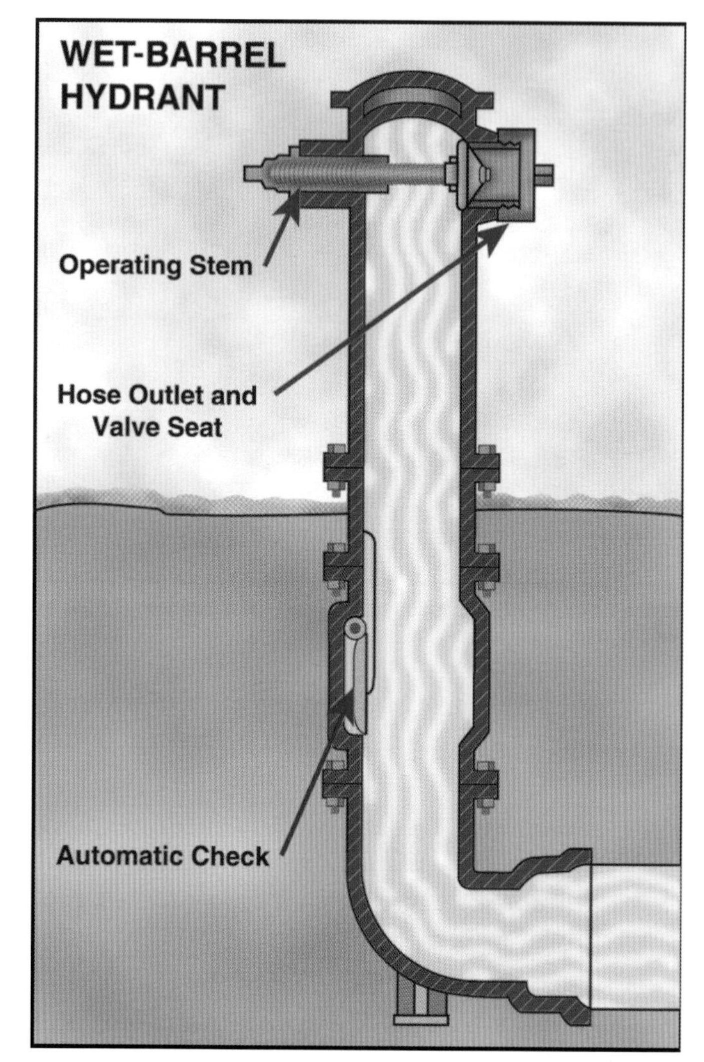

Figure 7.106 Wet-barrel hydrants have water right up to the discharge outlets when not in use.

The American Water Works Association has adopted specifications for a national standard hydrant for ordinary waterworks service in the United States. These specifications are designed to produce a hydrant that is free from difficulties such as trouble in opening and closing, interior mechanical parts that can work loose, leakage, excessive friction loss, failure to drain properly, and loose nipples. These specifications may be obtained from The American Water Works Association, 6666 West Quincy Avenue, Denver, Colorado 80235.

The actual flow of water from a hydrant may vary due to such conditions as feeder main location, encrustation, deposits, and totally or partially closed supply valves. Firefighters can make better tactical decisions if they know at least the relative available water flow of different hydrants in the vicinity. To address this problem, NFPA standardized a color code system (NFPA 291, *Recommended Practice for Fire Flow Testing and Marking of Hydrants*) to mark hydrants according to capacity (Table 7.5). The colors may vary due to geographic location, but the main intent of any color scheme is simplicity.

TABLE 7.5
Hydrant Color Codes

Hydrant Class	Color	Flow
Class AA	Light Blue	1,500 gpm (5 678 L/min) or greater
Class A	Green	1,000-1,499 gpm (3 785 L/min-5 677 L/min)
Class B	Orange	500-999 gpm (1 893 L/min-3 784 L/min)
Class C	Red	less than 500 gpm (1 893 L/min)

Figure 7.107 Nonindicating valves are usually buried in marked valve shafts or boxes.

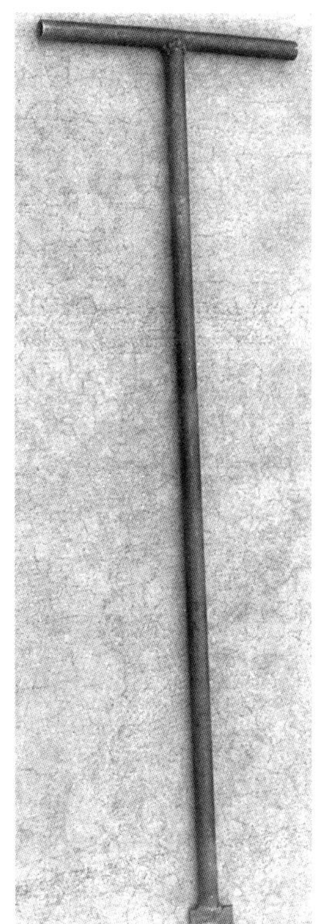

Figure 7.108 A water key tool is required for operating an underground, nonindicating valve.

Figure 7.109a A gate valve. **Figure 7.109b** A butterfly valve.

WATER MAIN VALVES

The main function of water main valves is to control water flow through the mains as circumstances dictate. Valves should be operated at least once a year to ensure that they are in good working condition. This test is commonly performed by municipal water department employees on public water systems and plant maintenance personnel on private systems.

Valve spacing should be planned so that a minimum length of the water distribution system will be out of service if a cutoff procedure is initiated. The maximum lengths for valve spacing should be 500 feet (150 m) in high-value districts and 800 feet (240 m) in other areas, as recommended by Commercial Risk Services, Inc. (formerly Insurance Services Office) engineers.

Valves for water systems are broadly divided into indicating and nonindicating types. An indicating valve shows whether the gate valve seat is open, closed, or partially closed. Valves in private fire protection systems are usually of the indicating type. Valves in public water systems are usually of the nonindicating type, except for some valves in treatment plants and pump stations. Two common indicator valves are the post indicator valve (PIV) and the outside screw and yoke valve (OS&Y). These were described and shown previously in this chapter in the section on automatic sprinkler systems.

Nonindicating valves in a water distribution system are usually buried or installed in manholes (Figure 7.107). If a buried valve is properly installed, the valve is operated through a valve box by a special valve key (Figure 7.108). This valve key may be carried on fire apparatus, but is more commonly kept by water department representatives. Control valves may be gate valves or butterfly valves (Figures 7.109 a and b). Gate valves are usually the nonrising stem type. As the valve nut is turned, the gate either rises or lowers to control the water flow. Butterfly valves are tight closing and they usually have rubber or a rubber composition seat that is bonded to the valve body. The valve disk rotates 90 degrees to open or shut the valve.

Inspectors should be aware of the consequences of stuck or partially closed valves. If a valve is partially closed, it would not be noticed during normal use of water; however, the high friction loss would prevent the fire department from obtaining sufficient water to combat a fire. This situation can be prevented by accurate and routine inspections.

Obstruction of Water Mains

After several years, fire flow tests may show a progressively inadequate flow through water

HYDRANT RECORD

LOCATION ____________ HYDRANT NO. ____________

POSITION ____________ MAKE ____________

INSTALLED ________ TYPE ________ TURNS TO OPEN ________ R. ________ L. ________

SIZE OF LEAD ____________ SIZE OF MAIN ____________

VALVE IN LEAD ________ FT. ________ TURNS TO OPEN ________ R. ________ L. ________

BENCH MARK ____________ ELEV. ____________

PRESSURE TESTS

DATE	STATIC PRESSURE	FLOW PRESSURE	GPM	DATE	STATIC PRESSURE	FLOW PRESSURE	GPM

REMARKS

RECORD OF MAINTENANCE

WORK PERFORMED ____________ DATE ____________

Flowed

Lubricated

Cap Gasket Replaced

Bonnet Gasket Replaced

Valve Leather Replaced

Drain Valve Replaced

Cap Replaced

Lead Valve Operated

Painted

Raised

Moved

Figure 7.116 A sample hydrant inspection/maintenance form.

given areas. Fire fighting defenses cannot be planned intelligently without this information. In order for fire service personnel to determine the quantity of water available for fire protection, it is necessary to conduct fire flow tests on the water distribution system. These tests include the actual measurement of static (normal operating) and residual pressures, and the formulas and calculations used to determine available water from these tests.

Fire flow tests are made to determine the rate of water flow available for fire fighting at various locations within the distribution system. By measuring the flow from hydrants and recording the pressures corresponding to this flow, the number of gallons available at any pressure or the pressure available at any flow can be determined through calculations or graphical analysis.

Before conducting a flow test, a responsible water department official should be notified because opening hydrants may upset the normal operating conditions in a water supply system. Notification is also important because water service personnel may be performing maintenance work in the immediate vicinity; therefore, the results of the flow test would not be typical for normal conditions. This practice of proper notification will also promote a better working relationship between the water department and fire service personnel.

Knowing the capacity of a water system is just as important as knowing the capacities of pumpers and water tanks. This knowledge is also essential when making pre-incident plans. The results of fire flow tests can be used to advantage by both the fire and water departments of a municipality. Fire officers familiar with fire flow test results are better qualified to locate pumpers at strong locations on a distribution system and avoid weak locations. Because test results indicate weak points in a water distribution system, they can be used by water works personnel to plan improvements in an existing system and to design extensions to newly developed areas. Tests that are repeated at the same locations year after year may reveal a loss in the carrying capacity of water mains and a need for strengthening certain arterial mains. Flow tests should be run after any extensive water main improvements, after extensions have been made, or at least every five years if there have been no changes.

USING THE PITOT TUBE AND GAUGE

Using a pitot tube and gauge to take a flow reading is not difficult, but it must be done properly to obtain accurate readings. A good method of holding a pitot tube and gauge in relation to a hydrant outlet or nozzle is illustrated in Figure 7.117. Note that the pitot tube is grasped just behind the blade with the first two fingers and thumb of the left hand while the right hand holds the air chamber. The little finger of the left hand rests upon the hydrant outlet or nozzle tip to steady the instrument. Unless some effort is made to steady the pitot tube, the movement of the water will make it difficult to get an accurate reading.

Figure 7.117 One method of holding the pitot tube and gauge involves resting the pinkie on the discharge outlet threads.

Another method of holding the pitot tube is illustrated in Figure 7.118. The left hand fingers are split around the gauge outlet and the left side of the fist is placed on the edge of the hydrant orifice or outlet. The blade can then be sliced into the stream in a counterclockwise direction (Figure 7.119). The right hand once again steadies the air chamber. The procedure for using a pitot tube and gauge is as follows:

Figure 7.118 Rest the fist on the side of the discharge outlet.

Figure 7.119 Rotate the blade into the stream.

Step 1: Open the petcock on the pitot tube and make certain the air chamber is drained. Then close the petcock.

Step 2: Edge the blade into the stream, with the small opening or point centered in the stream and held away from the butt or nozzle approximately one-half the diameter of the opening. For a 2½-inch (65 mm) hydrant butt, this distance is 1¼ inches (32 mm). The pitot tube blade should now be parallel to the outlet opening with the air chamber kept above the horizontal plane passing through the center of the stream. This increases the efficiency of the air chamber and helps avoid needle fluctuations.

Step 3: Take and record the velocity pressure reading from the gauge. If the needle is fluctuating, read and record the value located in the center between the high and low extremes.

Step 4: After the test is completed, open the petcock and be certain all water is drained from the assembly before storing.

COMPUTING HYDRANT FLOW

The easiest way to determine how much water is flowing from the hydrant outlet(s) is to refer to prepared tables for nozzle/outlet discharge (Tables 7.6 a and b). Jurisdictions may choose to develop their own tables based on the flow pressures that are common to their area. These tables are computed by using the following:

$$GPM = (29.83) \times C_d \times d^2 \times \sqrt{P}$$

$$L/min = (0.0667766) \times C_d \times d^2 \times \sqrt{P}$$

Where: C_d = The coefficient of discharge

d = The actual diameter of the hydrant or nozzle orifice in inches (mm)

P = The pressure in psi (kPa) as read at the orifice.

NOTE: 29.83 (.0667766) is a constant derived from the physical laws relating water velocity, pressure, and conversion factors that conveniently leave the answer in gallons per minute (liters per minute).

This formula was derived by assigning a coefficient of 1.0 for an ideal frictionless discharge orifice. An actual hydrant orifice or nozzle will have a lower coefficient of discharge, reflecting friction factors that slow the velocity of flow. The coefficient will vary with the type of hydrant outlet or nozzle used. When using a hydrant orifice, the operator will have to feel the inside contour of the hydrant to determine which one of the three types of hydrant outlets is being used (Figure 7.120 on page 217). When a nozzle is used, the coefficient of discharge depends on the type of nozzle. Refer to the manufacturer's recommendations for determining the coefficient of discharge for a specific nozzle.

The flow formula also depends on the actual internal diameter of the outlet or nozzle opening being used. A ruler with a scale that measures to at least sixteenths of an inch (mm) should be used to measure the diameter of the outlet or nozzle opening.

Assuming a 2½-inch (65 mm) hydrant outlet is used that has an actual diameter of 2⁷⁄₁₆ inches (2.44 inches [62 mm]) with a C factor of 0.80 and a flow pressure of 10 psi (69 kPa) read from the pitot gauge, the water flow equation would read:

$$GPM = 29.83 \times C_d \times d^2 \times \sqrt{P}$$

$$GPM = 29.83 \times 0.80 \times (2.44)^2 \times \sqrt{10}$$

$$GPM = 449.28 \text{ or } \approx 450$$

$$L/min = 0.0667766 \times C_d \times d^2 \times \sqrt{P}$$

$$L/min = 0.0667766 \times 0.80 \times (62)^2 \times \sqrt{69}$$

$$L/min = 1705.78 \text{ or } \approx 1700$$

Generally, 2½-inch (65 mm) outlets should be used to conduct hydrant flow tests. This is because the stream from a large hydrant outlet (4 to 4½ inches [100 mm to 115 mm]) contains voids, that is, the entire stream of water is not solid. For this reason, the listed formula alone will not give accurate results for flows using large outlets. If it is necessary to use the large outlets, a correction factor can be used to give more accurate results. The flow (as determined by gpm = $29.83 \times C_d \times d^2 \times \sqrt{P}$ or L/min - $0.0667766 \times C_d \times d^2 \times \sqrt{P}$) should be multiplied by one of the factors shown in Table 7.7 (on page 217), corresponding to the velocity pressure measured by the pitot tube and gauge.

TABLE 7.6a
Discharge Table for Circular Outlets* (U. S.)
Outlet Pressure Measured by Pitot Gauge

Outlet Pressure in lbs. per sq. inch	Outlet Diameter in Inches											
	2 3/8	2 1/2	2 5/8	2 3/4	2 7/8	3	3 1/8	3 7/8	4	4 3/8	4 1/2	4 5/8
	U.S. Gallons per Minute											
1	150	170	180	200	220	240	260	400	430	510	540	580
2	210	240	260	290	310	340	370	570	610	720	770	810
3	260	290	320	350	380	420	450	700	740	890	940	990
4	300	340	370	410	440	480	530	810	860	1030	1090	1150
5	340	380	410	450	500	540	590	900	960	1150	1220	1290
6	370	410	450	500	540	590	640	990	1050	1260	1340	1410
7	400	440	490	540	590	640	690	1070	1140	1360	1440	1520
8	430	480	520	570	630	680	740	1140	1220	1450	1540	1620
9	450	500	550	610	670	730	790	1210	1290	1540	1640	1720
10	480	530	580	640	700	760	830	1280	1360	1630	1730	1820
11	500	560	610	670	730	800	870	1340	1430	1710	1810	1910
12	520	580	640	700	770	840	910	1400	1490	1780	1890	1990
13	550	610	670	730	800	870	950	1450	1550	1850	1960	2070
14	570	630	690	760	830	900	980	1510	1610	1920	2040	2150
15	590	650	720	790	860	940	1020	1560	1660	1990	2110	2220
16	610	670	740	810	890	970	1050	1620	1720	2060	2180	2300
17	620	690	760	840	910	1000	1080	1660	1770	2120	2240	2370
18	640	710	780	860	940	1030	1110	1710	1820	2180	2310	2440
19	660	730	810	890	960	1050	1140	1760	1870	2240	2370	2510
20	680	750	830	910	990	1080	1170	1800	1920	2290	2430	2570
22	710	790	870	950	1040	1130	1230	1890	2020	2400	2550	2700
24	740	820	910	1000	1090	1180	1290	1970	2110	2510	2660	2810
26	770	860	940	1040	1130	1230	1340	2050	2190	2620	2770	2930
28	800	890	980	1070	1170	1280	1390	2130	2280	2720	2880	3040
30	830	920	1010	1110	1210	1320	1430	2210	2350	2820	2980	3150
32	860	950	1050	1150	1260	1370	1480	2280	2430	2910	3080	3250
34	880	980	1080	1180	1290	1410	1530	2350	2510	3000	3170	3350
36	910	1010	1110	1220	1330	1450	1580	2420	2580	3080	3260	3440
38	930	1040	1140	1250	1370	1490	1620	2480	2650	3170	3350	3540
40	960	1060	1170	1290	1400	1530	1660	2550	2720	3250	3440	3630

*Computed with Coefficient C = 0.90, to nearest 10 gallons per minute.

TABLE 7.6b
Discharge Table for Circular Outlets* (Metric)
Outlet Pressure Measured by Pitot Gauge

Outlet Pressure in kPa	Outlet Diameter in mm											
	60	64	67	70	73	76	79	98	102	111	114	117
	Liters per Minute											
5	484	550	603	658	716	776	839	1291	1398	1656	1746	1840
10	684	778	853	931	1012	1098	1186	1825	1977	2341	2470	2602
15	838	953	1045	1140	1240	1344	1453	2235	2422	2868	3025	3186
20	968	1101	1206	1317	1432	1552	1677	2581	2796	3312	3493	3679
25	1082	1231	1348	1472	1601	1736	1875	2886	3126	3702	3905	4113
30	1185	1348	1478	1613	1754	1901	2054	3161	3425	4056	4278	4506
35	1280	1456	1596	1742	1894	2054	2219	3415	3699	4381	4620	4867
40	1368	1557	1706	1862	2026	2195	2372	3650	3954	4683	4940	5203
45	1451	1651	1810	1975	2148	2328	2516	3871	4194	4967	5239	5519
50	1530	1741	1907	2082	2264	2455	2652	4081	4421	5236	5523	5817
55	1604	1826	2001	2184	2375	2574	2781	4281	4637	5492	5792	6101
60	1676	1907	2090	2281	2481	2689	2905	4471	4843	5736	6050	6373
65	1744	1985	2175	2374	2582	2799	3024	4653	5041	5970	6293	6633
70	1810	2060	2257	2463	2679	2904	3138	4829	5231	6195	6535	6883
75	1873	2132	2336	2550	2773	3006	3248	4999	5415	6413	6764	7125
80	1935	2202	2413	2634	2864	3105	3355	5162	5593	6623	6986	7358
85	1994	2270	2487	2715	2952	3200	3458	5321	5765	6827	7201	7589
90	2053	2335	2559	2794	3038	3293	3558	5476	5932	7025	7410	7805
95	2109	2399	2629	2870	3121	3383	3656	5625	6094	7217	7612	8019
100	2164	2462	2698	2945	3203	3471	3751	5772	6253	7405	7810	8227
105	2217	2522	2764	3017	3282	3557	3843	5914	6407	7589	8003	8430
110	2269	2582	2860	3089	3359	3640	3933	6054	6558	7766	8192	8628
115	2320	2640	2893	3158	3434	3722	4022	6190	6705	7940	8376	8822
120	2370	2697	2955	3225	3508	3803	4109	6323	6849	8112	8556	9012
125	2419	2752	3016	3292	3581	3881	4193	6453	6990	8279	8732	9198
130	2467	2807	3076	3358	3652	3958	4277	6581	7129	8443	8905	9380
135	2514	2860	3135	3422	3721	4033	4358	6706	7265	8604	9075	9559
140	2560	2913	3192	3484	3789	4107	4438	6829	7398	8761	9241	9734
145	2605	2964	3249	3546	3856	4180	4516	6950	7529	8917	9405	9907
150	2650	3015	3304	3607	3922	4251	4594	7069	7658	9069	9569	10076

*Computed with Coefficient C = 0.90, to nearest liter.

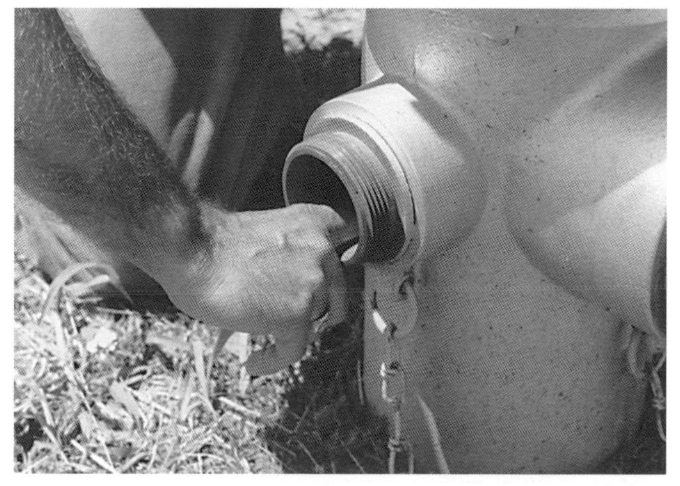

Figure 7.120 Determine the discharge orifice type by feeling the inside of the discharge.

TABLE 7.7
Correction Factors for Large Diameter Outlets

Velocity Pressure	Factor
2 psi (13.8 kPa)	0.97
3 psi (20.7 kPa)	0.92
4 psi (27.6 kPa)	0.89
5 psi (34.5 kPa)	0.86
6 psi (41.4 kPa)	0.84
7 psi (48.3 kPa) or over	0.83

From Table 7.6, a flow of 6 psi (42 kPa) through a 4-inch (100 mm) outlet is indicated as 1,050 gpm (3 974 L/min). However, tests have shown that only 84 percent of this quantity is actually flowing due to voids in the water stream. Accordingly, actual flow is 1,050 x 0.84 = 883 gpm (3 974 x 0.84 = 3 338 L/min).

These formulas allow the computation of total flow from the flowing hydrants when performing an area fire flow test. They also indicate the flow from the hydrant at the time of the test.

REQUIRED RESIDUAL PRESSURE

As a result of experience and water system analysis, fire protection engineers have established 20 psi (140 kPa) as the minimum required residual pressure when computing the available water for area flow test results. This residual pressure is considered enough to overcome friction loss in a short 6-inch (150 mm) branch, in the hydrant itself, and in the intake hose, as well as allowing a safety factor to compensate for gauge error. Many state health departments require this 20 psi (140 kPa) minimum to prevent the possibility of external water being drawn into the system at main connections. Pressure differentials can result in water main collapse or create cavitation, which is the implosion of air pockets drawn into pumps. A more common occurrence is that pumpers working at these low system pressures may be pumping near the water main's capacity. If a valve on the pumper is shut down too quickly, a water hammer is created. This sudden surge in pressure may be transferred to the water main, resulting in damaged or broken mains or connections.

FLOW TEST PROCEDURES

When testing the available water supply, determining the number of hydrants to be opened depends on an estimate of the flow available in the area. For example, a very strong probable flow requires several hydrants to be opened for a more accurate test. Enough hydrants should be opened to drop the static pressure by at least 10 percent. If more accurate results are required, the pressure drop should be as close as possible to 25 percent. For example, if the static pressure is 80 psi (560 kPa), then the residual pressure should be at least 72 psi (504 kPa). For more accurate results, the residual may be dropped 25 percent, which would be to 60 psi (420 kPa). The flow available at 20 psi (140 kPa) can then be determined by graphical analysis or mathematical calculations.

Another problem that may be encountered is that water mains may contain such low pressures that no flow pressure registers on the pitot gauge. If this occurs, straight stream nozzles with smaller than 2½-inch (65 mm) orifices must be placed on the hydrant outlet to increase the flow velocity to a point where the velocity pressure is measurable (Figure 7.121). It should be noted that using these straight stream nozzles will require an adjustment in the water flow calculation that must include the smaller diameter and the respective coefficient of friction.

Figure 7.121 It may be necessary to place a solid stream nozzle on the hydrant in order to get an accurate reading.

Flow tests are sometimes conducted in areas very close to the base of an elevated water storage tank or standpipe. This can result in flows that are quite large in gallons per minute (L/min). It should be realized that such large flows can only be sustained as long as there is sufficient water in the elevated tank or standpipe. It is advisable to conduct an additional flow test with the storage tank shut off to determine the quantity of water available when the storage has been depleted.

During a flow test, the static pressure and the residual pressure should be taken from a fire hydrant as close as possible to the location requiring the test results. This hydrant is commonly called the *test hydrant*. The *flow hydrants* are those hydrants where pitot readings are taken to find their individual flows. These readings are then added to find the total flow during the test.

In general, when flow testing a single hydrant, the test hydrant should be between the flow hydrant and the water supply source. In other words, the flow hydrant should be downstream from the test hydrant (Figures 7.122 a and b). The direction of flow can be determined by reviewing water maps supplied by the water department. When flowing multiple hydrants, the test hydrant should be centrally located relative to the flow hydrants. (**NOTE:** Water is actually never discharged from a test hydrant; rather, a capped gauge is placed on a discharge and the hydrant is opened fully.) The procedure for conducting an available water test is as follows:

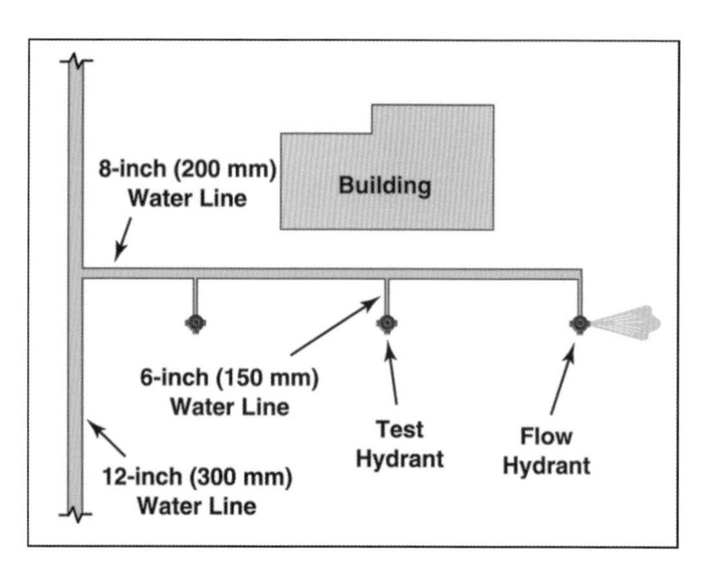

Figure 7.122a The flow hydrant should be downstream from the test hydrant.

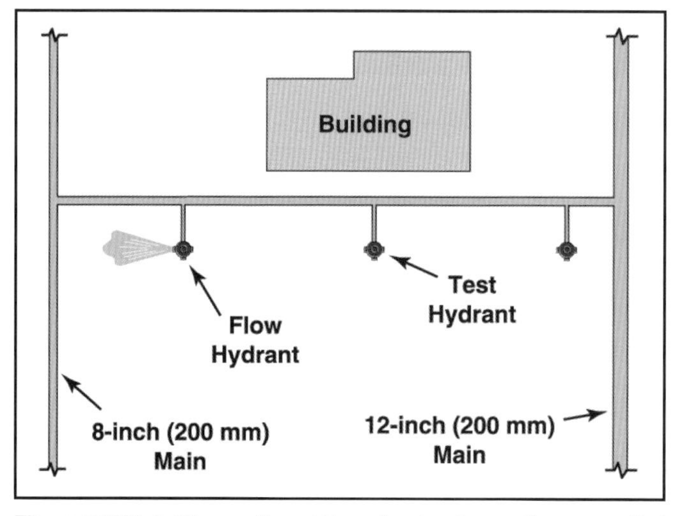

Figure 7.122b In this case it would be natural and correct to assume that the water is flowing in a direction away from the larger main and toward the smaller main.

Step 1: Locate personnel at the test hydrant and at all flow hydrants to be used.

Step 2: Remove a hydrant cap from the test hydrant and attach the pressure gauge cap with the petcock in the open position. After checking the other caps for tightness, slowly open the hydrant several turns. Once the air has escaped and a steady stream of water is flowing, close the petcock and fully open the hydrant (Figure 7.123).

Step 3: Read and record the static pressure as seen on the pressure gauge.

Step 4: The individual at the flow hydrant(s) removes the cap(s) from the outlet(s) to be flowed. When using a hydrant outlet, check and record the hydrant coefficient and the actual inside diameter of the orifice. If a nozzle is placed on the outlet, check and record its coefficient and diameter.

Step 5: Open flow hydrants as necessary and take and record the pitot reading of the velocity pressures (Figure 7.124). The individual at the test hydrant simultaneously reads and records the residual pressure. (**NOTE:** The residual pressure should not drop below 20 psi [140 kPa] during the test. If this happens, the number of flow hydrants must be reduced).

Step 6: Slowly close the flow hydrant to prevent water hammer in the mains. After checking for proper drainage, replace and secure all hydrant caps. Report any hydrant defects.

Figure 7.123 Record the static pressure after all the air has been bled from the test hydrant.

Figure 7.124 Use the pitot tube and gauge to record the pressure on the flow hydrant.

Step 7: Check the test hydrant for a return to normal operating pressure, then close the hydrant. Open the petcock valve to prevent a vacuum on the pressure gauge. Remove the pressure gauge. After checking for proper drainage, replace and secure the hydrant cap. Report any hydrant defects.

FLOW TEST PRECAUTIONS

Certain precautions must be observed before, during, and after flow tests to avoid injuries to those participating in the test or to passersby. Efforts must also be made to minimize damage to property from the flowing stream. Both pedestrian and automobile traffic must be controlled during all phases of the testing. This may require assistance from the local law enforcement agency. It may also be advisable to conduct flow tests in busy areas at off hours, such as very early in the morning.

Other safety measures include tightening caps on hydrant outlets not being used, not standing in front of closed caps, and not leaning over the top of the hydrant when operating it. Property damage control measures include opening and closing hydrants slowly to avoid water hammer, not flowing hydrants where drainage is inadequate, and always remembering to check downstream to see where the water will flow. Because flowing water across a busy street could cause an accident, take proper measures beforehand to slow or stop traffic. Do not flow water during freezing weather. A good rule to follow is: When in doubt, do not flow! If there are difficulties in conducting a flow test, give thought to their solutions so that the test can be completed without disruptions or property destruction.

COMPUTING AVAILABLE FIRE FLOW TEST RESULTS

There are two ways to compute fire flow test results: graphical analysis and mathematical computation.

Determining Available Water by Graphical Analysis

The water flow chart in Figure 7.125 is a logarithmic scale that has been developed to simplify the process of determining available water in an area. The chart is accurate to a reasonable degree

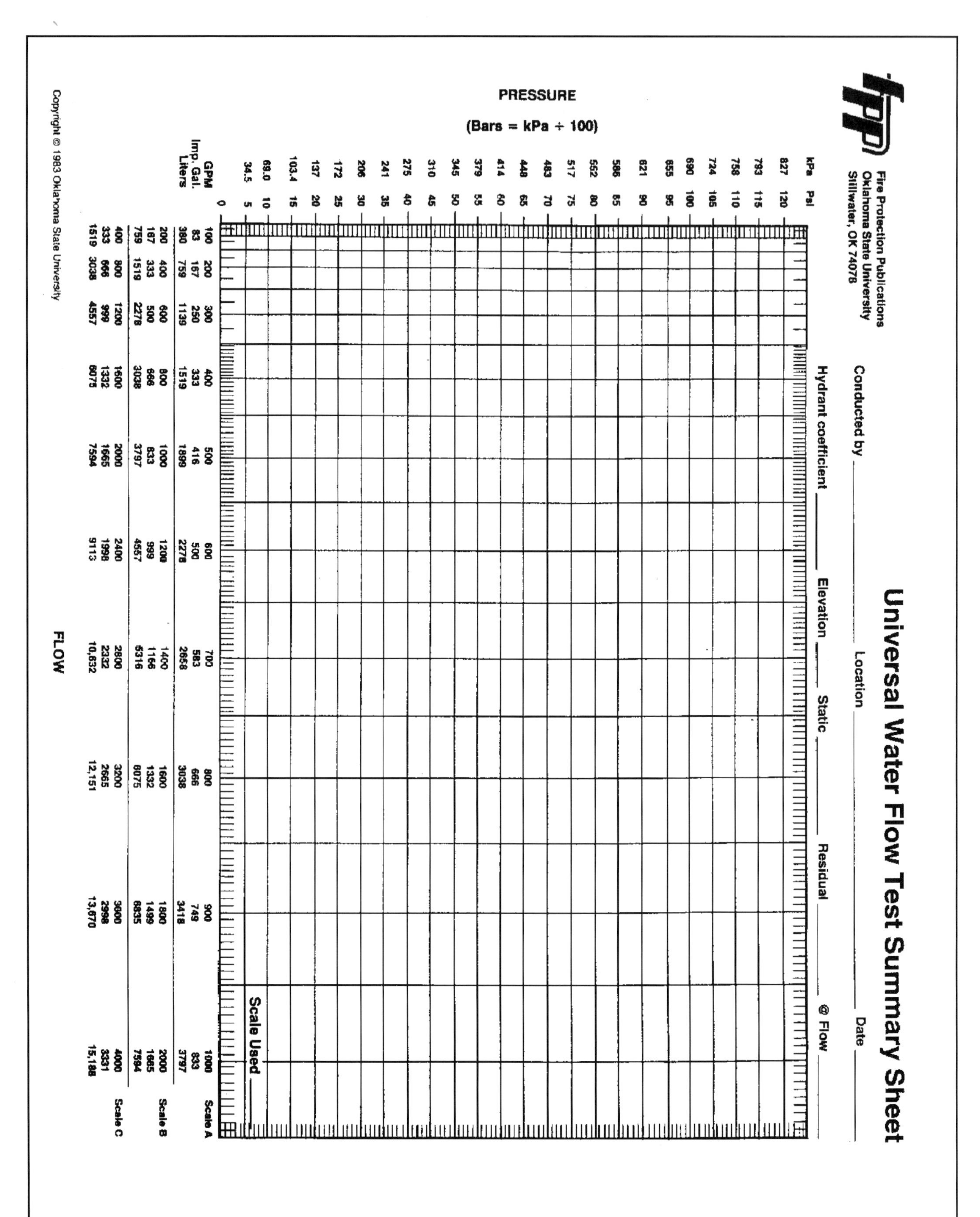

Figure 7.125 Available water supply can be calculated using special graph paper.

if one uses a fine-point pencil or pen when plotting results. The figures on the vertical and/or horizontal scales may be multiplied or divided by a constant, as may be necessary to fit any problem.

The procedure for graphical analysis is as follows:

Step 1: Determine which gpm (L/min) scale should be used.

Step 2: Locate and plot the static pressure on the vertical scale at 0 gpm (0 L/min).

Step 3: Locate the total water flow measured during the test on the chart.

Step 4: Locate the residual pressure noted during the test on the chart.

Step 5: Plot the residual pressure above the total water flow measured.

Step 6: Draw a straight line from the static pressure point through the residual pressure point on the water flow scale.

Step 7: Read the gpm available at 20 psi (140 kPa) and record the figure. This reading represents the total available water that can be relied upon.

The following are examples of graphical analysis for water flow tests using one and two outlets.

Example 1 (U.S.): One Outlet

Test Hydrant = 50 psi static and 25 psi residual

Flow Hydrant #1 = Using one 2½-inch outlet, with C = 0.9, pitot reading = 7 psi, and actual discharge diameter = 2.56 inches.

$(29.83)(0.9)(2.56)^2(\sqrt{7}) = 466$ gpm

Flow Hydrant #2 = Using one 2½-inch outlet, with C = 0.8, pitot reading = 9 psi, and actual discharge diameter = 2.44 inches.

$(29.83)(0.8)(2.44)^2(\sqrt{9}) = 426$ gpm

Total Water Flow = 466 + 426 = 892 gpm

Example 1 (Metric): One Outlet

Test Hydrant = 345 kPa static and 173 kPa residual

Flow Hydrant #1 = Using one 65 mm outlet, with C = 0.9, pitot reading = 48 kPa, and actual discharge diameter = 66.5 mm.

$(0.0667766)(0.9)(66.5)^2(\sqrt{48}) = 1841$ L/min

Flow Hydrant #2 = Using one 65 mm outlet, with C = 0.8, pitot reading = 62 kPa, and actual discharge diameter = 63.5

$(0.0667766)(0.8)(63.5)^2(\sqrt{62}) = 1\,691$ L/min

Total Flow = 1 841 + 1 691 = 3 532 L/min

Figure 7.126 shows the test results plotted for graphical analysis of the water supply. The static pressure of 50 psi (345 kPa) is plotted at 0 gpm (0 L/min). The residual pressure of 25 psi (173 kPa) is above the total measured flow of 892 gpm (3 532 L/min), Scale A. (**NOTE:** It is important to understand that pitot pressures are never plotted on the graph; only the flow that corresponds to the pitot pressures is used). A line drawn through the static and residual pressure points now represents the water supply at the test location. It is easy to note that approximately 978 gpm (4 000 L/min) would be available at 20 psi (140 kPa). This figure represents the minimum desired intake pressure.

Example 2 (U.S.): Two Outlets

Test Hydrant = 90 psi static and 50 psi residual

Flow Hydrant = Using two 2½-inch outlets, with each C = 0.9, pitot reading for each is 17 psi, and an actual diameter of 2.56 inches.

$(29.83)(0.9)(2.56)^2(\sqrt{17}) = 725$ gpm x two outlets = 1,450 gpm

Example 2: Two Outlets (Metric)

Test Hydrant = 621 kPa static and 345 kPa residual

Flow Hydrant = Using two 65 mm outlets, both with C = 0.9, pitot reading for each = 117 kPa, and the actual diameter = 66.5 mm.

$(0.0667766)(0.9)(66.5)^2(\sqrt{117}) = 2\,875$ L/min x two outlets = 5 750 L/min

This example shows that the water flow scale must be changed so that a line can be drawn down to the 20 psi (140 kPa) level (Figure 7.127 on page 223). The available water rate at 20 psi (140 kPa) in this case would be approximately 1,970 gpm (7 450 L/min).

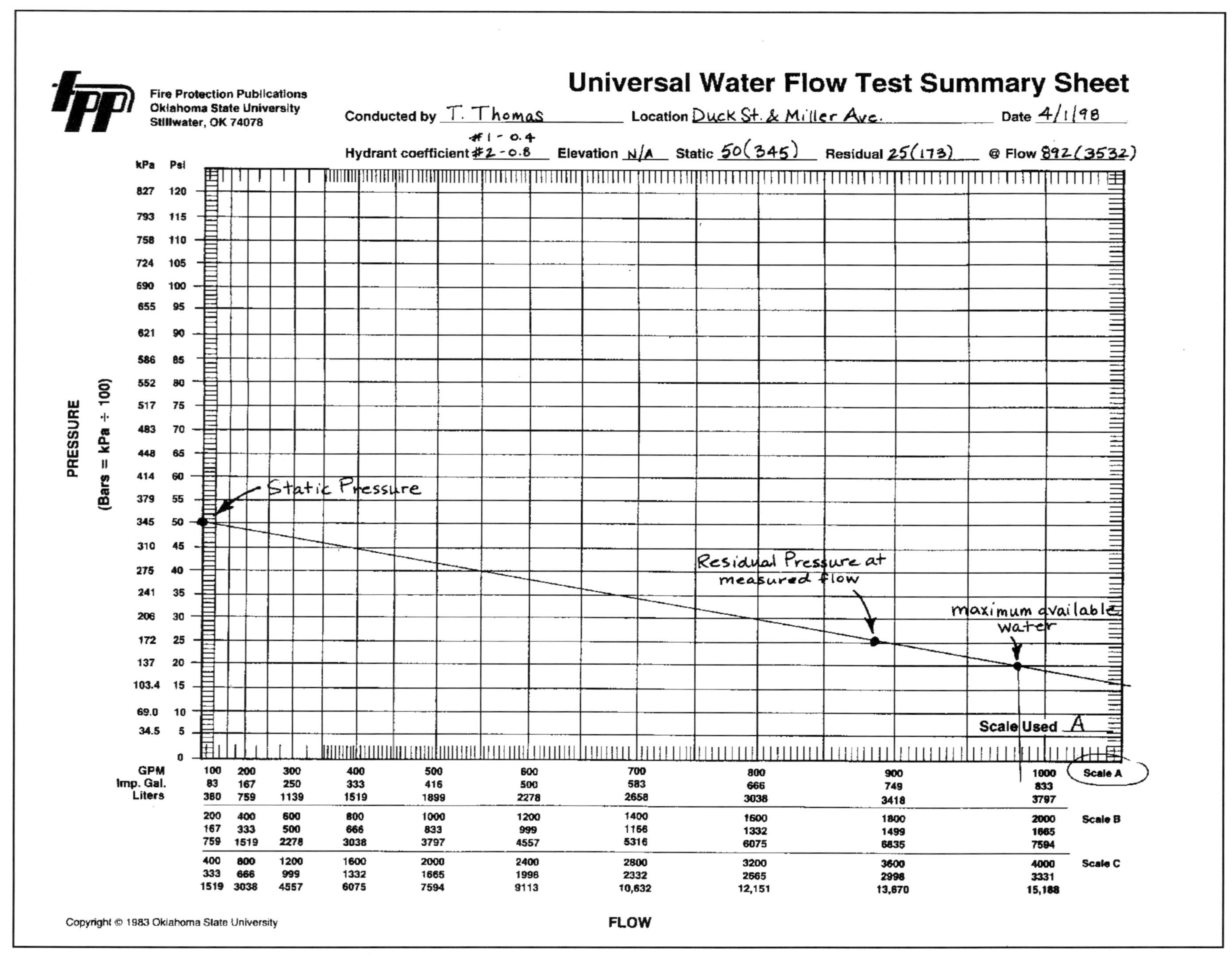

Figure 7.126 The maximum available water supply is considered the point where the line crosses 20 psi (137 kPa) on the chart.

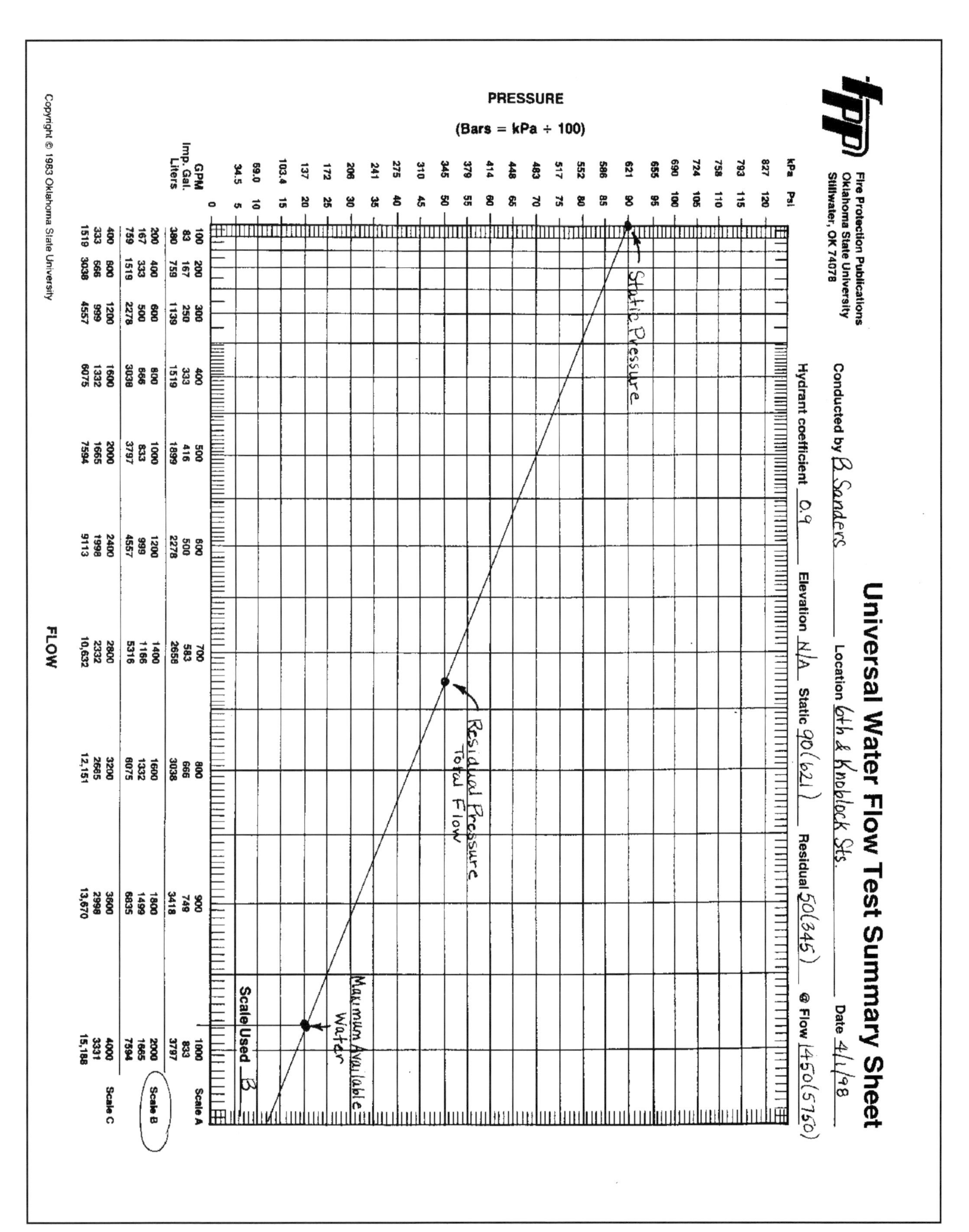

Figure 7.127 Note that Scale B is used on this example.

Determining Available Water by Mathematical Method

A variation of the Hazen-Williams formula for determining available water is written as follows:

$$Q_r = \frac{Q_f \times h_r^{0.54}}{h_f^{0.54}}$$

Where: Q_r = Flow available at desired residual pressure

Q_f = Flow during test

h_r = Pressure drop to residual pressure (normal operating pressure minus required residual pressure)

h_f = Pressure drop during test (normal operating pressure minus residual pressure during flow test)

The values for h_r or h_f to the 0.54 power are listed in Table 7.8.

Using the values from the first example, in addition to a normal operating pressure of 55 psi:

TABLE 7.8
Values for Computing Fire Flow Test

h	$h^{0.54}$	h	$h^{0.54}$	h	$h^{0.54}$	h	$h^{0.54}$	h	$h^{0.54}$	h	$h^{0.54}$	h	$h^{0.54}$
1	1.00	26	5.81	51	8.36	76	10.37	101	12.09	126	13.62	151	15.02
2	1.45	27	5.93	52	8.44	77	10.44	102	12.15	127	13.68	152	15.07
3	1.81	28	6.05	53	8.53	78	10.51	103	12.22	128	13.74	153	15.13
4	2.11	29	6.16	54	8.62	79	10.59	104	12.28	129	13.80	154	15.18
5	2.39	30	6.28	55	8.71	80	10.66	105	12.34	130	13.85	155	15.23
6	2.63	31	6.39	56	8.79	81	10.73	106	12.41	131	13.91	156	15.29
7	2.86	32	6.50	57	8.88	82	10.80	107	12.47	132	13.97	157	15.34
8	3.07	33	6.61	58	8.96	83	10.87	108	12.53	133	14.02	158	15.39
9	3.28	34	6.71	59	9.04	84	10.94	109	12.60	134	14.08	159	15.44
10	3.47	35	6.82	60	9.12	85	11.01	110	12.66	135	14.14	160	15.50
11	3.65	36	6.93	61	9.21	86	11.08	111	12.72	136	14.19	161	15.55
12	3.83	37	7.03	62	9.29	87	11.15	112	12.78	137	14.25	162	15.60
13	4.00	38	7.13	63	9.37	88	11.22	113	12.84	138	14.31	163	15.65
14	4.16	39	7.23	64	9.45	89	11.29	114	12.90	139	14.36	164	15.70
15	4.32	40	7.33	65	9.53	90	11.36	115	12.96	140	14.42	165	15.76
16	4.47	41	7.43	66	9.61	91	11.43	116	13.03	141	14.47	166	15.81
17	4.62	42	7.53	67	9.69	92	11.49	117	13.09	142	14.53	167	15.86
18	4.76	43	7.62	68	9.76	93	11.56	118	13.15	143	14.58	168	15.91
19	4.90	44	7.72	69	9.84	94	11.63	119	13.21	144	14.64	169	15.96
20	5.04	45	7.81	70	9.92	95	11.69	120	13.27	145	14.69	170	16.01
21	5.18	46	7.91	71	9.99	96	11.76	121	13.33	146	14.75	171	16.06
22	5.31	47	8.00	72	10.07	97	11.83	122	13.39	147	14.80	172	16.11
23	5.44	48	8.09	73	10.14	98	11.89	123	13.44	148	14.86	173	16.16
24	5.56	49	8.18	74	10.22	99	11.96	124	13.50	149	14.91	174	16.21
25	5.69	50	8.27	75	10.29	100	12.02	125	13.56	150	14.97	175	16.26

$Q_f = 892$ gpm

$h_r = 55$ psi – 20 psi = 35 psi

$h_f = 55$ psi – 25 psi = 30 psi

Under h at 35, $h^{0.54} = 6.82$. Under h at 30, $h^{0.54} = 6.28$.

So, $Q_r = \frac{892 \times 6.82}{6.28}$

$Q_r = 967$ gpm

Metric Example:

$Q_f = 3\,532$ L/min

$h_r = 380$ kPa – 138 kPa = 242

$h_f = 380$ kPa – 173 kPa = 207

NOTE: When doing these problems in metrics, quite often the figures obtained will be higher than those provided in Table 7.8. It will be necessary to use a calculator to determine $h^{0.54}$.

$Q_r = \frac{3\,532 \times 19.38}{17.81}$

$Q_r = 3\,843$ L/min

Although it is important for inspection personnel to understand these formulas and how the calculations are done, it is more common today for personnel to use computer programs to do these calculations. Inspection personnel simply enter the information from the flow tests into the computer and the available water supply is automatically determined. Several commercial water flow programs are available.

For more information on water supply analysis, see Fire Protection Publications' **Fire Protection Hydraulics and Water Supply Analysis** book by Pat D. Brock.

8

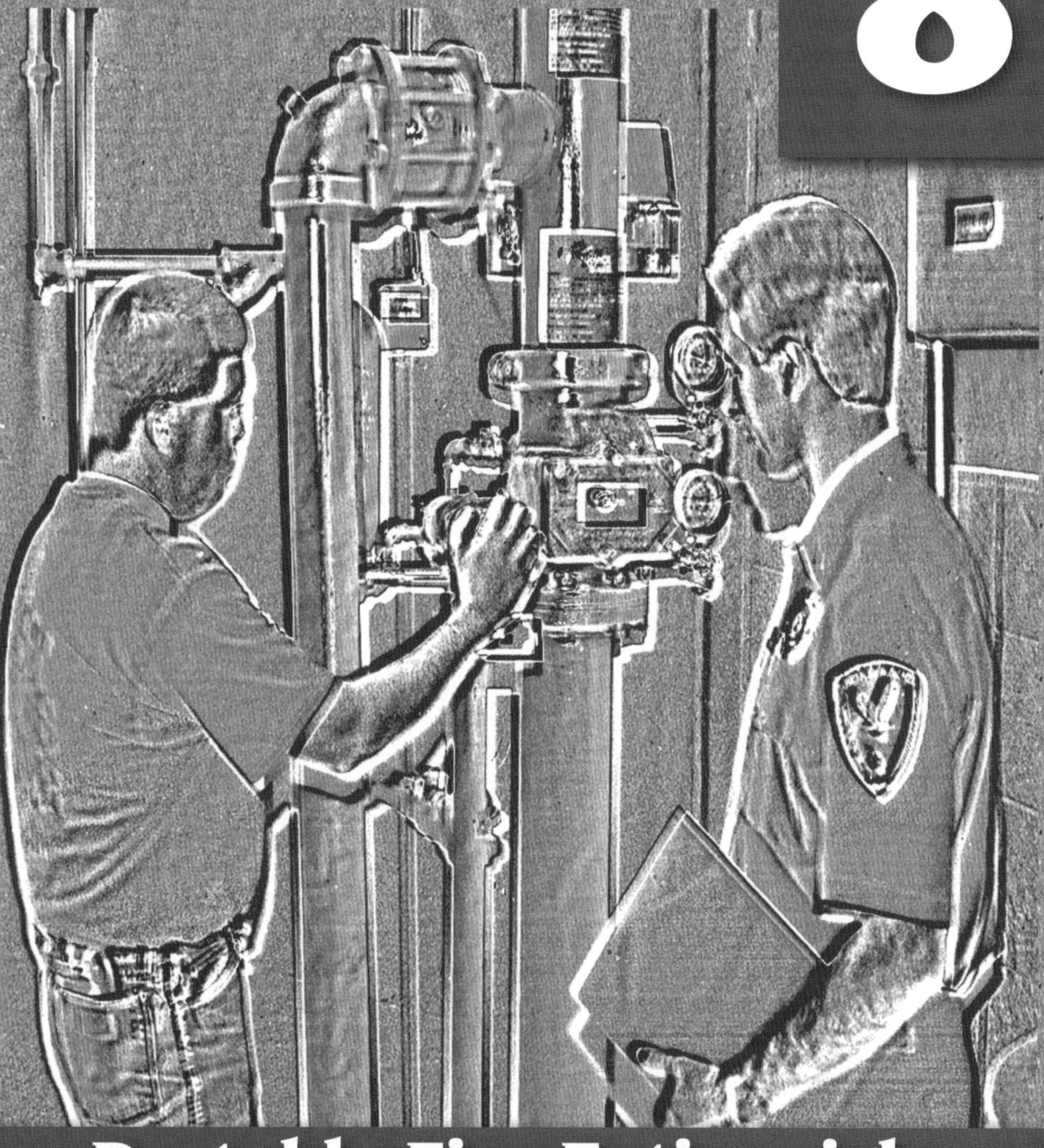

Portable Fire Extinguishers, Special Agent Fire Extinguishing Systems, and Fire Detection and Alarm Systems

This chapter contains information that will assist the reader in meeting the listed job performance requirements contained in NFPA 1031, ***Standard for Professional Qualifications for Fire Inspector and Plan Examiner*** (proposed 1998 edition).

Chapter 3 Fire Inspector I

3-3.5 Determine the operational readiness of fixed fire suppression systems, given test documentation and field observations, so that the system(s) is in an operational state, maintenance is documented, and all deficiencies are identified, documented, and reported in accordance with the policies of the jurisdiction.

3-3.6 Determine the operational readiness of existing fire detection and alarm systems, given test documentation and field observations, so that the systems are in an operational state, maintenance is documented, and all deficiencies are identified, documented, and reported in accordance with the policies of the jurisdiction.

3-3.7 Determine the operational readiness of existing portable fire extinguishers, given field observations and test documentation, so that the equipment is in an operational state, maintenance is documented, and all deficiencies are identified, documented, and reported in accordance with the policies of the jurisdiction.

Chapter 4 Fire Inspector II

4-3.4 Evaluate fire protection systems and equipment provided for the protection of a building or facility, given field observations of the facility and documentation, the hazards protected, and the system specifications, so that the fire protection systems provided are appropriate for the occupancy or hazard being protected and installed in compliance with applicable codes and standards, and all deficiencies are identified, documented, and reported in accordance with the policies of the jurisdiction.

4-4.3 Field verify the installation of a fire protection system, given shop drawings and system specifications for a process or operation, so that the system is reviewed for code compliance, installed in accordance with the approved drawings, and all deficiencies are identified, documented, and reported in accordance with the policies of the jurisdiction.

Chapter 8
Portable Fire Extinguishers, Special Agent Fire Extinguishing Systems, and Fire Detection and Alarm Systems

In addition to the water-based fire protection systems covered in the previous chapter, fire inspection personnel must be familiar with a variety of other types of fire protection equipment and systems. These include the following:

- Portable fire extinguishers
- Special agent fire extinguishing systems
- Fire detection and alarm systems

The most common private fire protection device is the portable fire extinguisher. Fire inspection personnel must have an intimate knowledge of the characteristics and applicability of each type of extinguisher. All employees of a facility should be familiar with the extinguishers in their facility and should know how to select and use the proper one in an emergency.

Special-agent, fixed-extinguishing systems are used in those situations where automatic water sprinkler systems are not desirable or compatible with the fire hazard. In these instances, protection must still be provided safely and effectively. This can be accomplished by the use of foam, carbon dioxide (CO_2), halogenated agent (or its replacement), or dry-chemical fixed systems. Inspectors must have a basic understanding of these systems so that their operability may be accurately evaluated during an inspection.

Fire detection and alarm systems are used to provide signals for alerting building occupants and/or organized fire protection units. They are also used to operate fire protection system components. All detection systems use some type of device that is sensitive to one or more products of combustion. Automatic fire detection and alarm systems with supplemental manual fire alarm pull stations should be installed in buildings for protection of life. These systems are especially important in isolated and/or high life-hazard facilities. With more emphasis on early detection and regulations that require detectors to be installed in dwellings, it becomes increasingly necessary for all fire service personnel to become well-versed in these devices. (**NOTE:** Automatic fire detection systems are not acceptable substitutes for automatic sprinkler systems.)

Appendix F contains sample inspection and testing forms for all of the systems described in this chapter.

PORTABLE FIRE EXTINGUISHERS

A portable fire extinguisher is viewed by some as the first line of defense against incipient fires of limited size. However, it should never be viewed as a substitute but rather as a complement to automatic fire detection and suppression systems. Requirements for extinguisher design and placement are contained in NFPA 10, *Standard for Portable Fire Extinguishers.*

The value of a fire extinguisher lies in the speed with which it can be used by personnel not trained as professional firefighters. For a portable fire extinguisher to be effective, the following requirements must be met:

- The extinguisher must be readily visible and accessible (Figure 8.1).
- The extinguisher must be suitable for the hazard being protected.
- The extinguisher must be in working order.

Figure 8.1 Portable fire extinguishers should be in a visible and accessible location.

- The extinguisher must be of sufficient size to control the fire.
- The person using the extinguisher must know how to operate it and be physically able to do so.

The following sections contain extensive information on how portable fire extinguishers are classified, rated, tested, used, and inspected. Also covered are the agents used in the extinguishers and the proper locations for extinguishers.

Rating of Portable Fire Extinguishers

No portable fire extinguisher is suitable for use on all fires. Therefore, portable fire extinguishers are designated with a letter or letters indicating the class or classes of fires they are designed to control. These labels are based on four classifications, or types, of fires:

- Class A: Fires involving ordinary combustibles such as wood, cloth, or paper (Figure 8.2).
- Class B: Fires involving flammable or combustible liquids, greases, and gases (Figure 8.3).
- Class C: Fires involving energized electrical equipment where the electrical nonconductivity of the extinguishing agent is of first importance. The materials involved are either Class A or B and can be

Figure 8.2 Class A fires include ordinary combustibles such as wood, paper, rubber, and plastic.

Figure 8.3 Class B fires involve flammable and combustible liquids and gases.

handled as such once the equipment is de-energized (Figure 8.4).

- Class D: Fires involving combustible metals such as magnesium, titanium, zirconium, and potassium (Figure 8.5). These fires may require special extinguishing agents or techniques. If there is doubt, the inspector should consult NFPA 325, *Guide to Fire Hazard Properties of Flammable Liquids, Gases, and Volatile Solids.*

Multiple letters or numerical-letter ratings are used on portable fire extinguishers that are effective on more than one class of fire. Class A and Class B extinguishers also receive a numerical rating that precedes the letter (Figure 8.6). This rating designates the size fire the extinguisher can be expected to suppress when used by an untrained operator. It is very important that the correct agent be used on a fire. Using the wrong agent can be dangerous and can result in a fire not being extinguished, a violent reaction, or both. The ratings for different types of extinguishers are as follows:

- Class A extinguishers: Ratings from 1-A through 40-A are designated for Class A fire extinguishers. A water-type extinguisher

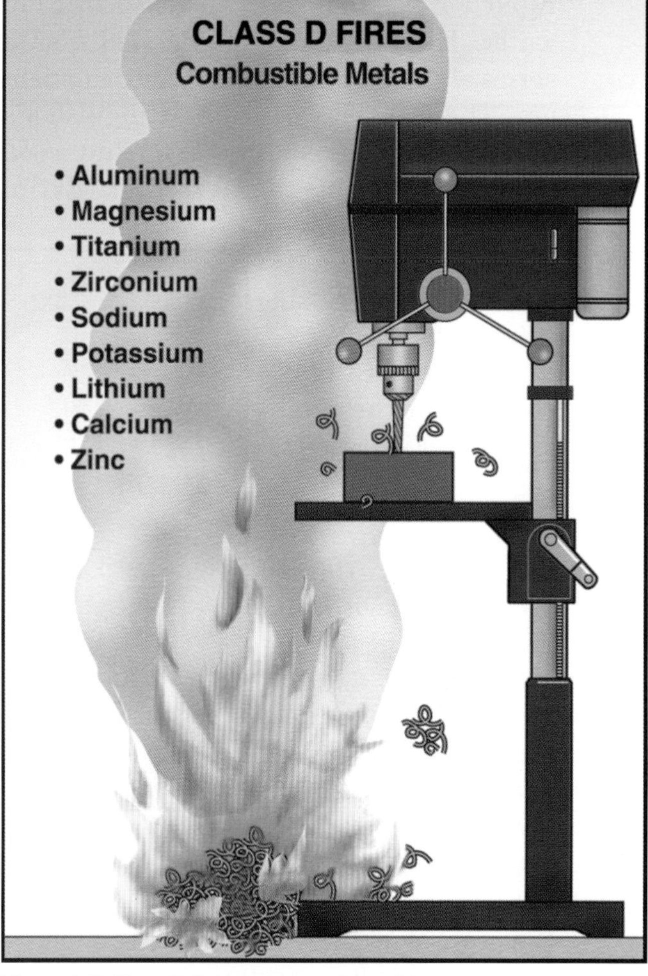

Figure 8.5 Class D fires occur in combustible metals.

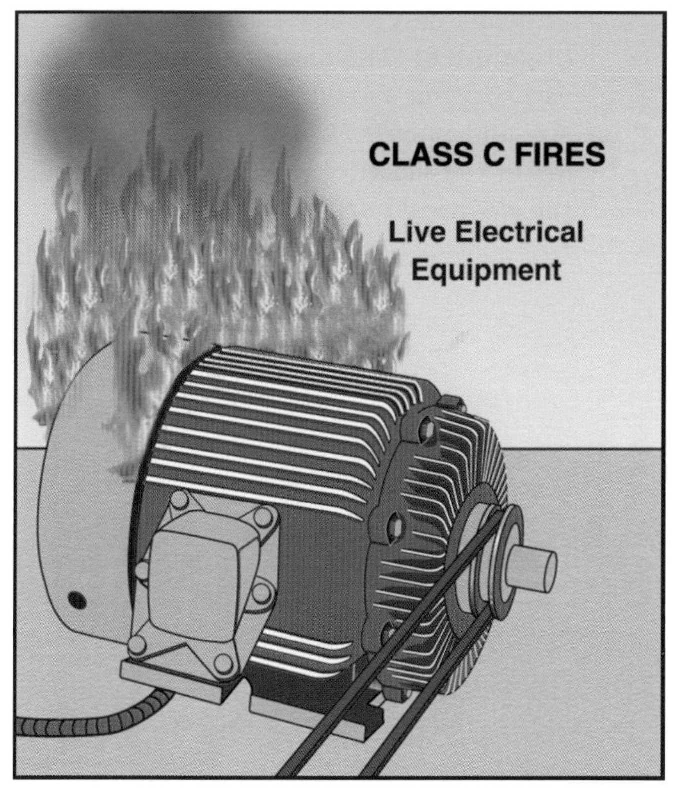

Figure 8.4 Class C fires involve energized electrical equipment.

Figure 8.6 The extinguisher's rated capabilities can be found on the faceplate.

rated 1-A requires 1¼ gallons (4.73 L) of water. These numbers are derived from a series of three tests for Class A extinguishers. These tests check the extinguisher's ability to extinguish fires involving wood cribs, wood panels, and excelsior. Extinguishers that are rated 1-A through 6-A are subjected to all three tests. Larger extinguishers are only subjected to the wood crib test (Figure 8.7).

- Class B extinguishers: Extinguishers for use on Class B fires are classified with numerical ratings from 1-B through 640-B. The number indicates the approximate area, in square feet (m^2), of fire involving a 2-inch (50 mm) layer of n-heptane in an 8-inch (200 mm) deep pan that can be extinguished. For example, a 10-B portable fire extinguisher can be expected to extinguish a fire of 10 square feet (0.9 m^2). This is the rating for an untrained operator. A trained operator should be able to extinguish 25 square feet (2.3 m^2) with a 10-B extinguisher (Figure 8.8).

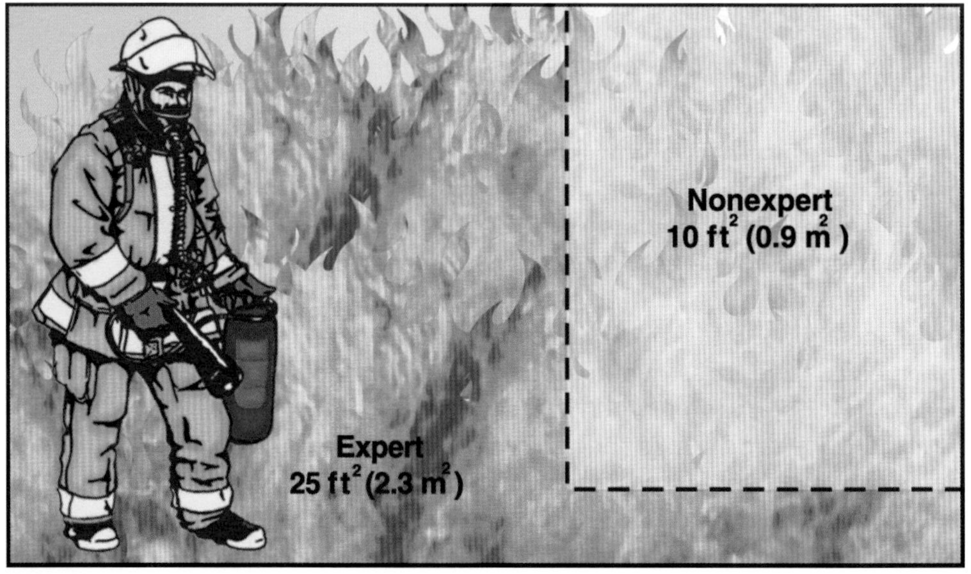

Figure 8.8 Expert users can extinguish considerably more fire than nonexperts.

Figure 8.7 The wood crib test is used to rate Class A fire extinguishers.

- Class C extinguishers: Extinguishers for Class C fires have no numerical rating and are tested only for electrical nonconductivity (Figure 8.9). The size of the portable fire extinguisher should be appropriate for the size and extent of Class A and/or B materials in the electrical equipment or around the electrical hazard.

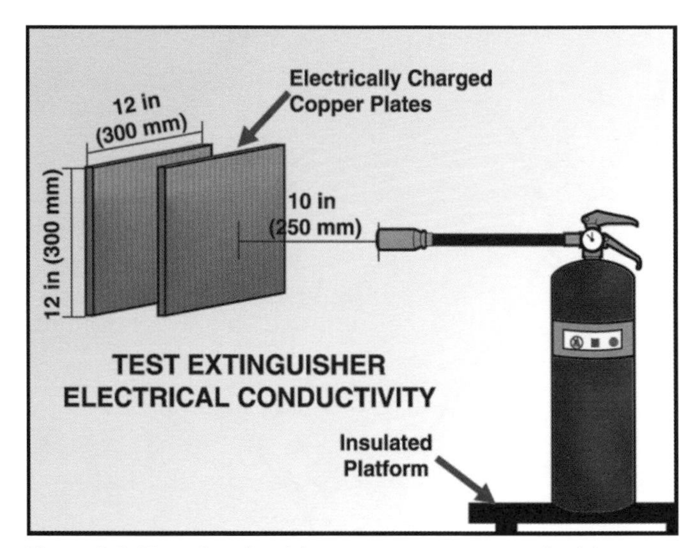

Figure 8.9 Class C extinguishers must pass a nonconductivity test.

- Class D extinguishers: Extinguishers for Class D fires have no numerical rating and the type of tests conducted vary depending upon the metals for which the extinguisher is intended. The faceplate of the extinguisher details the specific metals on which the extinguisher should be used, and how to use the extinguisher.

EXAMPLES OF EXTINGUISHER RATING LABELS

Foam Extinguisher Rated 4-A, 6-B. This extinguisher will extinguish four times the amount of Class A fire as a 1-A extinguisher, and six times as much Class B fire as a 1-B extinguisher.

Dry-Chemical Extinguisher Rated 10-B:C. This extinguisher will extinguish approximately ten times as much Class B fire as a 1-B unit and should extinguish a deep layer flammable liquid fire in a 10 square-foot (1 m^2) area. It is also safe to use on fires involving energized electrical equipment.

Multipurpose Extinguisher Rated 4-A, 20-B:C. This extinguisher has the Class A extinguishing equivalent of 5 gallons (20 L) of water, and approximately twenty times as much Class B fire extinguishing capability as a 1-B extinguisher, and should extinguish a deep-layer flammable liquid fire in a 20 square-foot (2 m^2) area. It is also safe to use on fires involving energized electrical equipment.

Extinguisher Symbols

To simplify the process of matching different types of extinguishers with types of fires, several methods for identifying extinguishers by using symbols have been developed. NFPA 10 recognizes two methods of extinguisher recognition: the pictorial system and the letter-symbol system.

PICTORIAL SYSTEM

The international "picture-symbol" labeling system is the most widely used identification system; it was designed by the National Association of Fire Equipment Distributors (NAFED). This system is designed to make the selection of fire extinguishers easier through the use of picture symbols (Figure 8.10). The symbols indicate the type of fire the extinguisher is capable of extinguishing. They also indicate when ***not*** to use an extinguisher on certain types of fires.

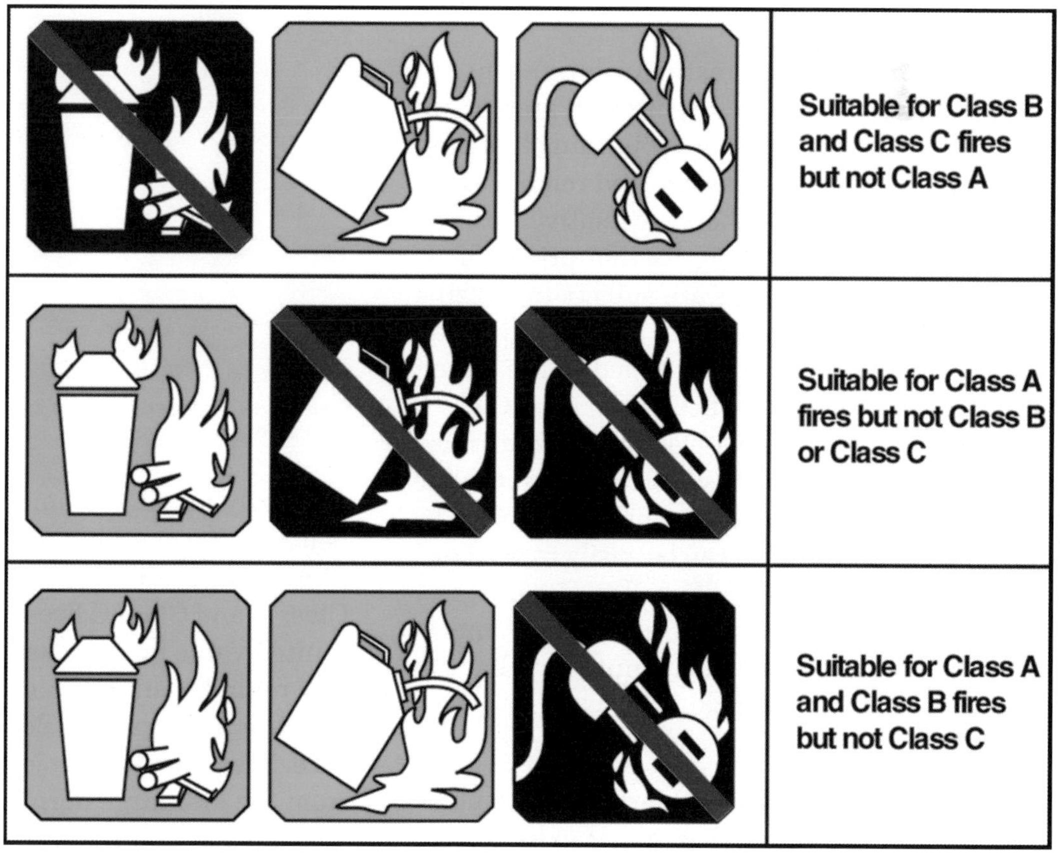

Suitable for Class B and Class C fires but not Class A

Suitable for Class A fires but not Class B or Class C

Suitable for Class A and Class B fires but not Class C

Figure 8.10 The pictorial method for marking portable fire extinguishers.

If an extinguisher is suitable for use on a particular class of fire, the picture-symbol background is light blue. If the extinguisher is not suitable for a particular class of fire, the picture symbol has a black background with a diagonal red line through the extinguisher symbol. Also, an extinguisher may be suitable for more than one class of fire.

LETTER-SYMBOL SYSTEM

The letter-symbol method of extinguisher identification is older than the picture-symbol method. In the letter-symbol method, each class of fire is represented by its appropriate letter: A, B, C, or D, which is enclosed by a particular geometric shape (Figure 8.11). In addition, the background of the geometric shape can be color-coded to further identify the extinguisher.

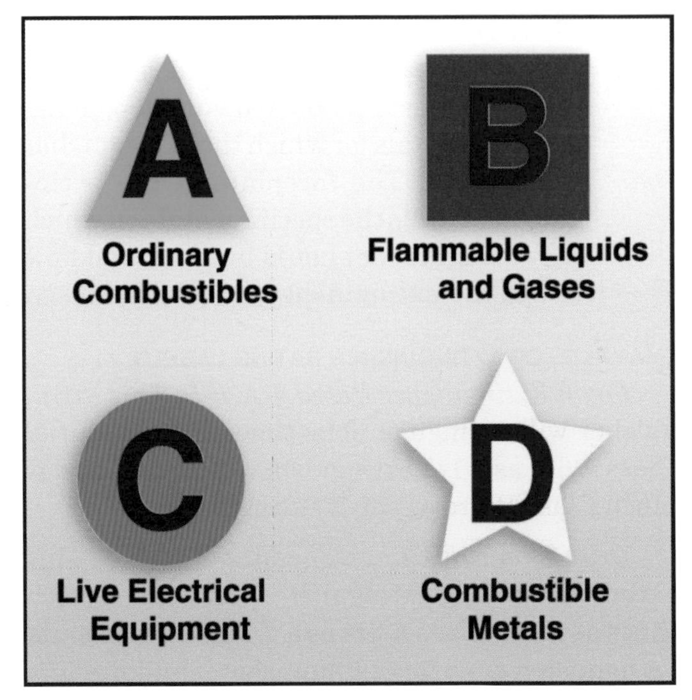

Figure 8.11 The letter-symbol method for marking portable fire extinguishers.

Extinguishing Agents

Portable fire extinguishers use many different types of extinguishing agents. Each agent may be able to control one or more classes of fire, but one agent cannot extinguish all classes of fire. The following sections highlight the more common extinguishing agents.

WATER

Water is a convenient extinguishing agent but is effective only on Class A fires. Water extinguishes primarily by cooling the burning fuel (Figure 8.12). Water is inexpensive and readily available, and water extinguishers are relatively easy to maintain. Water does have its limitations, however. Water extinguishers are subject to freezing and must be kept in a heated area, unless an approved antifreeze agent is added to the water. Water, in itself, is also ineffective on most Class B fires. Because it conducts electricity, water may not be used on Class C fires. Furthermore, the fire fighting capability of an extinguisher is limited by the amount of water that can be easily carried in a portable unit. Typically, 2½-gallon (10 L) water extinguishers are the most common size and 5-gallon (20 L) units are the maximum size that can be considered to be portable.

Figure 8.12 Water extinguishes fire primarily by cooling the fuel below its ignition temperature.

CARBON DIOXIDE

Carbon dioxide (CO_2) is a colorless, noncombustible gas that is heavier than air. It extinguishes primarily through a smothering action by establishing a gaseous blanket between the fuel and the surrounding air (Figure 8.13). It is suitable for Class B and Class C fires. Carbon dioxide has very limited value on deep-seated Class A fires, which can rekindle after the carbon dioxide dissipates into the atmosphere. Because of its gaseous nature, it is difficult to project carbon dioxide very far from the discharge horn of the extinguisher. Carbon dioxide extinguishers characteristically dis-

Figure 8.13 Carbon dioxide extinguishes fire by smothering it.

charge with a loud noise that may startle an untrained operator. When operated in areas of low humidity, there may also be a discharge of static electricity that can further startle an untrained operator.

The carbon dioxide is stored in the extinguisher in a liquid state at a pressure of about 840 psig (5 880 kPa). Storing the carbon dioxide as a liquid allows more agent to be stored in a given volume. When discharged from the extinguisher, the carbon dioxide has a white cloudy appearance. This is due to the small dry ice crystals formed by the condensation of surrounding water vapor that is carried along in the gas stream when it is discharged. Although the carbon dioxide is very cold when discharged, this has a minimal effect in cooling or extinguishing the fire.

Warning

Carbon dioxide is an asphyxiant. Do not use in a small, confined space such as a closet.

AQUEOUS FILM FORMING FOAM (AFFF)

Aqueous film forming foam (AFFF) produces both an air foam and a floating film on the surface of a liquid fuel (Figure 8.14). AFFF is suitable for both Class A and Class B fires. Most commonly, the AFFF concentrate is premixed with water in the extinguisher and discharged through a special

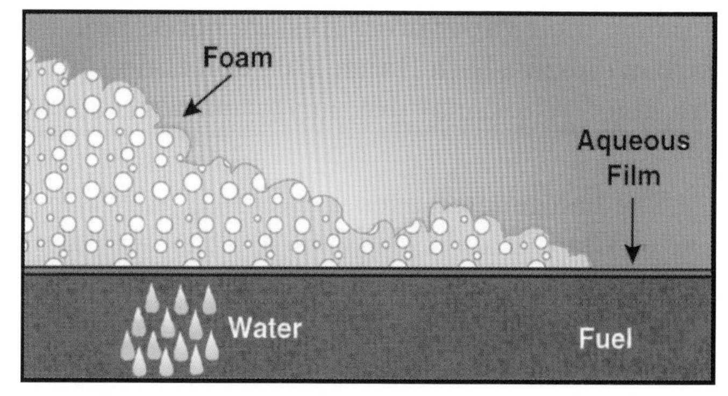

Figure 8.14 AFFF creates a film that floats on the surface of the fuel.

aerating nozzle. Because the foam agent is mixed with water, AFFF is effective on Class A fires by cooling and penetrating the fuel. This agent is very effective on flammable liquid fires because of the double effect of a foam blanket and a surface film to exclude air from the fuel. Note that the AFFF/water mix has all the same inherent limitations discussed earlier for plain water. (**NOTE:** AFFF is discussed in more detail later in this chapter).

HALOGENATED AGENTS

Halogenated extinguishing agents (herein referred to as halon agents) have been used for fire protection since the late 1940s. Their inherently clean nature made them ideal for protecting confined, occupied areas containing sensitive electronic equipment. The halon agents contain atoms from one of the halogen series of chemical elements: fluorine, chlorine, bromine, and iodine. While there have been a number of halon agents developed over the years, there are two that are most commonly used for fire protection purposes:

- Bromochlorodifluoromethane (CF_2BrCl) or Halon 1211 is most commonly used in portable fire extinguishers.
- Bromotrifluoromethane (CF_3Br) or Halon 1301 is most commonly used in fixed system applications.

Halon agents extinguish fire by a complicated, chemically dynamic process in which the halon molecule breaks down and inhibits the combustion reaction. Both Halon 1301 and Halon 1211 are gases at room temperature and are therefore clean agents. Halon 1211 is a vaporizing liquid. Halon 1301 is a liquefied gas that is stored under pressure in a manner similar to carbon dioxide. However, because their vapor pressures are not as high as carbon dioxide, both Halon 1301 and Halon 1211 are given additional pressurization with nitrogen.

The halogenated agents are principally effective on Class B and Class C fires, although Halon 1211 extinguishers also have Class A ratings. On a pound-for-pound basis, the halons are more effective than carbon dioxide.

There are two primary disadvantages of halon agents. The first is their toxicity. Exposure to modest concentrations of the halons (7 percent for Halon 1301 and 3 percent for Halon 1211) produces such undesirable effects as dizziness and reduction of physical and manual dexterity.

(**CAUTION:** Avoid prolonged exposure to the halons. In addition, the halons break down under fire conditions and liberate toxic substances such as chlorine, bromine, hydrogen chloride, and hydrogen bromide. These are irritating gases, and their presence is readily apparent to an extinguisher user.)

Another primary disadvantage of halon agents is their adverse effect on the earth's atmosphere, primarily the ozone layer. Concern over this issue led to a multi-national effort, called the United Nations Environmental Program (UNEP), to call for a treaty to limit the production of these chemicals. The treaty, known as the Montreal Protocol after the host city, was finalized in September of 1987. The basic elements of the Treaty were:

- Halon production would be frozen at 1986 levels beginning in 1992.
- There should be a 50 percent reduction in halon production by 1995.
- All halon use would be phased out by 2000.

Later on, the schedule was significantly moved up. New halon agents were not to be manufactured after January 1, 1994. However, limited production continues because there are some exceptions to the phase-out plan. Locations where halon agent use is deemed to be "essential" may be granted an exemption from the phase-out. The criteria for this exemption are:

- Halon agent use is necessary for human health and safety or critical for the functioning of society.
- There are no technically or economically feasible alternatives.
- All feasible actions must be taken to minimize emissions from use.
- The supply of halon agents from existing banks or recycled stocks is not sufficient to accommodate the need.

There has been considerable research and development on new clean agents to replace halon agents. These are commonly referred to as halon replacement or halon alternative agents. The intent of this research has been to develop agents that extinguish fires in the same manner as halon agents, but with no significant damage to the atmosphere. These agents can be put directly into an existing halon extinguishing system or portable extinguisher.

Several halon replacement agents, such as FM-110, FM-200, FE 25, and Inergen (IG-541), are commercially available. However, none of these agents are a "pound-for-pound" replacement for the halon agents. These replacement agents require anywhere from 2 to 20 times as much agent by weight or volume as halon agents to provide an equal amount of extinguishing capability.

For more information on halon replacement agents and their extinguishing systems, see NFPA 2001, *Standard on Clean Agent Fire Extinguishing Systems*.

DRY-CHEMICAL AGENTS

Several dry chemicals have proven very useful as extinguishing agents in portable fire extinguishers. These include sodium bicarbonate, potassium bicarbonate, urea potassium bicarbonate, and potassium chloride, which are usually referred to as "ordinary" dry-chemical agents. In addition, monoammonium phosphate is a "multipurpose" dry chemical. The three most frequently used agents are sodium bicarbonate, potassium bicarbonate, and monoammonium phosphate.

In physical form, the dry chemicals are very small solid particles — not gases or liquids (Figure

8.15). Because the dry chemicals are solid particles, they can be projected more effectively from the extinguisher nozzle than can gaseous agents. Therefore, dry chemicals do not dissipate into the atmosphere as rapidly as gases. Dry chemicals are especially suitable for controlling fires outdoors.

Figure 8.15 Dry chemical agents are very fine particles.

Sodium Bicarbonate

Sodium bicarbonate (also known as ordinary dry chemical) is effective on Class B and Class C fires. It is widely used for the protection of commercial food preparation equipment such as fryers and range hoods. When evaluated against an equal weight of carbon dioxide, sodium bicarbonate is twice as effective for Class B fires. Sodium bicarbonate has a very rapid knockdown capability against flaming combustion and has some effect on surface fires in Class A materials. In this connection, it has been used successfully on textile machinery where the fine textile fibers can produce a surface fire. Sodium bicarbonate used in fire extiguishers is chemically treated to be water repellent and free flowing.

Potassium Bicarbonate

Potassium bicarbonate (also known as Purple-K) has properties and applications similar to sodium bicarbonate. However, on a pound-for-pound basis, it is about twice as effective as sodium bicarbonate. That is, given a specific amount of potassium bicarbonate, one can extinguish a fire that is twice the size of a fire that is capable of being extinguished with the same amount of sodium bicarbonate. Like sodium bicarbonate, it is most effective on Class B and Class C fires. It is also treated to be water repellent and free flowing. Potassium bicarbonate is color-coded violet to differentiate it from other dry chemicals.

Monoammonium Phosphate

An ideal extinguishing agent would be equally effective on all classes of fire. The agent that comes closest to this goal is monoammonium phosphate (also known as multi-purpose dry chemical). It is effective on Class A, Class B, and Class C fires. It has an action similar to other dry chemicals on flammable liquid fires. Using a combination of extinguishing methods, it quickly knocks down flaming combustion. On Class A materials, the monoammonium phosphate melts, forming a solid coating and extinguishing the fire by a smothering action. This agent has a pale yellow color.

DRY POWDERS

Dry powder extinguishing agents are designed to extinguish Class D fires in combustible metals such as aluminum, magnesium, sodium, and potassium (Figure 8.16). There is no single agent that is effective on all combustible metals. In a given situation, the extinguishing agent must be carefully chosen for the hazard being protected. The following sections address just three of the more common Class D agents.

NA-X®

NA-X is a Class D extinguishing agent designed specifically for use on sodium, potassium,

Figure 8.16 Dry powders must be carefully applied to a Class D fire.

and sodium-potassium alloy fires. NA-X is not suitable for use on magnesium fires. Chemically, NA-X has a sodium carbonate base with additives to enhance flow. The extinguishing action is a crusting or caking on the burning material, causing an oxygen deficiency, and thereby extinguishing the fire. NA-X is listed by Underwriters Laboratories Inc. for use on burning materials at fuel temperatures up to 1,400°F (760°C). Application can be from portable extinguishers or by scoops from pails.

MET-L-X®

MET-L-X is a sodium chloride (salt) based extinguishing agent intended for use on magnesium, sodium, and potassium fires. Like other dry powders, it contains additives to enhance flowing and prevent caking in the extinguisher. It also extinguishes metal fires by forming a crust on the burning metal to exclude oxygen. The agent is applied from the extinguisher to first control the fire, and then the agent is applied more slowly to bury the fuel in a layer of the powder. The agent is stable when stored in sealed containers. It is nonabrasive and has no known toxic effects.

LITH-X®

Lith-X is an agent that can be used on several combustible metals. It was developed to control fires involving lithium but can also be used to extinguish magnesium, zirconium, and sodium fires. Lith-X consists of a graphite base that extinguishes fires by conducting heat away from the fuel after a layer of the powder has been applied to the fuel. Unlike other dry powders, it does not form a crust on the burning metal.

Types of Fire Extinguishers

Portable fire extinguishers use different methods to expel the extinguishing agent and can be broadly classified according to the method used. These include the following:

- Stored pressure
- Cartridge pressure
- Pump

STORED-PRESSURE MODELS

A stored-pressure model contains an expellant gas and extinguishing agent in a single chamber (Figure 8.17). The pressure of the gas forces the agent out through a siphon tube, valve, and nozzle assembly. The pressurizing gas can be a different gas from the agent itself. For example, dry-chemical extinguishers typically use nitrogen as an expellant gas. In other cases, the expellant gas can be the vapor phase of the agent itself, such as that in carbon dioxide extinguishers. Units that use a separate expelling gas have a pressure gauge that permits the user to see whether the extinguisher is ready for use. Air pressurized water (APW) extinguishers are among the most common type of stored-pressure extinguishers.

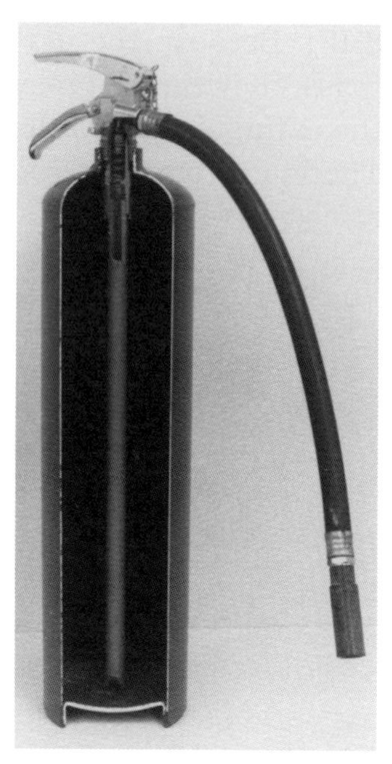

Figure 8.17 The inside of a stored-pressure fire extinguisher.

The stored-pressure extinguisher is simple to use. It usually requires only that the operator remove a safety pin and squeeze the valve handle. However, refilling the unit requires special charging equipment for pressurization and is normally serviced by distributors. This type of extinguisher may be found in such areas as office buildings or department stores where a high-use factor is not involved.

CARTRIDGE-OPERATED MODELS

The cartridge-operated extinguisher has the expellant gas stored in a separate cartridge, while the extinguishing agent is contained in an adjacent agent cylinder (Figure 8.18). To actuate the extinguisher, the expellant gas (carbon dioxide or nitrogen) is released into the agent chamber. The pressure of the gas forces the agent through the siphon tube into the hose. Discharge is controlled by a handheld nozzle. No pressure gauge is provided. During inspection, the expellant gas cartridge is weighed to ensure that it has adequate gas. The extinguisher is recharged by replacing the gas cartridge and refilling the agent chamber. This procedure may be performed in-house and does not require special equipment. These extinguishers are found in industrial operations, such as paint spraying or solvent manufacturing facilities, where they may be used frequently.

PUMP-OPERATED MODELS

A pump-operated extinguisher discharges its agent by the manual operation of a pump (Figure 8.19). This type of extinguisher is limited to the use of water as the extinguishing agent. Its primary advantage is that it can be refilled from any available water source in the course of extinguishing a fire. Maintenance is extremely simple, consisting mainly of ensuring that the extinguisher is full and has not suffered any mechanical damage.

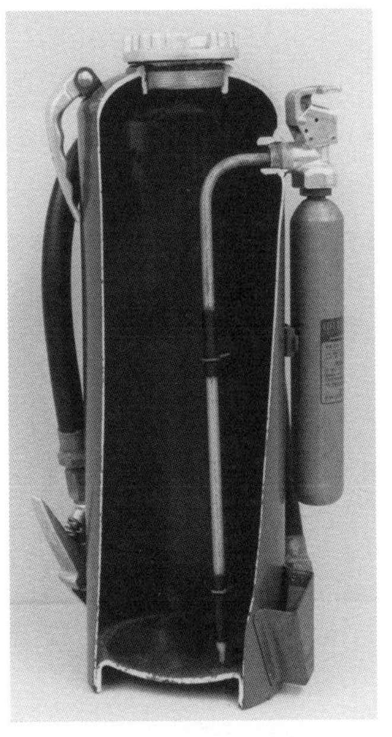

Figure 8.18 The inside of a cartridge-operated fire extinguisher.

Figure 8.19 A typical pump-operated portable fire extinguisher.

Auxiliary and Manual Fire Extinguishing Equipment

In addition to conventional portable extinguishers, a number of auxiliary devices are sometimes used. These devices include the following:

- Buckets of water
- Buckets of sand or salt
- Fire blankets
- Garden hoses

Although these devices cannot extinguish fire as efficiently as portable fire extinguishers, they can provide a measure of fire protection when problems are encountered with conventional extinguishers. These situations include such locations as construction sites where high rates of vandalism or theft exist.

A hanging, covered bucket containing 10 to 12 quarts (9.5 L to 11.4 L) of water is a common auxiliary extinguishing device. It is made of galvanized metal and has a rounded bottom. The rounded bottom makes the fire bucket unsuitable for normal use and deters its theft. Its use consists of simply throwing the contents over the fire. Hence, its control and effectiveness are highly dependent on the skill of the user. It is a good practice to provide the bucket with a cover to prevent its being used as a trash receptacle and to reduce evaporation.

Buckets filled with sand are occasionally used, although their value is very limited (Figure 8.20). They may be useful for the following types of situations:

- Spilled flammable liquid
- Certain combustible metals
- Cutting and welding operations

Fire blankets are made from several types of materials including flame-retardant fabrics, aluminized fabrics, and flame-resistant wool. To be of value, a fire blanket must be readily available and must be located within a few feet (meters) of the

Figure 8.20 Fire buckets filled with water or sand may still be found in some occupancies.

hazard. They are stored in wall cases or portable canvas bags (Figure 8.21). Treated blankets are also used around welding operations to protect nearby combustibles. Fire blankets are appropriate in situations where a person's clothing may catch fire, such as chemical laboratories.

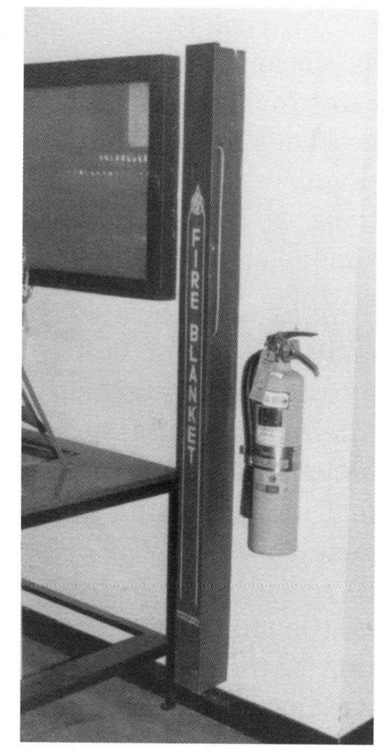

Figure 8.21 Fire blankets are commonly stored in wall cases that allow them to be unrolled and deployed quickly, even by the person needing it.

Small diameter hose (garden hose) can be very useful in combating Class A fires. Where a suitable outlet exists, such as a faucet, a small hose can provide a suitable amount of water. Small hoses are available with diameters from ½-inch to ¾-inch (13 mm to 19 mm). Very little operator skill is necessary to use these hoses. However, their useful range is limited by the length of the hose and nozzle. In addition, there is a natural tendency to use small hose for nonemergency purposes, such as lawn sprinkling, and it may be difficult to try and keep it available exclusively for emergency purposes. One should also keep in mind that this type of hose is only useful on incipient fires. As the fire grows in intensity, small hose cannot flow enough water to control it.

Obsolete Extinguishers

Fire inspection personnel must be aware of fire extinguishers that are out of date and no longer suitable for use. Employing these extinguishers, even when used as directed, could result in injuries or death to the user. All obsolete extinguishers should be removed from service and replaced with extinguishers that meet NFPA 10. The following sections examine some of the common types of obsolete extinguishers that are still frequently encountered.

In 1969, American fire extinguisher manufacturers ceased production of all inverting-type fire extinguishers (Figure 8.22). However, some of these obsolete extinguishers, including soda-acid, chemical foam, cartridge-operated water, and loaded stream, are still in use.

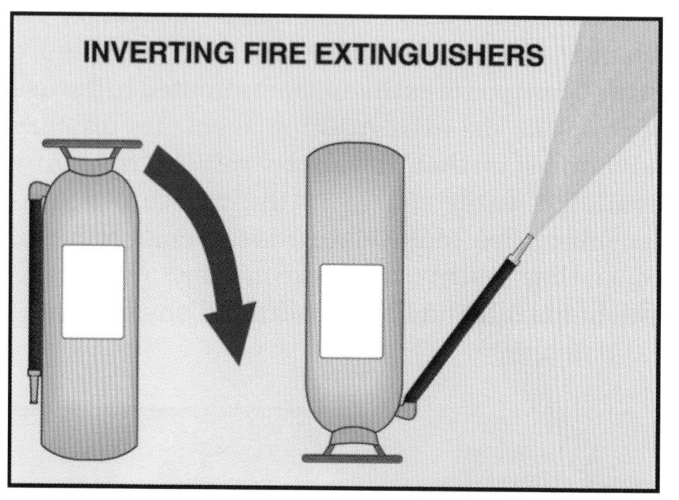

Figure 8.22 Inverting extinguishers became active when the user turned them upside down.

Soda-Acid Extinguisher. This is the most common of the obsolete extinguishers (Figure 8.23). When inverted, a bottle of acid mixes with the solution of sodium bicarbonate and water. This mixture results in a chemical reaction that produces carbon dioxide. The carbon dioxide produces the pressure to expel the liquid. This pressure increase, combined with an acid-corroded or otherwise weakened shell, could result in a violent failure of the unit. In warehouses and other areas, this type of extinguisher is subject to freezing. A 2½-gallon (10 L) unit has a rating of 2-A. Disadvantages associated with the inverted extinguishers include the following:

Figure 8.23 Soda-acid extinguishers are no longer suitable for use.

- Extinguishing agents are excellent conductors of electricity.

- Extinguisher cannot be turned off once it is activated.
- Extinguishing agent is more corrosive than water.
- Extinguisher is more costly to maintain.
- Extinguisher is potentially dangerous to the operator during use. If the discharge hose becomes blocked, it can build up pressures in excess of 300 psi (2 100 kPa), resulting in serious injury or death.
- Agent tank may have corroded over the years. This may result in a violent failure when the tank becomes pressurized, resulting in serious injury or death.

As of January 1, 1982, the Occupational Safety and Health Administration (OSHA) required that all soldered or riveted shell soda-acid extinguishers be removed from service in workplaces within the United States (Figure 8.24).

Figure 8.24 Older extinguishers had riveted tanks.

Cartridge-Operated Extinguisher. The cartridge-operated water extinguisher also requires inverting but also needs to be bumped on the floor to puncture a carbon dioxide cartridge that pressurizes the unit. This type of extinguisher can be supplied with an antifreeze solution. The 2½-gallon (10 L) unit has a rating of 2-A.

Inverting Foam Extinguisher. The inverting foam extinguisher looks like the soda-acid type. It also contains two solutions that when mixed together produce a foam as well as an expelling gas (Figure 8.25). The foam created has an expansion ratio of about 8 to 1; therefore, a 2½-gallon (10 L) unit could produce about 18 to 20 gallons (72 L to 80 L) of foam. Antifreeze cannot be added to the solution. This type of extinguisher has been replaced by those using AFFF.

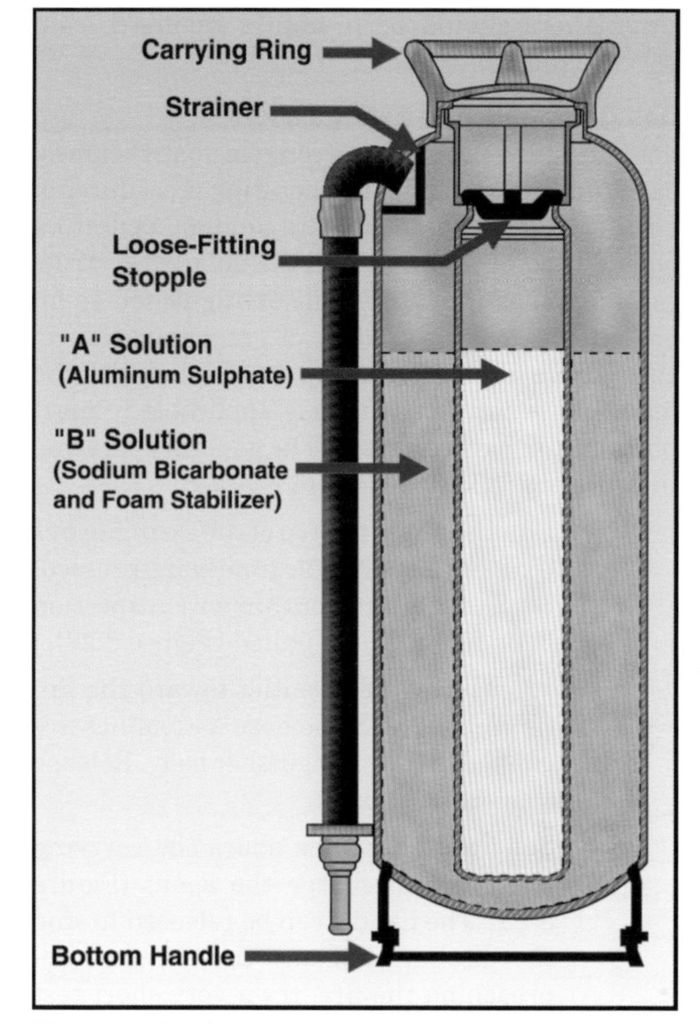

Figure 8.25 Chemical foam extinguishers operated on much the same principle as soda-acid extinguishers.

Vaporizing Liquid Extinguisher. This extinguisher, which became obsolete in the 1960s, used early halon agents such as carbon tetrachloride or chlorobromomethane. These agents produce toxic gases when applied to a fire. A common form of the extinguisher was the one-quart pump gun, which resembled a common bug sprayer (Figure 8.26). These ex-

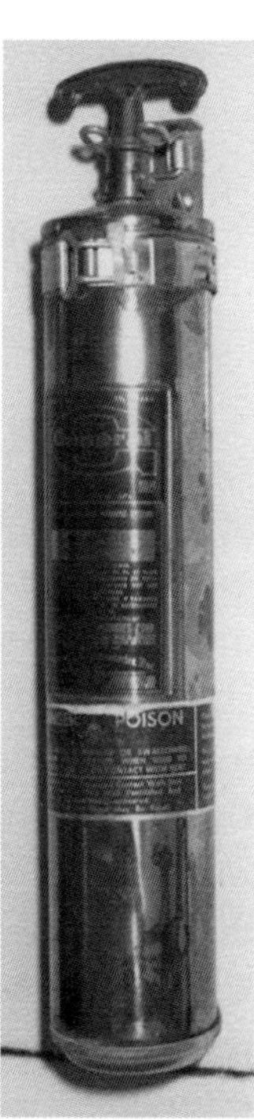

Figure 8.26 Early vaporizing liquid extinguishers were the predecessors to halon extinguishers and looked like bug sprayers.

tinguishers should be promptly removed from service and destroyed.

Using Portable Fire Extinguishers

Portable extinguishers come in many shapes, sizes, and types. While the operating procedures of each type of extinguisher are similar, operators should become familiar with the detailed instructions found on the label of the extinguisher. In an emergency, every second is of great importance; therefore, everyone should be acquainted with the following general instructions applicable to most portable fire extinguishers. The general operating instructions follow the letters P-A-S-S:

P — Pull the pin at the top of the extinguisher that keeps the handle from being pressed. Break the plastic or thin wire inspection band as the pin is pulled (Figure 8.27).

A — Aim the nozzle or outlet toward the fire (Figure 8.28). Some hose assemblies are clipped to the extinguisher body. Release the hose and point.

S — Squeeze the handle above the carrying handle to discharge the agent (Figure 8.29). The handle can be released to stop the discharge at any time. Before approaching the fire, try a very short test burst to ensure proper operation.

S — Sweep the nozzle back and forth at the base of the flames to disperse the extinguishing agent (Figure 8.30). After the fire is out, watch for remaining smoldering hot spots or possible flashback of flammable liquids. *Make sure that the fire is out.*

Modern extinguishers are designed to be carried to the fire in an upright position. When instructing the general public in the use of extinguishers, emphasize that extinguishers are operated in an upright position. (**NOTE:** Remember that only obsolete soda-acid extinguishers, foam extinguishers, and cartridge-operated water extinguishers are designed to be turned upside down.) Make sure that the fire is within range before discharging the extinguishing agent. Otherwise, the agent will be wasted. Smaller extinguishers require closer approach to the fire because they have less range than do larger extinguishers.

Figure 8.27 Pull the pin from between the two handle sections.

Figure 8.28 Attack the fire from a safe, yet effective distance.

Figure 8.29 Squeeze the handle sections together.

Figure 8.30 Sweep the agent over the fire.

Selection and Distribution of Portable Fire Extinguishers

In order to determine the number of fire extinguishers needed to adequately protect a property, fire inspectors must take into consideration a number of factors. The hazard or hazards present in the occupancy are probably the most important considerations in the selection of fire extinguishers; however, the following elements also should be considered during extinguisher selection:

- Chemical and physical characteristics of the combustibles that might be ignited
- Potential severity (size, intensity, and rate of advancement) of any resulting fire
- Location of the extinguisher
- Effectiveness of the extinguisher for the hazard in question
- Personnel available to operate the extinguisher, including their physical abilities, emotional characteristics, and any training they may have in the use of extinguishers
- Environmental conditions that may affect the use of the extinguisher (temperature, winds, presence of toxic gases or fumes)
- Any anticipated adverse chemical reactions between the extinguishing agent and the burning material
- Any health and occupational safety concerns such as exposure of the extinguisher operator to heat and products of combustion during fire fighting efforts
- Inspection and service required to maintain the extinguishers

The type, size, and number of extinguishers needed vary according to whether the occupancy is classified as light hazard, ordinary hazard, or extra hazard. The occupancy classifications are defined in NFPA 10 as follows:

Light Hazard Occupancy: An occupancy where the amount of combustibles or flammable liquids present is such that a fire of small size may be expected. Some examples are offices, school classrooms, churches, and assembly halls (Figure 8.31).

Ordinary Hazard Occupancy: An occupancy where the amount of combustible or flammable liquids present is such that fires of moderate size may be expected. Some examples are mercantile storage and display, automobile showrooms, parking garages, light manufacturing, school shop or laboratory areas, and warehouses not classified as extra hazard (Figure 8.32).

Extra Hazard Occupancy: An occupancy where the amount of combustibles or flammable liquids present is such that a fire of large size may be expected. Some examples are manufacturing processes such as painting, dipping, coating, flammable liquids handling, auto repair garages, restaurants with deep fat fryers, and aircraft and boat maintenance facilities (Figure 8.33).

Figure 8.31 Office buildings are considered to be light hazard occupancies.

Figure 8.32 Car dealerships are considered ordinary hazards.

Figure 8.33 Aircraft hangars are extra hazard occupancies.

For each hazard classification, NFPA 10 specifies the type of rated extinguisher needed, the maximum travel distance (the distance the operator has to travel to get the extinguisher), and the maximum areas that can be protected by each extinguisher. These specifications can be found in Tables 8.1 and 8.2.

CLASS A EXTINGUISHER DISTRIBUTION

In ordinary or low hazard occupancies, the authority having jurisdiction may approve the use of several lower-rated extinguishers in place of one higher-rated extinguisher. For example, two or more extinguishers may be used to fulfill a 6-A rating if there are enough individuals trained to use the extinguishers. When the weight of the extinguisher causes problems for those who will be operating it, two extinguishers of lesser weight may be used to replace the heavier extinguisher. If the area to be protected is less than 3,000 square feet (288 m²), at least one extinguisher of the minimum rating (2-A) should be provided.

TABLE 8.1
Maximum Area To Be Protected By Class A Extinguishers

	Light (Low) Hazard Occupancy	Ordinary (Moderate) Hazard Occupancy	Extra (High) Hazard Occupancy
Minimum rated single extinguisher	2-A	2-A	40A*
Maximum floor area per unit of A	3,000 sq ft	1,500 sq ft	1,000 sq ft
Maximum floor area for extinguisher	11,250 sq ft	11,250 sq ft	11,250 sq ft
Maximum travel distance to extinguisher	75 ft	75 ft	75 ft

*Two 2½-gal (9.46-L) water-type extinguishers can be used to fulfill the requirements of one 4-A rated extinguisher.

Table 8.1 shows the minimal sizes of fire extinguishers for the listed grades of hazards area that each Class A extinguisher can protect. These numbers can be used to determine the minimum number of extinguishers required to protect a particular area or building. For example, a building owner proposes to protect a 120,000 square foot (11 148 m²) light hazard occupancy with 2-A rated extinguishers. From the table we see that for light hazard occupancies, each unit of A can cover an area of 3,000 square feet (278.5 m²). Thus, a 2-A extinguisher will be able to cover twice that much area, or 6,000 square feet (557 m²). By division, the minimum number of extinguishers needed can be determined as follows:

$$\frac{120{,}000 \text{ ft}^2 \ (11\,148 \text{ m}^2)}{6{,}000 \text{ ft}^2 \ (557 \text{ m}^2)} = 20 \text{ extinguishers required}$$

Twenty extinguishers will be required, assuming that the 75-foot (23 m) maximum travel distance can be maintained. If a 75-foot travel distance cannot be maintained in certain areas, additional extinguishers will be required.

CLASS B EXTINGUISHER DISTRIBUTION

Determining the distribution of Class B extinguishers becomes more complicated and more critical than determining the distribution of extinguishers for Class A hazards. Flammable liquid fires develop very rapidly and occur in a variety of situations that are fundamentally different than Class A fires from a fire control standpoint. In providing extinguishers for Class B hazards, two situations may be encountered. One is a spill fire where the flammable liquid does not have depth, and the other involves flammable liquids with depth such as in dip tanks. NFPA 10 establishes ¼-inch (6.4 mm) deep as the criterion for a flammable liquid fire to be considered to be *with depth*. Anything less than this is considered to be *without depth*.

Flammable Liquid Fires Without Depth

In determining the proper distribution of extinguishers for flammable liquid fires without depth, the travel distances shown in Table 8.2 are used. Flammable liquids without depth are those situations where the liquids are used in a process but are not found in large, deep, open containers.

Notice that Table 8.2 specifies a travel distance only no area is specified. Because of the rapidly

TABLE 8.2
Travel Distances For Flammable Liquid Fires Without Depth

Hazard Class	Minimum Extinguisher Rating	Maximum Travel Distance
Light	5-B	30 feet (9.15 m)
	10-B	50 feet (15.25 m)
Ordinary	10-B	30 feet (9.15 m)
	20-B	50 feet (15.25 m)
Extra	40-B	30 feet (9.15 m)
	80-B	50 feet (15.25 m)

developing nature of a burning liquid fire, the speed with which the operator can begin using an extinguisher on the fire is extremely important. Also for this reason, the travel distance is less than permitted for Class A hazards.

In the case of flammable liquid fires, multiple extinguishers with lower ratings cannot be used to satisfy a requirement for a larger unit. This is because of the possibility of a flashback over the surface of the liquid if the fire is not completely extinguished with a small extinguisher. An exception to this is made for AFFF extinguishers where up to three extinguishers can be used to satisfy requirements in extra hazard occupancies. A foam extinguisher can establish a blanket of agent on the surface of the liquid to eliminate a flashback.

Flammable Liquid Fires With Depth

Individual hazards involving flammable liquids with depth should be protected by fixed extinguishing systems. This lessens the requirements for portable fire extinguishers in the area but will not eliminate the need for them. A spill fire could occur beyond the effective reach of a fixed system, and portable extinguishers would still be needed.

If there is no fixed system, an extinguisher must be provided that has a numerical classification equal to twice the surface area of the largest hazard in an occupancy. For example, an extinguisher with a 10-B rating would be needed to protect 5 square feet (0.464 m^2). It is possible to specify extinguishers in this manner because the surface area is usually known when flammable liquids are present in depth. An example would be an industrial occupancy with dip tanks or retention dikes. As with surface fires, smaller extinguishers cannot be used in lieu of a required larger extinguisher, although up to three AFFF extinguishers may be used to satisfy requirements, just as with surface fires.

Extinguishers for protection of flammable liquid hazards must be placed so that an operator is not endangered while attempting to reach an extinguisher. Extinguishers should not be placed over or behind a hazard.

CLASS C AND CLASS D EXTINGUISHER DISTRIBUTION FACTORS

There are no special spacing rules for Class C hazards because fires involving energized electrical equipment usually involve Class A or Class B fuels. The placement and distribution of extinguishers for combustible metals cannot be generalized. It involves analysis of the specific metal, the amount of metal present, the configuration of the metal (solid or particulate), and the characteristics of the extinguishing agent. NFPA 10 recommends only that the travel distance for Class D extinguishers not exceed 75 feet (23 m).

Installation and Placement

In addition to proper selection and distribution, effective use of fire extinguishers requires that they be readily visible and accessible. Therefore, proper extinguisher placement is an essential but often overlooked aspect of fire protection. Extinguishers must be mounted properly (for example, they should not protrude into traffic paths) to avoid injury to building occupants and to avoid damage to the device itself. Extinguishers are frequently placed in cabinets or wall recesses for this purpose (Figure 8.34). If an extinguisher cabinet is placed in a rated wall, then the cabinet must have the same fire rating as the wall assembly. Proper placement of extinguishers should provide for the following:

- The extinguisher should be near normal paths of travel, preferably near doors.
- The extinguisher must not be blocked by storage or equipment (Figure 8.35).
- The extinguisher should be near points of egress or ingress.
- The extinguisher must be visible.

Although an extinguisher must be properly mounted, it cannot be placed too high. It must be easily reached. The standard mounting heights specified for extinguishers are as follows:

- Extinguishers with a gross weight not exceeding 40 pounds (18 kg) should be installed so that the top of the extinguisher is not more than 5 feet (1.5 m) above the floor (Figure 8.36).
- Extinguishers with a gross weight greater than 40 pounds (18 kg), except wheeled types, should be installed so that the top of the extinguisher is not more than 3½ feet (1.1 m) above the floor.
- The clearance between the bottom of the extinguisher and the floor should never be less than 4 inches (100 mm).

Physical environment is very important to extinguisher reliability. The greatest concern is the temperature of the environment. Because testing laboratories evaluate water-based extinguishers at temperatures between 40°F and 120°F (4°C and 49°C), these extinguishers must be located where freezing is not possible. Other types of extinguishers can be installed where the temperature is as low as -40°F (-40°C) with specialized extinguishers available for temperatures as low as -65°F (-54°C). Extinguishers using plain water can be provided with an antifreeze recommended by the manufacturer. Care must be exercised in the use of an antifreeze. Ethylene glycol cannot be used, and calcium chloride cannot be used in stainless steel units. Antifreeze cannot be added to AFFF extinguishers.

Other environmental factors that may adversely affect an extinguisher are snow, rain, and corrosive fumes. A corrosive atmosphere can be encountered not only in an industrial environ-

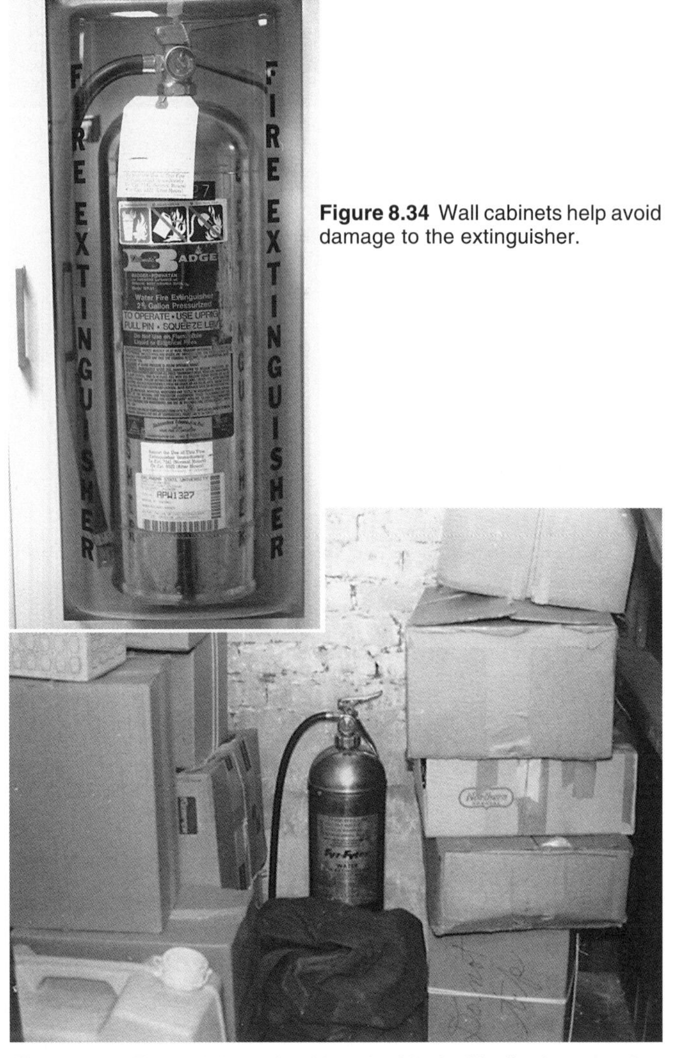

Figure 8.34 Wall cabinets help avoid damage to the extinguisher.

Figure 8.35 Extinguishers should not be blocked by furniture or other articles.

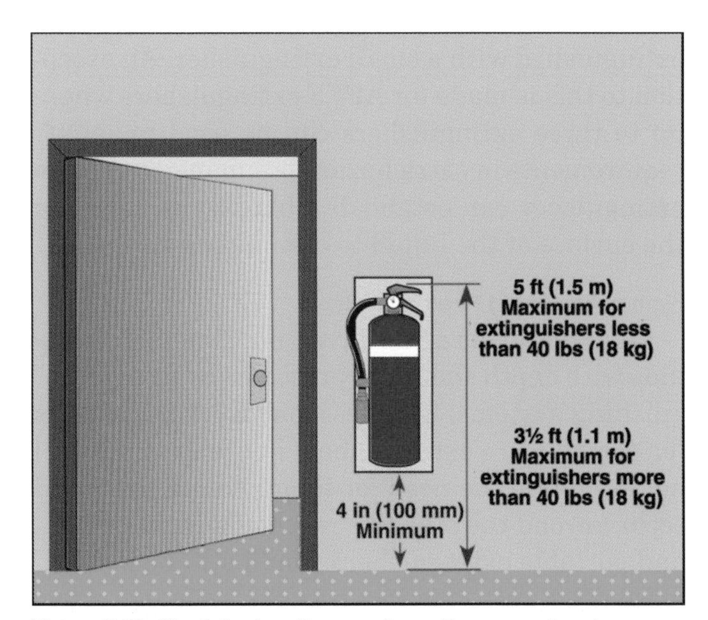

Figure 8.36 The following diagram shows the parameters for proper extinguisher placement.

ment but also in marine applications where extinguishers are exposed to salt spray. In the case of outdoor installations, the extinguisher can be protected with a plastic bag or placed in a cabinet (Figure 8.37). For marine applications, extinguishers are available that have been listed for use in a salt-water environment.

Figure 8.37 Outdoor extinguishers should be protected from the elements by a substantial cover or cabinet.

Inspecting Fire Extinguishers

Fire extinguishers must be inspected regularly to ensure that they are accessible and operable. In the United States, OSHA requires portable fire extinguishers in the workplaces to be inspected on a monthly basis. This is done by verifying that the extinguisher is in its designated location, that it has not been actuated or tampered with, and that there is no obvious physical damage or condition present that will prevent its operation. Inspection of portable fire extinguishers (or any other privately owned fire suppression or detection equipment) is the responsibility of the property owner or building occupant.

Although they are usually performed by the building owner or the owner's designate, fire inspectors should include extinguisher inspections in their building inspection and prefire planning programs. During the inspection, the inspector should remember that there are three important factors that determine the value of a fire extinguisher: its serviceability, its accessibility, and the user's ability to operate it.

The following items should be part of every fire extinguisher inspection:

- Check to ensure that the extinguisher is in a proper location and that it is accessible (Figure 8.38).
- Inspect the discharge nozzle or horn for obstructions. Check for cracks and dirt or grease accumulations.
- Check to see if the operating instructions on the extinguisher nameplate are legible.
- Check the lockpins and tamper seals to ensure that the extinguisher has not been tampered with (Figure 8.39).

Figure 8.38 Inspectors should check to make sure that all extinguishers are located where they should be.

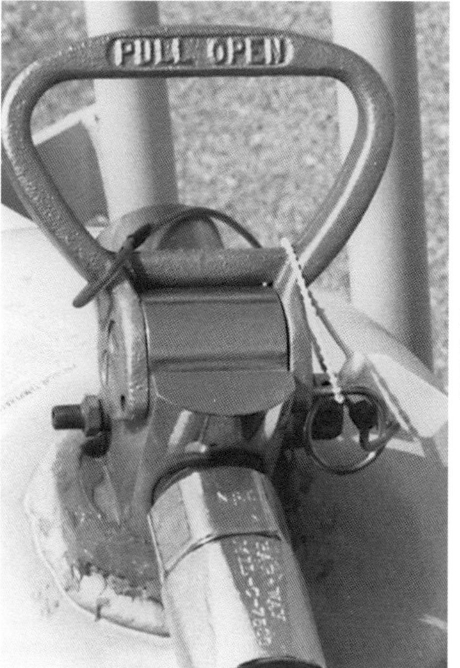

Figure 8.39 Pull pins should be held in place by a tamper seal.

- Determine if the extinguisher is full of agent and/or fully pressurized by checking the pressure gauge, weighing the extinguisher, or inspecting the agent level. If an extinguisher is found to be deficient in weight by 10 percent, it should be removed from service and replaced (Figure 8.40).
- Check the inspection tag for the date of the previous inspection, maintenance, or recharging (Figure 8.41).
- Examine the condition of the hose and its associated fittings.
- Check for signs of physical damage (Figure 8.42).

Figure 8.40 Make sure the extinguisher is charged to a proper operating pressure.

Figure 8.41 Check the inspection tag to make sure it is up to date.

Figure 8.42 Look for dents or other signs of damage on an extinguisher.

If any of the items listed are deficient, the extinguisher should be removed from service and repaired as required. The extinguisher should be replaced with an extinguisher that has an equal or greater rating.

Extinguisher Maintenance Requirements and Procedures

All maintenance procedures should include a thorough examination of the three basic parts of an extinguisher:

- Mechanical parts
- Extinguishing agents
- Expelling means

Building owners should keep accurate and complete records of all maintenance and inspections including the month, year, type of maintenance, and date of the last recharge. It is the inspector's responsibility to review these records.

NFPA 10 requires all fire extinguishers to be thoroughly inspected by qualified personnel at least once a year. Such an inspection is designed to provide maximum assurance that the extinguisher will operate effectively and safely. A thorough examination of the extinguisher will also determine if any repairs are necessary or if the extinguisher should be replaced.

Stored-pressure extinguishers containing a loaded stream agent should be disassembled for complete maintenance. Prior to disassembly, the extinguisher should be discharged to check the operation of the discharge valve and pressure gauge.

Stored-pressure extinguishers that require a twelve-year hydrostatic test must be emptied every six years for complete maintenance. Extinguishers having nonrefillable disposable containers are exempt.

All carbon dioxide hose assemblies should have a conductivity test. Hoses must be conductive because they act as bonding devices to prevent the generation of static electricity during carbon dioxide discharge. Hoses found to be nonconductive must be replaced.

For additional information regarding fire extinguishers, extinguishing agents, distribution, and applications, consult IFSTA's **Private Fire Protection and Detection** manual and NFPA 10.

SPECIAL AGENT FIRE EXTINGUISHING SYSTEMS

Special extinguishing systems are used in locations where automatic sprinklers may not be the best solution to fire problems. These locations include areas that contain flammable and combustible liquids, food preparation equipment, and highly sensitive computer or electronic equipment. A critical feature of most special extinguishing systems is that they have a limited amount of extinguishing agent; however, automatic sprinkler systems are usually supplied by an almost endless amount of water. An important distinction between the two systems is that an automatic sprinkler system is considered successful when it *controls* a fire; the specialized extinguishing system must *extinguish* the fire to be successful. The following sections address five specialized extinguishing systems: dry chemical, wet chemical, halogenated agents, carbon dioxide, and foam. More detailed information on inspecting and testing these types of systems may also be found in Appendix F of this manual.

Dry-Chemical Extinguishing Systems

A dry-chemical extinguishing system is used in locations where a rapid fire knockdown is required and where reignition is unlikely. The agents used in this system are the same as those described previously for portable fire extinguishers. This system is most commonly used to protect the following:

- Quenching operations
- Dip tanks
- Paint spray booths
- Commercial cooking operations (Figure 8.43)
- Exhaust duct systems

There are two main types of dry-chemical systems: local application and total flooding. The local application system, which is the most common type of dry-chemical system, is designed to discharge agent onto a specific surface such as the cooking area in a restaurant kitchen. A total flooding system is designed to introduce a thick concentration of agent into a closed area such as a paint booth. All dry-chemical systems must meet the requirements set forth in NFPA 17, *Standard for Dry Chemical Extinguishing Systems.*

Figure 8.43 Dry-chemical systems are commonly found in large-scale commercial cooking operations.

All of the agents discharge a cloud of chemical that leaves a residue. This residue creates a cleanup problem after system operation. A dry-chemical system is not recommended for an area that contains sensitive electronic equipment. The chemical residue has insulating characteristics that hinder the operation of the equipment unless extensive restorative cleanup is performed. The agent also becomes corrosive when exposed to moisture.

SYSTEM DESCRIPTION

Basically, all dry-chemical systems have the following components:

- Storage tank for expellant gas and agent
- Piping to carry the gas and agent
- Nozzles to disperse the agent
- Actuating mechanism

The storage container may contain both the agent and the pressurized expellant gas (stored pressure), or the agent and the gas may be stored separately (Figure 8.44). A pressure gauge attached to the container is an indication of a stored-pressure container (Figure 8.45). The expellant gas is either nitrogen or carbon dioxide. The containers are basically the same as those described

Figure 8.44 Dry-chemical system cylinders look much the same as those on portable extinguishers.

Figure 8.45 The dry-chemical system cylinder is equipped with a gauge much like a portable fire extinguisher.

for portable fire extinguishers except that the system containers are much larger. Although rare, some agent tanks may hold as much as 2,000 pounds (907 kg) of agent. However, most tanks are in the 30- to 100-pound (13.6 kg to 45.4 kg) range. The tanks must be located as close to the discharge point as possible. They also must be in an area that falls within the -40°F to 120°F (-40C to 49°C) temperature range.

Dry-chemical agent is delivered to the hazard through nozzles attached to a system of fixed piping (Figure 8.46). The piping is specially designed to account for the unique flow characteristics of dry-chemical agent. The proper size, the number of bends and fittings, and the pressure drop (friction loss) must be calculated into piping requirements. There are no standard nozzles. Each system manufacturer has its own nozzle designs.

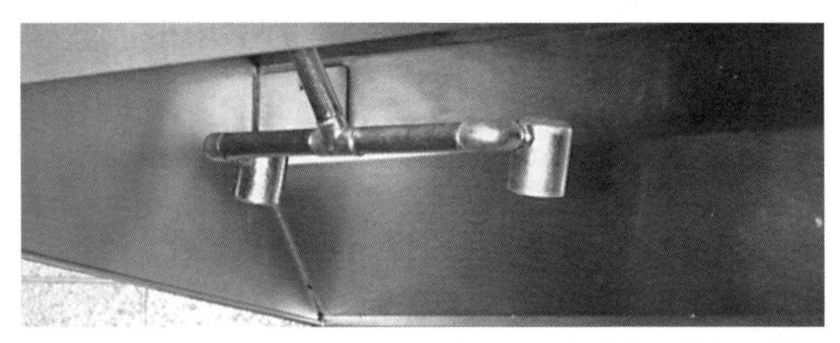

Figure 8.46 The dry-chemical nozzles are located directly above the protected area.

Dry chemical is released into the piping system in response to detection devices. Historically, actuation has most often been mechanical in response to the melting of fusible links (Figure 8.47). The fusible links then trigger a mechanical or electrical release that in turn triggers the flow of expellant gas and agent. Systems that have automatic actuation should also be equipped with audible warning signals that will ensure prompt evacuation of the area. This will lessen the chance of occupants suffering from reduced visibility or breathing difficulties as a result of the agent discharge. The majority of fixed systems must also be capable of manual release and must be equipped with automatic fuel or power shutdown (Figure 8.48).

Detailed maintenance of these systems should be left to reputable contractors. However, occupant representatives or fire inspectors should be able to inspect these systems for the following:

Figure 8.47 The system may be automatically operated by the melting of a fusible link.

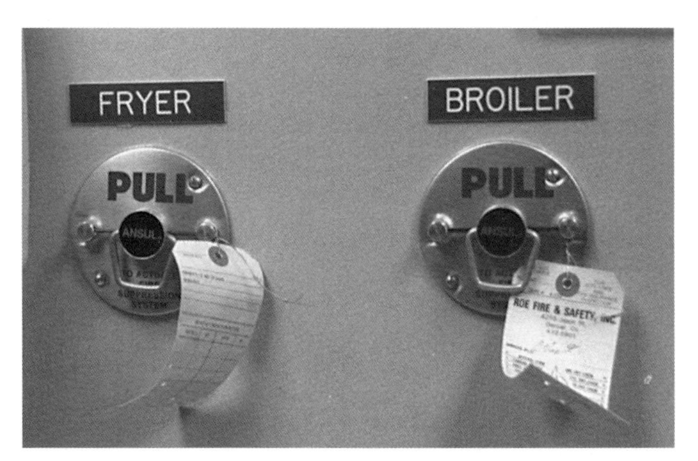

Figure 8.48 Dry-chemical systems have manual activation controls.

- Mechanical damage
- Aim of nozzles
- Change in hazard
- Proper pressures on stored-pressure containers

INSPECTION AND TEST PROCEDURES

The fire inspector should try to assure that building representatives inspect the dry-chemical system as required by NFPA 17. NFPA 17 recommends that the following procedures be performed monthly to ensure system readiness:

- Check all parts of the system to ensure that they are in their proper location and that they are connected.
- Inspect manual actuators for obstructions.
- Inspect the tamper indicators and seals to ensure that they are intact (Figure 8.49).
- Check the maintenance tag or certificate for proper placement.
- Examine the system for any obvious damage.
- Check the pressure gauges to ensure that they read within their operable ranges.

A record should be kept indicating that each inspection was made. Any problems that are noted should be corrected immediately. In most cases, this will require notification of the system service provider.

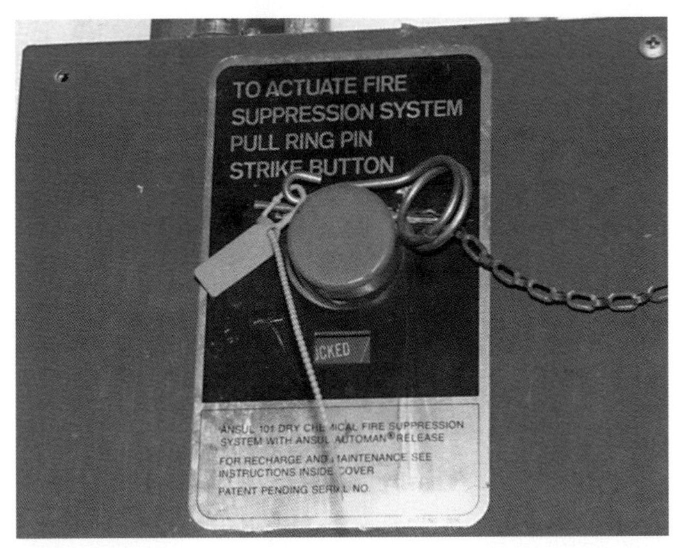

Figure 8.49 The tamper indicator should be in place on the manual release.

NFPA 17 recommends a more detailed inspection on a semiannual basis. The following procedures are recommended for this examination:

- Ensure that the size or nature of the hazard being protected has not changed.
- Examine all components thoroughly.
- Ensure that the piping contains no obstructions.
- Examine the agent to make sure that there is no caking or reduction of flow capabilities.
- Test all working components in accordance with the manufacturer's instructions. (**NOTE:** Normally, discharging the agent would not be required.)
- Correct any problems that are noted, and file a report with the occupant.

If the system actuator is controlled by a fusible link device, the fusible link should be replaced at least annually or as required by the manufacturer. If it appears to be distorted from frequent exposure to heat, it may need to be replaced more often. Dry-chemical cylinders that are less than 150 pounds (68 kg) must be hydrostatically tested every 12 years. Larger tanks have no hydrostatic test requirements.

Wet-Chemical Extinguishing Systems

A wet-chemical system is best suited for applications in commercial cooking hoods, plenums, ducts, and associated cooking appliances. The wet-chemical system operates and is designed similarly to a dry-chemical system (Figure 8.50). The wet-chemical agent is typically a solution that is composed of water and either potassium carbonate or potassium acetate and is delivered to the hazard area in the form of a spray. It is an excellent extinguishing agent for fires involving a flammable liquid, gas, grease, or ordinary combustibles such as paper and wood. It is not recommended for electrical fires because the spray may act as a conductor.

A wet-chemical system is most effective on fires caused by cooking hazards. The nature of the chemical is such that it reacts with animal or vegetable oils and forms a soapy foam. When using

Figure 8.50 Wet-chemical systems look much the same as dry-chemical systems.

a wet-chemical agent, grease or oil fires are extinguished by fuel removal, cooling, smothering, and flame inhibition.

For the most part, the components and actuation of wet-chemical systems are the same as those for dry-chemical systems. The inspection and testing procedures are also the same. Detailed requirements for wet-chemical systems can be found in NFPA 17A, *Standard for Wet Chemical Extinguishing Systems* or Underwriters Laboratories Standard UL 300, *Fire Testing of Fire Extinguishing Systems for Protection of Restaurant Cooking Areas.*

Halogenated Agent Extinguishing Systems

The properties of halogenated extinguishing agents (halon) were discussed in the previous section on portable fire extinguishers. Two of these agents, Halon 1211 and Halon 1301, are also widely used in fixed-system applications. Halon 1211 is primarily used in local application systems; however, it is used in some total flooding systems. Halon 1211 systems fall under the requirements set forth in NFPA 12B, *Standard on Halon 1211 Fire Extinguishing Systems*. Halon 1301 is primarily used in total flooding systems, and these systems are covered in NFPA 12A, *Standard on Halon 1301 Fire Extinguishing Systems*. The majority of halon systems in use today are of the total flooding variety.

A halon system is primarily used to protect against Class B and Class C fires because it has limited effectiveness on Class A fires. Halon systems are particularly useful in settings that require a "clean" agent. These settings usually contain processes or equipment of high value that would be damaged by other extinguishing agents such as water or dry chemical. Halon is not effective on most reactive metals or in locations containing self-oxidizing fuels.

SYSTEM DESCRIPTION

Components common to all halon systems are:

- Agent storage containers
- Actuators
- Nozzles and piping
- Detectors
- Manual releases
- Control panels

There is no standard container size, color, or shape. Tank capacities range from 5 to 600 pounds (2 kg to 275 kg) (Figure 8.51). The amount of agent in a container and the number of containers depend on the system design. Pressure gauges are used to indicate the pressure in the container. However, due to super pressurization with nitrogen, the pressure gauge reading may

Figure 8.51 Halon agent may be found in large cylinders.

not truly indicate whether the tank is completely full of halon or halon-replacement agent.

All containers have one or more valves to permit release of the agent into the hazard area. Although valves may be actuated mechanically, hydraulically, or pneumatically, most valves are actuated electrically in response to smoke or heat detection devices. Most systems also have a manual activation control (Figure 8.52). Although they are not recommended by the NFPA, many systems have one or more abort switches to stop system activation. Total flooding systems have a cross-zoned detection system to reduce the chance of accidental dumping of the system. A cross-zoned system consists of detectors that are connected in a double circuit. If a detector is actuated in either zone, local and central station alarms are activated. Only when detectors become active in both zones will the system be dumped. Actuation of the system must also be accompanied by visual and audible warning signals for occupants of the area (Figure 8.53).

Figure 8.52 Most halon systems have a manual release control.

The containers are connected to a system of fixed piping that terminates at the nozzles. There are no standard designs for piping or nozzles. Each manufacturer has its own favored style (Figure 8.54). There are a few minimum requirements, however. Piping must be able to withstand pressures of 620 psi (4 340 kPa) in low-pressure systems and 1,000 psi (7 000 kPa) in high-pressure systems. The piping must ***not*** be of the cast iron or nonmetallic variety.

A series of system controls and status indicators are contained on the system control panel. These controls and indicators are not standard from system to system. This control panel should be located in a highly accessible area.

Figure 8.53 Halon systems have visible and audible warning devices to signify a system discharge.

Figure 8.54 Halon system nozzles are engineered to work only on the system for which they were designed.

INSPECTION AND TEST PROCEDURES

The inspection and testing of a halon system should be performed *only by qualified individuals*. A halon system, along with its detection and control systems, is often too complex for the typical fire brigade member, fire department inspector, or maintenance staff to be familiar with. The expense of the agent and the environmental impact of accidental discharge also indicate that only properly trained individuals should attempt maintenance or testing of this system. Fire inspection personnel should be aware of the tests that are performed on these systems and should verify that they are regularly performed.

A halon system should be inspected semiannually. All components should be given a thorough review. Agent cylinders should be checked for loss of agent. This can be done by weighing the cylinder or inspecting pressure gauges (Figure 8.55). Any cylinder whose weight has decreased by 5 percent or whose pressure gauge has decreased by 10 percent (adjusted for temperature differences) must be replaced. (**NOTE:** The temperature at which the cylinder is stored will affect the pressure on the

Figure 8.55 The halon level may be checked by pressure gauges, sight gauges, or dipsticks.

gauge. The pressure will go down as the temperature goes down and vice versa; therefore, this must be taken into account.) All hoses on the system should be inspected for damage and tested if damage is suspected. Inspectors should also be alert for changes in room design or size that may affect the system's ability to provide an adequate amount of extinguishing agent.

Carbon Dioxide Extinguishing Systems

A carbon dioxide (CO_2) system is similar in design and application to a halon system. Although these systems were almost phased out of existence by the growth of halon in the 1970s and '80s, there has been a resurgence of new carbon dioxide systems in recent years. This is because of the previously discussed sanctions that have been placed on halon.

Both local application and total flooding carbon dioxide systems are in common. use. Handheld hose and standpipe systems are also in use, although they are not common and are not covered in this manual. All systems must adhere to the requirements set forth in NFPA 12, *Standard on Carbon Dioxide Extinguishing Systems*. Carbon dioxide is a "clean" smothering agent. Although some ice crystals accompany the carbon dioxide, the agent has little cooling effect on the fire.

The most serious problem involving carbon dioxide systems, especially total flooding systems, is personnel safety. The elimination of oxygen from a fire also eliminates oxygen from people. Total flooding systems are designed to deliver at least a 34 percent concentration of carbon dioxide into an enclosed area. This means that once the system has activated, carbon dioxide will make up at least 34 percent of the atmosphere. Concentrations this high are lethal, and people have been killed from these systems. For this reason, total flooding systems must be provided with predischarge alarms as well as discharge alarms (Figure 8.56). A predischarge alarm notifies those present that the system is about to activate. There is a delay before the system actually discharges the agent. However, alarms alone are not enough to ensure the safety of personnel. All affected personnel must be educated about the dangers of carbon dioxide. They must also be trained in proper emergency procedures relative to system discharge. Failure of alarms to operate or inadequate training could result in fatalities.

Local application systems are not as dangerous as total flooding systems. In local application systems, carbon dioxide is delivered directly onto the hazard as opposed to filling an enclosure with the gas. Danger to personnel is reduced if the local application system is located outdoors or in a large building.

Figure 8.56 The predischarge alarm for a carbon dioxide system gives occupants a chance to flee the area before discharge.

SYSTEM DESCRIPTION

The components of the carbon dioxide system are similar to those in the halon system. These components include actuation devices, agent storage containers, piping, and nozzles.

The following are three means of actuation for carbon dioxide systems:

- *Automatic Operation* — Operation that is triggered by a product-of-combustion detector.
- *Normal Manual Operation* — Operation that is triggered by a person manually operating a control device and putting the system through its complete cycle of operation, including predischarge alarms (Figure 8.57).
- *Emergency Manual Operation* — Operation to be used only when the other two actuation modes fail. This operation causes the system to discharge immediately and without any advance warning to individuals in the area (Figure 8.58).

Historically, fixed-temperature or rate-of-rise detection has been used for automatic actuation. However, modern systems may use smoke detectors or even flame-detection equipment to activate carbon dioxide systems. All of these actuation methods will trigger control valves on the carbon dioxide supply that allow agent to enter the system and discharge.

Carbon dioxide systems exist as either high-pressure systems or as low-pressure systems. In the high-pressure system, the carbon dioxide is stored in standard DOT cylinders at a pressure of about 850 psi (5 950 kPa) (Figure 8.59). The low-pressure system is for much larger hazards. The carbon dioxide in these systems is stored in large, refrigerated tanks where the liquefied carbon dioxide is stored at a temperature of 0°F (-18°C) (Figure 8.60). The carbon dioxide is stored at about 300 psi (2 100 kPa) at that temperature.

Figure 8.59 High-pressure carbon dioxide is stored in pressure cylinders.

Figure 8.57 The carbon dioxide system may be discharged by a manual pull station.

Figure 8.58 The carbon dioxide system will have a back-up manual control in case the main pull station does not activate the system.

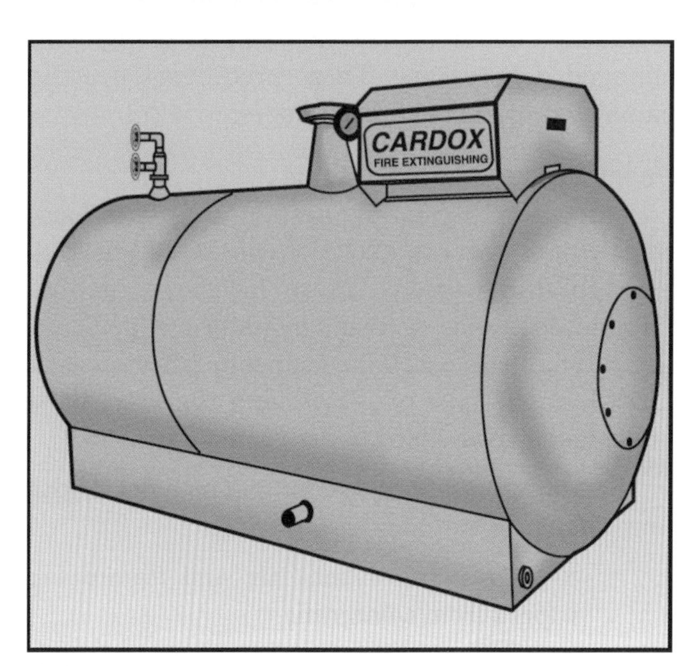

Figure 8.60 Low-pressure carbon dioxide may be stored in large, refrigerated tanks.

In either a low-pressure or a high-pressure system, the containers are connected to the discharge nozzles through a system of fixed piping. The requirements for piping are similar to those for halon systems. Nozzles for total flooding systems may be of the high- or low-velocity type (Figure 8.61). However, high-velocity nozzles promote better disbursement of the agent throughout the entire area. Local application nozzles are typically of the low-velocity type. This reduces the possibility of splashing the burning product.

Figure 8.61 Carbon dioxide nozzles have a hornlike appearance.

INSPECTION AND TEST PROCEDURES

Because of the relative complexity of carbon dioxide systems, routine inspection and testing is not normally performed by fire brigade members, fire inspectors, or facility maintenance personnel. Maintenance and testing should be performed annually by competent contractors, preferably licensed representatives of the system manufacturer. Fire service personnel can inspect for physical damage to components, check for excessive corrosion, and look for a change in hazard. Agent cylinders should be checked semiannually and changed if necessary. These procedures are the same as those for halon cylinders.

Foam Extinguishing Systems

A foam extinguishing system is used in locations where water in itself may not be an effective extinguishing agent. These locations include flammable liquids processing or storage facilities, aircraft hangars, and rolled paper or fabric storage facilities. The type of system and the foam used depends on the hazard being protected.

Foam extinguishes fire by one or more of the four following methods (Figure 8.62):

- *Smothering* prevents air and flammable vapors from combining.
- *Separating* intervenes between the fuel and the fire.

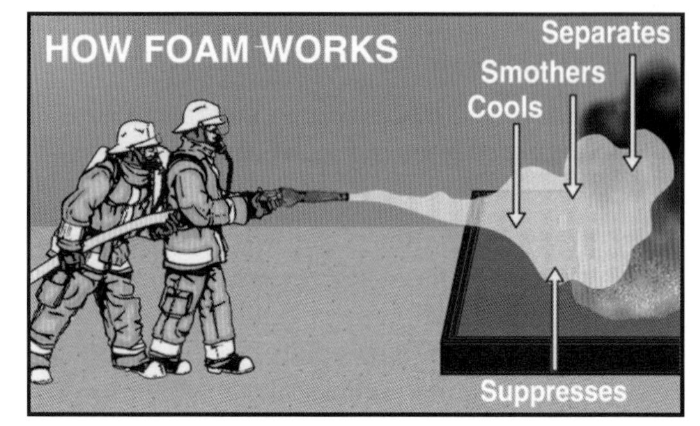

Figure 8.62 Foam works by a variety of methods.

- *Cooling* lowers the temperature of the fuel and adjacent surfaces.
- *Suppressing* prevents the release of flammable vapors.

In general, foam works by forming a blanket on the burning fuel. The foam blanket excludes oxygen, stops the burning process, and cools adjoining hot surfaces.

HOW FOAM IS GENERATED

Foams in use today are of the mechanical type; that is, they must be proportioned with water and aerated (mixed with air) before they can be used. Before discussing types of foams and the foam-making process, it is important to understand the following terms (Figure 8.63):

- *Foam Concentrate* — The raw foam liquid prior to the introduction of water and air. Foam concentrate is usually shipped in 5-gallon (20 L) pails or 55-gallon (220 L) drums. Foam concentrate for fixed systems is stored in large fixed tanks that hold 500 gallons (2 000 L) or more.
- *Foam Proportioner* — The device that introduces the correct amount of foam concentrate into the water stream to make the foam solution.
- *Foam Solution* — A homogeneous mixture of foam concentrate and water prior to the introduction of air.
- *Foam* — Once air is introduced into the foam solution, the completed product is called foam (also known as finished foam).

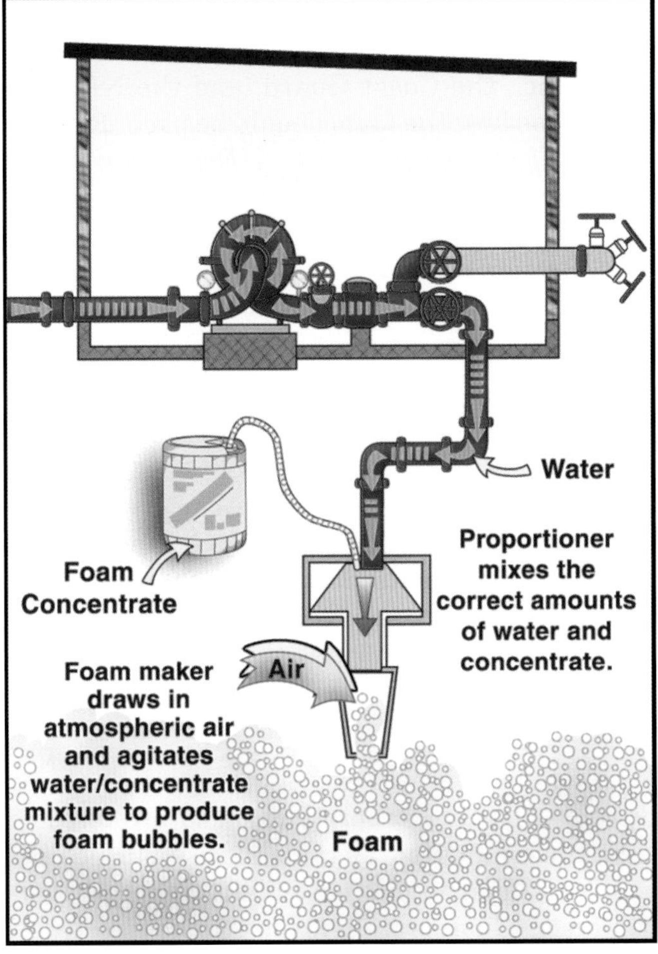

Figure 8.63 Foam is generated by mixing correct proportions of water, foam concentrate, and air.

Four elements are necessary to produce high-quality fire fighting foam: foam concentrate, water, air, and mechanical agitation (Figure 8.64). All of these elements must be present and blended in the correct ratios. Removing any element will result in either no foam or a poor quality foam.

There are two stages in the formation of foam. First, water is mixed with foam liquid concentrate to form a *foam solution*. This is the proportioning stage of foam production. Second, the foam solution passes through the piping or hoseline to a foam nozzle or sprinkler. The foam nozzle or sprinkler aerates the foam solution to form *(finished) foam*.

Proportioning equipment and foam nozzles or sprinklers are engineered to work together. Using a foam proportioner that is not hydraulically matched to the foam nozzle or sprinkler (even if the two are made by the same manufacturer) can result in unsatisfactory foam or no foam at all.

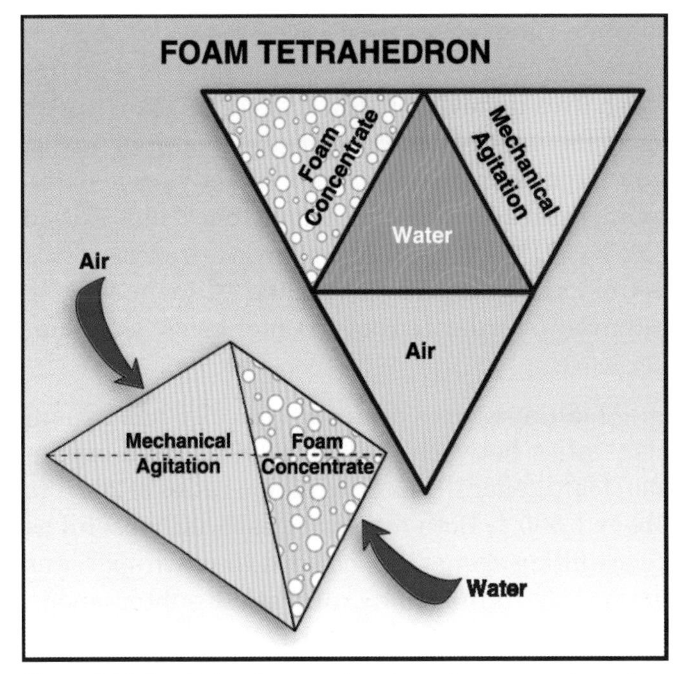

Figure 8.64 As with fire, there are four elements necessary to produce foam.

There are numerous appliances for making and applying foam. A number of these are discussed later in this chapter.

FOAM PROPORTIONING RATES

Finished foam is 94 to 99½ percent water. Class B low-expansion foams in use today are designed to be used at 1%, 3%, or 6% concentrations. In general, foams designed for hydrocarbon fires are used at 1% to 6% concentrations. Polar solvent fuels, such as lacquers or alcohols, require 3% or 6% concentrates, depending on the particular brand being used. Medium- and high-expansion foams are typically used at 1%, 1½%, 2%, or 3% concentrations.

To be effective, foam concentrates must also match the fuel to which they are applied. This is why it is extremely important to identify the type of fuel a system is protecting. Foams designed for hydrocarbon fires will not extinguish polar solvent fires, regardless of the concentration at which they are used. However, foams that are designed for polar solvent fires may be used on hydrocarbon fires.

EXPANSION RATES OF FOAM

Depending on its purpose, foam is designed for low, medium, or high expansion. Low-expansion foam, such as AFFF or FFFP, has a small air/

solution ratio, generally in the area of 7:1 to 20:1. Low-expansion foams are used primarily to extinguish fires involving liquid fuels. This foam is also used for vapor suppression on unignited spills. Low-expansion foam is most effective when the temperature of the fuel liquid does not exceed 212°F (121°C). If the fuel temperature exceeds this figure, much higher flow rates of foam will be required in order to cool the fuel below this temperature.

Medium-expansion foam typically has expansion ratios between 20:1 and 200:1. High-expansion foam generally has expansion ratios of 200:1 to about 1,000:1. Both foams are especially useful as space-filling agents in such hard-to-reach spaces as basements, mine shafts, and other subterranean areas. Steam dilution caused by the vaporization of the foam in heated areas displaces gas and smoke, thus cooling the environment and extinguishing confined space fires.

When used as a water additive in automatic sprinkler systems, high-expansion foam concentrate is an effective wetting agent that is capable of penetrating bulk-bailed commodities such as paper, rags, and cardboard. The foam moves quickly across the fuel surface and gives up its water rapidly.

When using high-expansion foam, it is important to check listings from Underwriters Laboratories Inc., the Coast Guard, and the NFPA to determine how the foam should be used. Refer to NFPA 11, *Standard for Low-Expansion Foam and Combined Agent Systems*, and NFPA 11A, *Standard for Medium- and High-Expansion Foam Systems*, for additional information on recommended uses. Contact the foam manufacturer for further information on its listed uses.

SPECIFIC FOAM CONCENTRATES

Fire fighting foam concentrate is manufactured with either a synthetic or protein base. Protein-based foams are derived from either plant or animal matter. Synthetic-based foam is made from a mixture of detergents. Some foams are a combination of the two. The following sections describe the types of foam concentrates that may be encountered in fixed systems.

Fluoroprotein Foams

Fluoroprotein foams are based on hydrolyzed protein solids. Fluoroprotein foams are fortified with fluoronated surfactants. These surfactants enable the foam to shed or separate from hydrocarbon fuels. Fluoroprotein foams can be injected at the base of a burning storage tank and allowed to surface and extinguish the fire (Figure 8.65). This

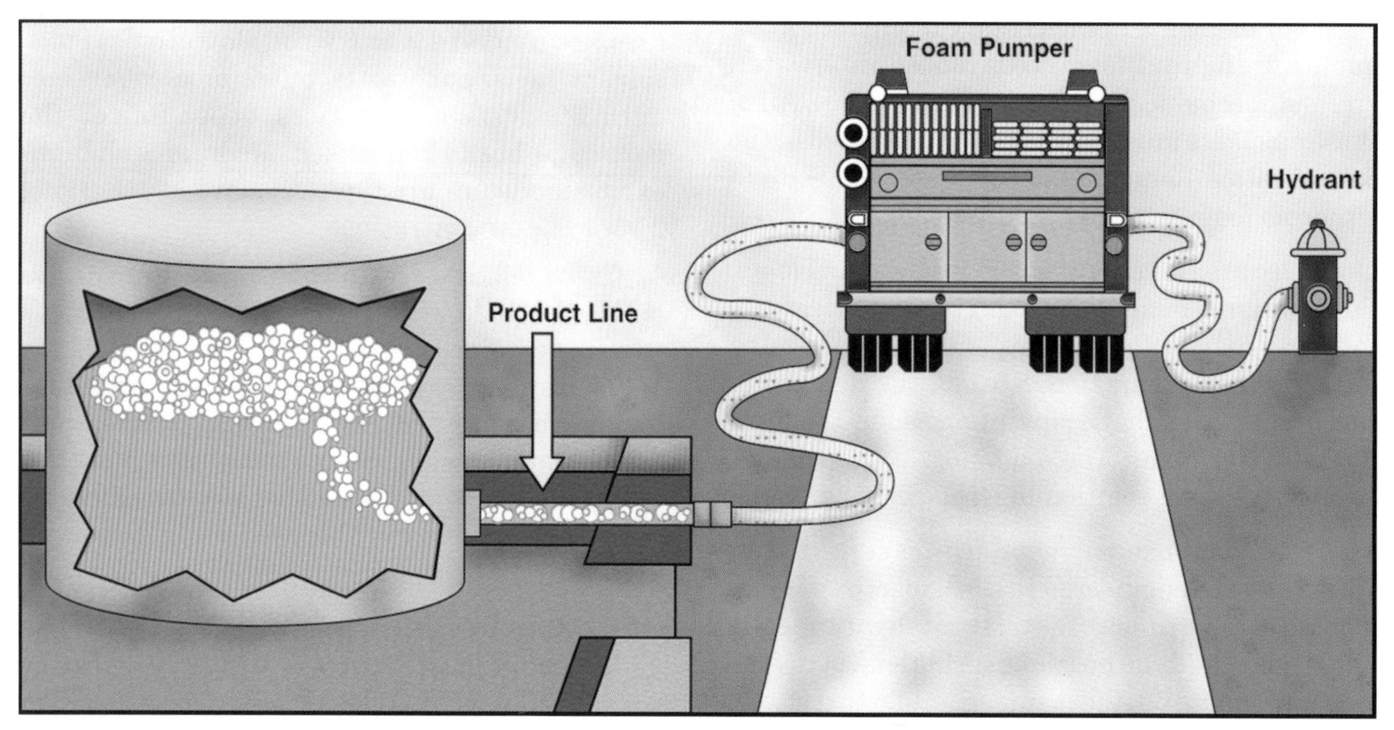

Figure 8.65 Fluoroprotein foams are commonly used for subsurface injection applications.

method is called subsurface injection. Fluoroprotein foams also provide a strong "security blanket" for long-term vapor suppression. Vapor suppression is especially critical with unignited spills.

Film Forming Fluoroprotein Foam (FFFP)

Film forming fluoroprotein foam (FFFP) concentrate is based on fluoroprotein foam technology with aqueous film forming foam (AFFF) capabilities. (**NOTE:** AFFF is discussed in the next section.) Film forming fluoroprotein foam incorporates the benefits of AFFF for fast fire knockdown and the benefits of fluoroprotein foam for long-lasting heat resistance. FFFP is available in an alcohol-resistant formulation.

Aqueous Film Forming Foam (AFFF)

Since synthetic foam's entry into the fire service some years earlier, the U.S. Navy was interested in using it as a flammable liquid fire fighting foam. The U.S. Navy discovered that when a fluoronated surfactant was added to detergent foam, the water that drained from the foam blanket actually floated on hydrocarbon fuel spills. This film is known as an aqueous film. The characteristics of AFFF are as follows:

- It is synthetic and premixable in portable fire extinguishers and apparatus water tanks.
- It can be used through low-expansion nonaerating nozzles, such as common fog nozzles, for Class B fuels.
- It can be stored at temperatures from 35°F to 120°F (2°C to 49°C). (**NOTE:** Some AFFFs are adversely affected by freezing and thawing, so it is important to consult the manufacturer.)
- It can be freeze-protected with a nonflammable antifreeze solution.
- It has good low-temperature viscosity.
- It has penetrating capabilities in baled storage fuels or high surface tension fuels such as treated wood.

When AFFF (as well as the previously mentioned FFFP) is applied to a hydrocarbon fire, three things occur:

- The foam solution drains from the foam blanket, which creates an air-excluding film. Excluding the air leads to a rapid fire knockdown.
- The rather fast-moving foam blanket then moves across the spill, which adds further insulation.
- As the aerated (7-20:1) foam blanket continues to drain its solution, more film is released. This gives AFFF the ability to "heal" over areas where the foam blanket is disturbed.

Today, AFFF is available in concentrations from 1% to 6% for use with fresh or salt water. Alcohol-resistant AFFF is also available from most foam manufacturers. Like other AFFF, alcohol-resistant AFFF is synthetic based. Two types of alcohol-resistant foams are available. One type is used on hydrocarbon fires at a 1% concentration and polar solvent fires at a 3% concentration. The other type is used on hydrocarbon fires at a 3% concentration and polar solvent fires at a 6% concentration.

When an alcohol-type AFFF is applied to polar solvent fuels, it creates a membrane rather than a film over the fuel. This membrane separates the foam blanket from the attack of the solvent. Then, the blanket acts in much the same way as a regular AFFF. Alcohol-resistant AFFF should be applied gently to the fuel so that the membrane can form first.

Medium- And High-Expansion Foams

Medium- and high-expansion foams are special-purpose foams. Because they have a low water content, they minimize water damage. Their low water content is also useful when runoff is undesirable. The major uses for these foams are the following:

- In pesticide fires
- For vapor suppression of fuming acids
- In coal mines and other subterranean spaces
- In concealed spaces such as basements
- In fixed-extinguishing systems for specific industrial uses, such as aircraft hangars (Figure 8.66)

Figure 8.66 Medium- and high-expansion foam systems are commonly found in large aircraft hangars.

FOAM PROPORTIONERS

There are several types of foam proportioners in common use for fixed systems. These proportioners include balanced pressure proportioners, around-the-pump proportioners, pressure proportioning tanks, and coupled water motor-pump proportioners. The following sections briefly describe each of these.

Balanced Pressure Proportioner

Using a balanced pressure proportioner is one of the most reliable methods of foam proportioning (Figure 8.67). The primary advantage of the balanced pressure proportioner is its ability to monitor the demand for foam concentrate and to adjust the amount of concentrate being supplied. Another major advantage of a balanced pressure proportioner is its ability to discharge foam from some outlets and plain water from others at the same time. Thus, a single fire pump can supply both foam and water lines. Supplying different types of lines is not possible with around-the-pump proportioners.

Figure 8.67 Balanced-pressure proportioners are commonly used in large, fixed foam systems.

Systems equipped with a balanced pressure proportioner have a foam concentrate line connected to each fire pump discharge outlet or the system riser (Figure 8.68). This line is supplied by a foam concentrate pump separate from the main fire pump. The foam concentrate pump draws the concentrate from a fixed tank. This pump is designed to supply foam concentrate to the outlet at the same pressure at which the fire pump is supplying water to the riser.

The orifice of the foam concentrate line is adjustable at the point where it connects to the riser. If 3% foam is used, the foam concentrate discharge orifice should be set to 3 percent of the total size of the water discharge outlet. If 6% foam is used, the foam concentrate discharge orifice is set to 6 percent of the total size of the riser and so on. Because the foam and water are being supplied at the same pressure and the sizes of the discharges are proportional, the foam is proportioned correctly.

Around-The-Pump Proportioner

The around-the-pump proportioner is another automatic proportioner. It is especially useful when there is low water pressure or when a motor is not available for a separate foam concentrate pump. This type of proportioner is the most common type of built-in proportioner installed in mobile fire apparatus today. This proportioner is also installed on some fixed systems.

This system runs a small return line from the discharge side of the pump back to the intake side of the pump (Figure 8.69). An in-line eductor is positioned on the pump bypass. This unit is rated for a specific flow and should be used at this rate, although it does have some flexibility. For example, a unit designed to flow 500 gpm (2 000 L/min) at a 6 % concentration will flow 1,000 gpm (4 000 L/min) at a 3 percent rate.

A major disadvantage of the around-the-pump proportioner is that the pump cannot take advantage of incoming pressure. If the inlet water supply is any greater than 10 psi (70 kPa), the foam

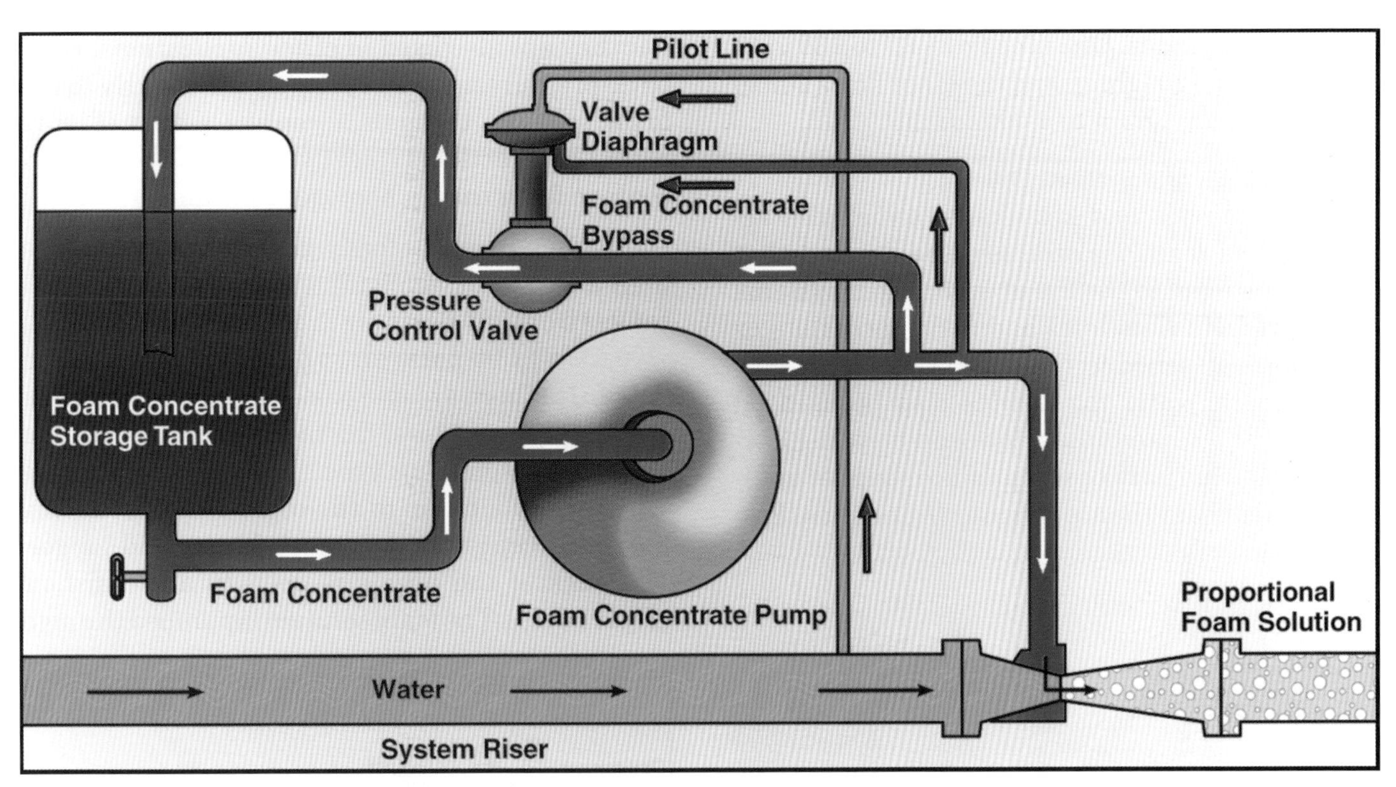

Figure 8.68 This diagram shows the basic design of a balanced pressure foam proportioning system.

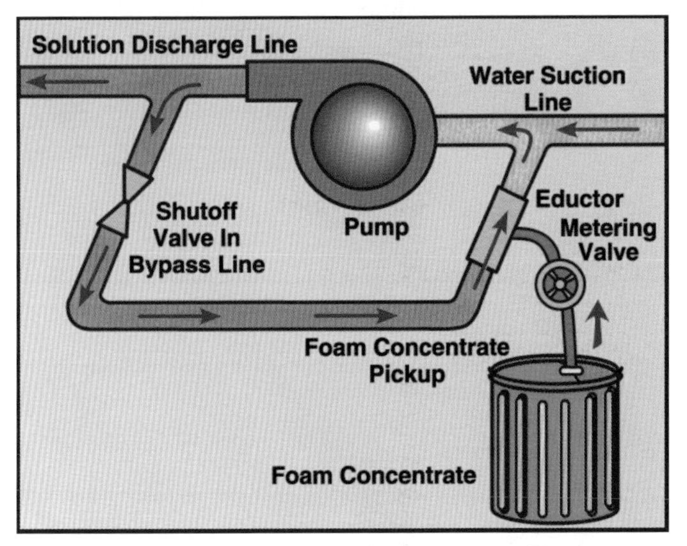

Figure 8.69 Around-the-pump proportioning systems utilize the energy generated by the fire pump to proportion foam concentrate into the stream.

concentrate will not be able to enter into the pump intake. Another disadvantage is that the pump must be dedicated solely to foam operation. An around-the-pump proportioner does not allow plain water and foam to be discharged from the pump at the same time.

Pressure Proportioning Tank System

This type of system consists of one or two foam concentrate tanks that connect both to the water supply and to foam solution lines of the overall system. This system is designed so that a small amount of water from the supply source is pumped into the concentrate tank(s). This water volumetrically displaces the concentrate into the foam solution line where it is mixed with discharge water. The system allows for automatic proportioning over a wide range of flows and pressures, and it does not depend on an external power source. However, the system is limited by the size of the concentrate tank. Once the concentrate is expended, the water must be drained from the tank before it can be refilled with concentrate.

Coupled Water Motor-Pump Proportioner

This proportioner consists of two positive-displacement rotary-gear pumps that are mounted on a common shaft. One pump is for water, and the other is for the foam concentrate. The water pump is proportionally larger than the foam pump. As water flows through the larger pump, it causes the smaller pump to turn and draft foam concentrate from the foam tank. Because the pumps are sized in proportion to each other, the correct foam/water solution is made. This type of proportioner is used in fixed-system applications. It is limited to only two sizes, and both are designed for 6 percent proportioning rates. One type flows 60 to 180 gpm (240 L/min to 720 L/min) and the other flows 200 to 1,000 gpm (800 L/min to 4 000 L/min).

FOAM NOZZLES AND SPRINKLERS

Foam nozzles and sprinklers, sometimes referred to as foam makers, are the devices that deliver the foam to the fire or spill. Fixed systems may have both handlines and foam sprinklers or generators attached to them. Standard fixed-flow or automatic water fog nozzles may be used on AFFF or FFFP handlines. However, aerating low-expansion foam nozzles will produce a higher quality finished foam than will water fog nozzles (Figure 8.70).

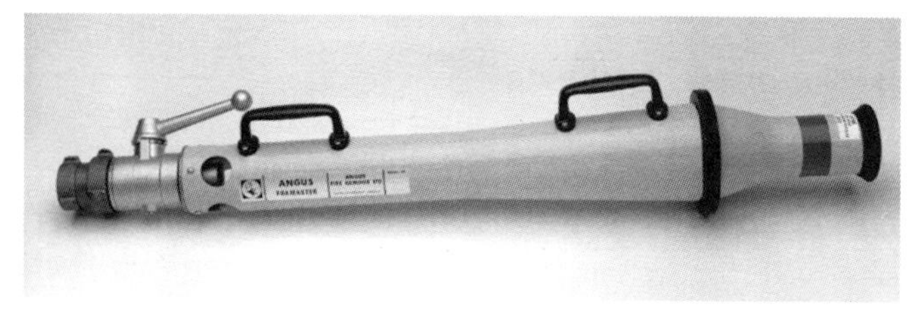

Figure 8.70 Low-expansion foam is applied through special air aspirating foam nozzles. *Courtesy of Angus Fire Armour, LTD.*

Foam/water sprinklers are found on fixed-foam deluge and foam/water systems. These systems use AFFF or FFFP foams. Many foam/water sprinklers resemble aspirating nozzles in that they use a venturi velocity to mix air into the foam solution (Figure 8.71). Some systems use standard sprinklers to form a less expanded foam through simple turbulence of the water droplets falling through the air. Foam/water sprinklers come in upright and pendant designs. Their deflectors must be adapted to meet the specific installation requirements.

Figure 8.71 Special sprinklers may be used on foam sprinkler systems.

High back-pressure foam aspirators, or forcing foam makers, are in-line aspirators used to deliver foam under pressure. These foam makers are most commonly used in subsurface injection systems that protect cone-roof hydrocarbon storage tanks, but they are also used in other applications. High back-pressure aspirators supply air directly to the foam solution through a venturi action (Figure 8.72). This action typically produces a low-air-content foam that is homogeneous and stable.

High-expansion foam generators produce a high-air-content, semistable foam. The air content ranges from 200 parts air to one part foam solution (200:1) to 1,000 parts air to one part foam solution (1,000:1). There are two basic types of high-expansion foam generators: the mechanical blower and the water-aspirating type. The water-aspirating type is very similar to the other foam-producing nozzles except that it is much larger and longer. The back of the nozzle is open to allow airflow (Figure 8.73). The foam solution is pumped through

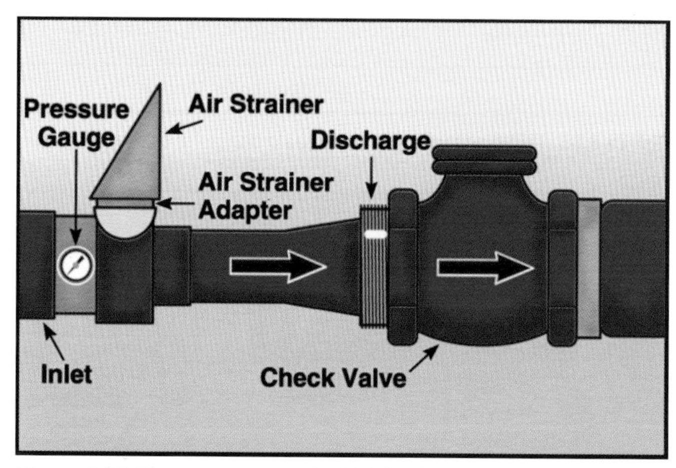

Figure 8.72 The components of a high back-pressure foam aspirator.

Figure 8.73 Medium- and high-expansion foam handlines use special nozzles to discharge foam.

the nozzle in a fine spray that mixes with air to form a moderate-expansion foam. The end of the nozzle has a screen, or series of screens, that breaks up the foam and further mixes it with air. These nozzles typically produce a lower-air-volume foam than do mechanical blower generators.

A mechanical blower generator is similar to a smoke ejector in appearance (Figure 8.74). It operates on the same principle as the water-aspirating nozzle except that the air is forced through the foam spray instead of being pulled through by the water movement. This device produces a higher-air-content foam and is typically associated with total flooding applications.

Figure 8.74 High-expansion foam may be discharged through mechanical foam generators. *Courtesy of Walter Kidde Inc.*

SYSTEM DESCRIPTION

A foam system must have an adequate water supply, foam concentrate supply, piping system, proportioning equipment, and foam makers (discharge devices). Basically, there are five types of foam extinguishing systems:

- Fixed
- Semifixed Type A
- Semifixed Type B
- High expansion
- Foam/water

Fixed System

A fixed foam extinguishing system is a complete installation that is piped from a central foam station. A fixed system automatically applies foam to the target hazard (Figures 8.75 a and b). The foam is discharged through fixed delivery outlets to

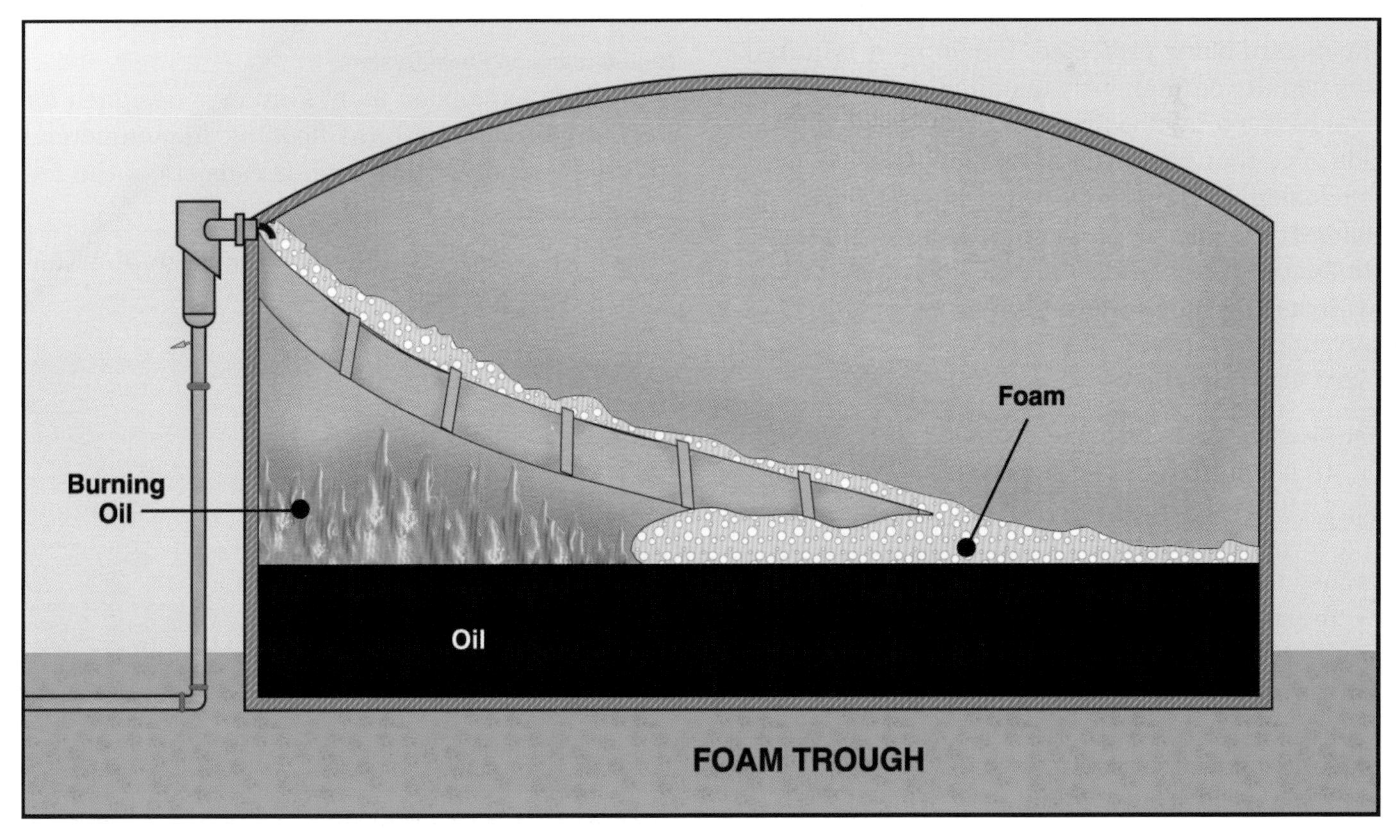

Figure 8.75a In this fixed system the foam trough gently applies foam to the surface of the fuel.

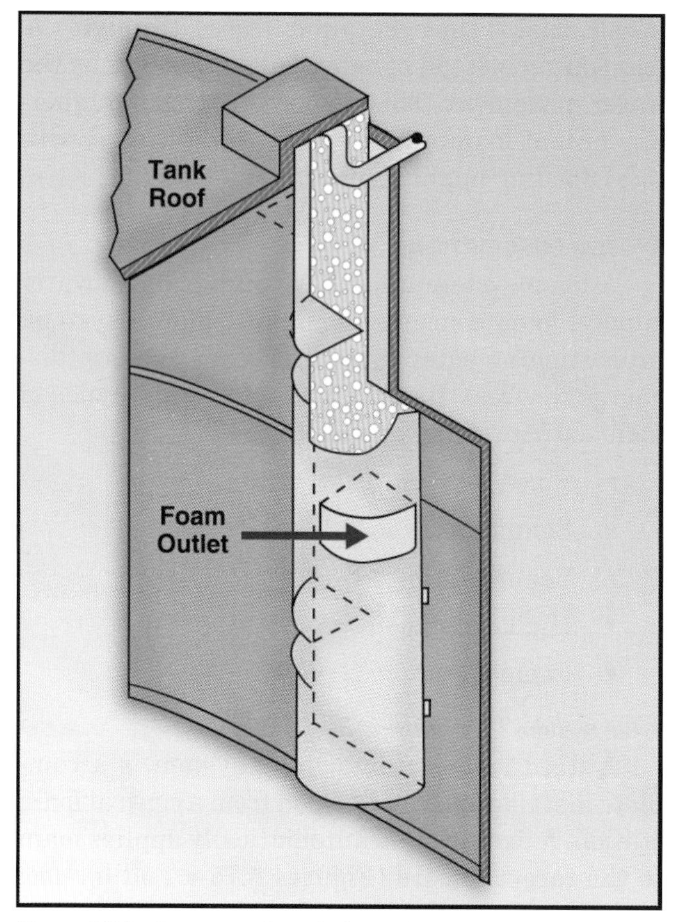

Figure 8.75b A cutaway view of a foam side dump.

the hazard being protected. If a pump is required, it is usually permanently installed. A fixed system may be the total flooding type or the local application type that uses either low-, medium-, or high-expansion foam. Most fixed systems are of the deluge type and require actuation by some sort of product-of-combustion detection system. Deluge systems have an unlimited water supply and may also have large foam supplies.

Semifixed Type A System

In a semifixed Type A system, the foam discharge piping is in place but is not attached to a permanent source of foam (Figure 8.76). The semifixed Type A system requires a separate source for foam solution (usually the fire brigade or fire department). This type of system is found in settings that involve several similar hazards. This system is primarily used on flammable liquid storage tanks. The foam source is a mobile foam solution apparatus (usually a fire truck). This system can be compared to dry standpipe systems.

Many subsurface injection systems are of this type. Subsurface injection is used in situations where topside application may not be effective because of wind or heavy fire conditions. Because the foam solution is lighter than the product in the tank, it will float to the top when introduced at the bottom of the tank. This is a highly effective method for extinguishing bulk tank fires.

Semifixed Type B System

A semifixed Type B system provides a foam solution source that is piped throughout a facility, much like a water distribution system. The foam solution is delivered to foam hydrants for connection to hoselines and portable foam application devices. The difference between a semifixed Type B system and a fixed system is that a fixed system actually applies foam to the hazard while this system merely provides foam capability to an area. From this point, the foam must then be applied manually.

High-Expansion Foam System

A high-expansion foam system is designed for local application or total flooding in commercial and industrial applications. It consists of the following primary components:

- An automatic detection or manual actuation system

Figure 8.76 The semifixed Type A system is basically a dry standpipe system used to protect oil storage tanks.

- A foam generator
- Piping from the water supply and foam concentrate storage tank to the generator

In the total flooding application, entire buildings can be filled to several feet (meters) above the highest storage area or equipment within a few minutes.

The high-expansion foam system can be actuated by any of the common fire detection devices, by a manual pull station, or by both. The foam generators are powered by electric or gasoline motors or by water. The generators should have a fresh-air intake to make sure that foam is not contaminated by products of combustion. Also, venting should be provided ahead of the foam to allow it to move through the area to be protected.

Foam/Water System

A foam/water system is basically a deluge sprinkler system with foam introduced into it (Figure 8.77). This system is used where there is a limited foam concentrate supply but an unlimited water supply. Thus, if the foam concentrate supply becomes depleted, the system will continue to operate as a straight deluge sprinkler system.

Typically, the foam/water system is an automatic system that operates in the same manner as a regular deluge system. The major differences are that the foam induction system and special aerating sprinklers are at the end of the piping. This type of system produces a lean foam solution that is expanded six to eight times when it is discharged from the sprinkler. This makes for a very fluid foam that will flow around obstructions after it is delivered.

The entire foam/water system may be divided into two parts: the water system and the foam system. The foam system contains a concentrate tank, pump, metering valve, strainer, piping, and the actuation unit. The water system components are the same as those described in Chapter 7 for deluge sprinkler systems.

Protein, fluoroprotein, and aqueous film forming foam (AFFF) concentrates may be used in these systems. AFFF may be discharged through regular water sprinklers with favorable results. When discharged through standard sprinklers, AFFF has greater velocity than when discharged through foam sprinklers. This tends to improve the spray and the penetration.

The system operates when the initiation devices sense the presence of fire and send an appropriate signal to the system control unit. This in turn triggers the deluge valve and the system. At the same time, the deluge valve on the foam side of the system opens, the foam pump starts, and the concentrate is introduced into the water flow. The foam/water solution flows to the sprinklers, where air is introduced to make foam. The foam is then delivered to the target area.

After the fire is out, the system must be shut down and thoroughly drained. The concentrate tanks must be refilled. The entire system must be thoroughly flushed to remove any foam residue. Once this is complete, the valves can be reset and the system restored to service.

FOAM SYSTEM INSPECTION AND TESTING

As with many of the other types of special extinguishing systems covered in this chapter, foam systems are highly complex systems that require specially trained personnel to inspect, service, and test them. All valves and alarms attached to the system should be checked semiannually. Foam concentrates, foam equipment, and foam proportioning systems should be checked annually. Qualitative tests should be performed on the concentrate to ensure that no contamination is

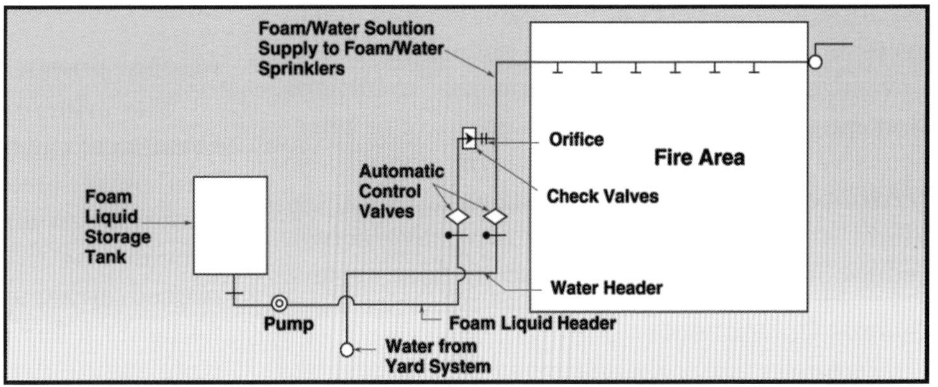

Figure 8.77 The components of a foam/water sprinkler system.

present. The concentrate tank should be checked for signs of sludge or deterioration (Figure 8.78). For more information on foam and foam systems, consult IFSTA's **Private Fire Protection and Detection** and **Principles of Foam Fire Fighting** manuals.

Figure 8.78 Fixed foam systems have large foam concentrate storage tanks.

FIRE DETECTION AND ALARM SYSTEMS

Fire detection and alarm systems used for fire protection are highly technical and include many types of equipment that are usually installed and maintained by specially trained individuals.

During inspections, fire inspectors should note the functional aspects of fire detection and alarm systems. Observant inspectors should be able to recognize physical and environmental conditions that may render the system inoperative or unresponsive to a hostile fire. Inspectors should also recognize conditions that may trigger an unwanted alarm and, by recommending corrective action, reduce fire department responses to false alarms.

System Components

Fire detection and signaling systems are highly technical. These systems include many forms of equipment that are usually installed and maintained by a specialist in fire detection and signaling systems. The components of the system should be listed by a nationally recognized testing laboratory, such as Underwriters Laboratory Inc. (UL) or the Factory Mutual System (FM), to ensure operational reliability (Figure 8.79). Testing reports may address either an entire system or individual components that may be used in interchangeable applications. The installation of the system shall conform to the applicable provisions of NFPA 70, *National Electrical Code*, and NFPA 72, *The National Fire Alarm Code*. The following sections highlight each of the basic components that can be found on all types of fire detection and signaling systems.

Figure 8.79 Fire alarm and detection equipment should be listed by a recognized testing authority.

SYSTEM CONTROL UNIT

The system control unit is essentially the "brain" of the system (Figure 8.80). This unit is responsible for processing alarm signals from actuating devices and transmitting them to the local or other signaling system. In actual installations, the system control panel is often referred to as the alarm or annunciator panel (Figure 8.81). All the controls for the system are located in the system control unit.

PRIMARY POWER SUPPLY

The primary electrical power supply usually consists of the building's main connection to the local public electric utility. An alternative power supply is an engine-driven generator that will provide electrical power. However, if a generator is used, either a trained operator must be on duty 24 hours a day or the system must contain multiple engine-driven generators. One of these generators must always be set for automatic start-

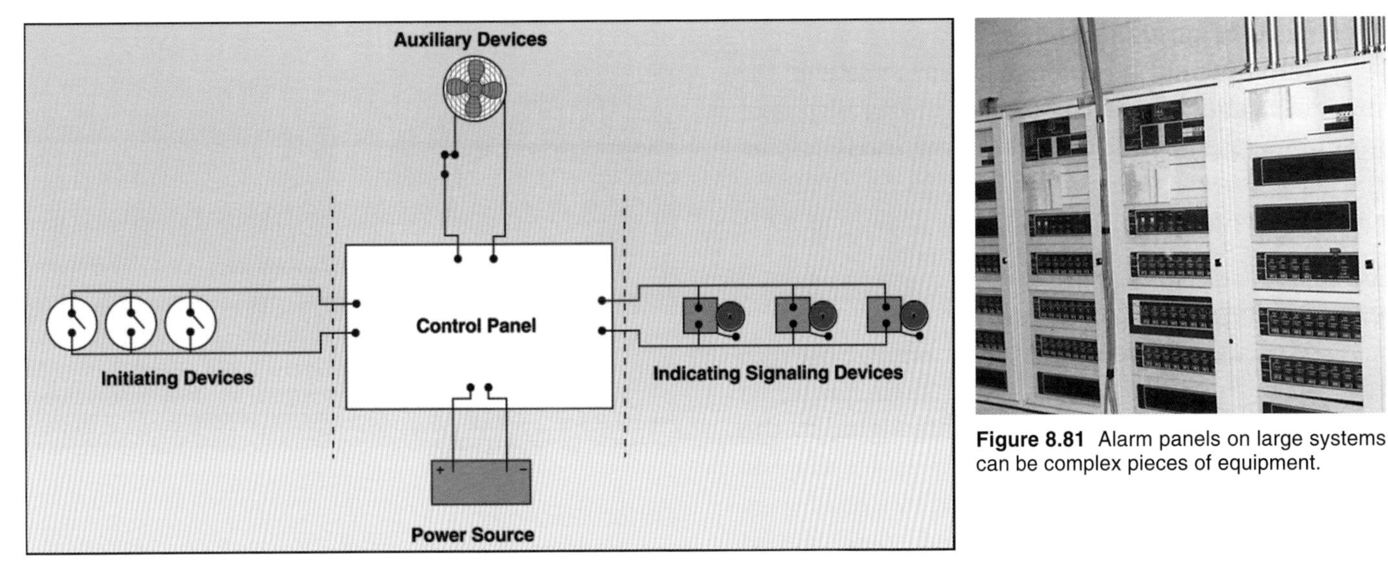

Figure 8.80 The system control panel is the brain center of the alarm and detection system.

Figure 8.81 Alarm panels on large systems can be complex pieces of equipment.

ing. Either power supply must be supervised and should signal an alarm if the power supply is interrupted.

SECONDARY POWER SUPPLY

A secondary power supply must be provided for the detection and signaling system. This will ensure that the system is operational even if the main power supply fails. The secondary system must be able to make the detection and signaling system fully operational within 30 seconds of the main power supply's failure. Table 8.3 shows the capabilities that the secondary power source must have.

TABLE 8.3
Secondary Power Supply Requirements

System Type	Maximum Normal Load	Alarm Load
Local systems	24 hours	5 minutes
Auxiliary systems	60 hours	5 minutes
Remote station systems	60 hours	5 minutes
Proprietary systems	24 hours of normal traffic	
Emergency voice/alarm communication systems	24 hours	2 hours

The secondary power source must consist of one of the following:

- Storage battery and charger (Figure 8.82)
- Engine-driven generator and a four-hour-capacity storage battery (Figure 8.83)
- Multiple engine-driven generators of which one must always be set for automatic starting

Figure 8.82 Fire alarm and detection systems are required to have a back-up power supply. This system utilizes a battery and charger system.

Figure 8.83 Diesel generators may be used for back-up electrical supply.

TROUBLE SIGNAL POWER SUPPLY

There must be a source of power available for the trouble signal indicator. This source of power may not be the primary power supply, but it may be the secondary power supply. In addition, it can be a totally independent power supply, as long as it does not entail the use of dry cell batteries.

INITIATING DEVICES

These are the manual and automatic devices that are either activated or that sense the presence of fire and then send an appropriate signal to the system control unit. The initiating device may be connected to the system control unit by a hard-wire system, or it may be radio controlled over a special frequency. Initiating devices include manual pull stations, heat detectors, smoke detectors, flame detectors, and combination detectors. These devices are covered in more detail later in this chapter.

ALARM-INDICATING DEVICE

Once an initiating device activates and sends a signal to the system control panel, that signal is processed by the control panel and appropriate action is taken. This may include the sounding and lighting of local alarms and the transmission of an emergency signal to a central station service or the fire department dispatch center. Local alarm devices include bells, buzzers, horns, recorded voice messages, strobe lights, and other warning lights (Figures 8.84 a through d). Depending on the design of the system, the local alarm may sound only in the area of the tripped detection device, or it may sound in the entire facility.

Figure 8.84a An alarm bell.

Figure 8.84b Buzzers are used for fire alarms.

Figure 8.84c Loudspeakers may be part of an alarm system.

Figure 8.84d Strobes provide warning for people who have hearing problems.

Other Services

Some occupancies have special requirements in the event of a possible fire condition. In these cases, the fire detection and alarm system can be designed to perform the following special functions:

- Shut down or reverse the heating, ventilation, and air-conditioning (HVAC) system for smoke control.
- Close fire doors and/or smoke or fire dampers (Figure 8.85).
- Pressurize stairwell for evacuation purposes.
- Override control of elevators and prevent them from opening on the fire floor (Figure 8.86).

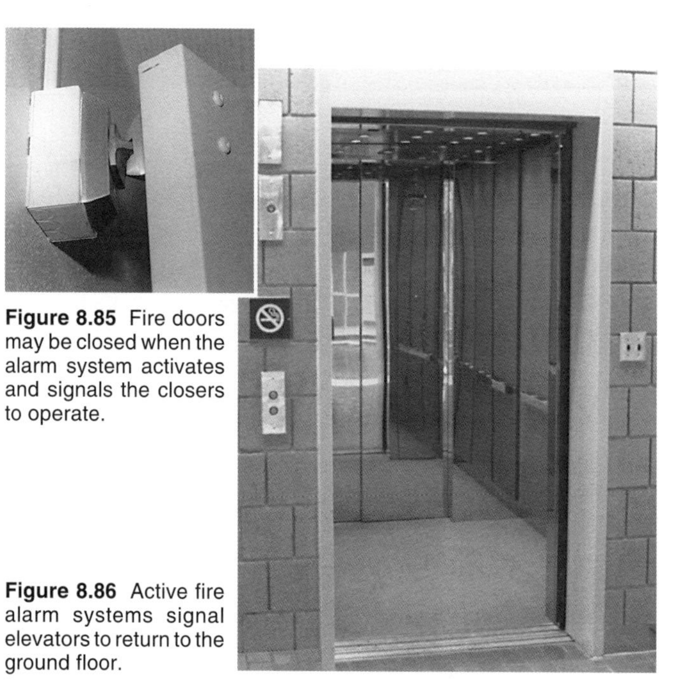

Figure 8.85 Fire doors may be closed when the alarm system activates and signals the closers to operate.

Figure 8.86 Active fire alarm systems signal elevators to return to the ground floor.

- Automatically return the elevator to the ground floor or other designated floor.
- Operate heat and smoke vents.
- Activate special fire extinguishing systems such as halon or halon-replacement systems or preaction and deluge sprinkler systems.

Emergency Voice/Alarm Communications Systems

An emergency voice/alarm communications system is a supplementary system that may be placed in a facility in addition to one of the other types of systems. The purpose of this system is to enable occupants and fire fighting personnel in the facility to communicate detailed information. This system may be a stand-alone system, or it may be integrated directly into the overall fire detection and signaling system.

There are two basic types of emergency voice/alarm communications systems: one-way and two-way. One-way systems are most commonly used to warn building occupants that action is needed and to tell them what actions need to be taken (Figure 8.87). People can be directed to move to a different portion of the building, they can be ordered to leave the building, or if in unaffected areas, they may be told to stay where they are. Two-way systems allow people at other locations in the building to communicate with the person at the main control station. This is accomplished by using either intercom controls or special telephones. The two-way system is most helpful to fire fighting personnel who are operating in the building, particularly in high-rises that interfere with portable radio transmissions (Figure 8.88). The emergency phones are connected in the stairwells and other locations as required by the fire department, and they allow fire fighting crews to communicate with the incident commander at the central control station. Most model building codes require these systems in high-rise structures.

Figure 8.87 Some systems use voice broadcasts to provide direction to occupants.

Figure 8.88 Firefighters will carry phone handsets with them that can be plugged into jacks that allow communication throughout the building.

Types of Signaling Systems

The expressed purpose of a protective signaling system is to give early warning of a fire. Signaling systems vary from the very simple to the very complex. A simple system may only sound a local evacuation alarm. A complex system may sound a local alarm, control building services, and notify outside agencies to respond. The type of system installed in any given occupancy depends on the following factors:

- The level of life safety hazard
- Structural features and size of the building
- Level of hazard presented by systems and contents in the building
- Local and state code requirements

The following sections examine each of the common types of signaling systems that inspection personnel are likely to encounter. The major types of systems include the following:

- Local
- Auxiliary
- Remote station
- Proprietary
- Central station

Previously, each of these types of systems had its own NFPA standard (NFPA 72A, 72B, 72C, etc.). However, as of May, 1993, the requirements for all fire alarm and protective signaling systems are contained in NFPA 72, *The National Fire Alarm Code*.

LOCAL SYSTEM

A local alarm system is designed to transmit both a visible and an audible alarm on the immediate premises served by the system. Its purpose is to alert the building's occupants and to ensure their life safety. The local system can be activated by manual means and/or by automatic detection devices. The local system may also be capable of supervision to make sure that no service interruptions go unnoticed. Local systems can be designed to activate the other services described at the end of the previous section. There are four basic types of local alarm systems: noncoded, zone noncoded, master coded, and zone coded.

Any of these types of local alarm systems may be equipped with a presignal alarm. Presignal alarms are employed in such locations as hospitals, where the potential for panic is high. The system responds initially with a presignal that alerts emergency personnel before the general occupancy is notified. This presignal is usually a discreet signal that is recognizable only by personnel who are familiar with the system. The presignal may be a recorded message over an intercom, a soft alarm signal, or a pager notification. The presignal provides emergency personnel with an opportunity to assist the general occupancy in evacuating. Depending on the policies of the occupancy and local code requirements, the emergency personnel may elect to handle the incident without sounding the general alarm. They may sound the general alarm after investigating the problem, or the general alarm may sound automatically after a certain amount of time has passed and the system control panel has not reset.

AUXILIARY SYSTEM

An auxiliary system is used only in those communities that are served by a municipal fire alarm box system. The auxiliary alarm system is a system within an occupancy that is attached directly to the municipal fire alarm box system. The municipal fire alarm box system may be of the hard-wired or radio-box type. When an alarm is activated in the protected occupancy, it is transmitted directly over the same communications lines as the alarms from the street boxes (Figure 8.89). This results in an alarm signal being received directly at the fire department dispatch center. The system is initiated by manual pull boxes, automatic fire detection devices, or waterflow indicating devices. Each community has its own requirements for these systems; some do not allow them at all. There are three basic types of auxiliary systems: local energy, shunt, and parallel telephone.

Figure 8.89 Auxiliary alarm systems are becoming less common in today's society.

Local Energy System

A local energy system has its own source of power and is not dependent on the supply source that powers the entire municipal fire alarm system. Initiating devices can be activated even when the power supply to the municipal system is interrupted. However, this may result in the alarm only being sounded locally and not being transmitted to the fire department. The ability to transmit alarms during power interruptions is dependent on the design of the municipal system.

Shunt System

A shunt system is an integral part of the municipal fire alarm system and is dependent on the municipal system's source of current for electric power. Should a ground fault occur in this type of system, an alarm indication (false alarm) will be sent to the fire department receiving point. NFPA 72 allows only manual pull boxes and waterflow detection devices to be used on shunt systems. Fire detection devices are not permitted on these systems.

Parallel Telephone System

A parallel telephone system is used on the type of municipal fire alarm system where each alarm box is connected to the fire department dispatch center by an individual circuit. The occupancy fire alarm system is also connected to the fire department dispatch center by an individual circuit.

REMOTE STATION SYSTEM

A remote station system is similar to an auxiliary system in that it is connected directly to the fire department dispatch center or other approved answering service. However, rather than being hooked to the dispatch center via a municipal fire alarm box system, the remote system is connected by some other form of connection, usually leased telephone lines (Figure 8.90). Where permitted, a radio signal over a dedicated fire department frequency may also be used. A remote station is common in localities that are not serviced by central station systems.

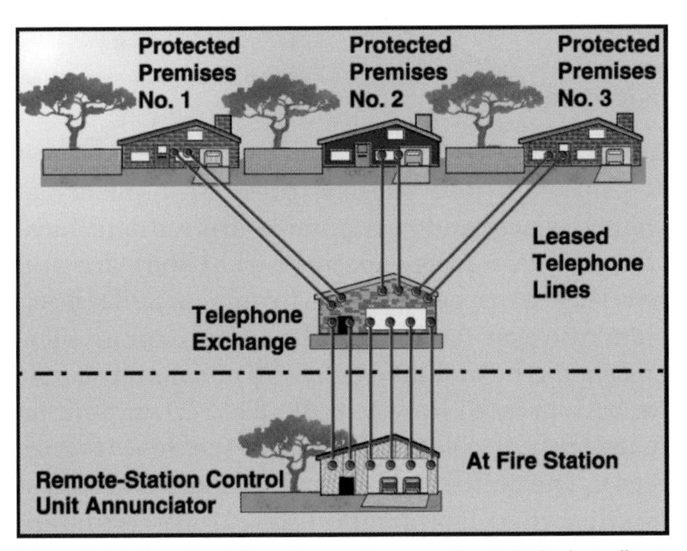

Figure 8.90 Remote station alarm systems use leased telephone lines.

A remote station system may be of either the coded or noncoded variety. A noncoded system is only allowable in situations where one occupancy is protected by the system. Coded systems may be used when up to five facilities are protected by the system. These five facilities may have a common connection to the remote station. A remote station system does not necessarily have local alarm capabilities because evacuation capabilities are not required by NFPA 72. The system will have the ability to transmit a trouble signal to the remote station when the system becomes impaired.

Depending on local preferences, the fire department may allow entities other than itself to monitor the remote station. In many small communities, it is monitored by the local police agency at its dispatch center. This is particularly common in communities that have volunteer fire departments whose stations are not constantly manned. In these cases, it is important that police-dispatch personnel are appropriately sensitized to the importance of these alarm signals and are trained in the actions that must be taken immediately upon receiving them.

PROPRIETARY SYSTEM

A proprietary system is used to protect large commercial and industrial buildings, high rises, and groups of commonly owned facilities in a single location, such as a college campus or industrial complex. Each building or area will have its own system that is wired into a common receiving point somewhere on the facility. The receiving point must be in a separate structure or in a protected part of the structure that is remote from any hazardous operations (Figure 8.91). The receiving station is constantly manned by representatives of the occupant who are trained in system operation and actions to take when an alarm is received. The operator should be able to automatically summon the fire department through the use of system controls or to do so manually by using the telephone.

Modern proprietary systems can be very complex systems that have a wide range of capabilities, including the following:

- Coded-alarm and trouble-signal indications

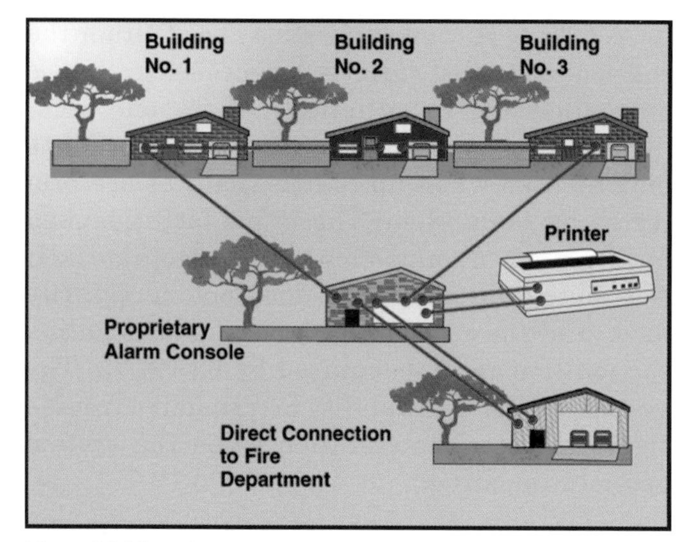

Figure 8.91 Proprietary alarm systems are common on large commercial or industrial sites where numerous buildings require monitoring.

- Building utility controls
- Elevator controls
- Fire and smoke damper controls

Many proprietary systems and receiving points are used to monitor security functions in addition to fire and life safety functions.

CENTRAL STATION SYSTEM

A central station system is basically the same as a proprietary system. The primary difference is that instead of having the receiving point for alarms on the protected premises and monitored by the occupant's representative, the receiving point is at an outside, contracted service point called a central station (Figure 8.92). Typically, the central station is a company that sells its services to many individual customers. When an alarm is activated at a particular client's location, central station employees take that information and initiate an emergency response. This involves calling the fire department and representatives of the occupancy. The alarm system at the protected property and the central station are connected by dedicated telephone lines or radio transmitters. All central station systems should meet the requirements set forth in NFPA 72.

Manual Alarm-Initiating Devices

Manual alarm-initiating devices, commonly called pull boxes or stations, are placed in structures to allow occupants to manually initiate the fire signaling system (Figure 8.93). Pull stations

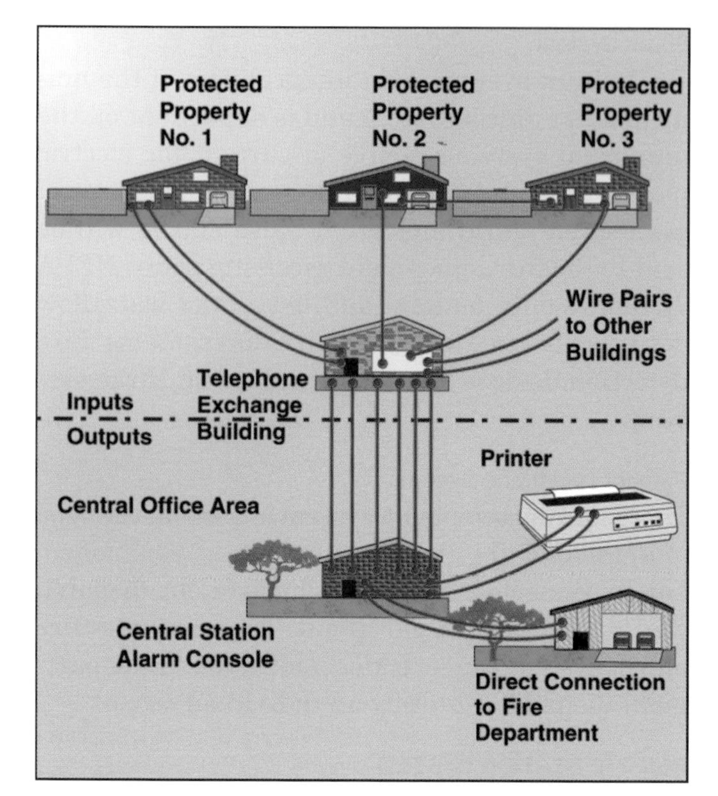

Figure 8.92 Central station alarm systems are among the most common in use today.

may be connected to systems that sound either local alarms, supervisory notification alarms, or both. Most codes do not require manual alarm-initiating devices in structures that are fully sprinklered with a system that triggers a local alarm when water begins to flow.

Figure 8.93 A typical manual pull box.

GENERAL REQUIREMENTS

Although pull stations come in a variety of shapes, sizes, and colors, most are red and have white lettering that specifies what they are and how they are to be used. The fire alarm pull station should be used for fire signaling purposes only. The box should be mounted on a wall or column so that its bottom is no less than 3½ feet (1.1 m) and no more than 4½ feet (1.3 m) from the floor (Figure 8.94). The pull station must be positioned in a location where it is plainly visible. Multistory buildings should have at least one pull station on each floor. In all cases, travel distances to the nearest pull station may not exceed 200 feet (60 m). It is wise to place pull stations at exits so that people can activate them on their way out of the building (Figure 8.95).Older-type pull stations that require the operator to break a small piece of glass with a provided mallet are no longer recommended (Figure 8.96). Originally, these types of devices were designed to cut down on false alarms and were somewhat effective. However, the broken glass presents an injury hazard to the operator. Plastic covers with tamper alarms reduce false alarms by sounding a loud alarm from the device if the cover is opened.

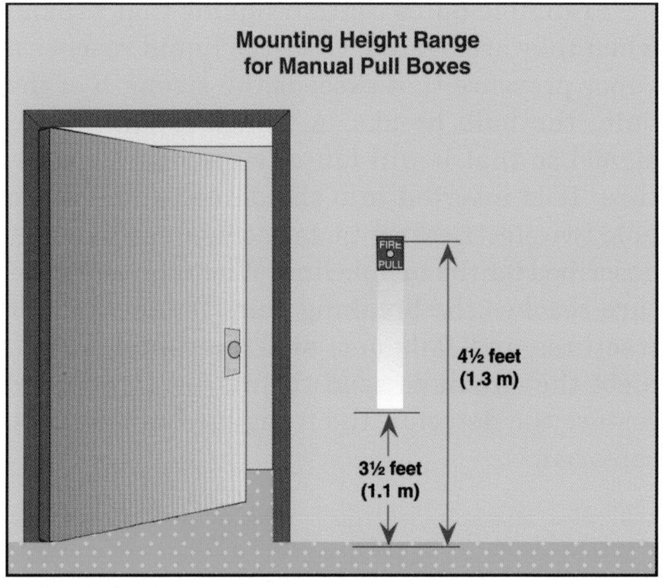

Figure 8.94 This diagram shows the mounting parameters for manual pull boxes.

Figure 8.95 It is wise to mount manual pull boxes near exits.

Figure 8.96 Some manual pull boxes require the user to break a thin sheet of glass before accessing the alarm.

Automatic Alarm-Initiating Devices

An *automatic alarm-initiating device*, sometimes simply called a detector, is a device that continuously monitors the atmosphere of a given area for the products of combustion (Figure 8.97). When such products are detected, the device sends a signal to the system control panel. This device is typically very accurate at sensing the presence of products of combustion; in many cases, however, other products that mimic fire conditions can be found in a given area, even when no emergency exists. For example, flame detectors may trip if a welder strikes an arc in a monitored area. These possibilities force fire protection system designers to take into account the normal activities that take place in a given area. System designers must then design a detection system that minimizes the chances of an accidental activation.

There are four basic types of automatic alarm-initiating devices. They include those that detect

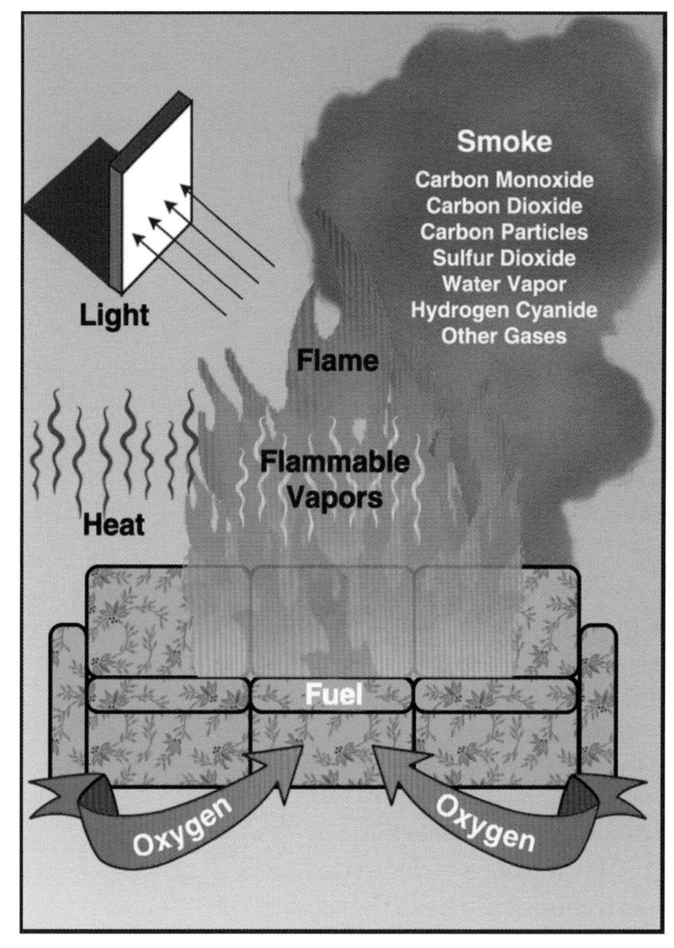

Figure 8.97 Detection devices are available to recognize any of the products of combustion.

heat, smoke, fire gases, and flame. The following sections describe the various types of devices in use.

FIXED-TEMPERATURE HEAT DETECTORS

Systems using heat detection devices are among the oldest types of fire detection systems in use. They are relatively inexpensive compared to other types of devices, and they are least prone to false activation. Heat detectors are limited, however, by the fact that they are typically the slowest type of system to activate under fire conditions.

Heat detectors must be placed in high portions of a room, usually on the ceiling, to be effective (Figure 8.98). Consideration must be given to the highest expected temperature in an area so that the appropriately rated heat detector can be selected.

Figure 8.98 Heat detectors are mounted on the ceiling.

Heat is an abundant product of combustion. It is detectable by certain devices using three primary principles of physics:

- Heat expands a material, closing a circuit.
- Heat melts a material, releasing stored mechanical energy.
- Heated materials have thermoelectric properties that are detectable.

All heat detection devices operate on one or more of these principles. The various types of fixed-temperature devices used in fire detection systems are covered in the following sections.

Fusible Links/Frangible Bulbs

While these two devices are more commonly associated with automatic sprinklers, they are also used in fire detection and signaling systems. The operating principles of links and bulbs that are used in fire detection and signaling systems are identical to the links and bulbs that are used with automatic sprinklers, which are described in Chapter 7 of this manual. It is their application that differs.

Fusible links are used to hold a spring device in the detector in the open position (Figures 8.99 a and b). When the melting point of the fusible link is reached, it melts and drops out. This causes the spring to release and touch an electrical contact that completes the circuit and sends an alarm signal. In order to restore the detector, the fusible link must be replaced.

Frangible bulbs contain liquids that expand when they are heated. Once the liquid reaches a vapor pressure that exceeds the strength of the bulb, the bulb breaks. A frangible bulb is designed so that it will fail at a specific temperature. It is inserted into the detection device to hold two electrical contacts apart, much like that described for the fusible link. When the temperature reaches the breaking point for the bulb, it fractures and falls out, and the contacts complete the circuit to send the alarm. In order to restore the detector, the frangible bulb must be replaced.

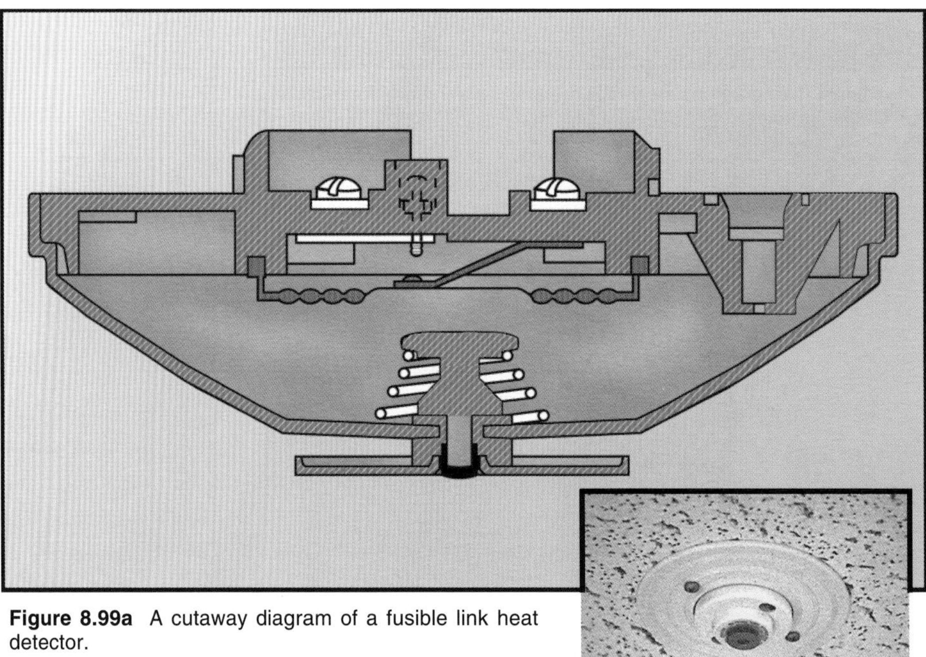

Figure 8.99a A cutaway diagram of a fusible link heat detector.

Figure 8.99b A typical fusible link heat detector.

Continuous Line Detector

Most of the detectors described in this chapter are the "spot" detector type; that is, they detect the conditions only at the specific spot where they are located. However, one type of heat detection device — the continuous line detection device — can be used to detect conditions over a wide area.

One type of continuous line detection device consists of conductive metal inner core cable sheathed within a stainless steel tubing (Figure 8.100). The inner core and the sheath are separated by an electrically insulating semiconductor material that keeps them from touching, but allows a small amount of current to flow between the two. This insulation is designed to lose some of its electrical resistance capabilities at a predetermined temperature anywhere along the line. These cables can be strung out over extremely large areas. When the heat at any given point reaches the resistance reduction point of the insulation, the current between the two components increases. This results in an alarm signal being sent to the system control panel. This detection device restores itself when the level of heat is reduced. These detectors are commonly found in cable trays, industrial locations, and aircraft repair hangars.

Another type uses two wires that are each insulated and bundled within an outer covering (Figure 8.101). When the melting temperature of each wire's insulation is reached, the insulation melts and allows the two wires to touch. This completes the circuit to send an alarm signal to the system control unit. To restore this type of line detector, the fused portion of the wires must be cut out and replaced with new wire.

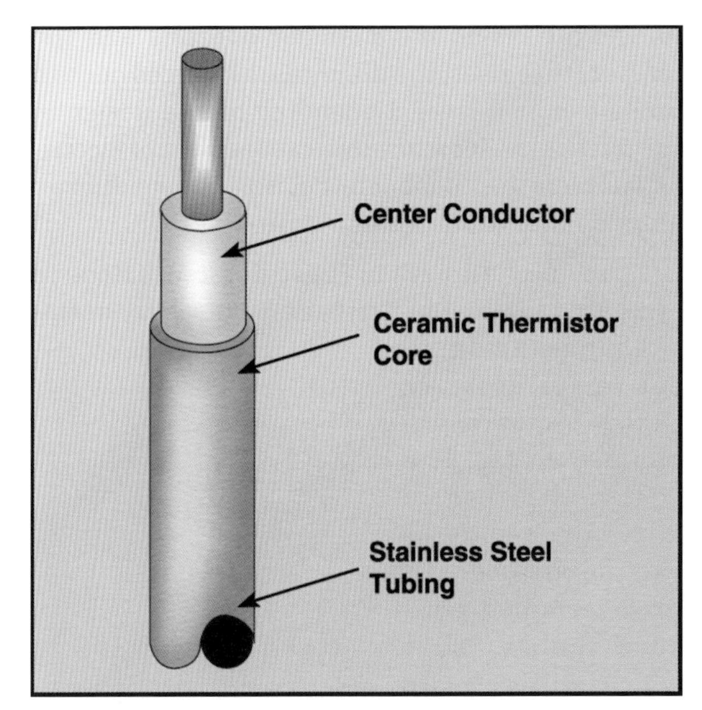

Figure 8.100 The basic components of a continuous line heat detector.

Bimetallic Detector

A bimetallic detector, sometimes referred to as a spot detector, uses two types of metals that have different heat-expansion ratios. These metals are

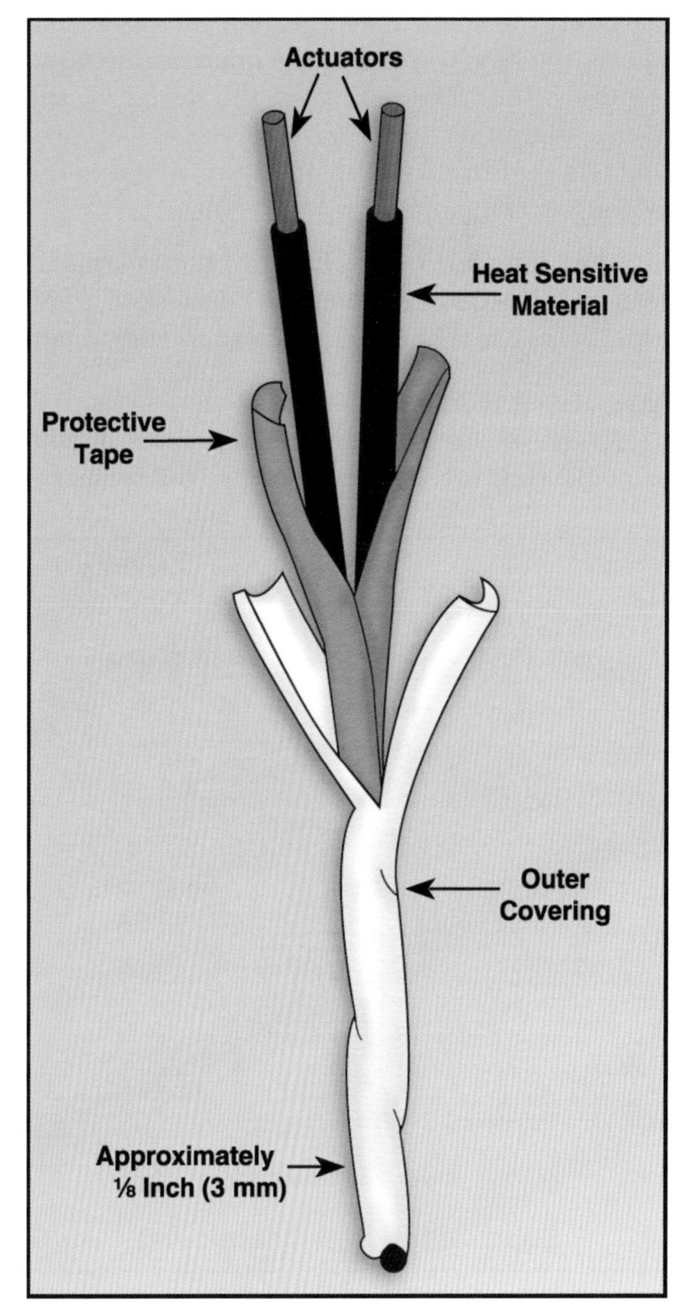

Figure 8.101 Some continuous line heat detectors utilize two wires.

each formed into thin strips, which are then bonded together. The fact that one metal expands faster than the other causes the combined strip to arch when subjected to heat. The amount that they arch depends on the characteristics of the metals, the amount of heat they are subjected to, and the degree of arch when in the normal position. All of these factors are calculated into the design of the detector.

The bimetallic strip may be positioned with either one or both ends secured in the device. When positioned with both ends secured, a slight bow is placed in the strip. When heated, the expansion causes the bow to snap in the opposite direction (Figure 8.102). Depending on the design of the device, this either opens or closes a set of electrical contacts that in turn send an alarm signal to the system control panel.

Most bimetallic detectors are the automatic resetting type. They do need to be checked, however, to ensure that they have not been damaged.

RATE-OF-RISE HEAT DETECTOR

A rate-of-rise heat detector operates on the principle that fires rapidly increase the temperature in a given area. The rate-of-rise detector detects these quick increases in temperature and responds at substantially lower temperatures than fixed-temperature detectors. Typically, rate-of-rise heat detectors are designed to send a signal when the rise in temperature exceeds 12°F to 15°F (7°C to 8°C) per minute. This is because temperature changes of this magnitude are not expected under normal, nonfire conditions.

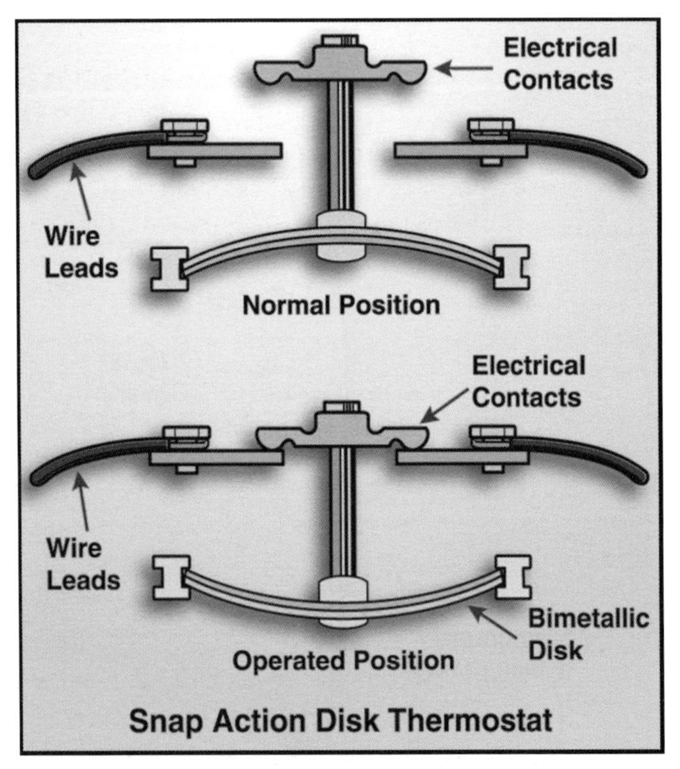

Figure 8.102 A bimetallic heat detector in the ready and operating position.

Most rate-of-rise heat detectors are reliable and not subject to false activations. However, they can occasionally be activated under nonfire conditions. An example would be when a rate-of-rise detector is placed near a garage door in an air-conditioned building. If the garage door is opened on a hot day, the influx of heated air will rapidly increase the temperature around the detector and cause it to activate. These situations can be avoided by careful placement.

There are several different types of rate-of-rise heat detectors in use; all are automatically reset. These are discussed in more detail in the following sections.

Pneumatic Rate-Of-Rise Spot Detector

A pneumatic spot detector is the most common type of rate-of-rise detector in use (Figures 8.103 a and b). It contains a small chamber filled with air and has a flexible metal diaphragm in the bottom. As the air inside the chamber expands and the temperature rises, the diaphragm is forced out to a predetermined level. Depending on the design of the particular detector, this causes a set of electrical contacts to either open or close, thus sending an alarm signal to the system control panel.

The air chambers in these detectors have to be vented to prevent activation caused by normal changes in ambient temperature or changes in barometric pressure. The vent is designed to allow air to enter or exit the chamber at a predetermined pressure below that which will activate the detector.

Figure 8.103a A pneumatic rate-of-rise heat detector.

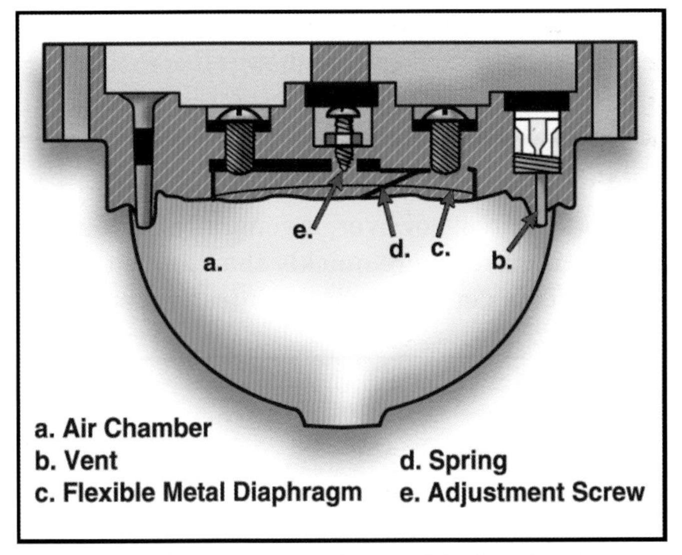

Figure 8.103b The components of a rate-of-rise heat detector.

Pneumatic Rate-Of-Rise Line Detector

While the spot detector monitors its exact area of location, a line detector can monitor large areas. Line detectors consist of a system of tubing arranged over a wide area of coverage (Figure 8.104). The space inside the tubing acts as the chamber (previously described in the section on spot detectors). These detectors also contain a diaphragm and are vented. When any area being served by the tubing experiences a temperature increase, the detector functions in the same manner as that described for the spot detector.

The tubing in these systems must be limited to about 1,000 feet (300 m) in length. The tubing should be arranged in rows that are not more than 30 feet (10 m) apart and 15 feet (5 m) from walls.

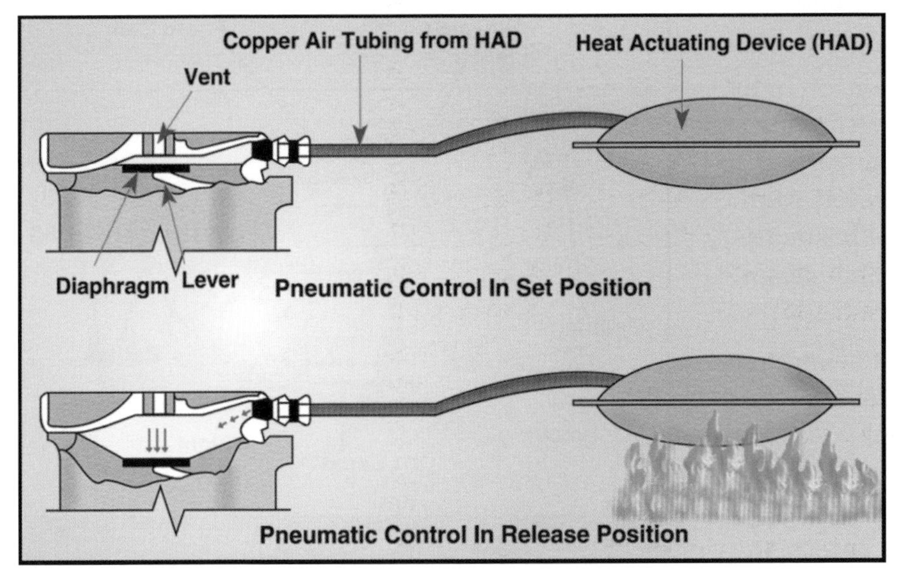

Figure 8.104 The operating principle of a pneumatic tube heat detector.

Rate Compensated Detector

This detector is designed for use in areas that are subject to regular — but slower than fire condition — temperature changes under normal conditions. It contains an outer bimetallic sleeve with a moderate expansion rate. This outer sleeve contains two bowed struts that have a slower expansion rate than the sleeve (Figure 8.105). The bowed struts have electrical contacts on them. In the normal position these contacts do not come together. When the detector is heated rapidly, the outer sleeve expands in length. This reduces the tension on the inner strips and allows the contacts to come together, thus sending an alarm signal to the system control panel.

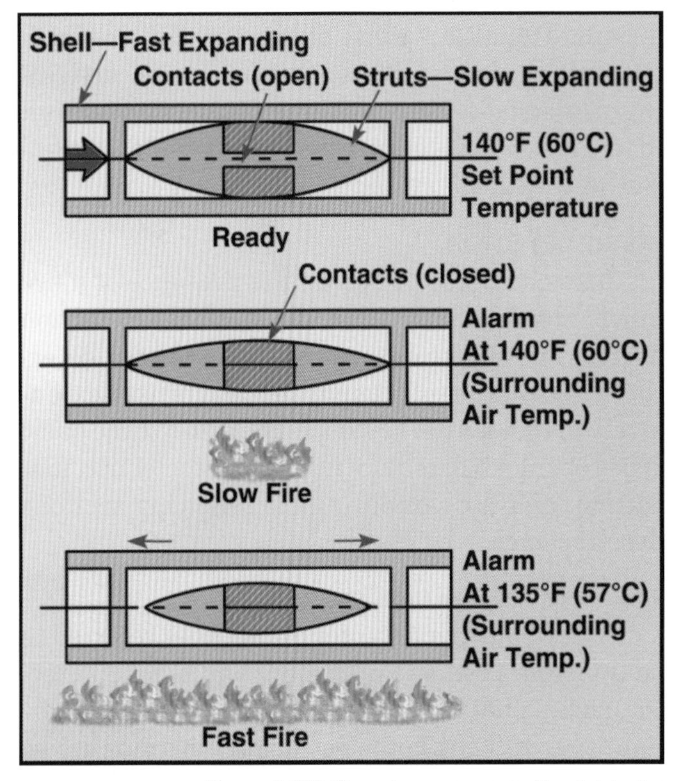

Figure 8.105 The rate compensated heat detector responds differently to varying levels of heat.

If the rate of temperature rise is fairly slow, such as 5°F (2°C to 3°C) per minute, the sleeve expands at a slow rate that will maintain tension on the inner strips. This prevents unnecessary system activation.

Thermoelectric Detector

This rate-of-rise detector operates on the principle that two wires made of dissimilar metals, when twisted together and heated at one end, will cause an electrical current to be generated at the other end. The rate at which the wires are heated determines the amount of current that is generated. These detectors are designed to bleed off or dissipate small amounts of current. This reduces the chance of a small temperature change activating an alarm. Large changes in temperature result in larger amounts of current flowing and in activation of the alarm system.

SMOKE DETECTOR

A smoke detector senses the presence of a fire much more quickly than does a heat detection device. The smoke detector is the preferred detector in many types of occupancies and is used extensively in residential settings. There are three basic types of smoke detectors in use: photoelectric, ionization, and air sampling. The following sections describe each.

Photoelectric Smoke Detector

A photoelectric detector, sometimes referred to as a visible products-of-combustion smoke detector, uses a photoelectric cell coupled with a specific light source. The photoelectric cell functions in two ways to detect smoke: beam application (obscuration) and refractory application.

The beam application uses a beam of light focused across the area being monitored onto a photoelectric cell. The cell constantly converts the beam into current, which keeps a switch open. When smoke blocks the path of the light beam, the current is no longer produced, the switch closes, and an alarm signal is initiated (Figure 8.106).

The refractory photocell uses a light beam that passes through a small chamber at a point away from the light source. Normally, the light does not strike the photocell, and no current is produced. When a current does not flow, a switch in the current remains open. When smoke enters the chamber, it causes the light beam to be refracted (scattered) in random directions. A portion of the scattered light strikes the photocell, causing current to flow. This current closes the switch and activates the alarm signal (Figure 8.107).

A photoelectric detector works satisfactorily on all types of fires; however, it generally responds to smoldering fires more quickly than ionization detectors. This detector is automatically reset.

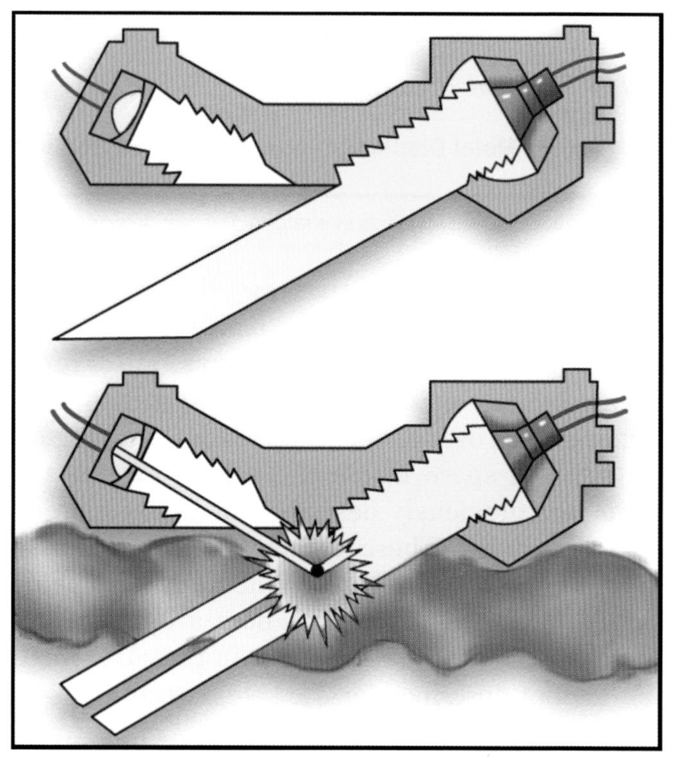

Figure 8.106 Photoelectric smoke detectors operate when smoke particles break a beam of light.

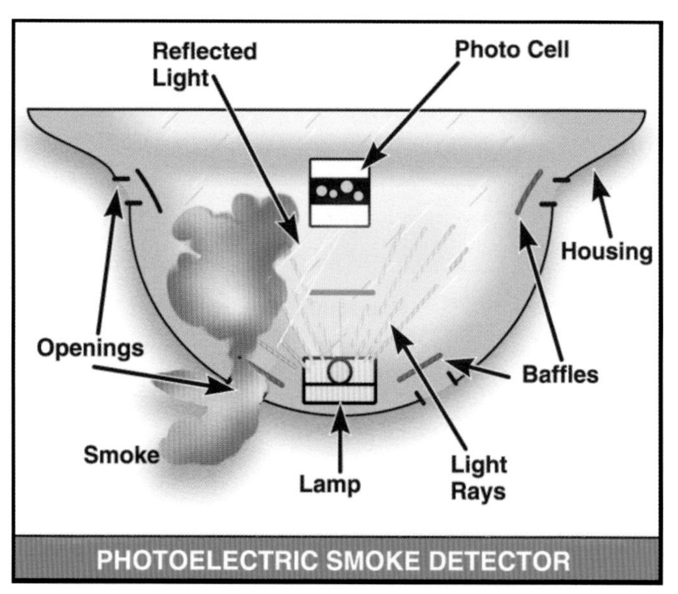

Figure 8.107 The components of a photoelectric smoke detector.

Ionization Smoke Detector

During a fire, molecules ionize as they undergo combustion. The ionized molecules have an electron imbalance and tend to steal electrons from other molecules. The operation of the ionization smoke detector, sometimes referred to as the invisible products-of-combustion detector, uses this phenomenon.

The detector has a sensing chamber that samples the air in a room. A small amount of radioactive material (usually americium) adjacent to the opening of the chamber ionizes the air particles as they enter. Inside the chamber are two electrical plates: one positively charged and one negatively charged. The ionized particles free electrons from the negative plate, and the electrons travel to the positive plate. Thus, a minute current normally flows between the two plates. When ionized products of combustion enter the chamber, they pick up the electrons freed by the radioactive ionization. The current between the plates ceases and an alarm signal is initiated (Figure 8.108).

An ionization detector responds satisfactorily to all types of fires; however, it generally responds quicker to flaming fires than do photoelectric detectors. However, ionization detectors are slower to detect smoldering fires. This detector is an automatic resetting type.

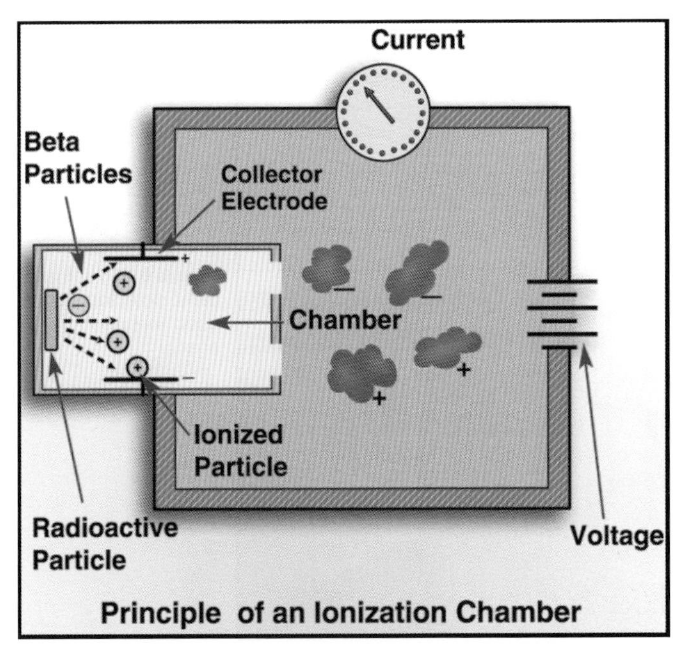

Figure 8.108 The principal features of an ionization smoke detector.

Air-Sampling Smoke Detectors

Air-sampling smoke detectors are designed to continuously monitor a small amount of air from the protected area for the presence of smoke particles. There are two basic types of air-sampling smoke detectors. The most common type is the cloud chamber type (Figure 8.109). This detector uses a small air pump to draw sample air into a high-humidity chamber within the detector. The detector then imparts the high humidity on the sample and lowers the pressure in the test chamber. Moisture condenses on any smoke particles in the test chamber. This creates a cloud in the chamber. The detector triggers an alarm signal when the density of this cloud exceeds a predetermined level.

The second type of air-sampling smoke detector is comprised of a system of pipes spread over the ceiling of the protected area (Figure 8.110). A fan in the detector/controller unit draws air through the pipes. The air is then sampled using a photoelectric sensor.

FLAME DETECTOR

A flame detector is sometimes called a light detector. There are three basic types:

- Those that detect light in the ultraviolet wave spectrum (UV detectors) (Figure 8.111)
- Those that detect light in the infrared wave spectrum (IR detectors) (Figure 8.112)
- Those that detect both types of light

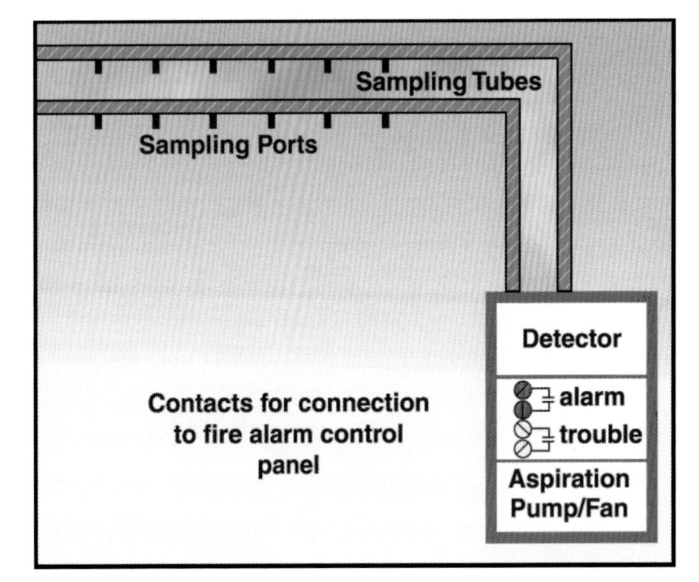

Figure 8.109 A cloud chamber air sampling smoke detector.

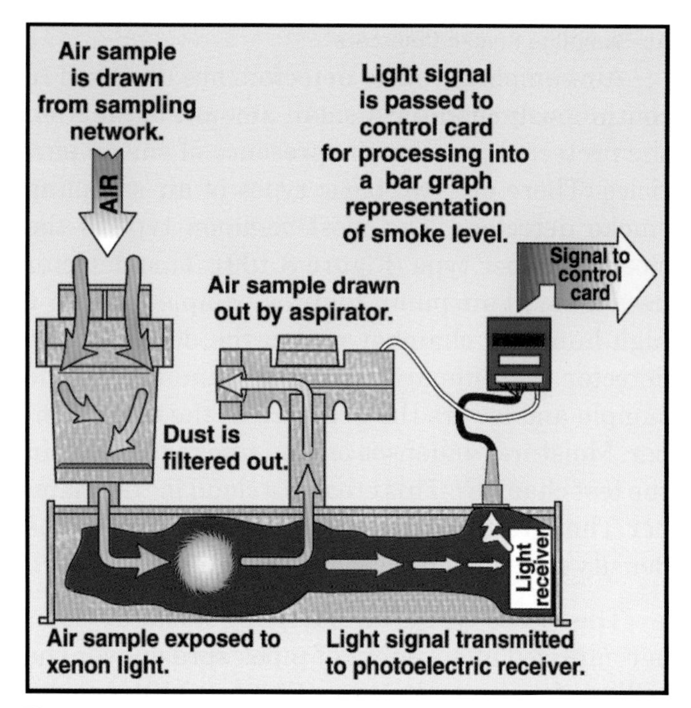

Figure 8.110 The operating principles of a tube-type air sampling smoke detector.

While these types of detectors are among the fastest to respond to fires, they can be easily tripped by nonfire conditions such as welding, sunlight, and other bright light sources. For maximum effectiveness, they must only be placed in areas where these possibilities can be avoided. They must also

Figure 8.111 An ultraviolet flame detector.

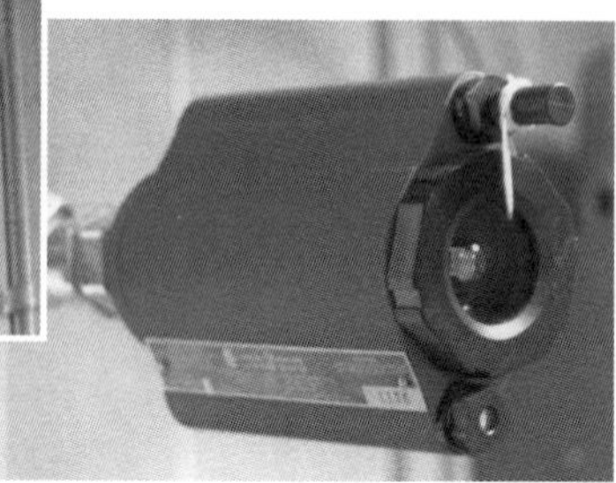

Figure 8.112 An infrared flame detector.

be positioned so that they have an unobstructed view of the protected area. If they are blocked, they cannot activate.

An ultraviolet detector can give false alarms when it is in contact with sunlight and arc welding. Therefore, its use is limited to areas where these and other sources of ultraviolet light can be eliminated. An infrared detector is effective in monitoring large areas (Figure 8.113). To prevent activation from infrared light sources other than fires, an infrared detector requires the flickering action of a flame before it activates to send an alarm. They are typically designed to respond to 1 square foot (0.9 m^2) of fire from a distance of 50 feet (15 m).

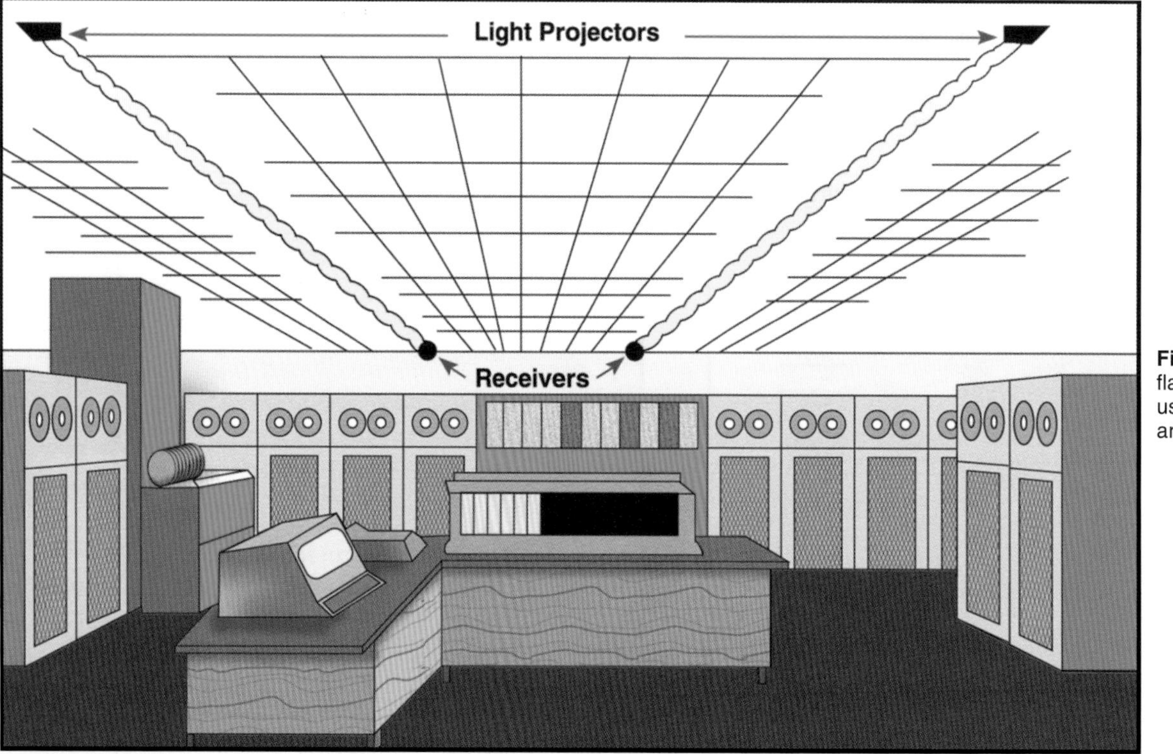

Figure 8.113 Infrared flame detectors can be used to protect large areas.

FIRE GAS DETECTOR

When fire breaks out in any confined area, it drastically changes the chemical gas content of the atmosphere in that area. Some of the gases released by a fire may include:

- Water vapor (H_2O)
- Carbon dioxide (CO_2)
- Carbon monoxide (CO)
- Hydrogen chloride (HCl)
- Hydrogen cyanide (HCN)
- Hydrogen fluoride (HF)
- Hydrogen sulfide (H_2S)

Only water, carbon dioxide, and carbon monoxide are released from all materials that burn. The other gases released are dependent on the specific chemical makeup of the fuel. Therefore, it is only practical to monitor the levels of carbon dioxide and carbon monoxide for fire detection purposes. A fire gas detector is somewhat faster than a heat detector, but not as fast as a smoke detector. It uses either semiconductors or catalytic elements to sense the gas and trigger the alarm. A fire gas detector is not as commonly used as other types of detectors.

COMBINATION DETECTOR

Depending on the design of the system, various combinations of the previously described detection devices may be used in a single device. These combinations include fixed rate/rate-of-rise detectors, heat/smoke detectors, and smoke/fire gas detectors. These combinations give the detector the benefit of both services and increase their responsiveness to fire conditions.

Inspecting and Testing Fire Detection and Signaling Systems

In order to ensure operational readiness and proper performance, fire detection and signaling systems must be tested when they are installed and then tested on a continuing basis. Tests conducted during installation are commonly called acceptance tests. Periodic testing is often referred to as service testing.

Fire department personnel who routinely conduct inspections will find it necessary to have a working knowledge of these systems (Figure 8.114). It is important to keep in mind, however, that fire department personnel are generally limited to visual inspections and to supervision of system tests. They will not have to operate or maintain these systems. In most cases, in-plant personnel or alarm system contractors actually perform system tests and maintenance. These personnel should be trained by representatives of their system's manufacturer.

Figure 8.114 Fire inspectors should have a working knowledge of alarm and detection systems.

ACCEPTANCE TESTING

Acceptance testing is performed soon after the system has been installed to make sure it meets the design criteria and functions properly. Acceptance tests may be required by the occupant's insurance carrier and/or local fire codes. This test should be witnessed by representatives of the occupant, the fire department, the system installer, and the system manufacturer (Figure 8.115). The fire department representative may be the fire marshal, an inspector, or a staff fire protection engineer. Some jurisdictions require a test certificate from the system manufacturer/installer indicating that the system has been thoroughly checked before the fire department inspection. This keeps the fire inspector from spending time checking a system that may not have been properly installed.

All functions of the fire detection and signaling system should be checked during the acceptance tests. Both the alarm and trouble modes of system operation should be checked. Actual wiring and circuitry should be checked against the system

Figure 8.115 The occupants should conduct all system tests.

drawing to ensure that all are connected properly. The system control panel should be thoroughly inspected to ensure that it is in proper working order. All interactive controls at the panel should be operated to ensure that they control the system as designed. All alarm initiation and indication devices and circuits must be checked for proper operation. The system should be operated on both the primary and secondary power supplies to make sure both will adequately supply the system.

All alarm-indicating and alarm-initiating devices, such as pull stations, detectors, bells, and strobe lights, should be checked to make sure they are operational. Each initiating device must be checked to ensure that it sends an appropriate signal and causes the system to go into the alarm mode.

Restorable heat detectors may be checked by subjecting them to heat generated by a hair dryer, heat lamp, or specialized testing appliance. Never allow an open flame to be used for testing detectors. Remember that some combination detectors have both restorable and nonrestorable elements. Exercise caution to avoid tripping the nonrestorable element. Restorable pneumatic heat detectors may be tested with a heating device (such as a common hair dryer) or with a pressure pump. Nonrestorable pneumatic detectors should be tested mechanically. Those detectors equipped with replaceable fusible links should have the links removed to see whether the contacts then touch and send an alarm signal. The links can then be replaced.

The manufacturers of smoke, flame, and fire gas detectors usually have specific instructions for testing their detectors. These instructions must be followed on both the acceptance and service tests. They may include the use of smoke-generating devices, aerosol sprays, or magnets. The use of nonapproved testing devices may result in voiding the manufacturer's warranty on the detector.

It is also important to check the response of outside entities who are responsible for monitoring the system. This is important on central station, auxiliary, remote station, and proprietary systems. The alarm-receiving capability and response of those involved must be guaranteed.

The results of all tests must be documented to the satisfaction of both the insurance carrier and the fire department. Only after all parts of the system have successfully passed the tests should a system certification certificate be issued. This is typically a preliminary step toward the issuance of a certificate of occupancy. NFPA 72 contains complete information on system acceptance tests.

SYSTEM SERVICE TESTING AND PERIODIC INSPECTIONS

A general inspection of fire detection and signaling equipment should be conducted on a routine basis. The inspections should be conducted by both the authority having jurisdiction and the occupant. Because fire departments may have many occupancies in their jurisdiction and testing these systems is time-consuming, it is not possible for the fire department to participate at every test. Occupants will have to test the systems on their own most of the time and document the results. At intervals specified by the fire department, fire department personnel will also witness the tests. The procedures covered in this section should be applied during occupant and/or fire department inspections and testing.

Inspect all wiring for proper support, wear, damage, or any other defects that may render the insulation ineffective. Where circuits are enclosed in conduit, inspect the conduit for solid connections and proper support. When batteries are used as an emergency power source, they should be checked for clean contacts and proper charge. Many batteries have floating-ball indicators that show whether the battery is properly charged.

All equipment, especially initiating and indicating devices, must be kept free of dust, dirt, paint, and other foreign materials. When either dust or dirt is found, it is recommended that the devices be cleaned with a vacuum cleaner rather than by wiping (Figure 8.116). Wiping tends to spread the debris around, causing it to settle on electrical contacts. This may inhibit the future operation of the system.

System control units, recording instruments, and other devices should not have objects stored on, in, or around them. Many system control panels that are designed with locking doors have storage areas for extra relays, light bulbs, and test equipment. If this space is not designed into the unit, these devices should be stored somewhere else. Otherwise, they may foul moving parts or cause electrical shorts that can result in system failure.

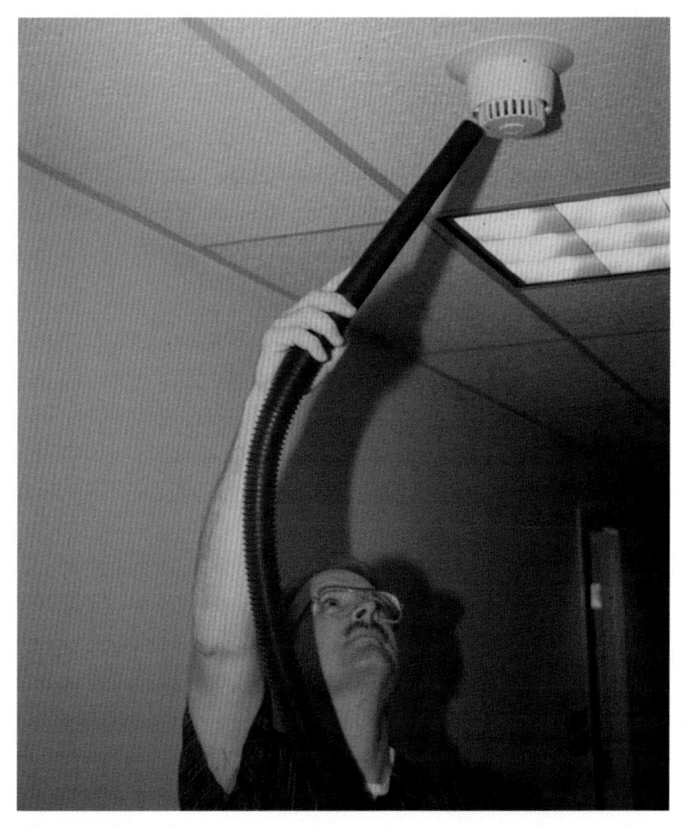

Figure 8.116 Ensure that proper maintenance functions, such as cleaning smoke detectors, are being performed within the occupancy.

INSPECTING AND SERVICE TESTING INITIATION DEVICES

Obviously, the system will not work at all unless the alarm-initiation devices are in proper working order and send the appropriate signal to the system control panel. The following sections highlight the procedures for checking these devices.

Testing Manual Alarm-Initiation Devices

Numerous items need to be checked when testing and inspecting a manual alarm-initiation device. Access to the device must be unobstructed, and each unit should be easy to operate. The housing should be tightly closed to prevent dust and moisture from entering the unit and disrupting service (Figure 8.117). Any chipped, cracked, or otherwise impaired glass should be removed and replaced. If the device is equipped with a cover or door, it should be checked to make sure it opens easily and all the components behind the door are in place and ready for service. Inspectors may wish to witness the activation of selected devices to ensure that the device and the system are operational.

Figure 8.117 Make sure that the manual pull box is still securely attached to the wall.

Testing Automatic Alarm-Initiation Devices

Without functional detection devices, the most elaborate wiring and signaling systems are virtually useless. The reliability of the entire system is, in fact, based mostly on the reliability of the detection devices. Automatic alarm-initiation devices should be checked after installation, after a fire, and periodically based on guidelines established by the authority having jurisdiction. All detector testing should be in accordance with

NFPA 72. This section highlights the procedures and issues related to detector inspections and testing.

Detectors must not be damaged or painted. Regardless of the type of detector in use, the following detectors should be replaced or sent to a recognized testing laboratory for testing:

- Detectors on systems that are being restored to service after a period of disuse
- Detectors that are obviously corroded
- Detectors that have been painted over, even if attempts were made to clean them
- Detectors that have been mechanically damaged or abused
- Detectors on circuits that were subjected to current surges, overvoltages, or lightning strikes
- Detectors subjected to foreign substances that might affect their operation

A permanent record of all detector tests must be maintained for a minimum of five years. The minimum information that should be included in the record are the date, the detector type, the location, the type of test, and the results of the test.

A nonrestorable fixed-temperature detector cannot be tested periodically. Testing would destroy the detector. For this reason, tests are not required until fifteen years after the detector has been installed. At this time, two percent of the detectors must be removed and replaced with new detectors. The old detectors must then be sent to an approved testing laboratory for testing. If one detector fails, several other detectors must be removed and sent for testing. Line detectors must have resistance testing performed semiannually.

A restorable heat detection device should be checked as described previously in the acceptance testing section. One detector on each signal circuit should be tested semiannually. A different detector should be selected each time.

A fusible-link detector with replaceable links should also be checked semiannually. This is done by removing the link and observing whether or not the contacts close (Figure 8.118). After the test, the fusible link should be replaced. It is recommended that the links be replaced at five-year intervals.

Figure 8.118 Test the system by pulling out a fusible link.

A pneumatic detector should be tested semiannually with a heating device or a pressure pump. If a pressure pump is used, the manufacturer's instructions must be followed closely.

A smoke detector should be tested semiannually in accordance with manufacturer's recommendations. The instruments required to carry out performance and sensitivity testing are usually provided by the manufacturer. Sensitivity testing should be performed after the detector's first year of service and every two years after that. Blowing cigarette smoke into the detector is not an acceptable method of testing smoke detectors.

Flame and gas detection devices are very complicated devices and should only be tested by highly trained individuals. Testing is typically performed by outside personnel operating on a contract basis.

INSPECTING SYSTEM CONTROL PANELS

Inspectors should check the system control panel to ensure that all parts are operating properly. All switches should perform their intended functions, and all indicators should light or sound when called for (Figure 8.119). When individual detectors are triggered, the system control panel should send out the appropriate signal, and the warning lamps should light. Auxiliary devices can also be checked at this time. The auxiliary devices include the following: local evacuation alarms, air-handling system shutdown controls, fire dampers, and the likc. All devices must be restored to proper operation after testing.

In connection with these tests, the receiving signals should also be checked. The proper signal and/or number of signals should be received and recorded. Signal impulses should be definite, clear, and evenly spaced to identify each coded signal. There should be no sticking, binding, or other irregularities. At least one complete round of printed signals should be clearly visible and unobstructed by the receiver at the end of the test. The time stamp should clearly indicate the time of signal and should not interfere in any way with the recording device.

Figure 8.119 Check the alarm panel to make sure everything appears to be in working order.

SYSTEM TESTING TIMETABLES

The following sections give a brief synopsis of the inspection and testing requirements for various types of systems. If any of these systems use backup electrical generators for emergency power, those generators should be run under load weekly for at least one-half hour.

Local Systems

Local alarm systems should be tested in accordance with guidelines established by the authority having jurisdiction and/or the occupant. There are no NFPA code requirements specifying the test period.

Central Station Systems

Central station signaling equipment should be tested on a monthly basis. Waterflow indicators, automatic fire detection systems, and supervisory equipment should be checked bimonthly. Manual fire alarm devices, water tank level devices, and other automatic sprinkler system supervisory devices should be checked semiannually. When these tests are to be conducted, it is important that both plant supervisory personnel and central station personnel be notified before the test. This will prevent them from dispatching fire units or evacuating occupants.

Auxiliary Systems

The system should be visually inspected monthly by the occupant to see that all parts are present and appear to be in working order. The operation of the system should be tested monthly to ensure that the signal is sent to the fire dispatch center. Noncoded manual fire alarm boxes should be tested semiannually.

Remote Station And Proprietary Systems

Most of the testing requirements for this type of system are established by the authority having jurisdiction. The fire detection components of these systems should be tested monthly. Waterflow indicators should be tested bimonthly.

Emergency Voice/Alarm Systems

A functional test of the various components in these systems should be conducted on a quarterly basis by the occupant. These tests can include selected parts of the system that are reflective of what may actually be used during an incident. However, all parts must be checked at least annually.

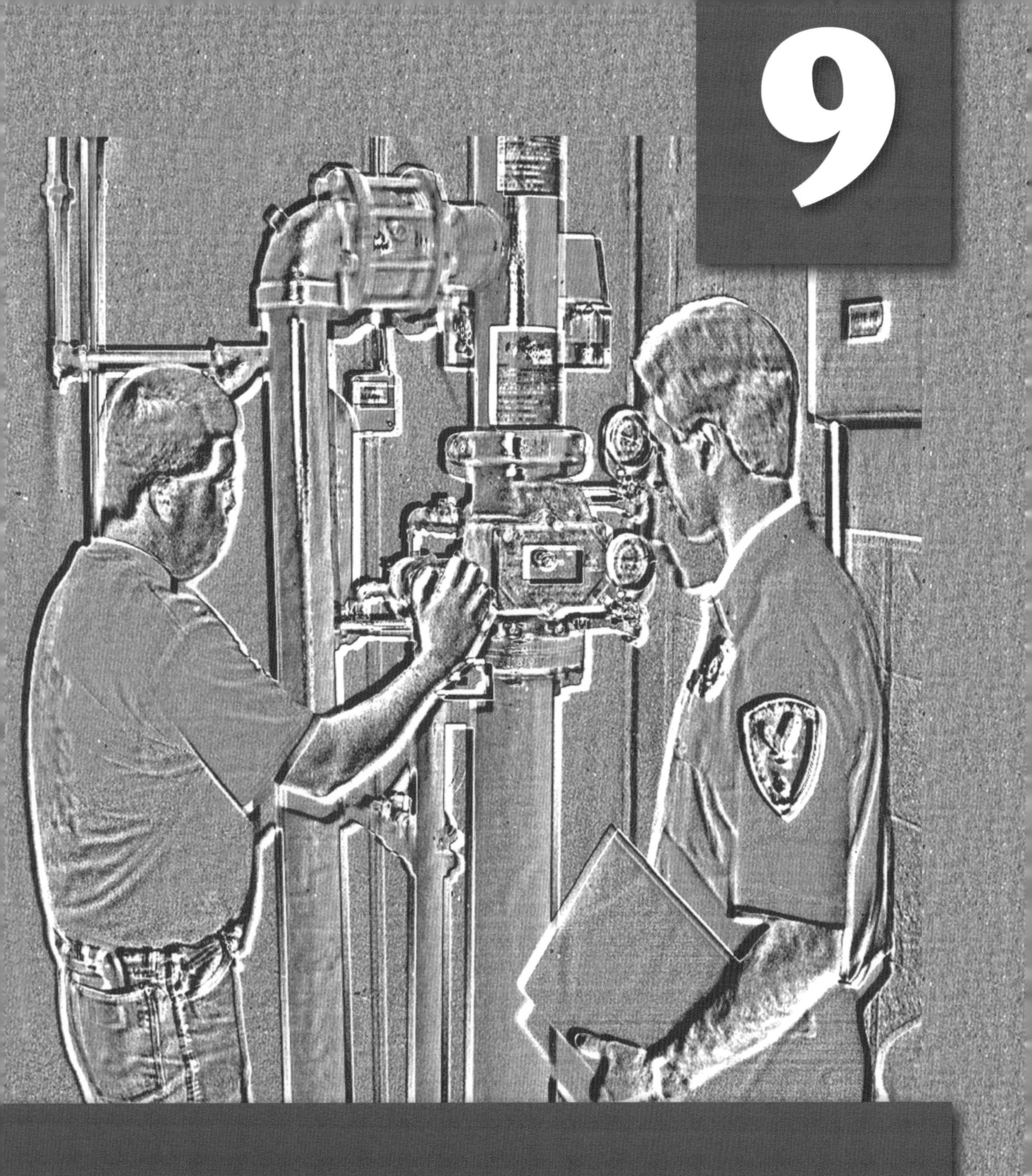

9

Plans Review

This chapter contains information that will assist the reader in meeting the listed job performance requirements contained in NFPA 1031, ***Standard for Professional Qualifications for Fire Inspector and Plan Examiner*** (proposed 1998 edition).

Chapter 3 Fire Inspector I

3-2.4 Recognize the need for a plan review, given a situation or condition, so that requirements for plan reviews are communicated in accordance with the policies of the jurisdiction.

3-3.9 Compare an approved plan to an installed fire protection system, given approved plans and field observations, so that any modifications to the system are identified, documented, and reported in accordance with the policies of the jurisdiction.

Chapter 4 Fire Inspector II

4-2.4 Process an plan review application, given a specific request, so that the application is evaluated and processed in accordance with the applicable policies and procedures of the jurisdiction.

4-3.3 Determine the type of construction in a new building, given plan review and observations or a description of the building's height, area, occupancy, and construction features, so that the construction type is properly classified according to applicable codes and standards.

4-4.1 Classify the occupancy type, given a set of plans, specifications, and a description of a building, so that the classification is made according to applicable codes and standards.

4-4.2 Compute an occupant load, given a floor plan of a building or portion of a building, so that the calculated occupant load is in accordance with applicable codes and standards.

4-4.3 Field verify the installation of a fire protection system, given shop drawings and system specifications for a process or operation, so that the system is reviewed for code compliance, installed in accordance with the approved drawings, and all deficiencies are identified, documented, and reported in accordance with the policies of the jurisdiction.

4-4.4 Verify that egress elements are provided, given a floor plan of a building or portion of a building, so that all elements are identified, checked against applicable codes and standards, and any deficiencies are discovered and communicated in accordance with the policies of the jurisdiction.

4-4.5 Verify the construction type of a building or portion thereof, given a set of approved plans and specifications, so that the construction type complies with the approved plans and applicable codes and standards.

Chapter 9
Plans Review

Many fire agencies do not have plans review authority; therefore, code enforcement often is a reactive process, beginning after a building is constructed and is ready for occupancy. Many of the problems encountered under this system can be avoided if code enforcement is begun *before* construction — at the plans review stage. To begin a proactive system of code enforcement, the fire department and the building inspection agency that has plans review authority must establish a cooperative relationship. With this type of system, the fire service plans review process can be incorporated into the process of applying for a building, hazardous materials, storage, use, or other type of permit.

The obvious advantage of establishing a plans review process is that it enables the fire service reviewer to point out discrepancies before construction begins. Correcting these problems before the start of construction improves the efficiency and cost effectiveness of the project. This will also improve the fire department's image and level of customer service with developers, which will help in forming harmonious relations between the two in future contacts.

When no fire protection engineer is employed for plans review, the task may be given to fire inspectors, who must be well prepared before attempting to examine any plans. This preparation requires that inspectors have a thorough knowledge of construction techniques, applicable codes, and knowledge of how to use them. Just as hoselines and a water supply are the basic tools of suppression forces, fire/building codes and standards and their application are the basic tools of the plans reviewer.

In addition to having a thorough knowledge of codes, fire inspectors must develop a good working relationship with all individuals and agencies involved in constructing the building. These persons and agencies may include governmental regulatory agencies, architects, engineers, contractors, insurance carriers, and the individuals for whom the building is being constructed.

Fire inspectors may also want to acquaint fire suppression forces with the proposed building project. By looking at the plans, firefighters can provide feedback about difficulties that may be encountered during fire suppression activities.

READING CONSTRUCTION DRAWINGS

The plans and specifications of buildings are shown in working drawings made by engineers, architects, or draftsmen. There are four main views of working drawings:

- Plan View
- Elevation View
- Sectional View
- Detailed View

Drawings should have a title block that contains specific information about the drawing and the project (Figure 9.1). The format, or location, of the information is determined by the designers. The title block should contain the following information:

1. The title of the drawing, such as **BASEMENT FLOOR PLAN**, the site plan, foundation plan, or floor plan.
2. The description of the project, such as **FIRE PROTECTION PUBLICATIONS OFFICE BUILDING**.

Figure 9.1 Each architectural firm has its own title block. *Courtesy of C.H. Guernsey & Company.*

3. The scale of the drawing. For example, ⅛" = 1'0" (one-eighth inch drawn on the paper is equal to 1 foot of actual construction). Scaling the drawing provides a view in exact proportion to the actual size of the building. Sometimes the scale is placed on the drawing rather than in the title block. Some computer-aided drawings (CAD) may not be drawn to scale.
4. The revisions block dates the drawing and any revisions (if applicable) for distribution to various people involved in the project.
5. The name of the firm producing the drawings. Such as **C. H. GUERNSEY & COMPANY**.
6. The name (usually initials) of the person performing the drafting, and the person responsible for checking the final drawings.
7. The sheet number in the set.
8. If the plans were drawn by a licensed architect or engineer, the title block should have the architect's/engineer's official stamp as issued by a state architectural/engineering registration board. However, on some drawings this stamp may be located elsewhere on the drawing.

Large construction projects are usually divided into four or more basic areas such as architectural, structural, mechanical, and electrical. Sheets containing a specific type of information are marked accordingly. For example, architectural information would be marked A1, A2, A3; structural sheets S1, S2, S3; mechanical sheets M1, M2, M3; and electrical sheets E1, E2, E3 and so on. Although architectural, structural, mechanical, and electrical sections are typical, many larger plans include sections for plumbing, fire protection, site plans, and landscaping.

The Plan View

The plan view is a two-dimensional view of the site or the building as seen from directly above the area. A plan view is extremely useful in giving information concerning the overall layout of the site or building. The fire inspector will commonly see two basic plan views: site plan and floor plan. On occasion, you will also be presented with the ceiling and roof plans.

THE SITE PLAN

The site, or plot, plan is usually one of the first sheets of a set of drawings. It identifies conditions currently existing on the site and relates information needed to locate the building (Figure 9.2). This information includes the north direction, lot dimensions, utility lines (gas, water, sewer, power, and telephone), grade level, contour lines, structures to be removed, concrete walks and driveways, and areas to be landscaped.

The north direction symbol usually points toward the top of the sheet. This conforms to universal standards used in the drawing of all maps.

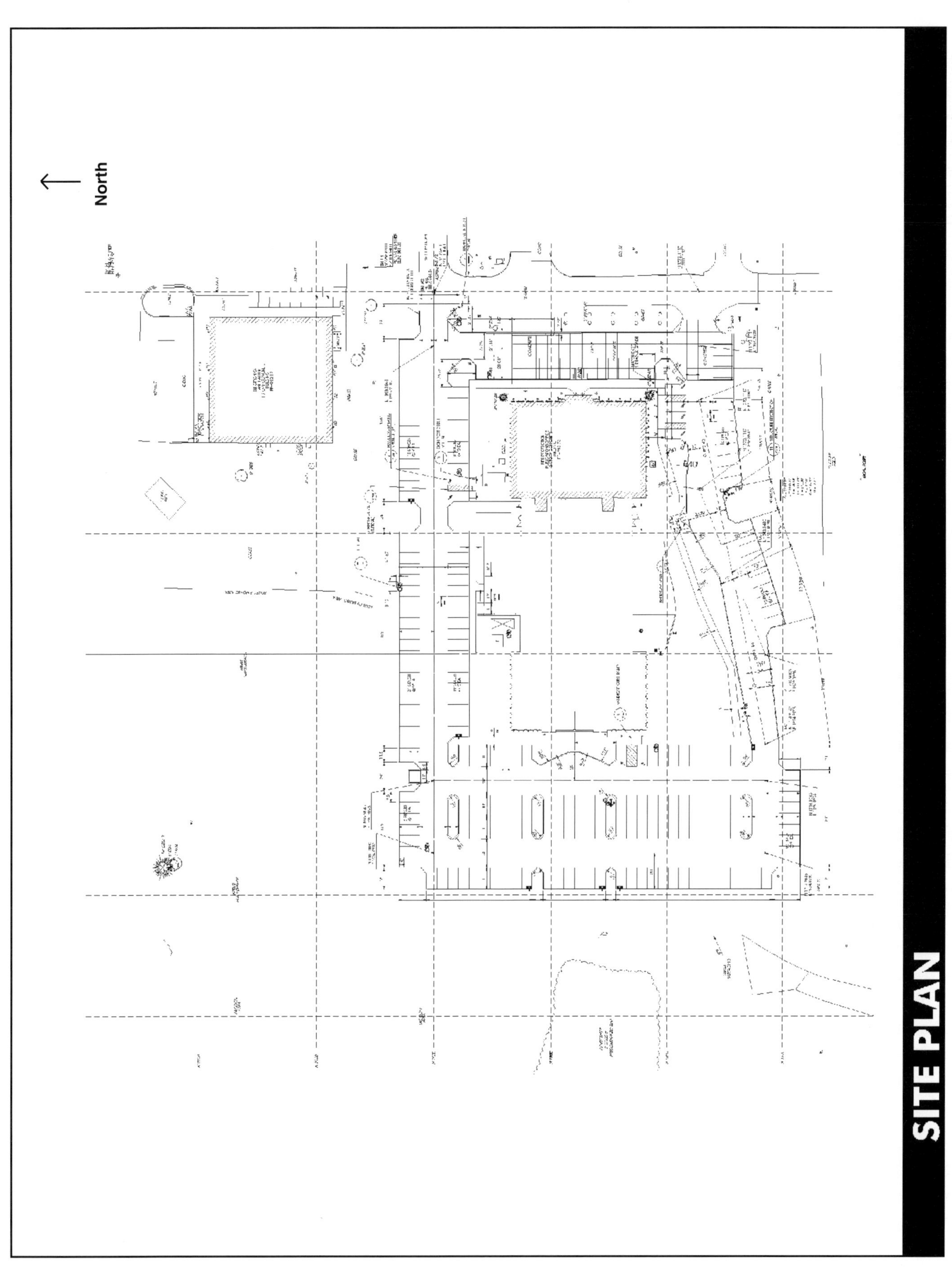

Figure 9.2 A site plan. *Courtesy of C.H. Guernsey & Company.*

The lot dimension is often shown as a broken line with the dimensions expressed in feet (meters) and decimal fractions of a foot (meter). The property lines are legal instruments that define the land boundaries for a particular site. Most building codes base fire separation on the location of the building in relation to that line.

The placing of the building on a site plan is called *dimensioning*. The objective of dimensioning is to ***clearly*** place the building on the site. This is usually accomplished by placing the dimensions some distance away from the shapes on the drawing through the use of extension lines. (**NOTE:** This can be accomplished without extension lines.) Dimensions are expressed in feet and inches (meters and millimeters).

The location of service lines (sewer, water, gas, and so on) is usually shown by a broken line marked at intervals with a letter. For example:

Water lines —w—w—w—w—w—w—w—w—w

Sewer lines —s—s—s—s—s—s—s—s—s

Power lines —p—p—p—p—p—p—p—p—p

Gas lines —g—g—g—g—g—g—g—g—g—g

In addition to the character in the broken line, a legend on the map also identifies the type of service line.

Structures or objects that must be removed from the area are also included on the site plan. These items are drawn with a broken line. Brief identification is also permitted.

Contour lines display the existing grade elevations and grade lines show the planned elevations after grading is completed. The numbers are given in feet (meters) and decimal fractions of a foot (meter) and are related to a benchmark. This information is particularly important when determining grade elevations for fire apparatus access to the site.

The locations of sidewalks, driveways, and graveled areas are usually shown on the plans. When they are included, their sizes and locations are relative to the lot lines or building dimensions. Other information that might be included is site access locations, width, and surface type.

Area landscaping is normally started when construction is complete or nearly so. Areas to be landscaped are often shown on site plans to illustrate the total concept created by the architect. The location of trees and other potential obstructions should be checked to ensure that landscaping does not block fire department connections or aerial ladder placement.

Other things that may be included on the site plan include:

- Location of fire hydrants and water mains
- Slope or grades that would affect the placement of fire department ground ladders
- Grades on the roadway that would affect the placement of fire department aerial apparatus
- Actual location of fire department connections

THE FLOOR PLAN

The floor plan provides information for constructing external walls, internal partitioning, doors and windows, ceiling and roof joists or trusses, cabinets, closets, shelving, electrical outlets, and fixtures (Figure 9.3).

The usual symbol used for exterior walls is a pair of parallel lines. The materials from which the wall will be constructed are sometimes listed on a floor plan; usually, however, construction materials are listed in a section view of the wall.

The interior walls (partitions) are also drawn with parallel lines. The function of each room may be listed on the sheet where the room is located or the rooms may just be numbered, as with proposed offices. When examining plans for room size, an allowance must be made for the thickness of an interior wall. Typically, the rooms are measured from the surface of a finished wall.

The symbol for a standard hinged door is a single line drawn from the hinge side of the doorway. An arc indicates the swing of the door. Codes also require doors to be installed in certain locations and in some cases swing in the direction of egress.

The windows and doors on the drawings are coded by letters or numbers enclosed in a circle or

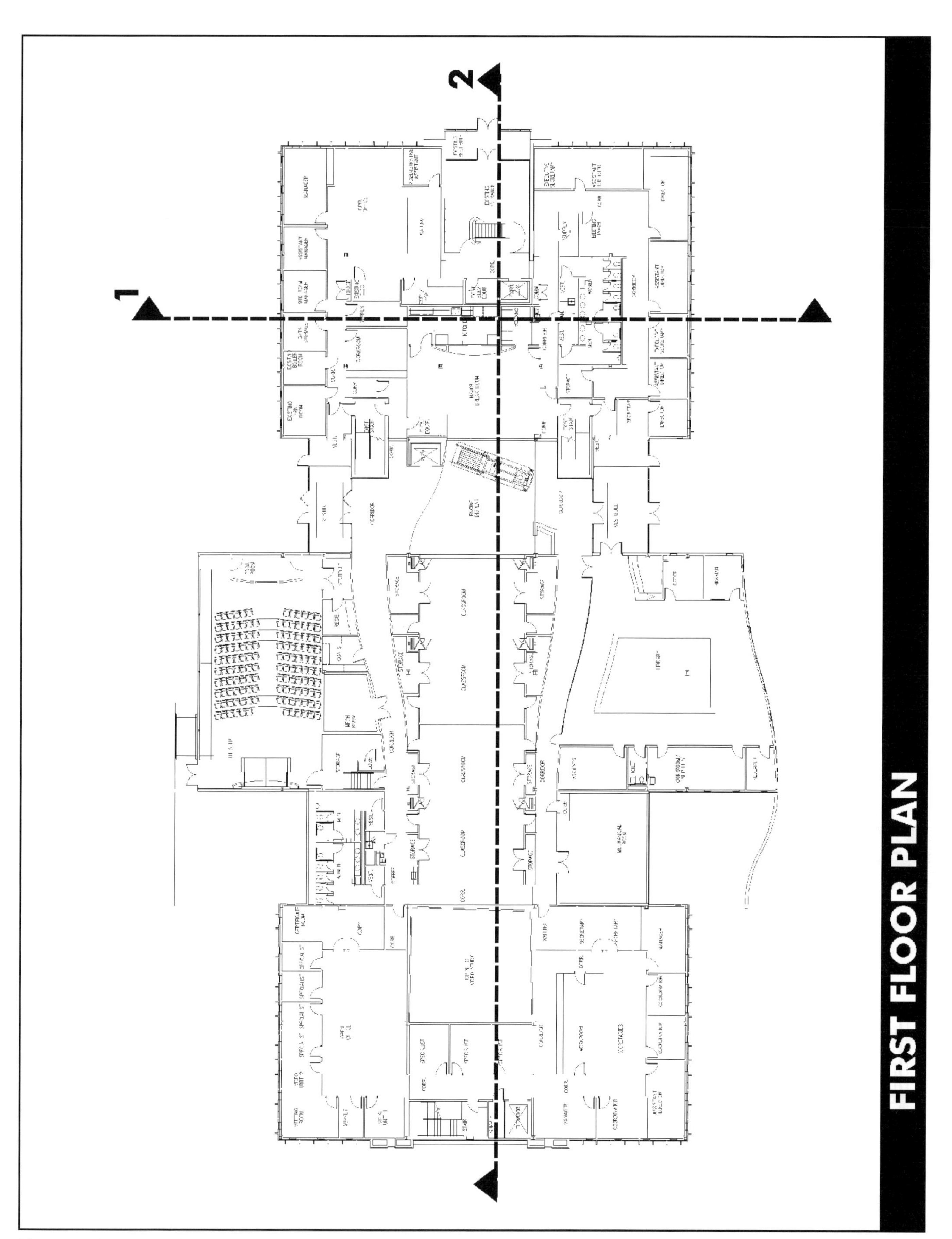

Figure 9.3 A floor plan. *Courtesy of C.H. Guernsey & Company.*

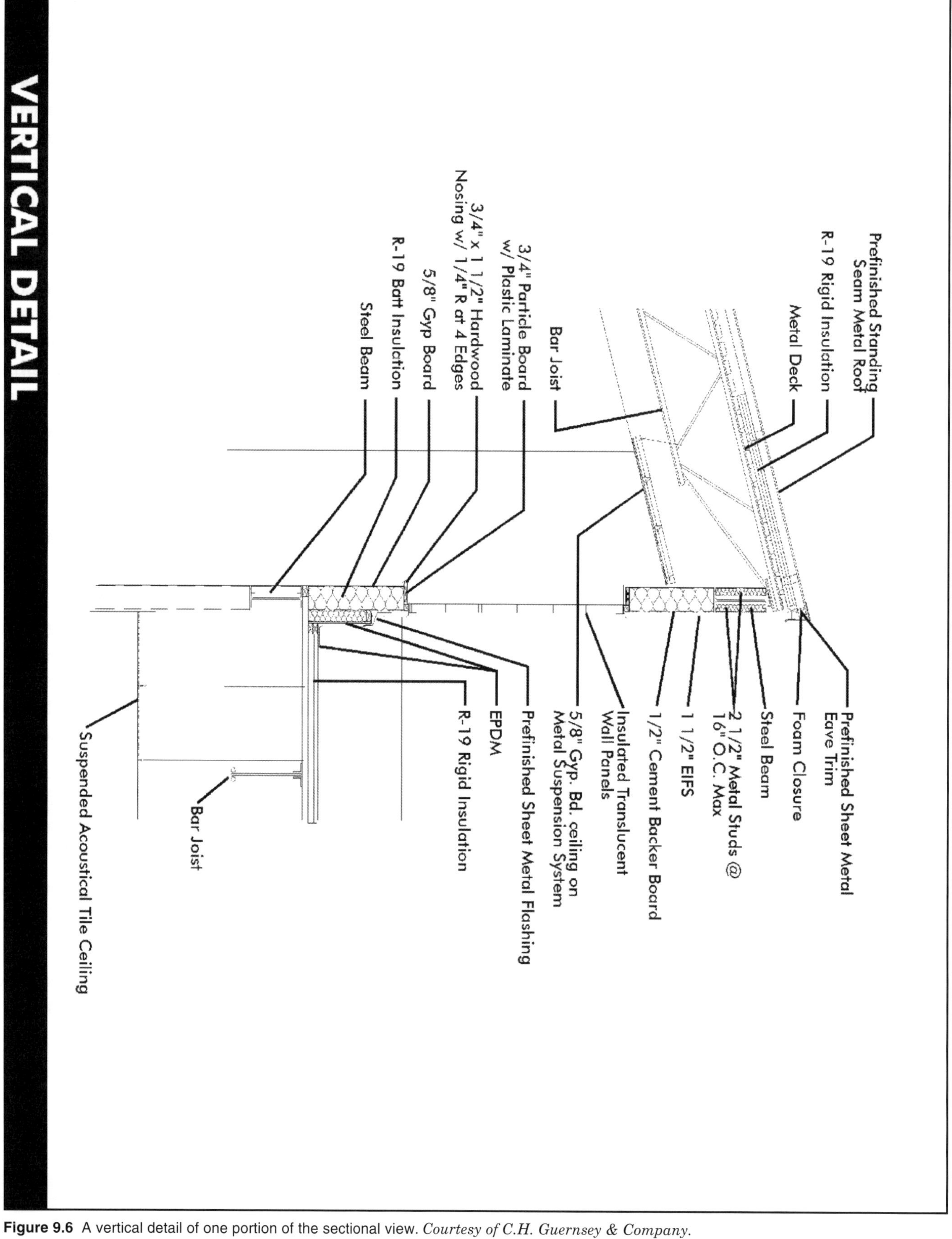

Figure 9.6 A vertical detail of one portion of the sectional view. *Courtesy of C.H. Guernsey & Company.*

The site plan review may also involve an assessment of the proposed water supply system. Fire inspectors may be required to determine if water is available in sufficient quantities and pressures to satisfy potential emergency demands. After inspectors have determined the minimum fire flows, they can verify the adequacy of the water supply by examining the type and location of hydrants (both public and private), pump systems, cross-connected water supplies, and the location and size of public water mains. Records of water flow testing performed in the area will be invaluable, assuming that they are recent and accurate. It is often necessary to use hydraulic calculations or flow tests to verify the adequacy of water supplies (Figure 9.7).

Fire inspectors should review the general building parameters: location, occupancy, fire and land-use zones, property line clearances, and special hazards. It is important for the building department to determine the use of the building in order to assign an occupancy classification. The occupancy classification dictates the total allowable area, maximum height, exposure protection, occupant load, necessity for fixed fire protection, and other factors unique to each building. Most cities have land-use zones; therefore, fire inspectors should check to see that the building is located in the proper zone. Finally, fire inspectors must identify any special hazards, such as tank installations or high-piled stock, and check for any corrections that are required.

Figure 9.7 Inspectors may be required to conduct water supply flow tests.

Reading Construction Documents

Construction drawings indicate the way in which the building is to be constructed and what materials are to be used. Building size, occupancy load and class, construction classification, exit systems, compartmentation, and other assorted considerations are all factors in determining if the building can be constructed and used as proposed. Because the inspector's task is to verify the information that the architect has provided, the construction documents should contain all the needed facts.

Reviewing the overall size of the building in terms of height and area should be the first step. This will provide information that is essential in carrying out later steps. The code requirements concerning building height limits were initially developed to ensure that fire hose streams could reach upper-story fires, that occupants would be able to exit the building swiftly, and that the building would not pose a hazardous exposure to surrounding buildings.

Today, building height restrictions are based on the occupancy classification and construction of the building. When determining the height of a building, fire inspectors must understand how the jurisdiction's code treats the following three factors:

- Lowest point to consider
- Highest point to consider
- Automatic sprinkler protection

Various codes handle sloping grades, basements, parapet walls, and penthouses differently when determining the lowest or highest points. In addition, many codes permit an additional story if complete sprinkler protection is provided.

Like height limitations, area limitations are intended to ensure effective fire suppression. Such factors as the capabilities of the fire department water supply, the community's layout, fire risk, and climatic conditions should be considered during an evaluation of area limitations. Each of the model building codes have different allowances for allowable area. Therefore, fire inspectors should use the definition from the appropriate code. The net floor area within the building will also be used in determining the maximum occupant load allowed in the structure or a certain part of the structure.

Allowances for additional areas also vary among codes. Area increases or decreases are based on automatic sprinkler protection, accessible building perimeter, and the distance between buildings. These increases and decreases are usually expressed in percentages.

The architect should determine an occupancy classification for the proposed building and identify the intended use of each room or section of the building. Fire inspectors should evaluate the fire and life hazards throughout the building to verify that the occupancy classification is correct. It is extremely important that the inspector be familiar with the applicable codes in order to make accurate judgments about mixed occupancies and special use requirements.

The various classifications for building construction consider only the factors necessary to define the building types. As with occupancy classification, the intended construction should be identified on the drawings. This is critical for required fire flow determination on some model codes. To verify the construction classification, fire inspectors should refer to the appropriate fire-resistance rating lists. (**NOTE:** Refer to Chapter 5 of this manual for more information about construction classifications.)

After verifying that the project will be built, inspectors should determine the occupant load before evaluating the life safety and fire protection features of the building. Fire inspectors can use the occupant density or occupant load factors given in the codes to determine the occupant load. Often, the architect is required to provide the occupant load calculations for the fire inspector's verification.

After fire inspectors have determined the occupant load, they should evaluate the means of egress. It is the inspector's responsibility to verify that various components of the exit system have been provided and that each means of egress provides a continuous path of travel to a place of safe refuge. Typically, codes require exits to be separated from all other parts of the building by construction having a specified fire-resistance rating. The architect should provide details of the separation walls as well as door assembly information. Fire inspectors should verify that there are no unnecessary openings, all penetrations are sealed, and that the fire-resistance ratings of the door assemblies are correct. Separation from the exit discharges and exit accesses must also be verified. For more information about both occupant load and means of egress, see Chapter 6 of this manual.

Fire inspectors should evaluate other related code requirements applying to means of egress, such as those concerned with interior finish, headroom, elevation changes, and obstructions. Means of egress should be highlighted on the architectural, mechanical, and electrical drawings so that inspectors can detect penetrations or other code violations easily.

Once the means of egress have been identified and the acceptability of each component has been verified, the reader should evaluate the capacity of the means of egress. Fire inspectors must evaluate the exit capacity according to the specifications of the appropriate codes. The architect may be required to provide the exit capacity calculations so that the inspector can verify that the exit capacity is sufficient. When determining the adequacy of exit signs, the inspector should examine the following components: location, spacing, color, illumination, design, and size. Individual codes will outline the requirements for exit signs in terms of these components.

Another aspect of fire protection to be examined during the architectural drawing review is building compartmentation. Building compartmentation includes fire barriers, smoke barriers, the protection of vertical openings and concealed spaces, and protection from hazards. The theory behind building compartmentation is to limit fire and resulting products of combustion, should they occur, to one area of the building.

The fire inspector should verify the fire-resistance rating of the construction separating the vertical openings, such as stairways and elevator shafts, from the remaining parts of the building. It may be helpful to highlight the vertical openings with a different color than the one used to highlight the exiting system. Atriums are vertical openings that are permitted if they are provided with adequate smoke control, automatic sprinkler protection, and fire barriers (Figure 9.8). Escalators may be permitted, although they may not serve as part of a means of egress. The escalator openings must be protected by rolling shutters, automatic sprinklers, partial enclosures, or a combination of sprinklers and mechanical ventilation. It may be necessary to refer to the mechanical and electrical plans to verify that the level of protection is adequate.

Fire inspectors should locate and check fire and smoke barriers for continuity, fire-resistance rating, and opening protection. It is often difficult to verify fire stopping in concealed spaces and penetrations. The architect must provide detailed plans that clearly show the fire stopping. When evaluating hazardous areas, fire inspectors should verify that adequate separation or automatic sprinkler protection has been provided.

Additional aspects that fire inspectors must consider are special hazards, interior finishes, furnishings and decorations, insulation materials, and materials that produce smoke and toxic gases.

Figure 9.8 Newer construction designs often feature large, irregularly shaped atriums.

Fire inspectors also must be aware that many new materials release harmful quantities of smoke and toxic gases when they burn. Therefore, there are regulations excluding the use of certain materials with high smoke production or smoke densities.

Interior finish requirements usually apply only to walls and ceilings because floor coverings are tested using a different standard. Fire inspectors must know what the applicable code requires for specific interior finishes such as cellular or foam plastic materials, incidental trim, and fire-retardant coatings. In addition, codes often permit less stringent requirements for interior finishes when automatic sprinklers are provided. If furnishings and decorations are included in the construction documents, fire inspectors should review the specifications and evaluate exit access.

The inspector should also verify that portable fire extinguishers, fire detection and signaling systems, or automatic extinguishing systems, if required, are located and installed according to the appropriate standards.

Understanding Mechanical Drawings

Mechanical systems are numerous and sometimes complex. It is therefore necessary that when fire inspectors review them, they gather as much information as possible to aid in evaluating the life safety hazards of the system (Figure 9.9).

It is important to understand the operation of the building's mechanical system and the factors that lead to the spread of smoke and fire throughout a building. While mechanical systems cannot prevent all smoke from filtering throughout a building, the quantity of smoke that does escape should be within human tolerances. The type of fuel to be used determines the type of heating, ventilating, and air conditioning (HVAC) equipment needed. The type of fuel also determines the type of venting needed, required clearances from combustibles, and quantities of air needed for proper combustion of the fuel. Most HVAC code requirements are based on the following NFPA standards:

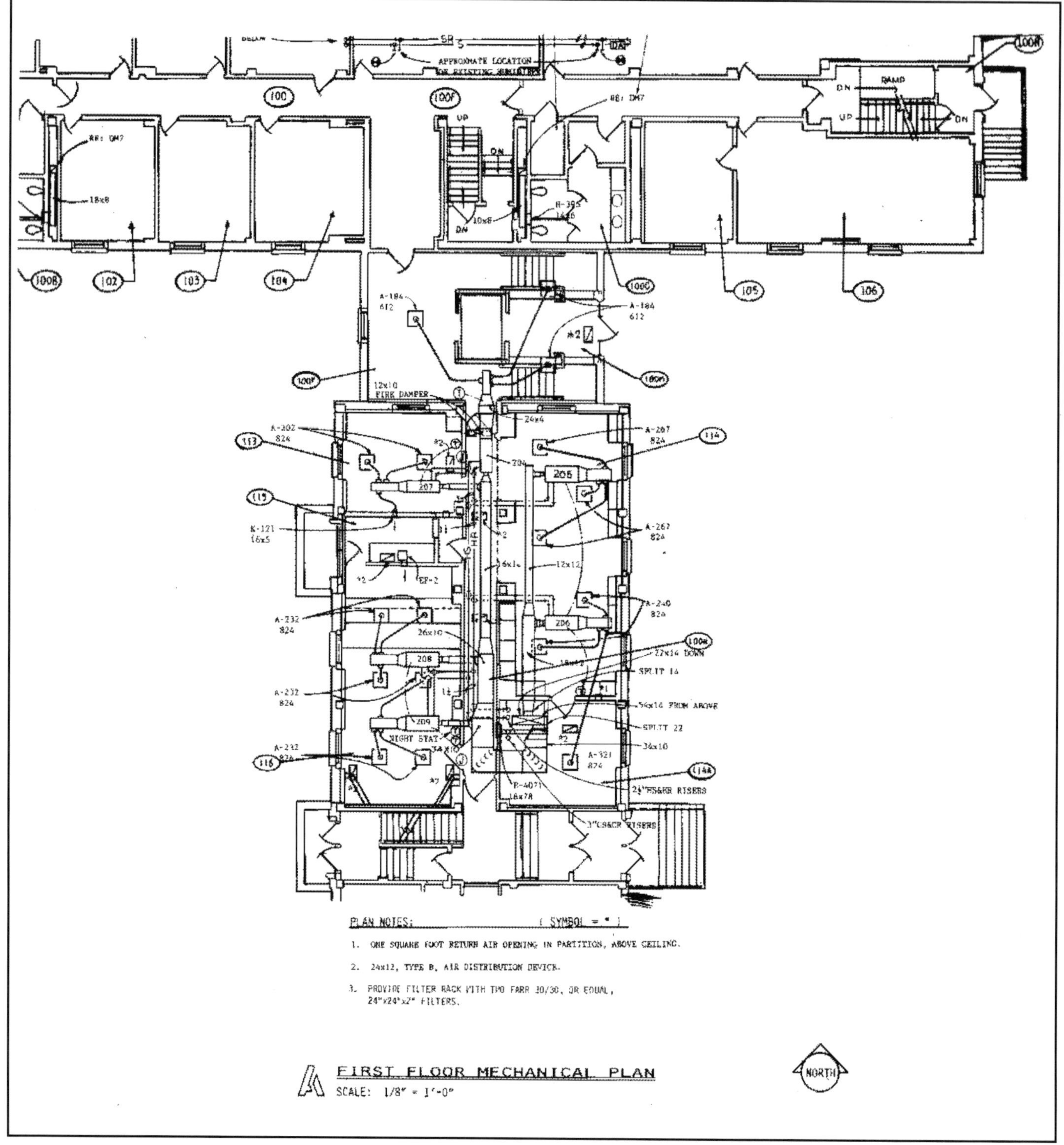

Figure 9.9 A typical mechanical plan.

- NFPA 31, *Standard for the Installation of Oil Burning Equipment*
- NFPA 54, *National Fuel Gas Code*
- NFPA 97, *Standard Glossary of Terms Relating to Chimneys, Vents, and Heat-Producing Appliances*
- NFPA 90A, *Standard for the Installation of Air Conditioning and Ventilation Systems*
- NFPA 90B, *Standard for the Installation of Warm Air Heating and Air Conditioning Systems*

- NFPA 92A, *Recommended Practice for Smoke-Control Systems*
- NFPA 92B, *Guide for Smoke Management Systems in Malls, Atria, and Large Areas*
- NFPA 96, *Standard for Ventilation Control and Fire Protection of Commercial Cooking Operations*
- NFPA 211, *Standard for Chimneys, Fireplaces, Vents, and Solid Fuel-Burning Appliances*

Fire inspectors can use these documents or the model code agency's mechanical codes, whichever are adopted by the AHJ, as a guide during their check of the mechanical plans.

During the review of the mechanical system, the inspector should verify the style and design of the HVAC system (Figure 9.10). The refrigerant used in the system (most commonly freon) should be checked for possible health or fire hazards. Fire inspectors should be familiar with code requirements concerning the automatic shutdown of heating systems to control the spread of fire and smoke. This shutdown can be accomplished with heat-sensing devices or smoke detectors inside the duct work. Inspectors must also determine if the system can be operated in a total exhaust mode. The total exhaust mode could be crucial in removing heat and smoke from a building during fire conditions. Fire and/or smoke dampers may be required where the ductwork penetrates fire-resistive walls or floors. The location of these features can be determined from the architectural drawings. Some codes prohibit exit areas from being penetrated by the ductwork. Exits are prohibited from being used for supply or return air. Additional references for damper placement and installation of ductwork are NFPA 90A, 90B, 92A, 92B, or the local mechanical code.

Figure 9.10 Inspectors must review the plans for large mechanical systems, such as HVAC systems.

Usually, smoke control systems are used in buildings involving either a large number of people or that have a large quantity of combustibles such as shopping malls, high rises, and buildings with open atriums (Figure 9.11). Smoke control systems must be engineered and tested and involve not only the mechanical systems but also the doors, partitions, windows, shafts, ducts, fan dampers, wires, controls, and pipes. For additional information about smoke control systems, consult NFPA 92A and 92B. Additional information can be obtained from the American Society of Heating, Refrigeration, and Air Conditioning Engineers (ASHRAE) manual.

Exit stairwell pressurization is another method of smoke control. The intent of stairwell pressurization is to maintain a smoke-free atmosphere for occupants to exit during a fire. Stairwell pressurization systems must also be engineered and tested to ensure that the applied pressure does not make it difficult for the average person to open stairwell doors.

Figure 9.11 Smoke control systems are used in large structures, such as shopping malls.

Understanding Electrical Drawings

General construction plans should include as much information about the electrical systems as possible (Figure 9.12). During the review of the electrical systems, the inspector must verify that

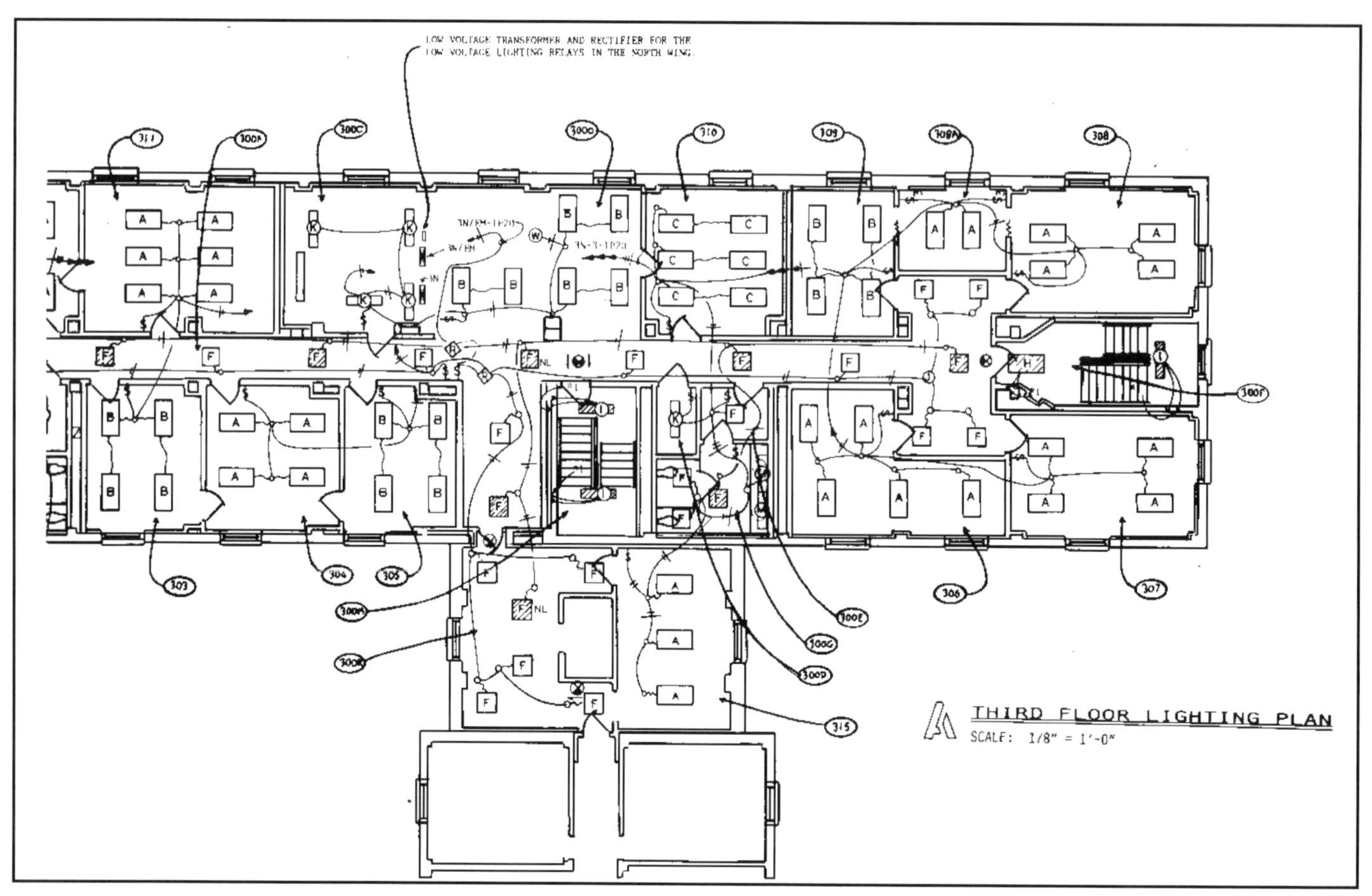

Figure 9.12 A common electrical system drawing.

illumination of exit ways is adequate. It is extremely important that fire inspectors take a detailed and systematic approach to the review of these systems. Exit illumination must be continuous during the time the building is occupied. The stairs, aisles, corridors, ramps, escalators, and passageways leading to an exit must be lit. Fire inspectors should evaluate such requirements as the level of illumination, power sources, emergency power, and special considerations for theaters or concert halls according to the appropriate code. General fire alarm requirements and emergency control systems must also be assessed.

Understanding Fire Protection Features

There are several fire protection feature drawings that fire inspectors may need to review. These include:

- Sprinkler systems
- Special agent extinguishing systems
- Fire detection and alarm systems

SPRINKLER SYSTEMS

A thorough, accurate review of sprinkler system plans is necessary to ensure code compliance (Figure 9.13). Inspectors need to be familiar with different types of sprinkler systems, their operation, and the specifications for each type of system. In addition to knowing national code requirements, fire inspectors need to be aware of local ordinances, amendments, and other requirements that may affect the installation of the sprinkler system. When required, the fire inspector needs to be able to develop a field inspection checklist of the specific requirements for sprinkler systems and the code sections that relate to the requirements. This checklist can be used as a guide during sprinkler plans review. See Appendix G for an example of a checklist.

Before the sprinkler review begins, the inspector determines which standard applies and what the basic requirements are. This involves classification of the hazard by occupancy (light, ordinary, or extra hazard) and by storage commod-

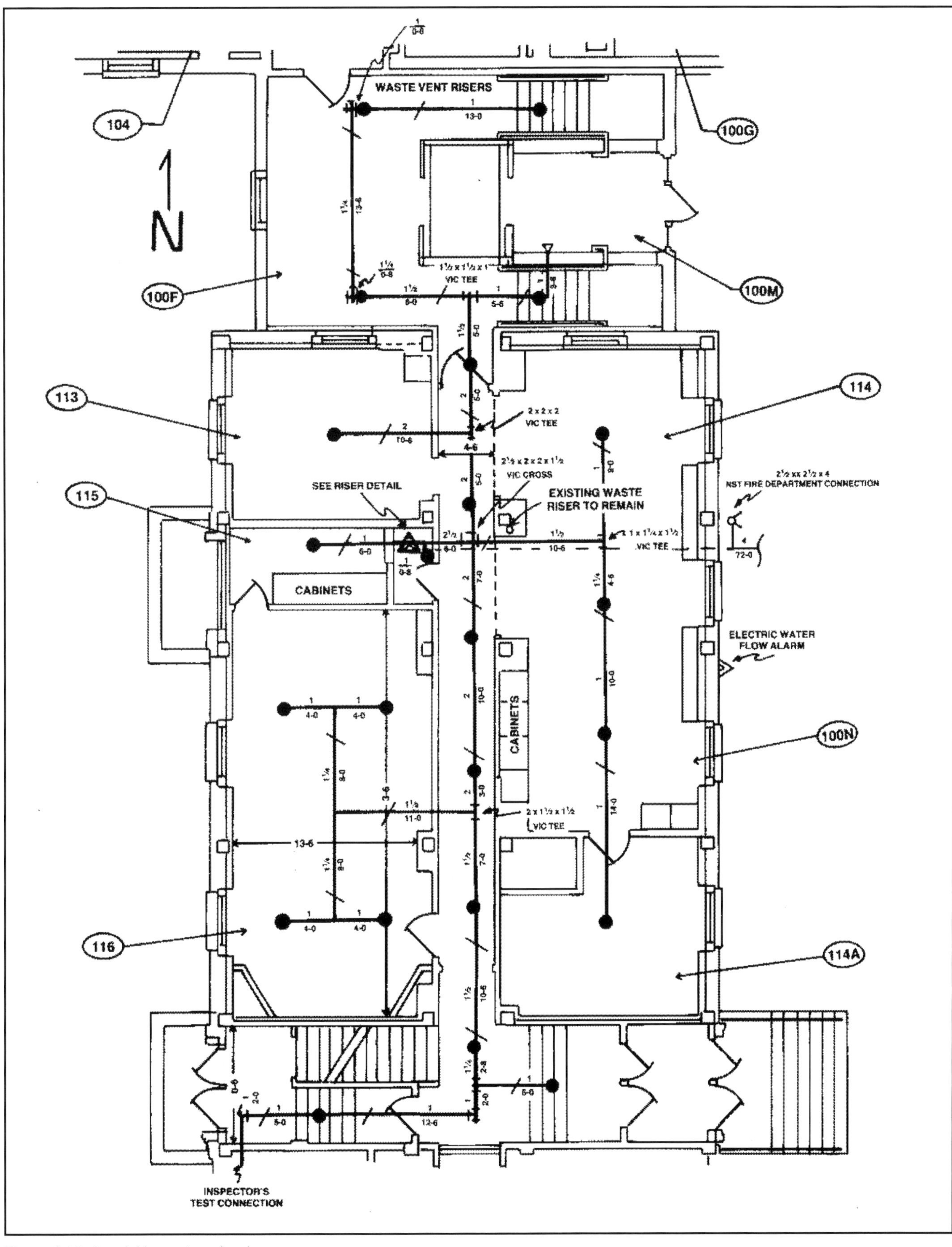

Figure 9.13 A sprinkler system drawing.

ity (shelf, high-piled, rack, or warehouse storage), or by both. The inspector must evaluate mixed occupancies very carefully to verify that the appropriate codes or standards have been followed. Drawings for hydraulically calculated systems should be accompanied by the calculations proving that the design complies with the code. The system designer provides all necessary information, including calculations and applicable data, on the drawings or in the specifications. All calculations, including computer printouts, must be reviewed carefully for accuracy.

Fire inspectors need to check several system design factors, including:

- Extent of coverage
- Type of system
- Size of system
- Water supply connections and valves
- Fire department connection
- Sprinkler types, temperature ratings, and locations
- Pipe sizes, lengths
- Number of pipe elbows and tees
- Type and number of pipe hangers

Rapidly changing sprinkler technology warrants a review of the sprinkler cut sheet to verify the sprinkler is installed in accordance with the UL listing of the sprinkler. Sprinklers should be installed as required in the standard to assure complete coverage of the building. This includes closets, stairwells, storage areas, walk-in freezers, and concealed spaces such as areas above suspended ceilings with combustible construction.

The size of each sprinkler system is based on the total square footage (square meters) of floor area protected by a single riser and control valve. Installation standards determine the size of the systems, the number of individual systems, and the number of system control valves needed. Fire inspectors should examine the specifications to verify the temperature rating, type, orifice size, area of coverage, use in special areas, spacing, and location of sprinklers. The drawings should clearly indicate all supports, bracing, connections, piping, valves, drains, and gauges throughout the system.

Water supplies for sprinklers must be of sufficient capacity and pressure to satisfy the calculated requirements for hydraulically designed systems. The sprinkler system designer should include a graph that shows the water demand compared to the available water supply (Figure 9.14). Flow tests are used to measure the pressure and flow available at the city main. Information indicating when, where, and by whom the last flow test was conducted should be provided. In addition to checking the pressure and the flow, fire inspectors need to check the total capacity of the water supply. It must be adequate for the expected duration of fire fighting operations listed in the codes.

SPECIAL AGENT EXTINGUISHING SYSTEMS

Fire inspectors should be familiar with each type of special agent extinguishing system and the applicable requirements. The architect's, engineer's, or contractor's specifications should be very detailed and include the following information:

- Definition of the area or equipment to be protected
- Type of system (local application or total flooding)
- Type of extinguishing agent being used
- Amount of agent required
- Concentration of extinguishing agent to be developed
- Storage container size
- Type of expellant gas
- Rate of discharge
- Duration of flow
- Layout and type of piping included, whether engineered or pre-engineered
- Location and type of discharge nozzles
- Method of actuation and auxiliary alarm functions such as shutting off ventilation
- Type of presignaling devices used, if required
- Area and volume of the protected space

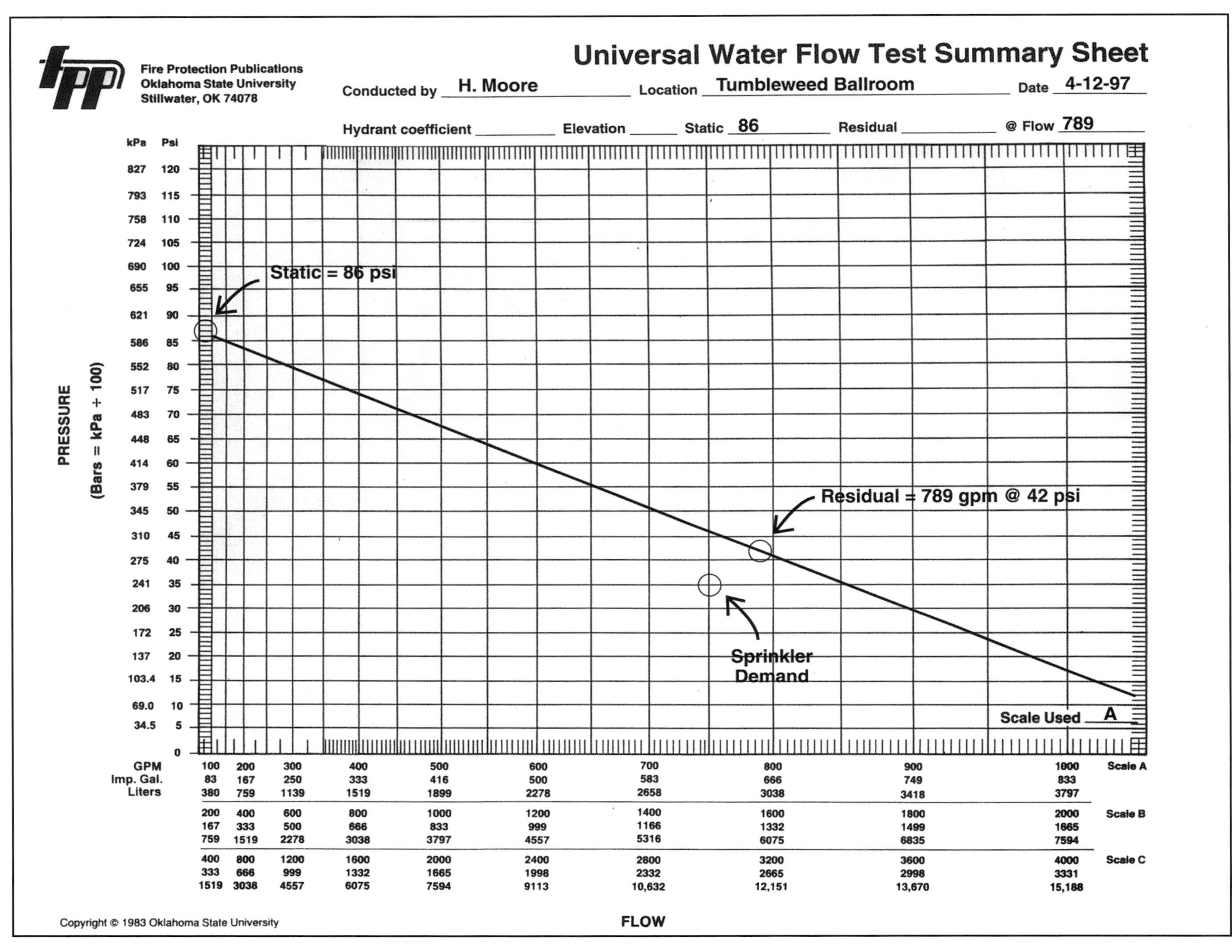

Figure 9.14 The available water supply must exceed the sprinkler demand value.

The architect or contractor should also include the calculation sheets used to design the system.

FIRE DETECTION AND ALARM SYSTEMS

Detailed information about fire alarm plans and communications systems is not usually provided on the general construction plans. Normally, separate plans must be submitted and approved before these systems can be installed. Plans for alarm systems should contain enough information for fire inspectors to evaluate the following:

- Signal initiation
- Signal notification
- Supervision of alarm systems
- Power supply
- Elevator control
- Automatic door closers
- Stair pressurization
- Smoke control
- Damper control
- Initiation of automatic extinguishing equipment
- Fire pumps
- Doors that unlock or close automatically when the alarm activates
- Lightning protection
- Battery load calculations
- Manufacturers' cut sheets
- Point-to-point wiring diagrams

There are numerous types of fire detection and alarm systems that can be employed for the protection of a building, its contents, and the people who are in the building. From the building occupants' standpoint, the most important alarm system is the local alarm system. Local alarm systems are used to alert the occupants of a building that it is necessary for them to leave the building. The three most common methods of initiating a signal are: manual operation of a pull box, automatic operation of heat or smoke detectors, or the actuation of sprinklers. All alarm equipment must be listed by UL or a similar testing agency for use as fire protective signaling devices.

The first method of actuating an alarm is by use of a manual pull box. Manual pull boxes must be highly visible and easily accessible. Instructions for activating an alarm should be printed directly on the box. Fire inspectors should be aware of specified maximum travel distances to pull boxes and code requirements concerning pull station placement for accessibility to the handicapped.

A signal can also be initiated by smoke or fire detectors that automatically activate the local alarm system. The operation of the detectors can be affected by several factors: ceiling construction, forced ventilation, and placement of the detectors. Therefore, fire inspectors should be familiar with the types of detectors specified and the code requirements for their use.

Local alarm signals can also be initiated by a flow switch in sprinkler water supply lines that operates when a sprinkler opens. When a flow switch operates, the alarm sounded in the building should sound the same as with any other alarm-initiating device on that system (i.e., the alarm should sound the same whether triggered by the flow switch or a smoke or heat detector). Some jurisdictions allow manual pull stations to be omitted in occupancies that are protected by sprinkler systems that sound an alarm when activated.

Once the initiating device has actuated, several types of devices are used to warn the occupants of the building or the persons in charge that a problem exists. These methods include audible indicating devices and visual signaling devices. Signaling devices used for alarm notification and evacuation purposes must be listed by UL or FM as fire protective signaling devices. Fire inspectors must verify the types of devices to be used and their locations. Audible devices may include bells, buzzers, electronic tones, recorded messages, or other recognizable sounds. The distinctiveness of the noise, as well as its level, are important factors in choosing acceptable audible warning devices. Visual warning devices include flashing or strobe lights. They are designed to complement audible devices to provide warning for the hearing impaired.

Trouble signals are nonemergency signals designed to warn those responsible for maintenance of the fire alarm system that there is a malfunction within the fire protection system. A supervisory circuit within the system monitors itself for problems, such as a power outage, problems with backup power supplies, or damaged wiring. Trouble signals should be distinct from alarm signals and should also operate a visual signal on the panel. The switches for silencing audible signals must not turn off the visual signal until the problem has been corrected.

Large, complex occupancies that need clarification as to what type of device is operating or the location of the device may require an annunciator panel. Examples of these types of buildings are institutions, hospitals, and high-rise buildings. The annunciator panel should be placed as close as possible to the entrance(s) of the building most likely to be entered by fire personnel. This panel may be located in the central command station within the building. This will allow fire personnel to quickly see which section of the building is involved upon their arrival. The annunciator panel may indicate other important information such as the shutdown of HVAC systems, the activation of the sprinkler system, or control of the elevators for use by fire personnel.

To review the alarm and detection system, fire inspectors need information from the specifications, floor plans, equipment list, and symbol list. The specifications should include the type and gauge of wire, protection provided for the wire (conduit or raceways), wiring methods, and methods of supervision. All fire alarm systems must have electrical supervision. Fire inspectors should use the floor plans to verify the location and number of alarm-actuating and signaling devices. By using the symbols list, they can determine if approved equipment is being specified. The fire inspectors should inspect the alarm system during construction and observe the final acceptance test to ensure code compliance.

10

Identification of Hazardous Materials and Storage, Handling, and Use of Flammable and Combustible Liquids

This chapter contains information that will assist the reader in meeting the listed job performance requirements contained in NFPA 1031, ***Standard for Professional Qualifications for Fire Inspector and Plan Examiner*** (proposed 1998 edition).

Chapter 3 Fire Inspector I

3-3.8 Recognize hazardous conditions involving equipment, processes, and operations, given field observations, so that the equipment, processes, or operations are conducted and maintained in accordance with applicable codes and standards, and all deficiencies are identified, documented, and reported in accordance with the policies of the jurisdiction.

3-3.12 Verify code compliance for incidental storage, handling, and use of flammable and combustible liquids and gases, given field observations and inspection guidelines from the authority having jurisdiction, so that applicable codes and standards are addressed and all deficiencies are identified, documented, and reported in accordance with the policies of the jurisdiction.

3-3.13 Verify code compliance for incidental storage, handling, and use of hazardous materials, given field observations, so that applicable codes and standards for each hazardous material encountered are properly addressed and all deficiencies are identified, documented, and reported in accordance with the policies of the jurisdiction.

Chapter 4 Fire Inspector II

4-3.6 Evaluate hazardous conditions involving equipment, processes, and operations, given field observations and appropriate documentation, so that the equipment, processes, or operations are installed in accordance with applicable codes and standards, and all deficiencies are identified, documented, and reported in accordance with the policies of the jurisdiction.

4-3.9 Verify code compliance for storage, handling, and use of flammable and combustible liquids and gases, given field observations and inspection guidelines from the authority having jurisdiction, so that applicable codes and standards are addressed and all deficiencies are identified, documented, and reported in accordance with the policies of the jurisdiction.

4-3.10 Verify code compliance for storage, handling, and use of hazardous materials, given field observations, so that applicable codes and standards for each hazardous material encountered are properly addressed and all deficiencies are identified, documented, and reported in accordance with the policies of the jurisdiction.

Chapter 10
Identification of Hazardous Materials and Storage, Handling, and Use of Flammable and Combustible Liquids

In today's society, it is inevitable that fire inspection personnel will have frequent contact with occupancies that use, handle, and store hazardous materials. A *hazardous material*, also called *dangerous goods* in Canada, is defined as any material that poses an unreasonable risk to the health and safety of persons or the environment if it is not properly controlled during handling, storage, manufacture, processing, packaging, use, disposal, or transportation.

Hazardous materials come in solid, liquid, and gaseous forms, and they pose risks to health and safety in many ways. They may be flammable, explosive, toxic, corrosive, radioactive, unstable, or reactive. Because of the dangers posed by hazardous materials, it is the inspector's job to see that they are stored and handled in accordance with the appropriate regulations.

By far, the most common types of hazardous materials that fire inspection personnel deal with are flammable and combustible liquids. These are found in many common occupancies within most jurisdictions, including service stations, manufacturing facilities, hardware stores, and lumberyards.

This chapter begins with a general discussion of how to recognize and identify hazardous materials. This is followed by an in-depth discussion of flammable and combustible liquids. Information on other types of hazardous materials with which inspection personnel must be familiar is contained in Chapter 11 of this manual.

NFPA 1031, *Standard for Professional Qualifications for Fire Inspector and Plan Examiner* requires that personnel certifying to the Inspector I level also meet the requirements for First Responders at the Awareness Level contained in NFPA 472, *Standard for Professional Competence of Responders to Hazardous Materials Incidents*. It is impossible to cover all the information necessary to meet that requirement in two chapters. The information necessary to meet the NFPA 472 awareness level requirements may be found in IFSTA's **Hazardous Materials for First Responders** or **Awareness Level Training for Hazardous Materials** manuals.

IDENTIFICATION OF HAZARDOUS MATERIALS

Fire inspection personnel must be very familiar with the methods for recognizing and identifying hazardous materials that they may encounter while performing their duties. This will allow them to accomplish the following:

- Record the necessary information on inspection forms and reports.
- Enforce the appropriate codes or ordinances applicable to the material(s).
- Notify fire or hazardous materials response crews of particular materials/problems that might be encountered during an incident at that facility (Figure 10.1).

Inspection personnel should be capable of identifying hazardous materials that are in both the transport and fixed use modes. Identifying hazardous materials in the transport mode is primarily accomplished through recognizing the placards and material numbering systems used by the U.S. Department of Transportation (DOT), Transport Canada, or other agencies. Inspectors should be familiar with the United Nations Classification

Figure 10.1 Most inspectors have occupancies within their jurisdictions that contain large quantities of hazardous materials. *Courtesy of Harvey Eisner.*

System and how to use the *North American Emergency Response Guidebook*.

Hazardous materials that are manufactured, stored, processed, or used at a particular site are not subject to regulations affecting materials during transport. Local agencies and county, city, and township governments may adopt their own identification system or use a widely recognized method such as that recommended by NFPA 704, *Standard System for the Identification of the Fire Hazards of Materials*. Some private companies and the military have created their own internal policies that regulate the marking of these materials. Inspectors should also be able to read and interpret Material Safety Data Sheets (MSDS), which provide information on the chemicals found in that occupancy.

United Nations Classification System

Both the United States and Canada have adopted the United Nations (UN) system for classifying and identifying hazardous materials transported both internationally and domestically. Under this system, nine hazard classes are used to categorize hazardous materials. In addition to these nine classes, a separate category exists for other regulated materials (ORM-D). The nine hazard classes used for categorizing hazardous materials are as follows:

Class 1 — Explosives

Class 2 — Gases

Class 3 — Flammable Liquids

Class 4 — Flammable Solids

Class 5 — Oxidizers

Class 6 — Poisons and Infectious Substances

Class 7 — Radioactive Substances

Class 8 — Corrosives

Class 9 — Miscellaneous

The UN system forms the basis for the DOT regulations (Table 10.1). The DOT classifies hazardous materials according to their primary danger and assigns standardized symbols to identify the classes; this is similar to what the UN system has done. DOT regulations cover several other types of substances in addition to the nine classes identified in the UN system. The major classes and a brief description of each are given in the following sections.

CLASS 1 — DIVISION 1.1, 1.2, 1.3, 1.4, 1.5, 1.6

An *explosive* is any substance or article, including a device, that is designed to function by explosion (i.e., an extremely rapid release of gas and heat) or that, by chemical reaction within itself, is able to function in a similar manner even if not designed to function by explosion, unless the substance or article is otherwise classed.

Explosives in Class 1 are divided into six divisions as follows:

- 1.1 — Explosives that have a mass explosion hazard. A mass explosion is one that affects almost the entire load instantaneously.
- 1.2 — Explosives that have a projection hazard but not a mass explosion hazard.
- 1.3 — Explosives that have a fire hazard and either a minor blast hazard or a minor projection hazard or both, but not a mass explosion hazard.
- 1.4 — Explosives that present a minor explosion hazard. The explosive effects are largely confined to the package, and no projection of fragments of appreciable size or range is to be expected. An external fire must not cause instantaneous explosion of almost the entire contents of the package.

TABLE 10.1
Examples of Department of Transportation Hazardous Materials Classes

Hazard Class	Product Example
1 Explosives	
1.1 Mass explosion hazard	Black powder
1.2 Projection hazard	Detonating cord
1.3 Fire hazard	Propellant explosives
1.4 No significant blast	Practice ammunition
1.5 Very insensitive	Prilled ammonium nitrate
1.6 Extremely insensitive	Fertilizer-fuel oil mixtures
2 Gases	
2.1 Flammable gas	Propane
2.2 Nonflammable gas	Anhydrous ammonia
2.3 Poisonous gas	Phosgene
2.4 Corrosive gas (Canada)	
3 Flammable Liquids	Gasoline, kerosene, diesel fuel
4 Flammable Solids, Spontaneously Combustible Materials, and Materials that are Dangerous When Wet	
4.1 Flammable solids	Magnesium
4.2 Spontaneously combustible	Phosphorus
4.3 Dangerous when wet	Calcium carbide
5 Oxidizers and Organic Peroxides	
5.1 Oxidizers	Ammonium nitrate
5.2 Organic peroxides	Ethyl ketone peroxide
6 Poisonous and Etiologic Materials	
6.1 Poisonous	Arsenic
6.2 Infectious (etiological agent)	Rabies, HIV, Hepatitis B
7 Radioactive Materials	Cobalt
8 Corrosives	Sulfuric acid, sodium hydroxide
9 Miscellaneous Hazardous Materials	
9.1 Miscellaneous (Canada Only)	PCBs, molten sulfur
9.2 Environmental Hazard (Canada Only)	PCB, asbestos
9.3 Dangerous Waste (Canada Only)	Fumaric acid
ORM-D (Other regulated materials)	Consumer commodities

- 1.5 — Very insensitive explosives. This division is composed of substances that have a mass explosion hazard but are so insensitive that there is very little probability of initiation or of transition from burning to detonation under normal conditions of transport.
- 1.6 — Extremely insensitive articles that do not have a mass explosive hazard. This division is composed of articles that contain

only extremely insensitive detonating substances and that demonstrate a negligible probability of accidental initiation or propagation.

CLASS 2 — DIVISION 2.1, 2.2, 2.3, (2.4 CANADA ONLY)

- 2.1 — *Flammable gas* is any material that is a gas at 68°F (20°C) or less at normal atmospheric pressure or a material that has a boiling point of 68°F (20°C) or less at normal atmospheric pressure and that:
  (1) Is ignitable at normal atmospheric pressure when in a mixture of 13 percent or less by volume with air or,
  (2) Has a flammable range at normal atmospheric pressure with at least 12 percent air, regardless of the lower limit.
- 2.2 — *Nonflammable compressed gas* (nonflammable, nonpoisonous compressed gas, liquefied gas, pressurized cryogenic gas, and compressed gas in solution) is any material (or mixture) that exerts, in the package, an absolute pressure of 41 psia (280 kPa) at 68°F (20°C) and does not meet the definition of Division 2.1 or 2.3.
- 2.3 — *Poisonous gas* (gas poisonous by inhalation) is a material that is a gas at 68°F (20°C) or less at normal atmospheric pressure, that has a boiling point of 68°F (20°C) or less at normal atmospheric pressure and that:
  (1) Is known to be so toxic to humans as to pose a hazard to health during transportation or,
  (2) In the absence of adequate data on human toxicity, is presumed to be toxic to humans because when tested on laboratory animals it has an LC_{50} value of not more than 5000 ml/m^3.
- 2.4 (Canada Only) — Corrosive Gases

CLASS 3

- *Flammable liquid* is:
  (1) A liquid having a flash point of not more than 141°F (61°C) or any material in a liquid phase with a flash point at or above 100°F (38°C) that is intentionally heated (mixtures and solutions) and offered for transportation or transported at or above its flash point in a bulk packaging.
  (2) A distilled spirit of 140 proof or lower is considered to have a flash point of no lower than 73°F (23°C).
- *Combustible liquid* is:
  (1) Any liquid that does not meet the definition of any other hazard class, except Class 9, and has a flash point above 141°F (61°C) and below 200°F (93°C).
  (2) A flammable liquid with a flash point at or above 100°F (38°C) that does not meet the definition of any other hazard class, except Class 9, may be reclassified as a combustible liquid.

NOTE: Specific flash points for the definition of a flammable liquid may vary depending on the regulation or standard referenced. For example, transportation regulations and codes for fixed facilities, such as NFPA 30, *Flammable and Combustible Liquids Code*, have slightly different definitions. These differences are important from both the inspection and enforcement perspectives.

CLASS 4 — DIVISION 4.1, 4.2, 4.3

- 4.1 — *Flammable solid* is any of the following three types of materials:
  (1) Wetted explosives
  (2) Self-reactive materials
  (3) Readily combustible solids
- 4.2 — *Spontaneously combustible materials* are the following:
  (1) A pyrophoric material, liquid or solid, that even in small quantities and without an external ignition source can ignite within five minutes after coming into contact with air.
  (2) A self-heating material that when in contact with air and without an energy supply has the potential to self-heat.

- 4.3 — *Dangerous when wet material* is a material that, by contact with water, is liable to become spontaneously flammable or to give off flammable or toxic gas at a rate greater than 1 liter per kilogram of the material per hour.

CLASS 5 — DIVISION 5.1, 5.2

- 5.1 — *Oxidizer* is a material that may, generally by yielding oxygen, cause or enhance the combustion of other materials.
- 5.2 — *Organic peroxide* is any organic compound containing oxygen in the bivalent -O-O- structure and that may be considered a derivative of hydrogen peroxide, where one or more of the hydrogen atoms have been replaced by organic radicals.

CLASS 6 — DIVISION 6.1, 6.2

- 6.1 — *Poisonous material* is a material, other than a gas, that is known to be so toxic to humans as to afford a hazard to health during transportation.
- 6.2 — *Infectious substance* (etiologic agent) is a viable microorganism, or its toxin, that causes or may cause disease in humans or animals and includes those agents listed in the regulations of the Department of Health and Human Services or any other agent that causes or may cause severe, disabling, or fatal disease.

CLASS 7

- *Radioactive material* is any material having a specific activity greater than 70 Becquerels per gram (Bq/g).

CLASS 8

- *Corrosive material* is a liquid or solid that causes visible destruction or irreversible alterations to human skin tissue at the site of contact or is a liquid that has a severe corrosion rate on steel or aluminum.

CLASS 9

- Miscellaneous hazardous materials

  (a) A material that has an anesthetic, noxious, or other similar property that could cause extreme annoyance or discomfort to a flight crew member so as to prevent the correct performance of assigned duties.

  (b) Meets the definitions in 49 CFR 171.8 for a hazardous substance or a hazardous waste.

  (c) Meets the definition in 49 CFR 171.8 for an elevated temperature material.
- In Canada, Class 9 has three divisions:

  9.1 — Miscellaneous Dangerous Goods

  9.2 — Environmental Hazard

  9.3 — Dangerous Waste

OTHER REGULATED MATERIALS — ORM-D

ORM-D is a material, such as a consumer commodity, that presents a limited hazard during transportation due to its form, quantity, and packaging.

Number Identification System

The United Nations (UN) has developed a system of numbers and reference materials that the United States and Canada use in conjunction with illustrated placards. The UN system provides uniformity in recognizing hazardous materials in international transport. By using the UN numbers and reference materials, the first responder can determine information on the general hazard class and the identity of certain predetermined commodities.

A haz mat identification number is a four-digit number assigned to each material listed in the Hazardous Materials Table appearing in the *North American Emergency Response Guidebook (NAERG)*. This four-digit number may appear on labels, placards, orange panels, or white square-on-point configurations in association with materials being transported in cargo tanks or portable tanks (Figure 10.2). It may also be found on fixed tanks at industrial or storage facilities. The identification number assists inspectors and first responders in identifying materials and in correctly referencing the material in the *NAERG*.

North American Emergency Response Guidebook (NAERG)

The *North American Emergency Response Guidebook (NAERG)* was primarily developed for

Figure 10.2 DOT numbers may be found on trailers, rail cars, and at some fixed facilities.

use by first responders such as firefighters, police, and emergency medical services personnel. The *NAERG* is a basic guide for initial actions to be taken by first responders to protect themselves and the general public when responding to incidents involving hazardous materials. The *NAERG* is intended to be carried in every public safety vehicle in the country. It is also useful for inspection personnel who are inspecting facilities where placarded containers are present. Prior to 1996, the United States Department of Transportation and Transport Canada developed their own books for each country. In 1996, they developed a common manual that is good for all of North America, including Mexico (Figure 10.3). Copies of the *NAERG* may be available through local or state/province emergency management offices.

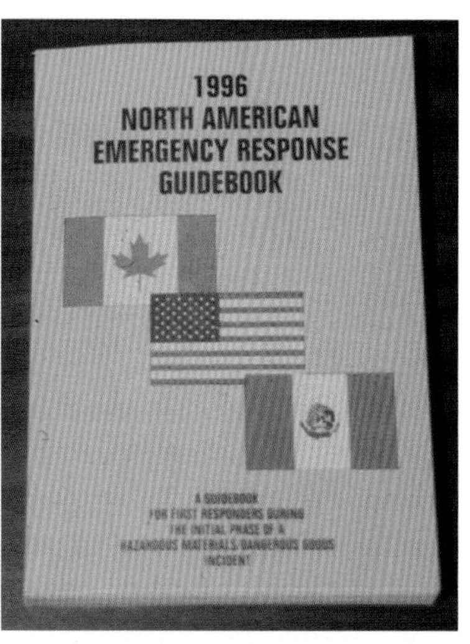

Figure 10.3 The NAERG is a useful tool for inspectors and responders.

The document provides several different means to identify hazardous materials. The inspector may look up the product either by name or UN number. Once the material is identified, the user is directed to a specific guide page that lists the hazards associated with the material and the initial actions that should be taken during an emergency involving that material (Figure 10.4). In the event the material cannot be specifically identified by the methods provided, the *ERG* provides guidance for initiating safety actions that are applicable to all materials until a specific decision can be made.

NFPA 704 System

NFPA 704 is an acceptable identification system that can be implemented in an area where materials are regularly stored and used. The NFPA 704 system is also an important aid in determining the actions and safety procedures to be used in the initial phases of a haz mat incident. The system is a widely recognized method for indicating the presence of hazardous materials at commercial, manufacturing, institutional, and other fixed-storage facilities. Use of this system is commonly required by local ordinances for all occupancies that contain hazardous materials. The NFPA 704 system is *not* designed to be used for the following:

- Transportation
- General public use
- Chronic (repeated, long-term) exposures
- Nonemergency occupational exposures

The NFPA 704 system offers the following:

- It provides the appropriate signal or alert to first responders that hazardous materials are present. The first-arriving responder who sees the NFPA 704 marker on a structure can determine the hazards of a single material in a marked container or can determine the relative combined hazard severity of the collection of numerous materials in the occupancy.
- It identifies the general hazards and the degree of severity for health, flammability, and reactivity.
- It provides immediate information necessary to protect the lives of both the public and emergency response personnel.

It is the fire inspector's responsibility to make sure that the correct NFPA 704 markings are appropriately displayed where required. The sys-

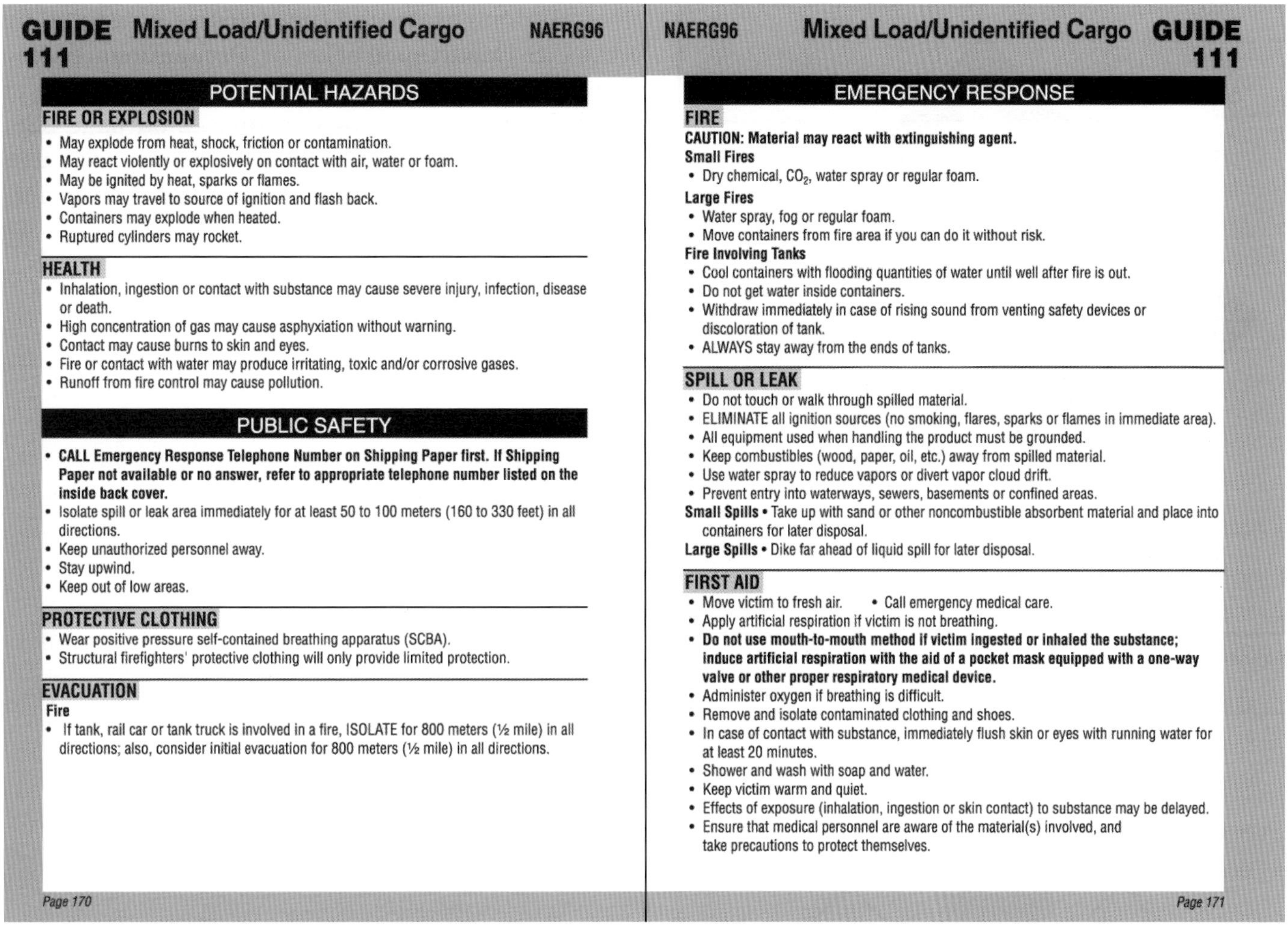

GUIDE 111 Mixed Load/Unidentified Cargo NAERG96

POTENTIAL HAZARDS

FIRE OR EXPLOSION

- May explode from heat, shock, friction or contamination.
- May react violently or explosively on contact with air, water or foam.
- May be ignited by heat, sparks or flames.
- Vapors may travel to source of ignition and flash back.
- Containers may explode when heated.
- Ruptured cylinders may rocket.

HEALTH

- Inhalation, ingestion or contact with substance may cause severe injury, infection, disease or death.
- High concentration of gas may cause asphyxiation without warning.
- Contact may cause burns to skin and eyes.
- Fire or contact with water may produce irritating, toxic and/or corrosive gases.
- Runoff from fire control may cause pollution.

PUBLIC SAFETY

- **CALL Emergency Response Telephone Number on Shipping Paper first. If Shipping Paper not available or no answer, refer to appropriate telephone number listed on the inside back cover.**
- Isolate spill or leak area immediately for at least 50 to 100 meters (160 to 330 feet) in all directions.
- Keep unauthorized personnel away.
- Stay upwind.
- Keep out of low areas.

PROTECTIVE CLOTHING

- Wear positive pressure self-contained breathing apparatus (SCBA).
- Structural firefighters' protective clothing will only provide limited protection.

EVACUATION

Fire

- If tank, rail car or tank truck is involved in a fire, ISOLATE for 800 meters (½ mile) in all directions; also, consider initial evacuation for 800 meters (½ mile) in all directions.

Page 170

NAERG96 Mixed Load/Unidentified Cargo **GUIDE 111**

EMERGENCY RESPONSE

FIRE

CAUTION: Material may react with extinguishing agent.

Small Fires

- Dry chemical, CO_2, water spray or regular foam.

Large Fires

- Water spray, fog or regular foam.
- Move containers from fire area if you can do it without risk.

Fire Involving Tanks

- Cool containers with flooding quantities of water until well after fire is out.
- Do not get water inside containers.
- Withdraw immediately in case of rising sound from venting safety devices or discoloration of tank.
- ALWAYS stay away from the ends of tanks.

SPILL OR LEAK

- Do not touch or walk through spilled material.
- ELIMINATE all ignition sources (no smoking, flares, sparks or flames in immediate area).
- All equipment used when handling the product must be grounded.
- Keep combustibles (wood, paper, oil, etc.) away from spilled material.
- Use water spray to reduce vapors or divert vapor cloud drift.
- Prevent entry into waterways, sewers, basements or confined areas.

Small Spills • Take up with sand or other noncombustible absorbent material and place into containers for later disposal.

Large Spills • Dike far ahead of liquid spill for later disposal.

FIRST AID

- Move victim to fresh air. • Call emergency medical care.
- Apply artificial respiration if victim is not breathing.
- **Do not use mouth-to-mouth method if victim ingested or inhaled the substance; induce artificial respiration with the aid of a pocket mask equipped with a one-way valve or other proper respiratory medical device.**
- Administer oxygen if breathing is difficult.
- Remove and isolate contaminated clothing and shoes.
- In case of contact with substance, immediately flush skin or eyes with running water for at least 20 minutes.
- Shower and wash with soap and water.
- Keep victim warm and quiet.
- Effects of exposure (inhalation, ingestion or skin contact) to substance may be delayed.
- Ensure that medical personnel are aware of the material(s) involved, and take precautions to protect themselves.

Page 171

Figure 10.4 A sample guide sheet from the NAERG.

tem does not identify the specific chemical or chemicals that may be present. Positive identification should be made through other means such as container markings, employee information, and company records. The inspector can then match the product with the recommended hazard levels established in NFPA 49, *Hazardous Chemicals Data*. NFPA 49 describes the properties and hazards of various materials and provides valuable information on assigning appropriate ratings to the NFPA 704 markers at facilities that contain listed chemicals.

Specifically, the NFPA 704 system uses a rating system of zero (0) to four (4). A zero indicates there is no hazard present, and a four represents a severe hazard. The rating is assigned to three categories: health, flammability, and reactivity. The rating numbers are arranged on a diamond-shaped marker or sign. The health rating is located on a blue background at the nine o'clock position. The flammability hazard rating is positioned on a red background at the twelve o'clock position. The reactivity hazard rating appears on a yellow background and is positioned at three o'clock (Figure 10.5). As an alternative, the backgrounds for each of these rating positions may be any contrasting color, and the numbers (0 to 4) may be represented by the appropriate color (blue, red, and yellow) (Figure 10.6). Special hazards are located in the six o'clock position and have no specified background color; however, white is most commonly used. The ratings for each hazard (health, flammability, and reactivity) are described in Table 10.2 on page 319. This table also describes the special hazards that may be indicated on the NFPA 704 marker.

Material Safety Data Sheet (MSDS)

The best source of information on a specific hazardous material is the manufacturer's data sheet known as a material safety data sheet (MSDS). State and federal legislation on hazard

Figure 10.5 The most common type of 704 symbol uses colored blocks.

Figure 10.6 An alternative 704 symbol design uses white blocks and colored numbers.

communication, right-to-know, and mandatory local notification on hazards make the MSDS a necessity at all occupancies that contain hazardous materials. Inspectors can acquire an MSDS from the manufacturer of the material, the supplier, the facility Hazard Communication Plan, or the Local Emergency Planning Committee.

Minimal content of the MSDS is mandated by the U.S. Department of Labor, Occupational Safety and Health Administration (OSHA). A sample of an MSDS is shown in Figure 10.7 on page 320. There is no set format for the MSDS, but each sheet has eight sections that contain the following information:

Section I

- Manufacturer's name and address
- Emergency telephone number
- Information telephone number
- Signature and date

Section II — *Hazardous Ingredients*

- Common name
- Chemical name
- Chemical Abstract Service (CAS) number
- OSHA Permissible Exposure Limit (PEL)
- ACGIH Threshold Limit Value (TLV)
- Other Exposure Limits

Section III — *Physical and Chemical Characteristics*

- Boiling point
- Specific gravity
- Vapor pressure
- Melting point
- Vapor density
- Evaporation rate
- Solubility in water
- Appearance and odor

Section IV — *Fire and Explosion Hazard Data*

- Flash point
- Flammable limits (LEL, UEL)
- Extinguishing media
- Special fire fighting procedures
- Unusual fire and explosion hazards

Section V — *Reactivity Data*

- Stability (stable/unstable, conditions to avoid)
- Incompatibility (materials to avoid)

TABLE 10.2
NFPA 704 Rating System

Identification Of Health Hazard		Identification Of Flammability		Identification Of Reactivity	
Type Of Possible Injury		Susceptibility Of Materials To Burning		Susceptibility To Release Of Energy	
Signal		Signal		Signal	
4	Materials that on very short exposure could cause death or major residual injury.	4	Materials that will rapidly or completely vaporize at atmospheric pressure and normal ambient temperature, or that are readily dispersed in air and that will burn readily.	4	Materials that in themselves are readily capable of detonation or of explosive decomposition or reaction at normal temperatures and pressures.
3	Materials that on short exposure could cause serious temporary or residual injury.	3	Liquids and solids that can be ignited under almost all ambient temperature conditions.	3	Materials that in themselves are capable of detonation or explosive decomposition or reaction but require a strong initiating source or which must be heated under confinement before initiation or which react explosively with water.
2	Materials that on intense or continued but not chronic exposure could cause temporary incapacitation or possible residual injury.	2	Materials that must be moderately heated or exposed to relatively high ambient temperatures before ignition can occur.	2	Materials that readily undergo violent chemical change at elevated temperatures and pressures or which react violently with water or which may form explosive mixtures with water.
1	Materials that on exposure would cause irritation but only minor residual injury.	1	Materials that must be preheated before ignition can occur.	1	Materials that in themselves are normally stable, but which can become unstable at elevated temperatures and pressures.
0	Materials that on exposure under fire conditions would offer no hazard beyond that of ordinary combustible material.	0	Materials that will not burn.	0	Materials that in themselves are normally stable, even under fire exposure conditions, and which are not reactive with water.

- Hazardous decomposition or by-products
- Hazardous polymerization (may or may not occur, conditions to avoid)

Section VI — *Health Hazard Data*

- Routes of entry
- Health hazards (acute or chronic)
- Carcinogenicity
- NTP (National Toxicological Program)
- IARC (International Agency for Research on Cancer) monographs
- OSHA regulated
- Signs and symptoms of exposure
- Medical conditions aggravated by exposure
- Emergency and first aid procedures

Material Safety Data Sheet

Hoechst Celanese

Chemical Group
Hoechst Celanese Corporation
*P.O. Box 819005/Dallas, Texas 75381-9005
*Information phone: 214 277 4000
Emergency phone: 800 424 9300 (CHEMTREC)

Ethylene oxide

Issued December 31, 1992 #40

Identification

Product name: Ethylene oxide
Chemical name: Ethylene oxide
Chemical family: Epoxide
Formula: $(CH_2)_2O$
Molecular weight: 44
CAS number: 75-21-8
CAS name: Oxirane
Synonyms: Dihydrooxirene; dimethylene oxide; 1,2-epoxyethane; oxiran; oxirane; oxacyclopropane; oxane; oxidoethane; alpha, beta-oxidoethane; EO; EtO.

***Transportation information**
Shipping name: Ethylene Oxide
Hazard class: 2.3, Poisonous Gas
Subsidiary hazard: 2.1, Flammable Gas
United Nations no.: UN1040
Packing group: 1
Emergency Response Guide no.: 69
DOT Reportable Quantity: 10 lb/4.54 kg

Physical data

Boiling point (760 mm Hg): 10.7°C (51°F)
Freezing point: -112.5°C (-171°F)
Specific gravity (H_2O=1 @ 20/20°C): 0.8711
Vapor pressure (20°C): 1094 mm Hg
Vapor density (Air =1 @ 20°C): 1.5
Solubility in water (% by WT @ 20°C): Complete
Percent volatiles by volume: 100
Appearance and odor: Colorless gas with sweet ether-like odor. Odor threshold: 500 ppm.

Fire and explosion hazard data

Flammable limits in air, % by volume
Upper: 100
Lower: 3.0

Flash point (test method):
Tag open cup (ASTM D1310): <0°F (<–18°C)
Tag closed cup (ASTM D56): –4°F (–20°C)

Extinguishing media:
Use water (flood with water), CO_2, dry chemical or alcohol-type aqueous film-forming foam. Allow to burn if flow cannot be shut off immediately.

Special fire-fighting procedures:
*If potential for exposure to vapors or products of combustion exists, wear complete personal protective equipment, including self-contained breathing apparatus with full facepiece operated in pressure-demand or other positive-pressure mode.

Dilution of ethylene oxide with 23 volumes of water renders it non-flammable. A ratio of 100 parts water to one part ethylene oxide may be required to control build-up of flammable vapors in a closed system. Water spray can be used to reduce intensity of flames and to dilute spills to nonflammable mixture. Use water spray to cool fire-exposed structures and vessels. Ethylene oxide is an NFPA Class 1A flammable liquid with a 51°F boiling point. Locations classified as hazardous because of the presence of ethylene oxide are designated Class 1.

Unusual fire and explosion hazards:
Rapid, uncontrolled polymerization can cause explosion under fire conditions. Vapor is heavier than air and can travel considerable distance to a source of ignition and flashback. Will burn without the presence of air or other oxidizers.

Special hazard designations

	HMIS	NFPA	Key
Health:	3	3	0 = Minimal
Flammability:	4	4	1 = Slight
Reactivity:	3	3	2 = Moderate
Personal protective equipment:	G	—	3 = Serious
			4 = Severe

SARA §311 hazard categories

Acute health:	Yes
Chronic health:	Yes
Fire:	Yes
Sudden release of pressure:	Yes
Reactive:	Yes

Component information (See Glossary at end of MSDS for definitions)

Component, wt. % (CAS number)	Exposure levels: OSHA PEL TWA	ACGIH TLV ®TWA	IDLH	Subject to SARA §313 reporting?
• Ethylene oxide, 99.95% (75-21-8)	1ppm[2]; 5 ppm excursion limit	1ppm[2]	800 ppm	Yes

(1) All components listed as required by federal, California, New Jersey and Pennsylvania regulations.
(2) Suspectd human carcinogen.

Reactivity data

Stability:
Potentially unstable

Hazardous polymerization:
Can occur.UNCONTROLLED POLYMERIZATION CAN CAUSE RAPID EVOLUTION OF HEAT AND INCREASED PRESSURE WHICH CAN RESULT IN VIOLENT RUPTURE OF STORAGE VESSELS OR CONTAINERS.

Conditions to avoid:
Heat, sparks, flame.

Materials to avoid:
Acetylide-forming metals (for example, copper, silver, mercury and their alloys): alcohols; amines; mercaptans; metallic chlorides; aqueous alkalis; mineral acids; oxides; strong oxidizing agents (for example, oxygen, hydrogen peroxide, or nitric acid).

Hazardous combustion or decomposition products:
Carbon monoxide.

Health data

Effects of exposure/toxicity data
Acute

Ingestion (swallowing): Can cause stomach irritation, also liver and kidney damage. Moderately toxic to animals (oral LD_{50}, rats: 0.1 g/kg).

Inhalation (breathing): Can cause irritation of nasal passages, throat and lungs; lung injury; nausea; vomiting; headache; diarrhea; shortness of breath; cyanosis (blue or purple coloring of the skin); and pulmonary edema (accumulation of fluid in the lungs) - signs and symptons can be delayed for several hours. Slightly toxic to animals (inhalation LC_{50}, rats, 4 hrs: 1460 ppm).

Skin contact: Can lead to severe reddening and swelling of the skin, with blisters.

*New or revised information; previous version dated October 1,1991.

Figure 10.7 A sample MSDS. *Courtesy of Hoechst Celanese.*

Ethylene oxide

page 2

#40

Issued December 31,1992

Sensitization (allergic reaction) possible. Large amounts evaporating from skin can cause frostbite.

Eye contact: Contact with liquid can cause severe injury to the cornea. High exposure to vapors is irritating.

Chronic

Mutagenicity: *In vitro*, mutagenic. *In vivo*, mutagenic. Human and animal inhalation studies show genetic material (DNA) damage, including hemoglobin alkylation, unscheduled DNA synthesis, sister chromatid chromosomal aberration, functional sperm abnormalities and dominant lethal effects.

Carcinogenicity: Carcinogenic to animals (inhaled, caused tumors and leukemia in rats at concentrations down to 10 ppm; caused tumors and lymphomas in mice at concentrations down to 50 ppm). Suspected of carcinogenic potential in humans (ACGIH). Listed as an experimental animal carcinogen and probable human carcinogen (IARC, OSHA, NTP).

Reproduction: No evidence of effect on female reproduction. Damages embryo and fetus. No evidence of malformed offspring. Causes dominant lethal effects (see "Mutagenicity" section).

Other: In animals, can damage the nervous system (for example, inhalation can cause paralysis of the hind legs of rats). Adversely affects blood and liver.

Medical conditions aggravated by exposure:
Significant exposure to this chemical may adversely affect people with chronic disease of the respiratory system, skin, blood, liver, central nervous system, kidneys and/or eyes.

Emergency and first aid procedures

Ingestion (swallowing): Patient should be made to drink large quantities of water. Then, induce vomiting by pressing finger down throat. Contact a physician immediately.

Inhalation (breathing): Remove patient from contaminated area. If breathing has stopped, give artificial respiration, then oxygen if needed. Contact a physician immediately.

Skin contact: Remove contaminated clothing. Wash contaminated skin with soap and water for 15 minutes. Contact a physician immediately.

Eye contact: Flush eyes with water for at least 15 minutes, lifting the upper and lower eyelids. Contact a physician immediately.

Spill or leak procedures

***Steps to be taken if material is released or spilled:**
Eliminate ignition sources. Ethylene oxide/air mixtures may detonate upon ignition. Avoid eye or skin contact; see "Special protection information" section for respirator information. Evacuate all personnel from the area except for those engaged in stopping the leak or in clean-up. Flood affected area with water. Use water spray to disperse vapors. Avoid runoff into storm sewers and ditches which lead to natural waterways. Call the National Response Center (800 424 8802) if the quantity spilled is equal to or greater than the reportable quantity (10 lb/day) under CERCLA "Superfund."

***Waste disposal method:**
All notification, clean-up and disposal should be carried out in accordance with federal, state and local regulations. Preferred methods of waste disposal are incineration or biological treatment in federal/state approved facility.

***Hazardous waste (40 CFR 261):**
Yes; hazardous waste codes U115, DOO1.

Special protection information

***Respiratory protection:**
Based on contamination level and working limits of the respirator, use a respirator approved by NIOSH/MSHA (the following are the minimum recommended equipment).

For ethylene oxide concentrations of:

≥1 ppm and ≤50 ppm — Air-purifying respirator with full facepiece and ethylene oxide approved canister.

>50 ppm and <800 ppm — Positive-pressure full facepiece supplied-air respirator, or continuous-flow full facepiece supplied-air respirator.

≥800 ppm or unknown concentration (such as in emergencies) — Positive-pressure self-contained breathing apparatus with full facepiece. Positive-pressure supplied-air respirator with full facepiece equipped with an auxiliary positive-pressure self-contained breathing apparatus escape system.

Ventilation

Local exhaust: Recommended as the sole means of controlling employee exposure.

Mechanical (general): Not recommended as the sole means of controlling employee exposure.

Protective gloves:
Butyl rubber.

Eye protection:
Chemical safety goggles. Contact lenses should not be worn if exposure to ethylene oxide is likely to occur.

***Additional protective equipment:**
For operations where spills or splashing can occur, use chemical protective clothing, including gloves and boots. A safety shower and eye bath should be readily available.

Special precautions

***Precautions to be taken in handling and storing:**
Storage vessels should be insulated and should have pressure relief valves. To avoid product contamination, install double-block valves on outlet of storage vessel. Electrical installations should be in accordance with Article 501 of the National Electrical Code. Do not incinerate ethylene oxide cartridges, tanks or other containers. Store in a cool, well-ventilated area. Replace or repair protective equipment that has been torn or otherwise damaged. Protective clothing wet with ethylene oxide should be immediately removed while under a safety shower. Decontaminate soiled clothing thoroughly before re-use. Contaminated leather articles should be destroyed. Do not expose to temperataures above 21°C (70°F). Keep away from heat, sparks and flame. Keep containers closed when not in use. Always open containers slowly to allow any excess pressure to vent. Use only DOT-approved containers. Use spark-resistant tools. Do not load into compartments adjacent to heated cargo. When transferring follow proper grounding procedures. Use with adequate ventilation. Do not store near combustible materials. Avoid breathing gas. Avoid contact with eyes, skin and clothing. Wash thoroughly with soap and water after handling. Do not enter storage area unless adequately ventilated.

*New or revised information; previous version dated October 1, 1991.

Glossary for Components information table

- **ACGIH** - American Conference of Governmental Industrial Hygienists
- **CAS** - Chemical Abstracts Service
- **Ceiling** - The concentration that should not be exceeded during any part of the working day.
- **IDLH** - Immediately Dangerous to Life or Health
- **OSHA** - Occupational Safety and Health Administration
- **PEL** - Permissible exposure limit
- **SARA** - Superfund Amendments and Reauthorization Act
- **Skin** - Potential contribution to overall exposure possible via skin absorption
- **STEL** - Short-term exposure limit; 15-min. time-weighted average
- **TLV** - Threshold limit value
- **TWA** - 8-hour time-weighted average

Chemical Group
Hoechst Celanese Corporation
*P.O. Box 819005/Dallas, Texas 75381-9005
*Information phone: 214 277 4000
Emergency phone: 800 424 9300 (CHEMTREC)

Figure 10.7 Continued

Section VII — *Precautions for Safe Handling and Use*

- Steps to be taken in case material has been released or spilled
- Waste disposal methods
- Handling and storing precautions
- Other precautions

Section VIII — *Control Measures*

- Respiratory protection
- Ventilation (local, mechanical, special, other)
- Protective gloves
- Eye protection
- Other protective clothing or equipment
- Work/hygienic practices

U.S. Material Safety Data Sheets are not accepted in Canada because they are slightly different from the Canadian MSDS.

For more detailed information on recognizing and identifying hazardous materials, see IFSTA's **Hazardous Materials for First Responders** or **Awareness Level Training for Hazardous Materials** manuals.

FLAMMABLE AND COMBUSTIBLE LIQUIDS

As stated previously in this chapter, flammable and combustible liquids are the most common types of hazardous materials fire inspection personnel encounter. Because of this fact, the remainder of this chapter is dedicated to a detailed look at the things which inspectors should know about flammable and combustible liquids. Included is information on the characteristics, storage, handling, and use of flammable and combustible liquids.

Characteristics of Flammable and Combustible Liquids

In general, NFPA 30 states that *flammable liquids* are any liquids having a flash point below 100°F (38°C) and having a vapor pressure not exceeding 40 psi (256 kPa). **Note that these figures differ from those stated previously for the U.S. DOT**. NFPA 30 divides flammable liquids into Class IA, IB, and IC liquids:

- Class IA flammable liquids have flash points below 73°F (23°C) and boiling points below 100°F (38°C).
- Class IB flammable liquids have flash points below 73°F (23°C) and boiling points at or above 100°F (38°C).
- Class IC flammable liquids have flash points at or above 73°F (23°C) and below 100°F (38°C).

Combustible liquids have flash points at or above 100°F (38°C). Combustible liquids are further divided into Class II, IIIA, and IIIB liquids:

- Class II combustible liquids have flash points at or above 100°F (38°C) and below 140°F (60°C).
- Class IIIA combustible liquids have flash points at or above 140°F (60°C) and below 200°F (93°C).
- Class IIIB combustible liquids have flash points at or above 200°F (93°C).

The *flash point* is the minimum temperature at which a liquid fuel gives off sufficient vapors to form an ignitable mixture with the air near the surface (Figure 10.8). Although a flash fire may occur at this temperature, not enough vapors are being released to support continued combustion. Flash points depend on the atmospheric pressure at the time of the test, the oxygen concentration, the type of test used, and the skill of the individual performing the test. Therefore, if the atmospheric pressure is actually lower than the pressure was during the test, it is possible that flammable vapors will be emitted at a much lower temperature. Listed flash points are determined by one of two methods: the open cup test or the closed cup test. Results of studies have indicated that the flash points determined by the open cup method are approximately 10°F to 15°F (3°C to 4°C) higher than those determined by the closed cup method for the same material.

For liquids, the *fire point* refers to the temperature at which a liquid fuel, once ignited, will continue to burn. For all other fuels, the ignition temperature is the temperature at which the fuel will continue to burn. The *autoignition temperature*, sometimes simply referred to as the ignition

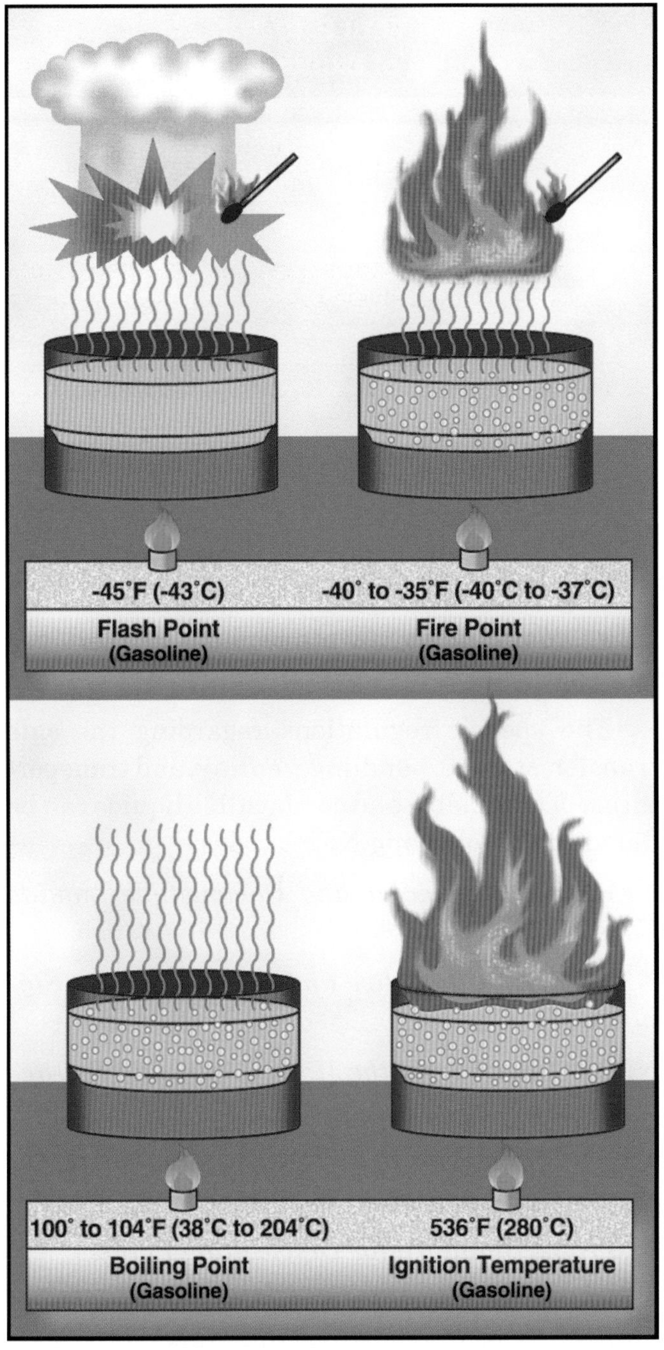

Figure 10.8 There are a variety of temperature terms that are important for inspectors to know.

temperature, refers to the temperature at which the fuel will ignite independent of another ignition source.

The *boiling point* of a liquid is that temperature at which the vapor pressure of the liquid is equal to the external pressure applied to it. In many cases, this external pressure is simply the atmospheric pressure at the given location. When this state of equilibrium exists, boiling occurs. The bubbles produced during boiling are bubbles of vapor that have formed within the liquid and are rising to the surface, where they can escape into the atmosphere.

The *specific gravity* of a flammable or combustible liquid refers to the ratio of the weight of the liquid to an equal volume of water (Figure 10.9). Because water is the standard, its value is 1.0. Liquids with a specific gravity of less than 1.0 are therefore lighter than water and will float on water. Most hydrocarbons have a specific gravity less than 1.0. Liquids with a specific gravity greater than 1.0 will allow water to float on their surface.

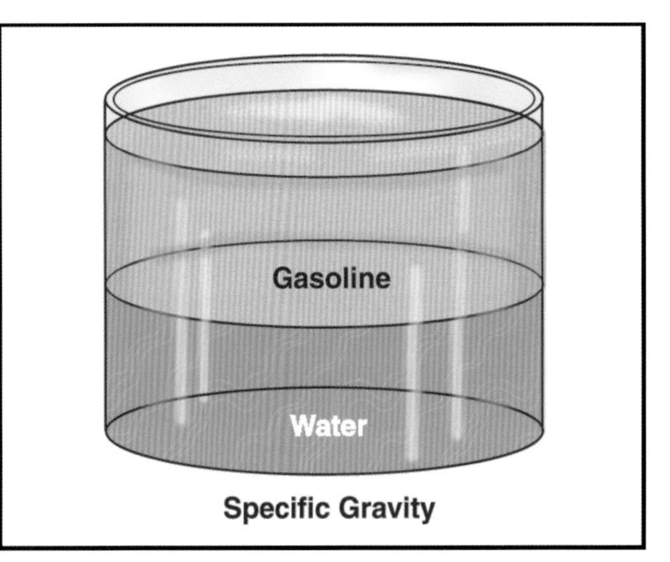

Figure 10.9 Gasoline's specific gravity is less than that of water. Thus, gasoline will float on water.

Before any flammable or combustible liquid ignites, it must first emit a certain amount of vapors. These vapors then mix with an oxidizing agent (air) to form a flammable vapor concentration. The *flammable and explosive limits* of flammable or combustible liquids are the upper and lower concentrations of the vapor in air that will produce a flame at a given pressure and temperature (Figure 10.10). These limits are expressed as a percent of the vapors in the mixture with an oxidizer. For example, 16 to 25 percent are the flammable limits of anhydrous ammonia. The lowest concentration of vapors in air that will ignite is known as the lower explosive limit (LEL) or lower flammable limit (LFL). Mixtures below the LEL/LFL are said to be too lean to burn. The highest concentration of vapors

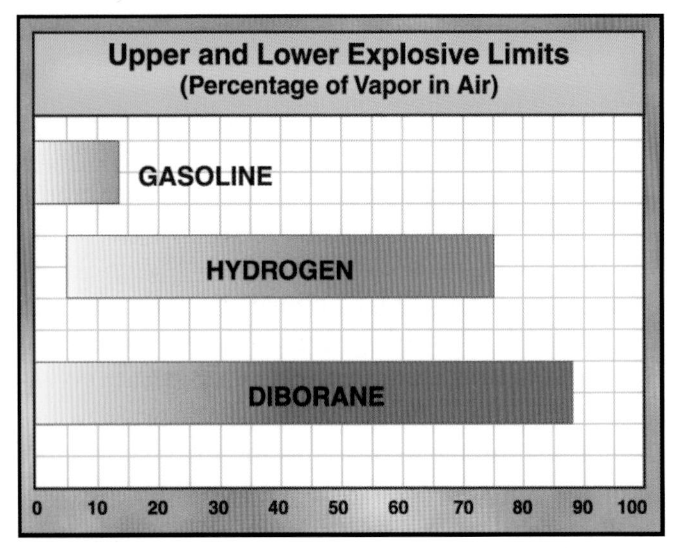

Figure 10.10 The flammable limits for several common substances.

in air that will ignite is known as the upper explosive limit (UEL) or upper flammable limit (UFL). Vapor concentrations above this point are said to be too rich to burn. Variations in temperature affect explosive limits: when the temperature increases, the range increases; when the temperature decreases, the range decreases. Examples of explosive ranges are shown in Table 10.3.

When the temperature of a flammable or combustible liquid is raised, more vapors are emitted from the liquid. If the liquid is in a closed container, the pressure inside the container will increase. If the temperature rise is significant, the resulting pressure rise inside the container may be sufficient to cause container failure, often with explosive force. Such explosions are known as Boiling Liquid Expanding Vapor Explosions (BLEVEs).

INHERENT HAZARDS AND IGNITION SOURCES

The hazards of flammable and combustible liquids range from accidental pollution (due to leaks) to fires to explosions. Fire prevention measures are based on one or more of the following techniques or principles:

- Eliminating or excluding sources of ignition such as electrical, mechanical or frictional sparks; static electricity; open flames; hot surfaces; and incompatible materials
- Excluding air (oxygen)

TABLE 10.3
Examples of Explosive Ranges

	Lower Explosive Limit	Upper Explosive Limit
Butyl alcohol	1.4	11.2
Xylene	1.1	7.0
Gasoline	1.4	7.6

- Storing liquids in closed containers or systems
- Ventilating to prevent the accumulation of vapors within the flammable range
- Maintaining an atmosphere of inert gas instead of air

The specific regulations regarding the safe transfer, storage, handling, venting, and transportation of flammable and combustible liquids can be found in the following NFPA codes:

NFPA 30, *Flammable and Combustible Liquids Code*

NFPA 30A, *Automotive and Marine Service Station Code*

NFPA 30B, *Code for the Manufacture and Storage of Aerosol Products*

NFPA 31, *Standard for the Installation of Oil Burning Equipment*

NFPA 32, *Standard for Drycleaning Plants*

NFPA 33, *Standard for Spray Application Using Flammable or Combustible Materials*

NFPA 34, *Standard for Dipping and Coating Processes Using Flammable or Combustible Materials*

NFPA 36, *Standard for Solvent Extraction Plants*

NFPA 325, *Guide to Fire Hazard Properties of Flammable Liquids, Gases, and Volatile Solids*

NFPA 329, *Recommended Practice for Handling Underground Releases of Flammable and Combustible Liquids*

NFPA 385, *Standard for Tank Vehicles for Flammable and Combustible Liquids*

NFPA 386, *Standard for Portable Shipping Tanks for Flammable and Combustible Liquids*

NFPA 395, *Standard for the Storage of Flammable and Combustible Liquids at Farms and Isolated Sites*

CONDITIONS CONDUCIVE TO EXPLOSIVE ATMOSPHERES

Flammable vapors, which are released by flammable liquids, are responsible for flame propagation and explosions. Therefore, storage and handling procedures that prevent vapor release and minimize the amount of liquid exposed to the air are of prime importance in preventing explosive atmospheres.

When flammable vapor-air mixtures explode, they usually do so in a confined space such as a building, room, or container. The confined space is required to allow a high enough concentration of vapor to collect in order for an explosion to occur. The violence of the explosion depends upon the quantity of the vapor-air mixture, the concentrations of the vapors, and the type of enclosure containing the mixture. Explosions occurring with the vapor-air mixture near the extreme limits of the flammable range have been found to be less severe than those occurring near the middle of the flammable range.

If flammable and combustible liquids should ignite, extinguishing techniques involve shutting off the fuel supply, excluding the air, cooling the liquid to prevent evaporation, or applying a combination of these techniques. These extinguishing techniques are basically the same as preventive measures.

Storage of Flammable and Combustible Liquids

Standards and specifications concerning proper storage practices are set forth by such organizations as the American Petroleum Institute (API), the National Fire Protection Association, Underwriters Laboratories, Inc., and model building or fire codes. When evaluating such requirements, fire inspectors must remember that the size of the tank is relatively unimportant compared to the characteristics of the liquid being stored, the design of the tank, foundations and supports, size and location of vents, and piping and connections used throughout the installation.

There are several different types of storage receptacles used for flammable and combustible liquids. *Containers* have a storage capacity of 60 gallons (240 L) or less (Figure 10.11). *Portable tanks* are larger than 60 gallons (240 L) and are not intended for a fixed installation (Figure 10.12). The term *storage tank* refers to a vessel greater than 60 gallons (240 L) and in a fixed location (Figure 10.13). All portable tanks that have a

Figure 10.11 Flammable liquids come in a variety of containers.

Figure 10.12 A common style of portable flammable liquid tank.

Figure 10.13 Large flammable and combustible liquids storage tanks can hold several million gallons (liters) of fuel.

storage capacity greater than 660 gallons (2 640 L) should be treated according to the same standards as fixed storage tanks.

STORAGE TANKS

Storage tanks can range from small tanks holding several hundred gallons (liters) to huge tanks holding millions of gallons (liters) of liquid. There are three basic classifications of storage tanks: aboveground, underground, and those located inside buildings. All types of tanks should be installed in accordance with NFPA 30.

Aboveground Storage Tanks

Aboveground storage tanks for flammable and combustible liquids are classified in three categories, depending upon the pressures for which they are designed. Atmospheric tanks are designed for pressures of 0 to 0.5 psi (0 to 3 kPa), low-pressure storage tanks for pressures of 0.5 to 15 psi (3 to 103 kPa), and high-pressure vessels for pressures greater than 15 psi (103 kPa) (Figures 10.14 a through c).

Aboveground tanks should be constructed of steel or concrete unless conditions warrant using some special material. The thickness of the tank walls depends upon the weight and corrosiveness of the liquids the tank will hold. If corrosion is a severe problem, the tank may have a special coating or lining. Some tanks are also constructed to be especially resistant to bullets or vandalism.

The inspector should check to ensure that aboveground tanks are placed on sturdy foundations and are provided with adequate supports. Exposed supports under fixed tanks must be protected by fire-resistant materials with ratings of two hours or more (Figure 10.15). Tanks should be located slightly above the normal ground level in order to keep water from accumulating around

Figure 10.14a An open-top floating roof tank is one example of an atmospheric pressure tank.

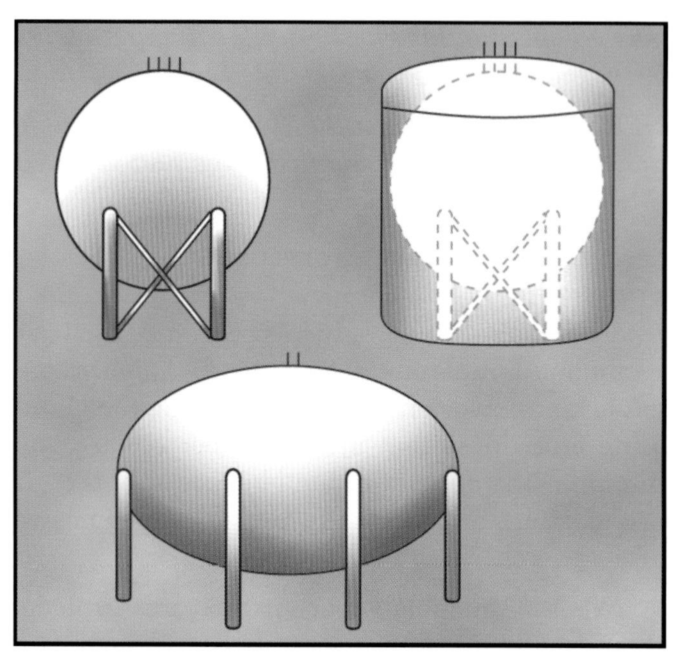

Figure 10.14b Several examples of low-pressure flammable liquids storage tanks.

Figure 10.14c A high-pressure storage tank.

Figure 10.15 In many cases raised horizontal tanks rest on exposed steel supports. *Courtesy of Conoco Oil Co.*

them. The specified distances from tanks to items such as property lines and public ways depend upon the size and design of the tanks, their protection from exposure, available fire control resources, the stability of the liquids that the tanks contain, and the internal pressure of the tanks. These distances should be in accordance with the requirements set forth by the local building codes, fire codes, planning codes, or the authority having jurisdiction.

Fire inspectors must also carefully consider the spacing between tanks. The spacing between two tanks should be at least 3 feet (1 m) (Figure 10.16). For tanks containing liquids classified as Class I, Class II, or Class IIIA, the distances between the tanks must be at least one-sixth the sum of their diameters. The distances will vary depending on the stability of the liquid. For those tanks containing unstable flammable or combustible liquids, the distances must be at least one-half the sum of their diameters.

Certain tanks should have additional spacing so that inside tanks are accessible for fire fighting operations. These tanks include those that are arranged in an irregular fashion or in rows of three or more, tanks that contain Class I or Class II liquids, and tanks that are in the drainage route of Class I or Class II liquids.

When liquefied petroleum gas (LPG) containers are located in the same area as flammable or combustible liquid storage tanks, the gas containers and storage tanks must be at least 20 feet (6 m) apart (Figure 10.17). Fire inspectors should also make sure that measures are taken to prevent the flammable or combustible liquids from accumulating under the gas containers. If the liquid storage tanks are enclosed by dikes, any nearby liquefied petroleum gas containers must be on the outside of the dike and at least 10 feet (3 m) from the wall of the dike.

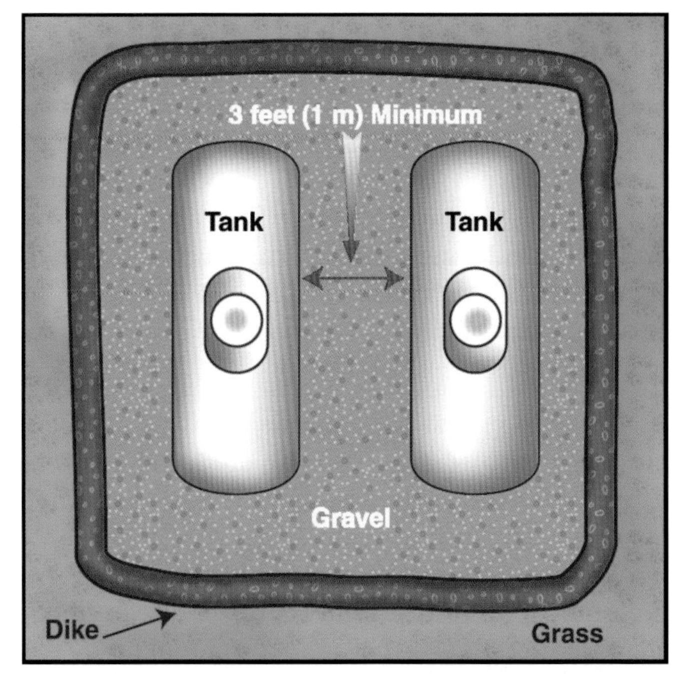

Figure 10.16 Horizontal tanks should be at least 3 feet (1 m) apart.

Figure 10.17 Large LPG tanks must be at least 20 feet (6 m) away from other flammable and combustible liquid tanks.

In large storage applications, one of the most important considerations is the properly designed storage vessel. The vessel must be a container that is liquidtight and designed to carefully control the release of vapors. In addition to normal venting, most aboveground tanks require emergency venting to prevent a BLEVE.

A BLEVE occurs when excessive pressure inside a container causes the container to explode and rupture. A fire around or under a tank can cause the liquid in the tank to vaporize, resulting in high internal pressures. If the tank does not have adequate venting to allow the vapor to escape and burn at the vents, it will rupture and explode (Figure 10.18). It is also possible for a BLEVE to result if the steel in the vapor space is softened by heat and fails.

All tanks must have normal venting to allow air or vapors to flow into or out of the tanks during filling or emptying operations. Vents that are clogged, damaged, or otherwise too small may cause the tank to rupture because of internal pressure. An internal vacuum can even cause the tank to collapse. The inside nominal diameter of a vent is 1¼ inches (32 mm) minimum, with the size dependent on the size of the tank. Containers that have more than one fill or discharge connection must have vents whose sizes are based on maximum possible simultaneous flow. Vent discharges must be arranged to prevent flame impingement on any part of the tank or nearby tanks in case the vapors exiting the vent ignite. Vents located near buildings or public places must be 12 feet (4 m) from the ground and arranged so that vapors are released outside and in a safe area. Vent pipe manifolding should be avoided unless it is required for recovering vapors, conserving vapors, or controlling air pollution. If vent pipe manifolding is used, pipe sizes must be large enough to allow for the maximum possible simultaneous flow from all connected tanks.

Figure 10.18 Overpressurization of a tank could lead to a BLEVE.

All aboveground storage tanks must have some device or inherent design that will relieve any excess pressure caused by nearby fires (Figure 10.19). The only exceptions to this requirement are tanks storing Class IIIB liquids in quantities exceeding 12,000 gallons (48 000 L) that are not in a diked area with, or in the path of, Class I or Class II liquids. Fire inspectors should make sure that the tanks have emergency relief provisions such as loose manhole covers that will rise under pressure, rupture disks, a weak seam between the roof and shell of the tank, or conventional emergency relief devices designed for the specific application. If a tank has emergency relief provisions that only relieve pressure, it must have normal and emergency vents whose total venting capacity is sufficient to prevent the tank from rupturing. All commercial venting devices should be stamped with the opening pressure, the pressure at which the valve reaches the full open position, and the accompanying flow capacity in cubic feet per hour (cubic meters per hour).

Figure 10.19 Storage tanks must have adequate relief vents.

Besides emergency relief provisions, all tanks or tank areas storing Class I, II, or IIIA liquids must have dikes or some form of drainage to prevent any leaks or accidental discharges from endangering nearby facilities, property, or waterways (Figure 10.20). If dikes are used, the volume of the diked area should be large enough to contain the entire amount that could be released from the largest tank. The walls of the dike should be liquidtight and constructed of steel, concrete, solid masonry, or earth. Dikes that are made of earth and that are 3 feet (1 m) high or more must have a 2-foot (0.6 m) wide flat section at the top (Figure 10.21). Loose combustible materials, empty or full drums, and other items must be kept out of the diked area.

If drainage is used, there should be a slope of at least one percent toward the drainage system, and the basin of the system should be large enough to hold the entire contents of the largest tank. Furthermore, the drainage system must not empty onto nearby property or into water supplies, public drains, or public sewers.

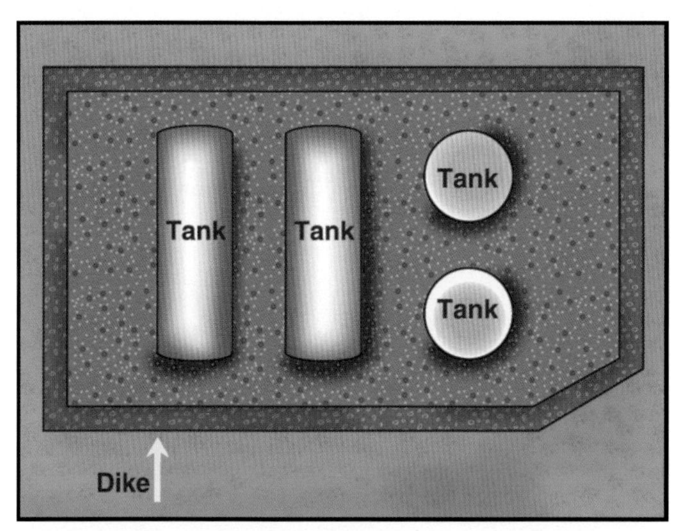

Figure 10.20 Dikes should completely surround storage tanks.

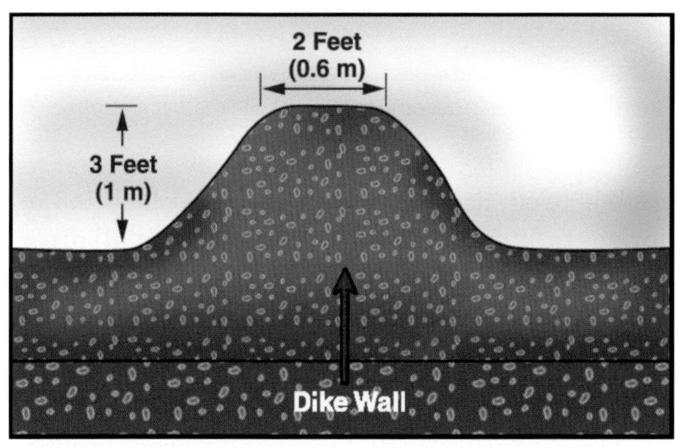

Figure 10.21 The dike should be at least 3 feet (1 m) tall and 2 feet (0.6 m) wide at the top.

Fire inspectors must also check the connections for the tanks. All connections through which liquids may flow should have valves located as closely as possible to the tanks (Figure 10.22). Those connections that are located below the liquid level of the tanks and through which liquids normally do not flow should have a liquidtight closure. For tanks containing Class I liquids, gauging openings that must be made and broken must be located outside buildings and away from any ignition source. The connections should be at least 5 feet (2 m) from any building opening.

Figure 10.22 Valves should be located as close to the tanks as possible.

Underground Storage Tanks

From a fire protection standpoint, underground tank storage is perhaps the safest form of storage for flammable and combustible liquids if the tanks are designed correctly, installed safely, and maintained properly. This type of storage is most commonly found in bulk and retail fuel storage facilities, car rental agencies, and truck depots (Figure 10.23).

Underground tanks must be designed for their intended use. The designer must consider how much pressure the surrounding earth or pavement will exert and whether vehicles will cross over the tanks. The designer/installer must also consider the level of the water table in the area where the

Figure 10.23 Underground tanks are commonly found at service stations.

Figure 10.24 Tanks must be gently placed in the ground to avoid damage.

tank will be buried. Tanks that are not properly installed in high water table areas have been known to float up and out of the ground. This causes damage to the tank and surrounding area, and usually results in a fuel leak. The tanks may be made of metal, fiberglass, unlined concrete (for liquids with a specific gravity of 40° API or greater), or lined concrete (for liquids with a specific gravity lighter than 40° API).

If proper precautions are taken, underground storage tanks may be buried almost anywhere. The tanks must be placed on a firm foundation and surrounded with at least 6 inches (150 mm) of noncorrosive, inert materials, such as clean sand, earth, or gravel, that have been well tamped into place. The tank must be placed in the ground carefully because dropping or rolling the tank can puncture it, break a weld, or scrape off the protective coating if it has one (Figure 10.24). Underground tanks that are not subject to vehicular traffic must be covered with at least 2 feet (0.6 m) of earth, or at least 1 foot (0.3 m) of earth plus 4 inches (100 mm) of reinforced concrete.

If underground tanks are likely to have vehicles passing over them, they must be protected against damage by at least 3 feet (1 m) of earth, or 1.5 feet (0.5 m) of well-tamped earth plus 6 inches (150 mm) of reinforced concrete or 8 inches (200 mm) of asphaltic concrete (Figure 10.25). If concrete is used, it must extend at least 1 foot (0.3 m) horizontally from the edge of the tank in all directions (Figure 10.26). Any piping that may be damaged must be protected with sleeves, casings, or flexible connections that will absorb any vibrations or impacts.

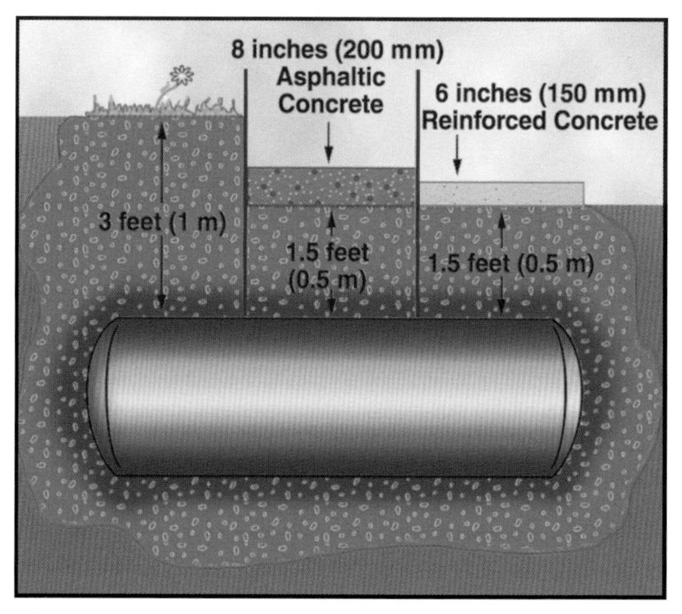

Figure 10.25 The required depth of the tank is dependent on the material that will cover it.

The vents used in underground storage must be large enough to prevent any liquid or vapor from

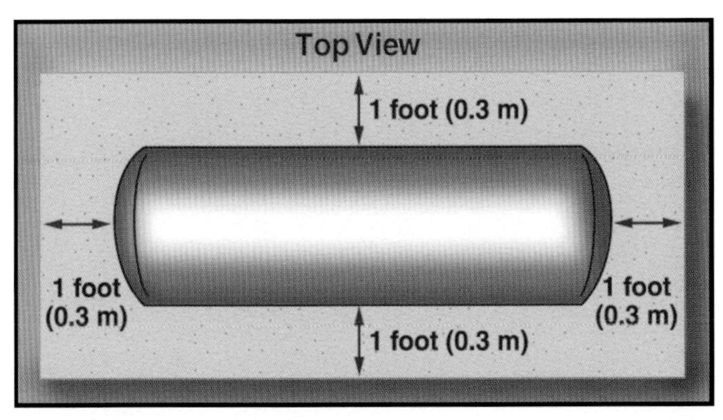

Figure 10.26 The concrete pad above the tank must extend at least 1 foot (0.3 m) beyond all sides of the tank.

being blown back toward the opening where the tank is filled. The vents should have at least a 1¼-inch (32 mm) nominal inside diameter. Equipment for recovering vapors must not obstruct or restrict the piping for vents unless the tanks and associated components have protection that prevents back pressure from developing. Vents should also be protected so that dirt, debris, and insect nests will not block them. The vents shall be equipped with flame arresters and must be located so that vapors are not discharged to an unsafe area.

Vents for Class I liquids must discharge outside buildings, must be higher than the fill pipe opening, and must be at least 12 feet (4 m) above the adjacent ground level (Figure 10.27). Vents for Class II and IIIA liquids also must discharge outside buildings, be higher than the fill pipe opening, and be placed above the normal snow level.

Vent pipes should be protected from physical damage. This can be accomplished by placing them in an inaccessible location or by constructing a barrier around them. The vent pipes must enter the tanks through the top and must be sloped so that they drain toward the tanks. When the vents are manifolded, the pipe sizes must be large enough to allow a discharge within the pressure limitations of the system. Storage facilities for Class I liquids may not have manifolding vents with storage facilities for Class II or Class III liquids unless provisions are made to keep vapors from Class I liquids from contaminating the Class II or Class III tanks.

In addition to vent openings, underground tanks will have openings for fill and discharge lines, gauging, and vapor recovery (Figure 10.28). All

Figure 10.27 Underground tanks require vent pipes.

Figure 10.28 The location of the underground tank is usually signified by the presence of fill openings.

connections for these openings must be vaportight. Any openings to allow the tank to be gauged manually must have liquidtight caps or covers if the openings are separate from the fill pipes. Fill pipes for Class IB and IC liquids must terminate within 6 inches (150 mm) of the tank bottom to minimize the possibility of generating static electricity. Filling, discharge, and vapor recovery connections that are made and broken for Class I, II, and IIIA liquids must be located outside buildings at least 5 feet (1.5 m) from the nearest building openings and in an area free of ignition sources.

Leakage from underground tanks is a major concern because water supplies may be contaminated, the environment damaged, and explosions may occur as a result of accumulated vapors. Leaks as small as one drop per second can result in a loss of 34 gallons (136 liters) in a month. Furthermore, a 1⁄16-inch (1.5 mm) stream can leak 2,520 gallons (10 800 L) per month, and a 1⁄4-inch (6 mm) stream can leak 28,080 gallons (106 300 L) per month.

Perhaps the simplest and most economical method of detecting leaks is to be sure that inventory is stringently controlled, taking into account unavoidable losses from evaporation, product shrinkage, and improper meter calibration. Testing of tanks and related piping shall be done in accordance with NFPA 30 and EPA regulations.

Soils containing fill such as cinders, shale, construction debris, or other materials may be highly corrosive. Tanks installed in these types of soils should have protective coatings, wrappings, cathodic protection, or be made of materials that resist corrosion.

Problems can occur if different metals are used for tanks and piping. In this case, a chemical reaction known as electrolytic corrosion occurs where the two metals are joined. This corrosion will result in a failure of the connection. Tanks and piping should be constructed of the same materials.

Abandoned tanks can be safeguarded in one of two ways. The first method for safeguarding tanks involves removing them from the ground. All liquid must be removed from the tank lines. All inlets, outlets, and underground piping must be capped. When tanks are to be removed, an approved method of inerting, removal, and disposal must be used. The second method, inerting, involves leaving the tank in the ground permanently. In this case, the fill line, gauge opening, pump suction, and vent lines are disconnected. The underground piping must be capped. The tank is then filled with an inert, solid material.

Inside Storage Tanks

The most common types of tanks found inside buildings are those used to store fuel oil and drain or waste oil (Figure 10.29). Tanks of less than 660 gallons (2 640 L) require no special fire protection features. If amounts greater than 660 gallons (2 640 L) are to be stored inside the structure, they must be placed in a fire-resistant room or enclosure so that they are isolated from the rest of the building. This enclosure must be in accordance with NFPA 30. A raised noncombustible sill or liquidtight ramp is needed at the opening to prevent the liquid from flowing into the structure if a leak should occur.

Figure 10.29 Fuel may be stored inside buildings in simple metal barrels.

Tanks containing Class I, Class II, or Class IIIA liquids must have overflow prevention devices such as float valves, fill-line meters, low heat pumps, or liquidtight overflow pipes that are at least one pipe size larger than the fill pipe. The vents and fill pipes must terminate on the outside of the building. Openings provided for vapor recovery equipment must be protected against vapor release by dry break connections or spring-loaded check valves.

Tanks used to store flammable or combustible liquids inside a building have the same wall thickness and design requirements as tanks used for outside storage. Inside tanks do require several additional safeguards, however. They must have automatic-closing, heat-actuated valves for each connection that is used to withdraw liquid and that is located below the liquid level. Such valves prevent liquids from flowing out of the tanks in the event of pipe rupture. Inside tanks must also have ventilation equipment to remove flammable vapors from within the enclosure. Using automatic sprinkler systems to protect the tanks and associated piping should be carefully considered. Manual gauging openings and other tank openings must have vaportight caps or covers.

CONTAINER AND PORTABLE TANK STORAGE

The most common form of flammable and combustible liquid storage vessels encountered are containers and portable tanks. Virtually every occupancy that the inspector enters has some type of these vessels in it. The following sections will take a look at the vessels themselves as well as proper methods of storage.

Containers for storing flammable and combustible liquids come in several forms, including glass containers, metal drums, safety cans, and polyethylene containers (Figures 10.30 a through d). Safety cans should be UL or FM approved. Portable tanks must conform to Chapter 1, Title 49 of the *Code of Federal Regulations* (DOT Regulations) or NFPA 386, *Standard on Portable Shipping Tanks for Flammable and Combustible Liquids*. Table 10.4 highlights the maximum amounts of each

Figure 10.30a Some flammable and combustible liquids are stored in glass containers.

Figure 10.30b A flammable liquid drum.

Figure 10.30c Safety cans are one of the safest types of storage containers.

Figure 10.30d Some fuel cans are constructed of plastic.

TABLE 10.4
Maximum Allowable Size of Containers and Portable Tanks

Liquids	Flammable Liquids			Combustible	
Container Type	Class IA	Class IB	Class IC	Class II	Class III
Glass	1 pt	1 qt	1 gal	1 gal	5 gal
Metal (other than DOT drums) or Approved Plastic	1 gal	5 gal	5 gal	5 gal	5 gal
Safety Cans	2 gal	5 gal	5 gal	5 gal	5 gal
Metal Drum (DOT Specification)	60 gal	60 gal	60 gal	60 gal	60 gal
Approved Metal Portable Tanks and IBCs	660 gal	660 gal	660 gal	660 gal	660 gal
Rigid Plastic IBCs (UN 31H1 or 31H2) and Composite IBCs (UN 31HZ1)	Not Permitted	Not Permitted	Not Permitted	660 gal	660 gal
Polyethylene DOT Specification 34, UN 1H1, or as authorized by DOT Exemption	1 gal	5 gal	5 gal	60 gal	60 gal
Fibre Drum NMFC or UFC Type 2A; Types 3A, 3B-H, or 3B-L; or Type 4A	Not Permitted	Not Permitted	Not Permitted	60 gal	60 gal

classification of liquid that can be stored in each type of container.

The inspector will most often encounter flammable or combustible liquids stored in containers that are 5 gallons (20 L) or less in size. Although there are other acceptable methods of storing these small amounts, as the chart indicates, the safest containers are approved safety cans. The maximum allowable size for approved safety cans is 5 gallons (20 L). Safety cans are constructed to reduce the chance of leakage or container failure. They are also designed to virtually eliminate vapor release from the container under normal conditions. Safety cans use self-closing lids with vapor seals and contain a flame arrester in the dispenser opening (Figures 10.31 a and b). The self-closing lid also acts as a pressure relief device when the can is heated. The inspector should be aware that not all containers that are painted red are approved safety cans; nonapproved safety cans are sometimes red as well.

Portable tanks must be designed with an emergency venting device to limit internal pressures to 10 psig (70 kPa), or 30 percent of the bursting pressure of the tank, whichever is greater. The total venting capacity of both normal and emergency vents should prevent rupture of the tank. If fusible vents are used, they must be designed to operate at a temperature not exceeding 300°F (149°C).

Figure 10.31a Safety cans often have a pouring spout.

Figure 10.31b Spark arresters are located within safety can openings.

Inside Storage of Containers and Portable Tanks

The amount of flammable and combustible liquids that can be stored in an occupancy depends upon the occupancy classification and storage conditions. Dwelling occupancies and buildings containing three or fewer dwelling units should not store more than 25 gallons (100 L) of Class I or Class II liquids, or 60 gallons (240 L) of Class IIIA liquids. Assembly occupancies and buildings containing more than three dwelling units may store no more than 10 gallons (40 L) of Class I and Class II liquids, or 60 gallons (240 L) of Class IIIA liquids. All liquids must be stored in approved cabinets and safety cans. Office, educational, and health care occupancies should store only the amount of flammable or combustible liquids needed for operation, maintenance, demonstrations, or treatment. In these facilities, the containers for Class I or Class II liquids must be approved safety cans holding no more than 2 gallons (8 L).

Mercantile occupancies and retail stores storing flammable or combustible liquids in areas accessible to the general public should store no more than the amount needed for displays or operating purposes (Figure 10.32). If the quantity of liquids

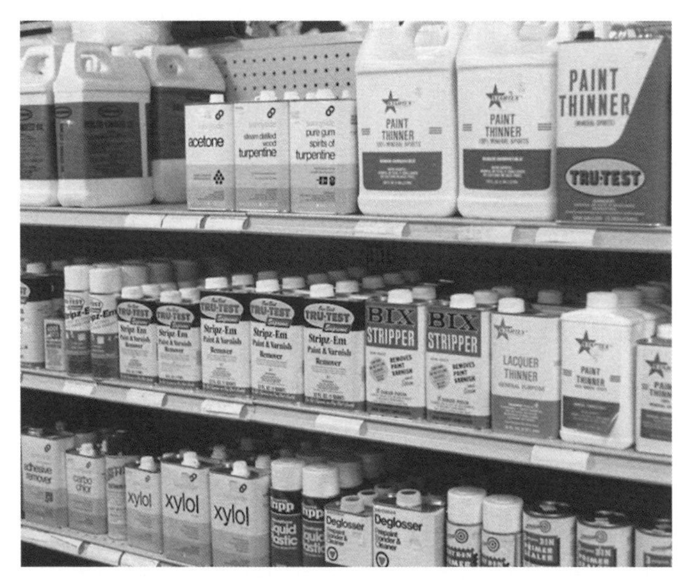

Figure 10.32 Mercantile occupancies should store only a reasonable amount of flammable and combustible liquids.

exceeds certain limits, the excess must be stored in a room approved for storing flammable liquids.

In order to further guard against accidents involving the storage of small quantities of flammable and combustible liquids, it is important that containers and portable tanks be stored in appropriate storage cabinets or rooms. Storage in these places will further protect the liquids from dangerous exposure to fire conditions should they occur.

Storage Cabinets

Small quantities of flammable or combustible liquids in normal operating areas should be stored in approved storage cabinets (Figure 10.33). Storage cabinets shall be designed in accordance with NFPA 30. They must be able to limit the internal temperature at the center of the cabinet — 1 inch (25 mm) from the top — to not more than 325°F (163°C) when subjected to a 10-minute fire test. This fire test employs burners simulating a room fire exposure using the standard time-temperature curve. All joints and seams should remain tight and the door should remain securely closed during the fire test. The cabinet is not required to be vented for fire protection purposes.

Figure 10.33 Flammable liquid containers may be stored inside a flammable liquids storage cabinet.

Storage Rooms

To be safe, facilities storing containers of flammable or combustible liquids should have rooms that are specially constructed of fire-resistive materials in order to protect the containers from any nearby fire exposure. Any openings from these rooms into other areas should have a raised sill that is liquidtight and composed of noncombustible materials in accordance with NFPA 30 (Figure 10.34). These openings also must have approved, self-closing fire doors. The joint where the walls meet the floor must be liquidtight to prevent any leakage of spilled liquids into the main facility. Low-pressure steam, hot water, or electrical heating units should be used to heat the facilities. In areas where Class I liquids are stored, electrical installations must be approved for Class I, Division 2 locations as defined by NFPA 70. The electrical installations in areas containing strictly Class II or Class III liquids may be approved for general use. Inside storage rooms must have some form of ventilation, either gravity or mechanical (Figure 10.35).

The liquids must be dispensed through approved and listed pumps or self-closing faucets. If the Class I or II storage containers have a capacity

Figure 10.34 Rooms containing flammable liquids should have raised sills at the doors.

Figure 10.35 Ventilation systems are important in flammable and combustible liquids storage rooms. Also, note the explosionproof electrical fixtures in this room.

of 30 gallons (120 L) or more, they should not be stacked on each other. A clear aisle approximately 3 feet (1 m) wide must be maintained at all times.

If Class I liquids are being dispensed in a room, the room must have mechanical ventilation. This ventilation must provide at least one cubic foot per minute of exhaust per square foot of floor area (1 m^3 per m^2) but not less than 150 cubic feet per minute (4 m^3). Air must be moved across the room to prevent any accumulation of flammable vapors. Furthermore, the exhaust from the room must exit the building completely and may not be recirculated. Recirculation is permitted only when the exhaust is continuously monitored by a fail-safe system that automatically sounds an alarm, stops recirculation, and provides full exhaust to the outside of the building if the vapor-air mixture exceeds one-fourth of the lower flammable limit. The dispensing area of a mechanical ventilation system must have an audible alarm to alert personnel if the ventilation system fails. The exhaust installation should follow the guidelines established in NFPA 91, *Standard for Exhaust Systems for Air Conveying of Materials*.

According to NFPA 30, areas where Class 1A or unstable liquids are stored or processed, explosion venting through one or more of the following methods should be in place:

- Open air construction
- Lightweight walls and/or roof
- Lightweight wall panels and roof hatches
- Explosion-venting windows

Outside Container Storage

Flammable and combustible liquids are often stored outside in portable containers such as drums. Fire inspectors should make sure the containers are located so that the threat of fire spreading from these containers is minimal. The storage area must have a fence or security guard to protect it against vandalism (Figure 10.36). The facilities must be kept free of weeds, debris, and other combustible materials that could ignite or spread an outside fire to the storage area. When storage facilities are located next to buildings, the exterior walls of the building must be constructed of noncombustible or approved limited-combus-

Figure 10.36 Outdoor storage of flammable and combustible liquid containers should be protected from vandalism and theft. *Courtesy of Ed Prendergast.*

tible materials having a fire-resistance rating of at least two hours. No opening is permitted within 10 feet (3 m) of the storage area. Storage areas must be graded so that spills have at least a 6-inch (150 mm) curb. If the storage area has a curb, there must be some provision for removing rain or spilled liquids that accumulate.

Fire inspectors should allow only containers that have been approved for the liquids that are placed in them. The containers should have one or more devices that provide sufficient venting capacity to limit the internal pressure of the containers to 10 psi (70 kPa) or 30 percent of the bursting pressure of the container. Each tank should have a pressure-actuated vent that operates at approximately 5 psi (35 kPa). If fusible venting devices are used, they should be designed to operate at a temperature less than 300°F (149°C). Drains in the storage area must end in a safe location and be accessible during fires. Quantities of more than 1,100 gallons (4 400 L) may not be located adjacent to buildings. Local fire or building codes usually contain tables giving the required spacing distances.

Handling, Transferring, and Transporting Flammable and Combustible Liquids

Assuming that equipment containing flammable and combustible liquids is properly maintained and that the tanks or containers do not allow the vapors to escape, fire inspectors must next be concerned with the handling, transfer, and transport of the liquids. It is through these actions that most problems occur. The inspector must always be observant for situations or conditions that pose a hazard. The occupants may demonstrate excellent practices when the inspector is present, but revert to their normal, less safe ways of doing things as soon as the inspector leaves. With a certain amount of experience, the inspector will be able to determine by observing the layout of the facility how normal operations are conducted.

There are several simple rules to follow to ensure safe handling of flammable and combustible liquids within a facility. Most of them apply to handling small amounts of the liquids because they are the most common amounts handled. Class I and Class II liquids should be kept in covered containers when not actually in use. When handling any flammable or combustible liquid that is not in a closed container, a way must be provided for safe disposal in case a leak or spill occurs. Class I liquids must not be used in the presence of any possible ignition source such as open flames, electrical arcs, or heating elements. In addition, Class I liquids cannot be stored in containers that are pressurized with air. In some circumstances, they may be stored in containers that are pressurized with an inert gas. Appropriate bonding and grounding procedures should always be followed.

Finishing operations are one of the most common processes inspectors assess. Many of the more common finishing processes use materials that can form flammable mixtures if they combine with oxygen. Such materials are usually applied by spraying, flow coating, hand brushing, dipping, or by using electrostatic equipment. Smoking *must* be prohibited in such areas. The electrical wiring and equipment must be of the appropriate classification for the particular hazard. Whenever possible, solvents used for cleaning equipment should have flash points greater than 100°F (38°C).

Several additional safeguards are necessary for paint finishing operations. First, adequate ventilation is needed to prevent the residues and vapors of the flammable liquids from accumulating. Fire inspectors should make sure that all potential ignition sources, including open flames and equipment that produce sparks, are removed. Only paints and solvents required for the immediate operation should be allowed in the area. Other paints and solvents may be stored in metal cabinets or special flammable liquids storage rooms. Because overspray residues may be a spontaneous heating hazard, they should be cleaned up regularly. See NFPA 33 and the building code for further fire suppression requirements related to paint finishing operations.

Operations that use spray booths also require some special safety procedures (Figure 10.37). Each spray booth should have an exhaust duct leading to the outside by the most direct route. These ducts should be constructed of steel and should not pass through combustible floors. Both the booths and

Figure 10.37 A typical paint spray operation. *Courtesy of Ed Prendergast.*

ducts should be separated from combustible walls and roof material by clearances that are equal to those required for metal stacks. The booths should be used to apply only combusible types of finishes because different finishes may chemically react and cause a fire or explosion. Overspray residues should be routinely removed from the booths and duct systems. All tools that are used for cleaning should be nonsparking.

Fire inspectors should consider the inherent hazards associated with dip tanks. All dip tank operations must be located on the ground floor of noncombustible buildings. The covers on the tanks should be kept closed when not in use. The covers should also be the self-closing type that activate if there is a fire. In addition, the tanks must have overflow pipes located approximately 6 inches (150 mm) below the top of the tank because tanks may overflow during a fire. Tanks must also have bottom drains so that the tank can be drained both manually and automatically if there is a fire. The use of nonflammable solvents should be recommended for consideration by the dip tank operation management whenever it is possible.

Accidents frequently occur during the transfer of liquids from one vessel to another. Although transferring can be hazardous regardless of the amount of liquid being transferred, the inspector should be particularly concerned with loading and unloading in bulk-handling operations. Loading and unloading stations for Class I liquids should be located no closer than 25 feet (8 m) from storage tanks, property lines, or adjacent buildings (Figure 10.38). Loading and unloading stations for Class II and Class III liquids should be located no closer than 15 feet (5 m) from these same objects. These stations should be constructed on level ground. Curbs, drains, natural ground slope, or other means are needed to keep any spills in the original area (Figure 10.39). Adequate ventilation, natural or mechanical, must be maintained.

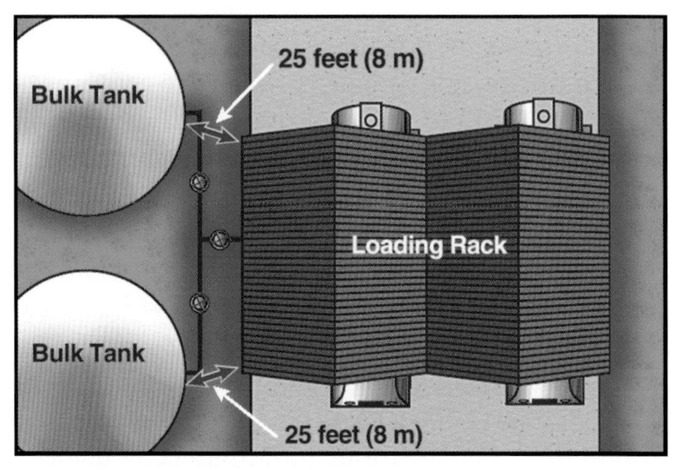

Figure 10.38 Loading racks should be at least 25 feet (8 m) from storage tanks.

Figure 10.39 Curbing around loading rack areas should help limit the spread of spilled fuels. *Courtesy of Conoco Oil Co.*

Several liquids, including light fuel oils, toluene, gasoline, and jet fuels can develop static electrical charges on their surfaces. If a static discharge occurs at the same time that a flammable mixture is near, a fire or explosion can result. To protect against static discharge, tanks must be bonded together with a metal chain or strap (Figure 10.40). Tanks also need to be grounded to neutralize static charges when Class I liquids are loaded and unloaded (Figure 10.41). Bonding and

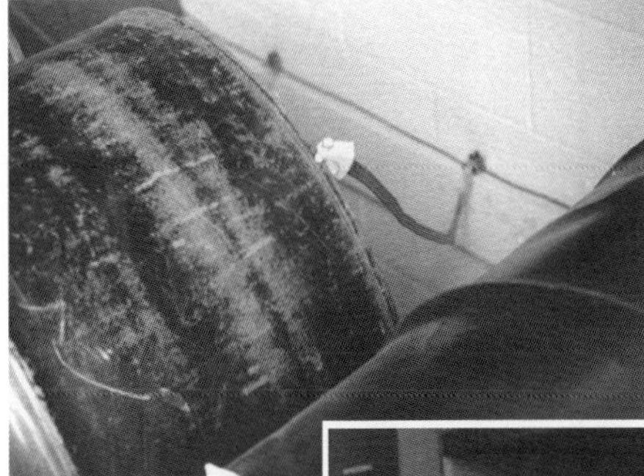

Figure 10.40 Containers should be bonded together in some manner.

Figure 10.41 Grounding is important during liquid transfer operations.

grounding are also necessary when Class II or Class III liquids are loaded into containers that have previously contained Class I liquids. All bonding connections must be completed before dome covers, lids, or caps are removed. Bonding is also needed between the dispensing device and the receiving containers.

Perhaps the safest procedure for handling flammable or combustible liquids, from a fire protection standpoint, is to pump them from underground tanks through a piping system to dispensing equipment. The dispensing equipment must be kept in specially designed rooms or outside. Piping materials must resist any corrosion from the material being transferred. Piping must also be able to withstand the maximum service pressure and temperature as well as thermal shock and physical damage.

All transfer systems must be designed so that liquids will not continue to flow by gravity or by siphoning if a pipe breaks. Control valves should be located at easily accessible points along the piping to control the flow. Fire inspectors should be sure that all piping and connections are inspected on a regular basis for signs of deterioration or leakage and replaced as necessary.

GASOLINE STATION VAPOR RECOVERY SYSTEMS (VRS)

Another common transferring system is the gasoline station vapor recovery system (VRS). Gasoline vapors are released into the atmosphere each time gasoline is dispensed into motor vehicle fuel tanks or service station bulk storage tanks. These vapors, when mixed in the right proportions with air, create a mixture that can ignite with explosive force. To reduce this fire hazard and to comply with the air pollution standards developed by the Environmental Protection Agency (EPA), many gas stations have begun using vapor recovery systems.

The gasoline vapors in a gasoline tank occupy whatever space is not filled with liquid. If additional gasoline is put into a tank, it will displace an equal volume of gasoline vapor. The displaced gasoline vapor will be released into the atmosphere (Figure 10.42).

As gasoline is pumped from a storage tank to a vehicle tank, a vacuum is created. Air then enters the storage tank through the tank vent and occupies the space formerly filled by the gasoline. The gasoline entering the vehicle tank displaces the vapors in the tank and the vapors are released into the surrounding atmosphere.

In a VRS, the top of the vehicle tank is connected to the top of the storage tank, creating a closed loop exchange of gasoline and vapor (Figure 10.43). A similar technique is used when the gasoline transport truck refills the storage tank (Figure 10.44).

Two types of VRS for gasoline service stations are currently in operation: balance systems and vacuum assist systems. Both systems work on the principle of a closed loop exchange of gasoline and vapors. The difference between the systems is the way in which the closed loop is accomplished.

The balance system has a highly specialized nozzle that provides a tight seal around the fill

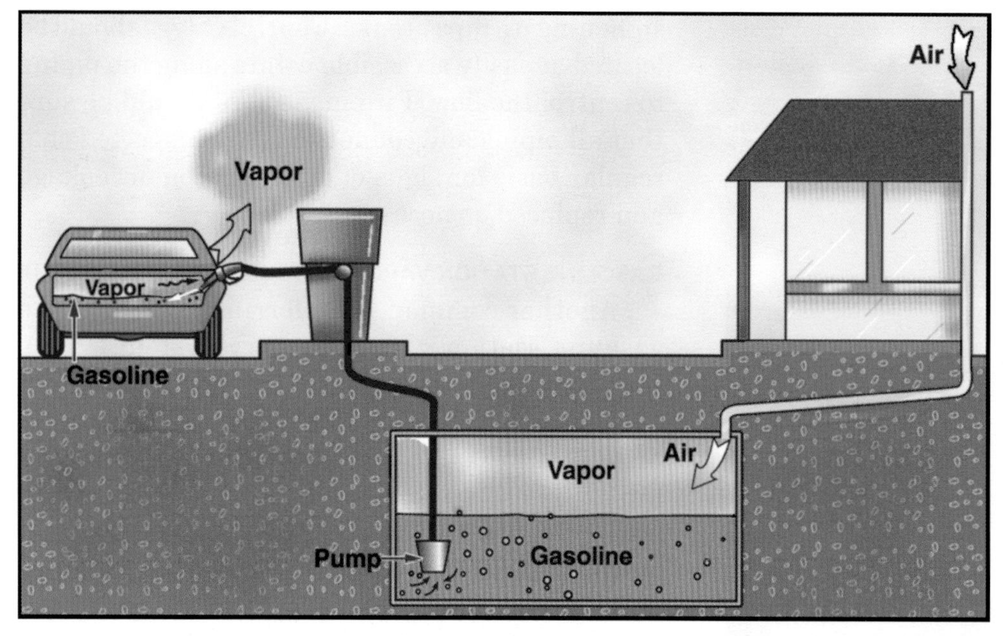

Figure 10.42 Systems not equipped with vapor recovery systems lose vapors to the atmosphere.

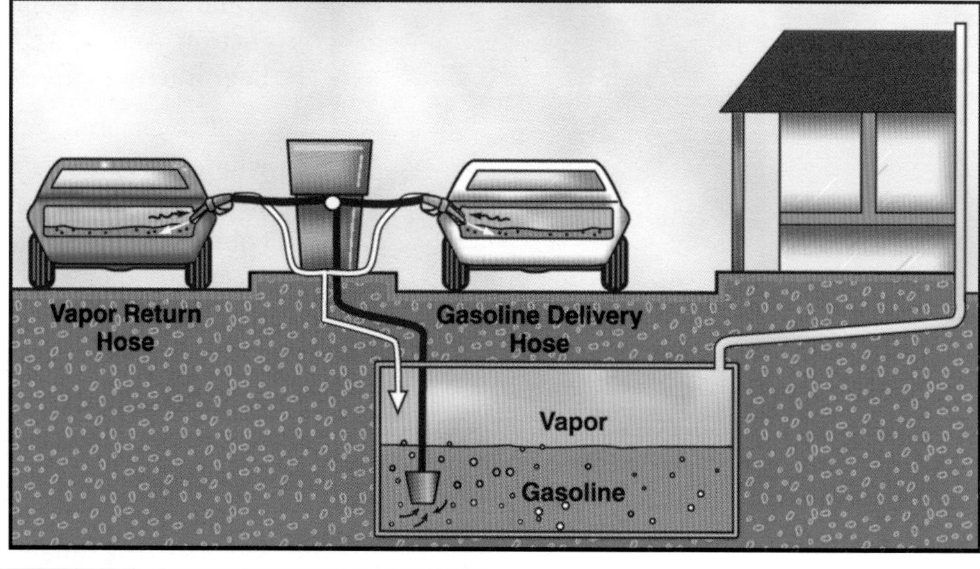

Figure 10.43 The VRS systems send vapors back to the storage tank.

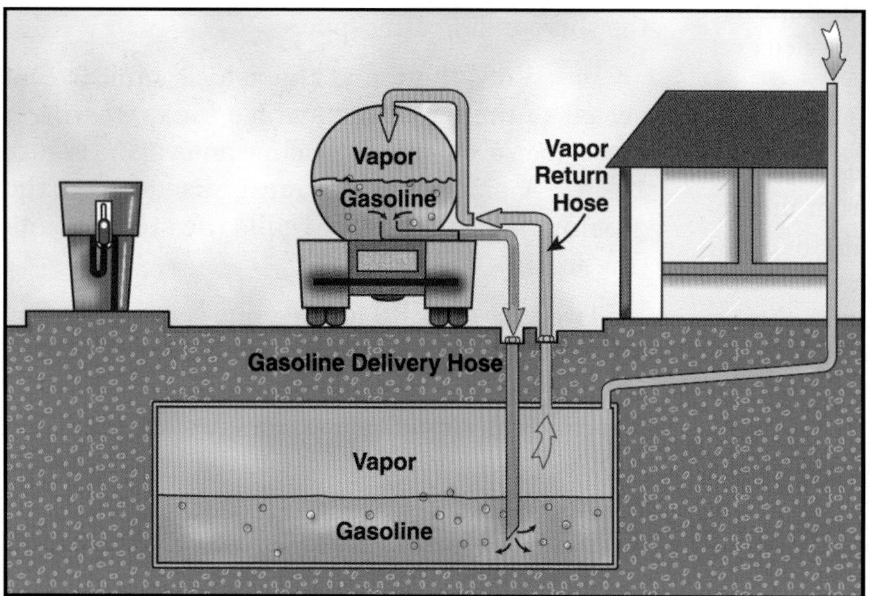

Figure 10.44 The VRS system is also used during underground tank filling.

neck of the vehicle's fuel tank. This special nozzle, which is larger, heavier, and more complex than an ordinary dispensing nozzle, has an interlock so that the nozzle must be pushed with force into the gasoline tank fill neck before the gasoline will flow (Figure 10.45). This interlock maintains a tight seal while the gasoline is being dispensed. Because of this tight seal, flammable vapors are not released into the atmosphere.

The vacuum assist system uses a small vacuum generator, which keeps the vapor return hose and piping under a slight negative pressure (Figure 10.46). If the dispensing nozzle and the gasoline tank fill neck are not tightly sealed, air is drawn into the vapor return hose. This keeps vapors from being released into the atmosphere. The dispensing nozzle used in the vacuum assist system is smaller, lighter, and less complicated than that used in the balance system because the tightness of the seal between the nozzle and the fill neck is not as crucial. Most vacuum assist systems have a small processor that eliminates vapors when the amount of vapors being released is greater than the amount of gasoline being exchanged.

VRS Inspections

Inspections made during the installation of a VRS are as important as inspections made while the VRS is in service. Moreover, a complete inspection during installation may save the service station owner time and money that would later be required to correct previously undetected problems.

During any inspection of a VRS, special emphasis should be placed on connections, piping, and nozzles and equipment. The dangerous accumulation of flammable vapors can be avoided by ensuring that connections and other fittings are secured and that no sags are present in the vapor return line. Due to the complexity of VRS, the inspector may find it necessary to research the particular systems in the area. Technical data about individual VRS may be obtained from the VRS manufacturer or the appropriate government agency.

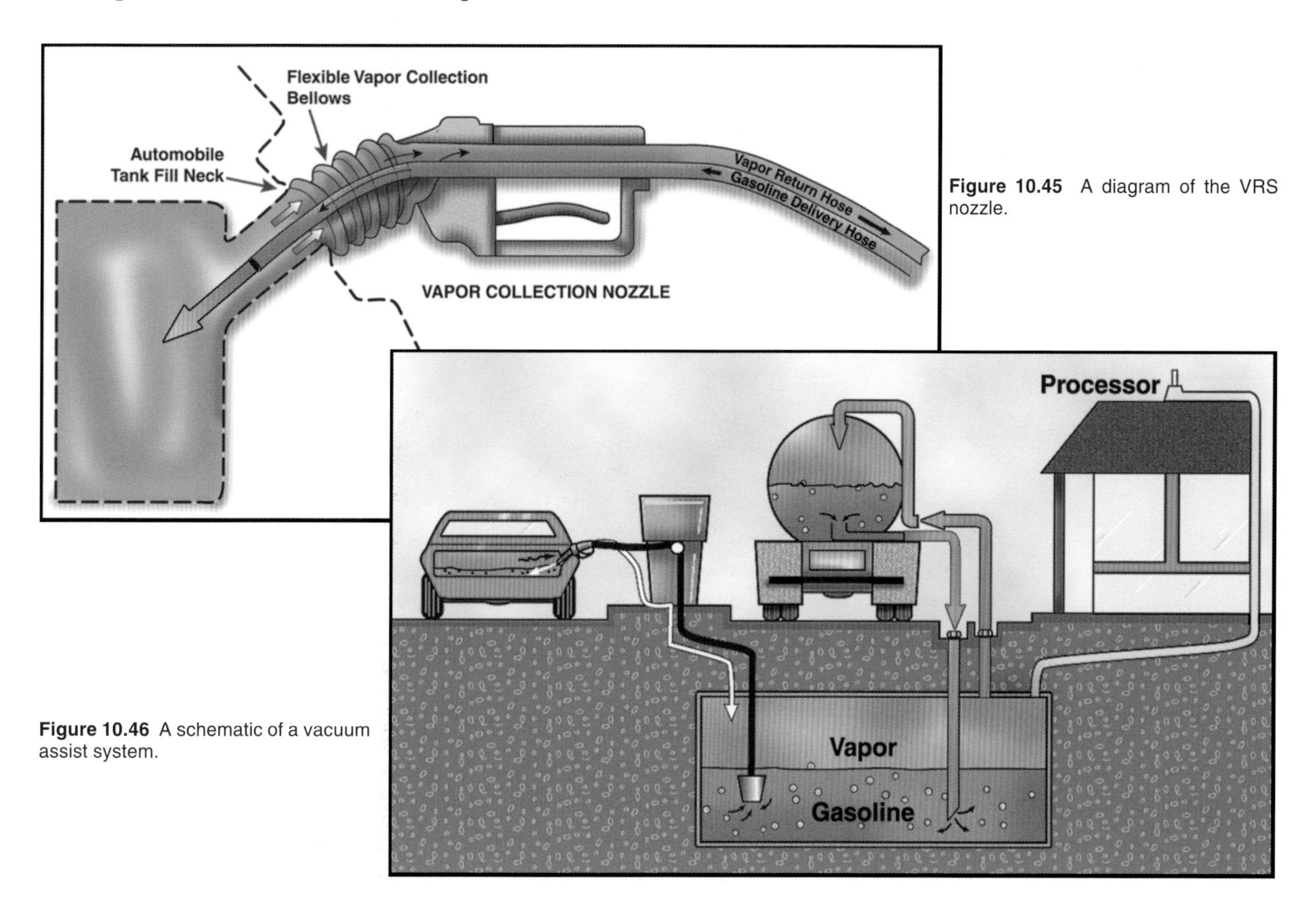

Figure 10.45 A diagram of the VRS nozzle.

Figure 10.46 A schematic of a vacuum assist system.

11

Storage, Handling, and Use of Other Hazardous Materials

This chapter contains information that will assist the reader in meeting the listed job performance requirements contained in NFPA 1031, ***Standard for Professional Qualifications for Fire Inspector and Plan Examiner*** (proposed 1998 edition).

Chapter 3 Fire Inspector I

3-3.8 Recognize hazardous conditions involving equipment, processes, and operations, given field observations, so that the equipment, processes, or operations are conducted and maintained in accordance with applicable codes and standards, and all deficiencies are identified, documented, and reported in accordance with the policies of the jurisdiction.

3-3.13 Verify code compliance for incidental storage, handling, and use of hazardous materials, given field observations, so that applicable codes and standards for each hazardous material encountered are properly addressed and all deficiencies are identified, documented, and reported in accordance with the policies of the jurisdiction.

Chapter 4 Fire Inspector II

4-3.6 Evaluate hazardous conditions involving equipment, processes, and operations, given field observations and appropriate documentation, so that the equipment, processes, or operations are installed in accordance with applicable codes and standards, and all deficiencies are identified, documented, and reported in accordance with the policies of the jurisdiction.

4-3.10 Verify code compliance for storage, handling, and use of hazardous materials, given field observations, so that applicable codes and standards for each hazardous material encountered are properly addressed and all deficiencies are identified, documented, and reported in accordance with the policies of the jurisdiction.

Chapter 11
Storage, Handling, and Use of Other Hazardous Materials

In addition to flammable and combustible liquids, fire inspectors must have a working knowledge of the full range of hazardous materials they may encounter. Inspectors should have at least a general knowledge of the materials in each of the U.N. classifications. Inspectors who are responsible for a jurisdiction that has a larger-than-usual presence of a particular material should take it upon themselves to become more familiar with that material.

This chapter is intended to provide the inspector with basic information on the characteristics, storage, handling, and use of hazardous materials other than the flammable and combustible liquids that were covered in Chapter 10. For more detailed information on hazardous materials and handling emergencies in which they are involved, see IFSTA's **Hazardous Materials for First Responders** or **Hazardous Materials: Managing the Incident** by Hildebrand, Noll, and Yvorra.

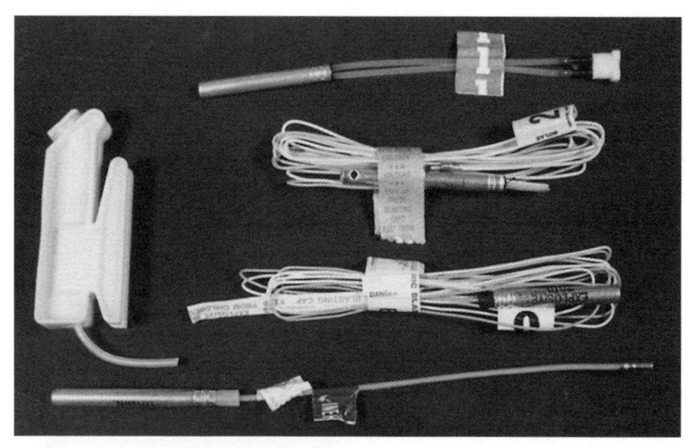

Figure 11.1a Detonation devices are one type of explosive. *Courtesy of the Institute of Makers of Explosives.*

Figure 11.1b Common types of explosives. *Courtesy of Atlas Powder Co.*

EXPLOSIVES

Explosive materials (U.N. Class 1) are used in military applications and in mining, logging, construction, fireworks displays, and demolition operations (Figures 11.1a and b). In the normal storage and transport states, explosive materials are relatively stable and can be handled safely. Explosives (except for old or damaged explosives) will not detonate if they are stored and handled properly. However, they must be protected from open flame, excessive heat, friction, impact, electrical shock, and chemical contamination. Explosives, particularly dynamite, may detonate if not handled properly after they have been stored improperly or abandoned for many years and have started to decompose. Decomposition is indicated by either a crystallized residue on the explosive or by the internal contents leaking through the exterior container. Explosives in this condition are very motion sensitive and require special handling. Only properly trained personnel, such as bomb disposal technicians, should handle, neutralize, or remove damaged or decomposed explosives. If inspectors encounter explosives in this condition, they should evacuate the area or structure immediately and notify the appropriate authority.

When fire inspectors consider storage provided for explosives, they must ensure that the explosives are protected from external sources of energy and that employees are protected in case of explosion. Regulations concerning the manufacture, distribution, and storage of explosives are issued by the Bureau of Alcohol, Tobacco, and Firearms (BATF), which is part of the U.S. Department of the Treasury. Additional regulations are set forth by NFPA 495, *Explosive Materials Code*, and may be enforced at the local level if they have been adopted as an ordinance.

General Storage Regulations for Explosives

There are some general storage regulations that apply to all explosives. No detonators, flammable or combustible liquids, or spark-producing tools or other metal implements shall be stored with other explosives. Piled oxidizers, such as ammonium nitrate, must be separated from readily combustible fuels (Figure 11.2). Cases of dynamite must be stored flat and rotated periodically (Figure 11.3). Powder kegs must be stored either on end (bungs down) or on the side (seams down). Containers of explosives must be stored according to corresponding grades and brands. They must be located so that the brand and grade marks are visible. At no time may small arms ammunition be stored in the same magazine (storage facility) with Class A or B explosives.

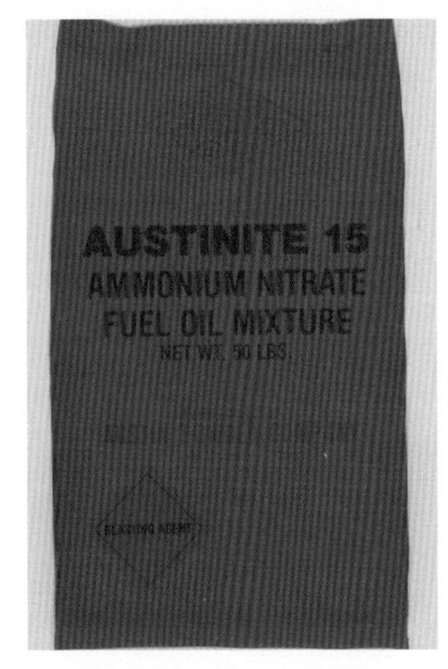

Figure 11.2 Ammonium nitrate and fuel oil mixture is one type of explosive. *Courtesy of Atlas Powder Co.*

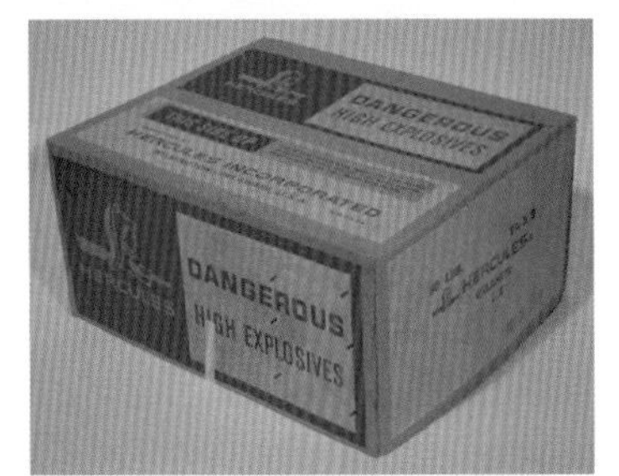

Figure 11.3 A case of dynamite.

Older stocks of explosives should be used or shipped first. Packages must be opened or repackaged at least 50 feet (15 m) from the magazine. Loose explosives and open packages are not permitted inside the magazine. Where artificial lighting is necessary, only approved safety flashlights or electric lanterns may be used. When magazines need to be heated, hot water radiant heat, forced air directed over hot water, or low pressure steam coils should be used. Smoking must be prohibited in or near the magazine, and employees should not carry matches, lighters, or other flame-producing devices. The interior of the magazine must be kept clean and the surrounding area should be clear of grass, undergrowth, leaves, and other flammable debris. Empty dynamite cases and powder kegs must be removed from the magazine area.

The BATF regulations and NFPA 495 contain the requirements for the five types of storage facilities or magazines. These requirements are described in the following sections.

TYPE 1 MAGAZINE

A *Type 1 magazine* is a permanent facility for the storage of explosive materials sensitive to initiation by a number 8 test blasting cap. These materials will mass detonate. These magazines are fire resistant, weather resistant, theft resistant, and bullet resistant (Figure 11.4). Materials stored in Type I magazines include dynamite or nonelectric blasting caps. Walls must be constructed of 8-inch (200 mm) blocks with the hollow spaces filled with well-tamped sand; brick or solid cement block construction 8 inches (200 mm) thick; wood construction covered with 26-gauge metal having ¾-inch (19 mm) plywood or wood sheathing with a 6-inch (150 mm) space between the exterior and interior sheathing and the space filled with dry sand; or 14-gauge metal lined with 4-inches (100 mm) of brick, solid cement block, or hardwood; or walls filled with 6 inches (150 mm) of sand. Doors must be constructed of steel plate ⅜-inch (10 mm) thick and lined with four layers of ¾-inch (19 mm) tongue and groove hardwood flooring or be constructed of a 14-gauge metal plate lined with 4

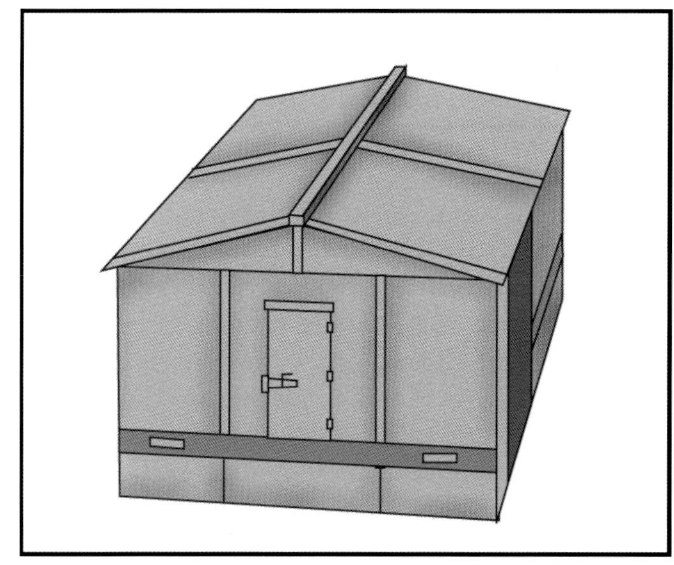

Figure 11.4 A Type 1 magazine.

Figure 11.5 A Type 2 magazine. *Courtesy of TREAD Corp.*

inches (100 mm) of hardwood. Roofs must be constructed of 14-gauge metal or ¾-inch (19 mm) wood sheathing covered by not less than 26-gauge metal or other noncombustible roofing material. All exposed wood must be covered by similar material.

TYPE 2 MAGAZINE

A *Type 2 magazine* is a portable or mobile magazine used for outdoor or indoor storage of the same materials found in a Type 1 magazine (Figure 11.5). These magazines are fire resistant, theft resistant, weather resistant, and bullet resistant. Magazines used strictly for indoor storage need not be bullet resistant. The construction requirements vary based on the end use of the magazine. Generally, the top, sides, bottom, and doors must be constructed of metal and lined with at least 4 inches (100 mm) of hardwood or similar bullet-resistive construction. Outdoor magazines must be supported to prevent direct contact between the floor and ground.

TYPE 3 MAGAZINE

A *Type 3 magazine* is a portable facility for the temporary storage of explosive materials under the constant attendance of a qualified employee. These facilities are bullet resistant, fire resistant, theft resistant, and weather resistant. Type 3 magazines must be secured by a five-tumbler padlock at all times. If the magazine is constructed of wood, the sides, bottom, and doors must be made of 4-inch (100 mm) hardwood and should be braced at all corners. The wood must then be covered with sheet metal of at least 20 gauge. If the magazine is constructed of metal, the sides, bottom, top, and doors must be made of 12-gauge metal lined with a nonsparking material.

TYPE 4 MAGAZINE

A *Type 4 magazine* is a permanent, portable, or mobile magazine for storing explosives that will not detonate when initiated by a number 8 test blasting cap. These materials will not mass detonate. These facilities are fire resistant, theft resistant, and weather resistant. Materials likely to be found in these magazines include blasting agents, various water gels, black powder, and smokeless powder. This type of magazine must be constructed of masonry, wood covered with metal, or fabricated metal. Doors must be metal or wood covered with metal.

TYPE 5 MAGAZINE

A *Type 5 magazine* is a permanent, portable, or mobile magazine for storing explosive materials that will not detonate when initiated by a number 8 test blasting cap. These facilities include tanks, tank trailers, tank trucks, semitrailers, bulk trucks, and bins. These magazines are theft resistant, and outdoor magazines are also weather resistant. They must be secured with at least one case-hardened five-tumbler lock. If the magazine is vehicular, it must be immobilized when unattended.

SECURITY MEASURES FOR EXPLOSIVES STORAGE

The Bureau of Alcohol, Tobacco, and Firearms (BATF) is responsible for enforcing Federal Explosive Laws under the Organized Crime Act of 1970.

Prior to purchasing and using explosives in the United States, individuals must complete a form (4710) from the BATF and have it processed under penalty of perjury. Unfortunately, when manufactured explosives are unavailable, individuals can easily obtain the necessary ingredients to compose homemade explosives, or they can steal explosives. This situation presents a real problem because these explosives may be used in acts of arson (incendiary bombs), terrorism, murder, or illegal entry. Thus, businesses that manufacture, store, or use explosives must have storage security programs.

Magazines should be supervised at all times by competent individuals who are responsible for enforcing all safety regulations. Someone should open and inspect the magazine at least every three days to make sure that nothing has been damaged or removed from the magazine unless authorized. The security officer should report any theft, loss, or unauthorized removal of explosive materials within 24 hours to the BATF, the police department, and the Bureau of Fire Prevention.

Handling and Use of Explosives

Explosives must never be abandoned or left unattended in the open. The tools used to open explosives packages should be approved, nonsparking tools. Empty packing materials, such as boxes, paper, or fibers, should be disposed of by remote burning.

When blasting is to be performed in congested areas or close to structures, railways, or highways, the safety of the general public and workers should be assured by using warning precautions that may include sirens or other signals. Before initiating an explosion, the person in charge should give a loud warning signal.

All operations involving the transportation of explosive materials must be conducted in accordance with the regulations of the Department of Transportation (DOT). Personnel involved in transporting explosives must not drive, load, or unload the vehicle recklessly. Further, individuals performing these duties may not smoke or carry matches or other flame-producing devices.

Before transferring explosives from one vehicle to another within a municipality, the appropriate personnel should notify the fire and police departments. Vehicles used to transport explosives should have a closed body or the load should be covered with a flame-retardant and moistureproof tarpaulin (Figure 11.6). Two fire extinguishers with a combined rating of 2-A:10-B:C should be carried on the vehicle. Explosives must not be transported over or through any prohibited tunnels, subways, bridges, roadways, or elevated highways. At no time should vehicles transporting explosives be left unattended, nor should they be driven within 300 feet (90 m) of each other. The drivers of these vehicles should also avoid congested traffic and densely populated areas when possible. In some localities, specified routes will be named for the carrier.

Upon receipt of explosives at a terminal, the carrier shall notify the fire inspection office of the completed shipment. The carrier shall also notify the consignees of the shipment's arrival. Once the explosives have reached the terminal, the consignee must remove them within 48 hours.

Figure 11.6 A specially designed explosives transport vehicle. *Courtesy of TREAD Corp.*

COMPRESSED AND LIQUEFIED GASES

Fire inspection personnel will find compressed and liquefied gases in numerous types of occupancies within their jurisdiction. These include:

- Retail stores
- Laboratories in educational facilities
- Medical facilities
- Welding shops or supply stores
- Industrial facilities
- Research labs

Depending on the type, compressed or liquefied gases can present a significant fire and safety hazard for the occupants and fire suppression crews. It is important for fire inspection personnel to ensure that gases are stored and handled in a proper manner to assure maximum safety for everyone involved.

Closely related to compressed and liquefied gases are cryogenic liquids. These special substances are covered later in this chapter.

Characteristics of Compressed and Liquefied Gases

Compressed gases are those materials which, at a normal temperature, exist as gases when pressurized within a container. In contrast, liquefied gases are ones which, at normal temperatures, exist in both liquid and gaseous states when pressurized in a container. Furthermore, this condition will continue as long as some amount of the liquid remains in the container.

Gases are often classified by their principal use. Fuel gases (which are, of course, flammable) are generally used to produce heat, light, or power. Examples of fuel gases are liquefied natural gas and propane (Figure 11.7). Industrial gases are used in heat treating, chemical processing, refrigeration, water treatment, welding, and cutting (Figure 11.8). Examples of industrial gases are methyl-ethyl-propane (MEP) gas, methyl acetylene propane (MAPP) gas, acetylene, freon, argon, oxygen, nitrogen, and chlorine. Medical gases are used in doctors' offices and hospitals for providing basic life support and anesthesia. Medical gases include oxygen, cyclopropane, and nitrous oxide. Some of the most toxic gases found in modern society are used in the manufacture of semiconductor materials. Examples of gases used in the semiconductor industry include arsine, phosphine, and diborane.

Figure 11.7 LPG and propane are commonly used for barbecues and portable stoves.

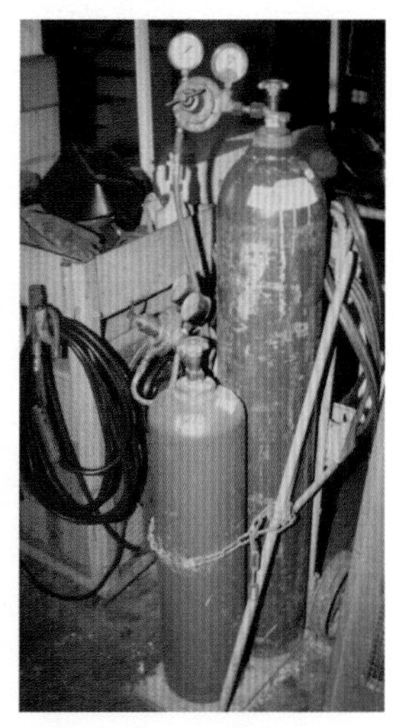

Figure 11.8 Acetylene is an industrial gas used for welding and cutting.

VAPOR DENSITY

Vapor density is used to evaluate the relative weights of gases in much the same way as specific gravity is used to evaluate liquids. Vapor density is the weight of a gas as compared to the weight of an equal volume of air. Air has a vapor density of 1.0. Gases with vapor densities less than 1.0 will rise and can be expected to concentrate near the ceiling or to form vapor clouds before they dissipate. Gases with a vapor density greater than 1.0 can be expected to lay near the ground or lowest point of a room, structure, or area (Figure 11.9). Materials with a vapor density close to 1.0 will mix with air and be found at all levels within the room. Table 11.1 shows the vapor densities of some common gases inspection personnel might encounter.

Figure 11.9 Substances with vapor densities greater than air will stay close to the ground.

TABLE 11.1
Common Vapor Densities

Product	Vapor Density
Ethyl Ether	2.6
Methane	0.6
Propane	1.6
Ethyl Alcohol	1.6
Acetone	2.0

TEMPERATURE/PRESSURE RELATIONSHIPS

According to *Charles' Law*, if the volume of a gas is kept constant and the temperature is raised, the volume will increase in direct proportion to the increase in temperature. This relationship is given by the formula:

$$\frac{V_1}{T_1} = \frac{V_2}{T_2}$$

where: V_1 = Original volume

T_1 = Original temperature

V_2 = Final volume

T_2 = Final temperature

A second important relationship is known as *Boyle's Law*. This law states that the volume of gas varies inversely with the applied pressure, if the gas is confined. This relationship can be shown by the formula:

$$P_1V_1 = P_2V_2$$

where: P_1 = original pressure

V_1 = original volume

P_2 = final pressure

V_2 = final volume

Therefore, doubling the applied pressure reduces the volume by one-half. Conversely, reducing the pressure by one-half results in the volume doubling. Thus, if a gas in a cylinder or in any other type of unfired pressure vessel is heated, the pressure within the vessel will also increase, causing pressure relief devices to activate. If heating continues and pressure is not relieved, the container will rupture.

Storage of Compressed and Liquefied Gases

Flammable gases are stored in a variety of cylinders and tanks. Flammable gases are also handled through gas pipelines. All these types of containers must be carefully designed, built, and maintained. Cylinders for flammable gases must be built according to specifications and regulations set forth by the U.S. Department of Transportation (DOT) in the United States or the Canadian Transport Commission (CTC) in Canada (Figure 11.10). These specifications cover metal composition, joining methods, wall thickness, heat treatments, marking, proof testing, type of openings required, and cylinder testing. The regulations also govern safety devices, in-service transportation, design pressure, and the gases that the cylinders may contain. Very small containers and those holding nonflammable cryogenics below 40 psi (280 kPa) are exempt.

Tanks used to store relatively small quantities of gas at moderate pressures are designed according to Section VIII (Unfired Pressure Vessels) of the *Boiler and Pressure Vessel Code* developed by the American Society of Mechanical Engineers (ASME). Tanks designed to hold large quantities of low-pressure gas are built according to standards set forth by the American Petroleum Institute (API) (Figure 11.11).

Inspectors should be aware of the techniques for safe installation and storage of compressed and liquefied gases. Careless handling of these materials may result in serious accident or death. Compressed and liquefied gases are most commonly stored in metal cylinders. The contents should be marked in large letters near the top of the cylinder (Figure 11.12).

Figure 11.10 There are government requirements for the construction of gas pressure cylinders.

Figure 11.11 A low-pressure storage tank.

Cylinders should be stored in a safe, dry, well-ventilated area specifically designed for storing compressed or liquefied gases. Cylinders should be stored valve end up on a level, noncombustible floor. This is to prevent materials from leaking from the cylinder and also because cylinders have thicker metal on the bottoms and are less susceptible to corrosion. Empty cylinders should be stored in a separate area from full cylinders. Incompatible materials should not be stored close to each other (for example, oxygen and flammable gases).

Figure 11.12 The contents of the cylinder may be stenciled near the top of the container.

If the cylinders are stored outside, they should be protected from adverse weather conditions such as accumulations of ice and snow in the winter and continuous direct sunlight in the summer. Gas cylinders are designed for temperatures no greater than 130°F (54°C).

Many jurisdictions have occupancies on which bulk storage of compressed and liquefied gases may be found. These include industrial facilities where the gases are used and storage facilities from which the gases may be supplied to other customers (Figure 11.13). At these locations, compressed and liquefied gases may be found in large storage tanks that range in size from 300 to 2,000,000 gallons (1 200 L to 8 000 000 L). Depending on the contents of the tank, specific codes may limit the size of the tank. For example, NFPA 1, *Fire Prevention Code,* limits the size of bulk LPG tanks to 120,000 gallons (480 000 L).

Figure 11.13 Many jurisdictions have occupancies on which bulk storage of compressed and liquefied gases may be found.

Inspectors should make sure that all bulk tanks are in good condition and have functioning overpressurization devices. Bulk tanks should have steel supports. Depending on the circumstances, these supports may need to be fire protected. Grass, weeds, and other combustible materials should not be within 10 feet (3 m) of the tank. Tanks that contain flammable or combustible products should be within a diked area and at least 10 feet (3 m) from the dike walls.

Inspectors should ensure that the following additional safety precautions are observed:

- *Absolutely No Smoking* in the area where cylinders are stored or used.
- Adequate ventilation is provided to prevent accumulation of flammable gas vapors.
- Electric lights are in a fixed position, enclosed in glass to prevent contact with gas, and the switch to the light is located outside the storage room.
- Glass light enclosures are equipped with a guard to prevent breakage.
- Cylinder valves are covered with safety caps when cylinders are not in use (Figure 11.14).
- Stored cylinders are anchored with chains around the tank so that they cannot accidentally be knocked over (Figure 11.15).
- Bulk cylinders may be nested, as opposed to anchored and chained (Figure 11.16).

HAZARDS ASSOCIATED WITH COMPRESSED AND LIQUEFIED GAS STORAGE

The primary hazard associated with the storage of compressed and liquefied gases is the high pressure caused by the storage itself. Compressing a gas into a container creates a tremendous source of potential energy; this energy is prevented from being released by the container itself. Normally, this stored energy is released in a controlled manner through a valve, dispensing instrument, or a

Figure 11.14 Cylinders that are not in use should be stored with their caps in place.

Figure 11.15 Cylinders should be secured upright by chains or racks.

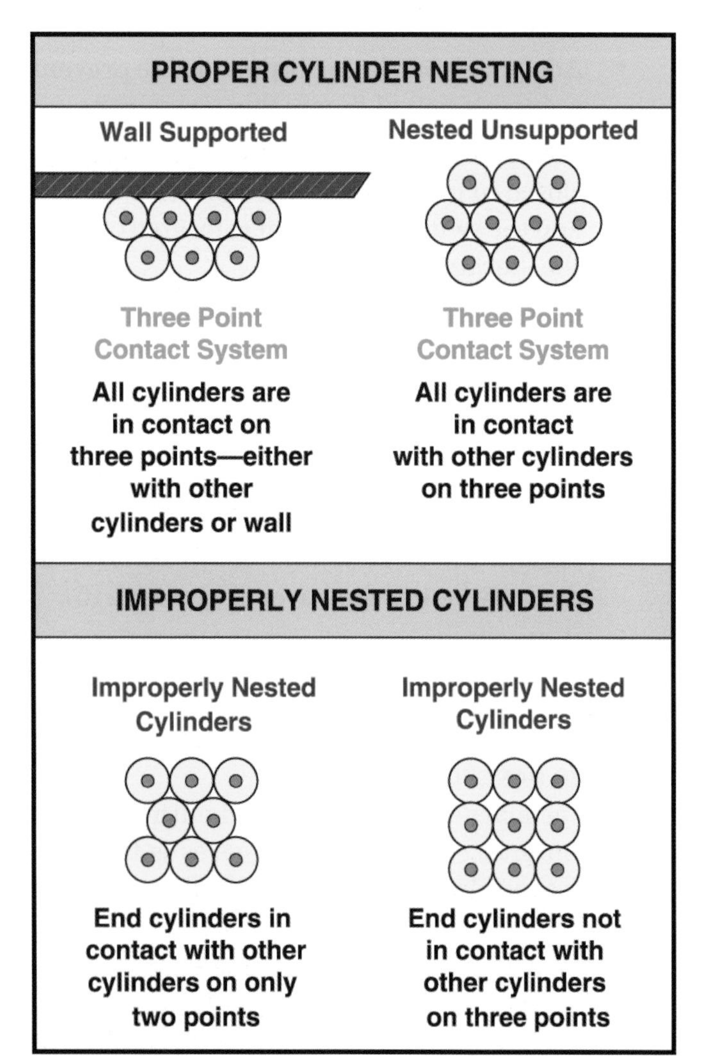

Figure 11.16 The proper and improper methods of nesting cylinders.

pressure release device. The *uncontrolled* release of compressed and liquefied gases may ultimately result in a fire and/or explosion of tremendous proportions. (**NOTE:** A container may be a large storage tank, small storage bottle, pipeline, or any other device designed to handle compressed and liquefied gases.)

There are four basic causes of container failure that lead to uncontrolled release of energy:

- Excessive pressures caused by the expansion of gases when they are heated
- Actual combustion of the gases within the container
- Structural failure of the container caused by flame impingement on the container
- Mechanical damage to the container such as collision with another object or shearing of the valve stem

The uncontrolled release of gas from a container will result in pieces of the tank being violently hurled about and the formation of a vapor cloud. If the gas is flammable or combustible, any ignition source may result in an explosion or fire. If the container fails under fire conditions, this explosion or fire will most likely occur immediately upon failure of the container.

IGNITION SOURCES

The process of forming a combustible mixture of flammable gas and air occurs in a series of steps. First, a flammable gas is released from the equipment, container, or piping holding it. The gas then mixes with the air and diffuses throughout the building or enclosure. How quickly the gas accumulates within the enclosure depends upon the rate of gas release, the density of the gas, and the ventilation features of the enclosure. Eventually, the proportions of gas and air become such that the mixture falls within its flammable limits and can be ignited. If the mixture is ignited, the amount of heat increases rapidly and is absorbed primarily by nearby combustibles. If the heated air cannot expand, the pressure will rise in the enclosure, very likely causing structural damage.

Fire inspectors must control ignition sources to prevent fires and explosions. To do so, they must analyze potential ignition sources, including electrical arcs, hot surfaces, open flames, as well as operations involving cutting, welding, grinding, and the use of impact tools. All electrical fixtures within the hazard area must be explosionproof (Figure 11.17). Lift trucks used within the hazard area should be electrically powered. The electrical equipment on the trucks should be designed, constructed, and assembled for use in atmospheres containing flammable vapors. Lift trucks rated for this service have the symbol EX stamped on the unit.

Figure 11.17 Electrical equipment, including fire alarm equipment, should be explosionproof.

Static electricity is a potential ignition source where gases are transferred. Flowing gas that is contaminated with metallic oxides, scale, or liquid particles may develop a static electrical charge. If electrically charged gas contacts an ungrounded conductive material, the charge will be transferred to that body. Bonding the two receptacles together with a metal chain or cord and grounding them will prevent the static discharge. If the gas is transferred in a completely closed system, however, no charges can be transferred and the system need not be electrically grounded or bonded.

Handling Compressed and Liquefied Gases

When inspectors are evaluating handling procedures for gas cylinders, they should ensure that workers are properly trained and are under competent supervision. The following general rules will help control hazards when accepting and handling compressed gas cylinders:

- Accept only cylinders approved for use in interstate commerce for transportation of compressed or liquefied gases.
- Do not allow numbers or marks stamped on cylinders to be removed or changed, with the exception of the marking of new hydrostatic testing numbers.
- Do not allow cylinders to be dragged; however, they may be rolled on their bottom edge (Figure 11.18).
- Protect cylinders from cuts or other physical damage.
- Do not lift cylinders with an electromagnet.
- Do not drop cylinders or let them strike each other violently.
- Do not use cylinders for rollers, supports, or any purpose other than to contain gas.
- Do not tamper with safety devices on valves of cylinders.
- When in doubt about the proper handling of a compressed gas cylinder or its contents, consult the supplier.
- When cylinders are empty, mark them "empty" or "MT." Replace valve caps.

Figure 11.18 Cylinders should be rolled on their bottoms.

GAS TRANSFER OPERATIONS AND SAFETY PRECAUTIONS

One of the major problems associated with transferring a gas from one container to another is that the gas, which may be flammable or toxic, may be released into the atmosphere. Often, these gases are released as a result of inadequate or improperly maintained equipment or operator negligence.

The suppliers of flammable gases should provide information about safe handling of gases. This information should include data about transporting, storing, setting up, and using cylinders safely. Information should also be supplied about safely using the hardware for the associated pressure system.

Bulk gas is usually transferred by replacing storage tank cylinders or by refilling cylinders or tanks. There should be a standard procedure for all disconnect and transfer operations. When deliveries are made, someone should verify that the cylinders are in good condition and are labeled properly.

All cylinders on loading stations must be clearly labeled with the name of the gas being used in order to reduce the possibility that someone will connect the wrong containers. Color coding can be useful, but should not be relied upon totally. The loading hose furnished by the receiver must be properly maintained and should be replaced at least every five years. When not in use, the hose should be stored away from direct sunlight. Bonding is generally not required if the hose has vaportight connections.

The drivers of the trucks should be warned when unloading operations are in progress so that they will not attempt to drive away while the hoses are still connected. When the transfer operation is complete, nonreturnable containers must be disposed of properly. This includes having someone remove the plugs, puncture the shells so that they cannot be reused, decontaminate them, and finally dispose of them as nonhazardous waste.

Fire inspectors should look for the following poor handling and storage practices involving gases:

- Storage and/or use of cylinders below ground level or in nonventilated or poorly ventilated areas.
- "Locking up" a connected (to cylinder) cutting torch head or instrument in a tool box cabinet.
- The presence of oil or grease on oxygen cylinders or equipment, which can cause an explosion.
- Not securing cylinders properly or not using a cap.
- Modifying a cylinder by welding legs, hooks, and so on (Figure 11.19).
- Storing incompatible gases in the same area.
- Storage or setup of cylinders in an area where poor housekeeping practices have allowed an accumulation of combustible materials (Figure 11.20).

Figure 11.19 Be alert for cylinders that have been improperly modified.

Figure 11.20 Combustible materials must not be stored close to flammable gas cylinders.

- Using cylinders that have been abused, dented, arc struck, in a fire, have broken or bent valves, or are severely corroded.
- Failing to provide and/or maintain pressure relief devices with appropriate settings and adequate pressure relief capacity.

If any of these problems are noted, the occupant should be required to correct them as soon as possible.

Pipelines are commonly used for transferring natural gas, LP gas, and some industrial gases such as hydrogen, oxygen, ethylene, and ammonia (Figure 11.21). The U.S. Department of Transportation regulates design pressures, pipe materials, valves, meters, service lines, corrosion control, and testing requirements. NFPA 54, *National Fuel Gas Code*, provides additional recommendations and requirements.

Figure 11.21 Pipeline systems have pumps located along the system.

TRANSPORTATION OF COMPRESSED AND LIQUEFIED GASES

Compressed and liquefied gases, like flammable and combustible liquids, are under strict federal regulation. The *Code of Federal Regulations, Title 49 Transportation* details these regulations, including the requirements of the Department of Transportation (DOT) for transporting these gases by air, motor vehicle, rail, and to some extent, vessel (Figures 11.22 a and b). Fire inspectors should have a copy of this code available for reference.

Much of the material in CFR 49 is adopted by reference from nationally recognized organizations. States also regulate transportation, and local jurisdictions may enact additional regulations. For example, a jurisdiction may prohibit propane cylinders from being transported through tunnels. Fire inspectors must research the local, county, or state regulations that affect their particular jurisdiction.

Figure 11.22a Compressed gas vehicles may be found at some occupancies.

Figure 11.22b Large industrial facilities may have compressed gas rail cars on site.

Cryogenic Liquids (Refrigerated Liquids)

Gases can be converted to liquid form through refrigeration. This process is known as cryogenics. Cryogenic liquids are produced and stored at extremely low temperatures (-130°F [-90°C] and colder). (**NOTE:** Cryogenic liquids are also known as refrigerated liquids, especially when in transit.) One advantage of using cryogenic liquids is the ability to modify a material's liquid-to-gas volume ratio. For example, 862 volumes of gaseous oxygen can be liquefied to a single volume by reducing the gas temperature below its boiling point. A cryogenic cylinder of liquid oxygen can hold twelve times more gas than a pressurized cylinder of oxygen.

Besides the savings in storage space and weight they offer, cryogenic liquids are valuable simply because they are so cold. For example, liquid nitrogen is used to freeze liquids for emergency pipeline repairs; to solidify gumlike materials, such as plastics and cosmetics, before they are ground; and for medical purposes.

HAZARDS OF CRYOGENIC LIQUIDS

Basically, the hazards of cryogenic liquids can be reduced to three categories:

1. The inherent hazard of the particular gas, which may be intensified when it is in liquid form. Liquid hydrogen and methane are still flammable. Liquid oxygen will support combustion with great intensity. Liquid fluorine is both reactive and toxic.
2. The tremendous liquid-to-vapor ratio.
3. The extremely low temperatures.

All cryogenic liquids, except oxygen, are either asphyxiants or are toxic. There is also the possibility that a gas with a lower boiling point than oxygen will extract the oxygen from the air exposed to it, reducing concentrations to levels that will not support life.

Because of their extremely low temperatures, all cryogenic liquids can inflict severe burns (similar to severe frostbite) to exposed flesh. When small amounts are spilled on skin, they tend to move across the flesh quickly. In large amounts, they cling to skin because of their low surface tension. The gases produced by cryogenic liquids are also extremely cold. Inhaling these gases can severely damage the respiratory tract. Vapors can damage the eyes by causing the water in the eyes to freeze.

Cryogenic liquids will refrigerate the moisture in the air and create a visible fog. The fog normally extends over the entire area containing cryogenic vapors.

STORAGE AND HANDLING OF CRYOGENIC LIQUIDS

Because of the temperature and vapor ratio, cryogenic liquids must be stored in special containers. These containers are typically double-walled and have a vacuum in the area between the walls. An insulating material, such as perlite, or a multilayered, aluminized Mylar® film, stored under vacuum pressure, fills the space between the inner shell and the outer vessel jacket. The inner tank is built of corrosion-resistant stainless steel, aluminum, or various copper alloys. Carbon steel is not used because it becomes too brittle at these temperatures.

There are several types of cryogenic containers: spherical tanks, cylindrical tanks (either vertical or horizontal), and dewars (Figures 11.23 a and b). Heat from the atmosphere will penetrate the most well-insulated tank. The temperature differential between ambient air and the product can be several hundred degrees. Some of the cryogenic liquid will absorb heat and evaporate, but most of the fluid remains at the same temperature. Through this process of self-refrigeration, the temperature and vapor pressure inside a cryogenic container remain uniform. If the container remains unused for a long time, a slow rise in temperature can be expected.

Several pressure relief devices are used to control pressure in cryogenic containers. The most common types used are pressure relief valves,

Figure 11.23a Some facilities have large cryogenic tanks on site.

Figure 11.23b Be alert for occupancies that are storing large amounts of cryogenics on site in tank trailers.

frangible disks, and a safety vent in the insulation space. A pressure relief valve is a device that opens well below any pressure that would threaten the integrity of the tank. A frangible disk ruptures if the pressure rise is excessive or if the pressure relief valve malfunctions. The insulation space also guards against leakage from the inner vessel.

If cryogenic liquids are trapped anywhere and expand, they can cause a violent pressure explosion. Their liquid-to-vapor ratios can create havoc when the product is heated in a confined space. This is particularly true in piping. A pressure relief device should be installed on every length of pipe between two shutoff valves. All pipes should be sloped up from the container to avoid the possibility of trapped fluids. Pipe and its associated fittings should be manufactured from stainless steel, aluminum, copper, or Monel®.

FLAMMABLE SOLIDS

There are a variety of materials that are loosely categorized under the term flammable solids. Metal powders, readily combustible solids that ignite by friction, and self-reactive materials that undergo a strong exothermic decomposition fall under the description of flammable solids. Also included are explosives that are wetted to suppress their explosive properties.

Spontaneously combustible materials is another type of flammable solid with which inspectors may need to be familiar. There are two primary types of spontaneously combustible materials:

- *Pyrophoric materials*, liquid, solid or gaseous, that even in small quantities and without an external ignition source can ignite within five minutes after coming into contact with air.
- *Self-heating materials* that when in contact with air and without an energy supply have the potential to self-heat.

Dangerous-when-wet materials are those materials that, on contact with water, are likely to become spontaneously flammable or to give off flammable or toxic gas at a rate greater than 1 liter per kilogram of the material per hour. Magnesium phosphide is an example of a dangerous-when-wet material.

Storage and Handling of Flammable Solids

A variety of containers are used for packaging Class 4 materials. Tubes, pails, steel and fiberboard drums, cardboard boxes, and bags are used for nonbulk packaging of flammable solids (Figures 11.24 a and b). Depending on the materials being stored, the packaging may be designed for any one of a number of strategies, including:

- Containers with tightly secured lids to prevent contact of the material with moisture.
- Containers that are filled with an inert medium that excludes air from the material.
- Containers that are filled with a liquid (oil or water) to prevent a reaction from occurring.

Figure 11.24a Small flammable solids containers.

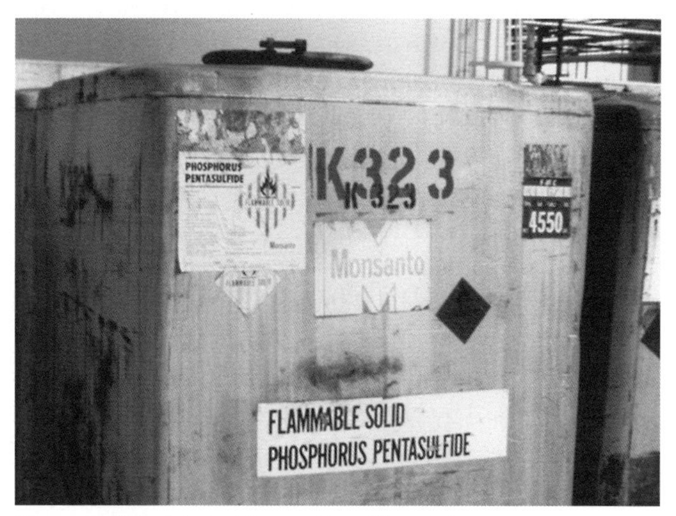

Figure 11.24b A typical flammable solid container.

Inspectors should check to make sure that the appropriate materials are stored in the appropriate containers. Furthermore, incompatible materials, such as a water-reactive material and another that must be stored in water, should not be stored or handled close to each other. If the flammable solids are metals, the appropriate size and type Class D fire extinguisher or extinguishing agent should be located close to the hazard. NFPA 10, *Standard on Portable Fire Extinguishers* provides requirements on extinguisher sizes and travel distances.

TOXIC AND HIGHLY TOXIC MATERIALS

Toxic and highly toxic materials are commonplace in industrial and residential occupancies. They include such common items as disinfectants, weed killer, rodent killers, insecticides, and various cleaning solvents. These materials are further classified as irritants, sensitizers, or other hazards. They include such chemicals as morphine, strychnine, atropine, aniline, arsenic acid, benzene, toluene, xylene, creosote, sodium fluoride, hydrogen cyanide, and cyanogen chloride.

Materials classified as toxic or highly toxic include those substances that are capable of producing serious illness or death once they enter the bloodstream. Anesthetics reduce the muscular powers of vital bodily functions and result in loss of consciousness, suppressed breathing, or even death. Irritants produce some type of local inflammation or irritate the skin, respiratory membranes, or eyes.

Toxic or highly toxic materials may enter the body through absorption, ingestion, injection, or inhalation (Figure 11.25). It is imperative that fire inspectors be aware of the possible toxic effects of a material and make sure that sufficient personal protective equipment is available. They must also consider the threshold limit value (TLV). This value is the concentration of a given toxic material that generally may be tolerated without ill effects. Threshold limit values (TLVs) are published by the American Conference of Governmental Industrial Hygienists and by NIOSH. These values may serve as guides for fire inspectors in determining the maximum allowable concentration of a substance to which an individual may be repeatedly exposed without harmful effects. The inspector must check for proper ventilation in these areas, along with the adequacy of exit requirements. The action of these products under fire conditions and the products of combustion will also be noteworthy.

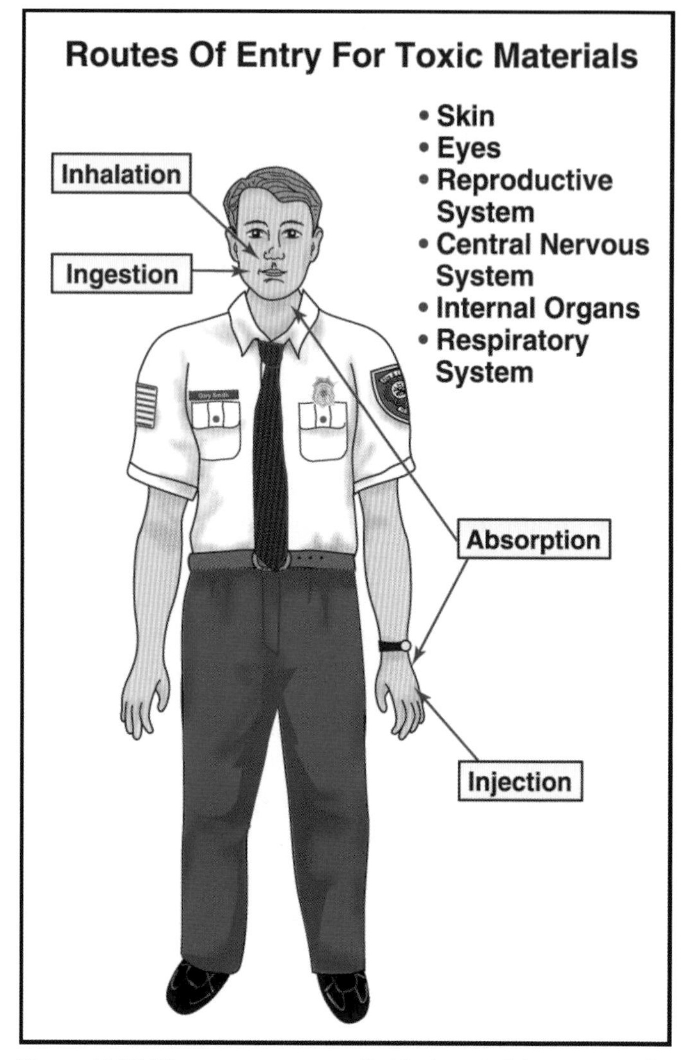

Figure 11.25 There are many ways that toxic materials may enter the human body.

The toxicity of a material refers to the ability of that substance to do bodily harm by causing some form of chemical action within the body. Toxic materials commonly enter the body through the skin, eyes, lungs, or digestive tract, eventually entering the bloodstream. Because of the many avenues of entry, the body organs affected by the substance may be remote from the point of contact.

From an exposure standpoint, toxicity may be divided into acute or chronic forms. Acute toxicity refers to the effects caused by a single dose or exposure. Depending upon the inherent concentrations and rates of exposure, the effects may be as minor as a simple headache or as serious as death. Acute toxicity is expressed in terms of LD_{50} and LC_{50} for purposes of hazard evaluation. *LD_{50}* refers to the ingested dose of a given substance that was lethal to fifty percent or more of the animals tested when they swallowed or ate the substance. *LC_{50}* refers to the concentration in the air of a given substance that killed fifty percent of more of the test animals when they inhaled or absorbed the vapors, fumes, or mists of the substance.

Chronic toxicity refers to the effects of substances upon repeated exposure over a long period of time. Chronic exposure is most often found in industrial situations where toxic substances are part of a daily or regular process.

Storage and Handling of Toxic Materials

Because they come in gaseous, liquid, or solid forms, there are a wide variety of acceptable methods for storing toxic materials. They may be found in small quantities or bulk storage. The containers used to store them can range from small jars to large tanks (Figures 11.26 a and b). Fire inspectors should make sure that whatever container is used is in good condition and shows no signs of leakage. The fire inspector should also assure that appropriate warning signs are located near the stored toxins. These signs should indicate the nature of the hazard and the actions to be taken in the event of a release. The inspector should make sure that any detection and/or warning systems that are used in that facility are in proper working condition.

The particular fire code used within a jurisdiction will have specific requirements regarding the storage of toxic materials. Most codes require that toxic materials stored inside a building be stored in either a room or structure of noncombustible construction *or* inside approved hazardous material storage cabinets. Special spill control/drainage systems and ventilation systems should also be in place. Both of these should be designed to limit or eliminate any outside contamination if an uncontrolled release occurs. Most codes also require that toxic material be separated from other hazardous materials by at least a one-hour-rated fire separation. Toxic materials that are stored outside should be protected from the weather by a canopy or some other type of enclosure.

Figure 11.26a Toxic materials are found in many retail stores.

Figure 11.26b Toxic materials may be stored in fairly benign looking containers.

Another extremely important factor that inspectors should consider, but one that is hard to check on by simply walking through a facility, is the level of training and preparedness of the employees who work with the toxic materials. Inspectors should ask to see verification that the employees have been trained properly. This might include checking training records, certificates, or program materials.

OXIDIZERS AND ORGANIC PEROXIDES

Oxidizers are chemicals, other than a blasting agent or explosives, that initiate or promote combustion in other materials thereby causing fire either of itself or through the release of oxygen or other gases. *Organic peroxides* are organic compounds that contain the bivalent -O-O- structure which may be considered to be a structural derivative of hydrogen peroxide in which one or both of the hydrogen atoms have been replaced by an organic radical. Organic peroxides can present an explosion hazard (detonation or deflagration) or they can be shock sensitive. They can also decompose into various unstable compounds over an extended period of time. Oxidizers are not necessarily combustible in themselves; however, by releasing oxygen, they support combustion and increase the intensity of burning. When in contact with materials such as water, oils, greases, hydrocarbons, and various cleaning agents, oxidizers react chemically to ignite the material. This reaction can result in an explosion. When exposed to heat, oxidizers decompose, releasing oxygen that accelerates the fire. The accelerated fire produces more heat, causing the oxidizer to further decompose, again adding to the intensity of the fire. Some oxidizers decompose spontaneously and may do so with violent or explosive results. Oxidizers that inspectors are most likely to come in contact with include:

- Sodium nitrate
- Potassium nitrate
- Ammonium nitrate
- Cellulose nitrate
- Nitric acid
- Potassium chlorate
- Calcium hypochloride
- Perchloric acid

Storage and Handling of Oxidizers

The storage and packaging of oxidizers runs a wide spectrum that is typical of hazardous materials. A common package is the plastic-lined, multiply paper bag (Figure 11.27). Metal tins and fiberboard, plastic, and metal drums are also used for packaging oxidizers. Portable, stationary, and applicator tanks contain oxidizers, which are made in slurry form for agricultural use. Liquid or slurry-form oxidizers are shipped in stainless

Figure 11.27 Oxidizers are stored in multi-layered sacks.

steel tank trucks. Dry-bulk tank trucks and railcars are used to move bulk loads of dry oxidizers. Many organic peroxides are limited in the quantity that can be shipped. The containers range in size from a few ounces (ml) up to 55-gallon (220 L) drums. Unlike similar containers used with other hazard class materials, the small organic peroxide containers are vented. Tank trucks may move some peroxides, but special permits are required by DOT. Hydrogen peroxide solutions, however, are allowed to be shipped in tank cars made of aluminum.

The particular fire code used within a jurisdiction will have specific requirements regarding the storage of oxidizers. Most specify the maximum amount of oxidizers, by weight, that may be stored in any single pile (Figure 11.28). There will also be a minimum specified distance between each pile. The requirements may vary depending on whether

Figure 11.28 Plain ammonium nitrate may be found stacked in piles.

the oxidizers are stored inside or outside. They may also vary depending on the specific type of oxidizer in question. Oxidizers stored inside generally must be stored in a noncombustible structure that is detached from other structures. The building should be insulated to provide temperature control. Other issues that may be contained in the code include:

- Spill and drainage control
- Heat and smoke venting
- Explosion control
- Standby power
- Automatic fire detection systems

The codes generally specify a minimum clearance from exposed structures for outside storage of oxidizers. Aboveground tanks of liquid oxidizers should be surrounded by a secondary containment dike to confine a spill should the tank fail.

RADIOACTIVE MATERIALS

The use and transportation of radioactive materials has been relatively commonplace for several decades. Radioactive materials are typically used in medicine, smoke detectors, power generation, weapons, nondestructive testing, manufacturing, food preservation, and at many universities and laboratories. Radioactive materials are routinely transported by roadway, rail, air, and are generally stored near where they are used. Fire inspectors should be able to recognize radioactive material markings and labels, and use standard techniques to prevent fires in and around such material.

Radioactive materials are generally no more unsafe or hazardous in a fire situation than other hazardous chemicals or substances, and are often less hazardous in a fire situation than these other materials. The property of a material that makes it radioactive is of no consequence in a fire and does not affect the behavior of the material in a fire. As with other hazardous materials, the behavior of a radioactive material in a fire is dependent on the chemical make-up of the material, not the radioactive properties. (**NOTE**: X-ray machines used in medical, dental, and veterinary clinics do not contain any radioactive material and present no radiation hazard when they are turned off.)

There are four basic types of radiation emitted from radioactive materials. *Alpha radiation* is the least penetrating and can be completely stopped by a piece of paper or cloth (Figure 11.29). *Beta radiation* is more penetrating than alpha, but can be stopped by a fairly thin layer of metal or plastic. *Neutron radiation* is more penetrating than either beta or alpha, but can be stopped by water or polyethylene. *Gamma radiation*, which is identical to medical x-rays, is the most penetrating form and requires lead or concrete to shield it. Neutron and alpha radiation are rarely encountered; beta and gamma radiation are relatively common in the applications previously listed.

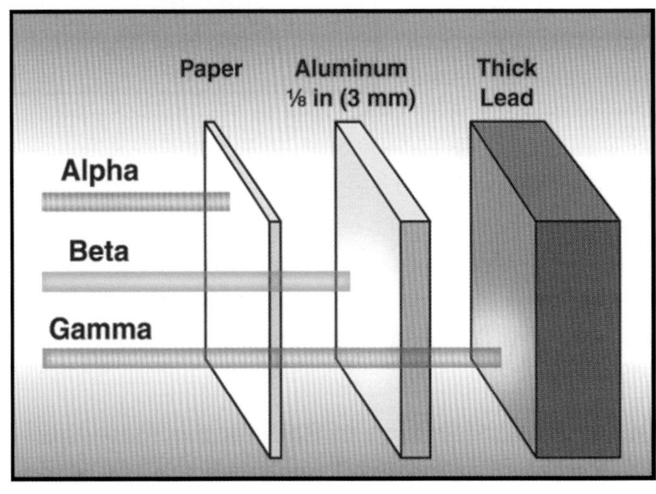

Figure 11.29 The different types of radiation have differing penetration powers.

Radiation doses are measured in Roentgens (R) or rems; these two units are essentially equal. These units are relatively large, thus milli-Roentgens or millirems are commonly used. A millirem is one-thousandth (1/1000) of a rem. In the United States, current government regulations do not allow personnel working with radioactive materials to receive more than 5,000 millirems per year. Every person receives about 400 millirems per year from background sources, such as outer space, the earth, medical sources, and food. Various methods and instruments are available to measure radiation. Most facilities using radioactive materials possess radiation measuring equipment and personnel trained to use it. Facilities with radioactive materials will have a person designated as the Radiation Safety Officer (RSO), or an equivalent title. This is the person most knowledgeable about radioactive materials at a particular site or location, and the person who should be contacted regarding any issues with the materials. In the United States, regulation of the possession, use, and storage of radioactive material is done by the Nuclear Regulatory Commission (NRC) or by state government.

Storage of radioactive materials is similar in many ways to the storage of chemicals and other hazardous materials. Warning signs and labels are required by state and federal laws (Figure 11.30). Storage should be in shielded containers. Lead is the most common material used for shielding, which in itself makes the container relatively fire resistant. When not in use, the radioactive material should be stored in the proper container.

Figure 11.30 The standard symbol for radiation is readily recognizable.

Storage and Handling of Radioactive Materials

Radioactive material packages are the strongest containers used to store and transport hazardous materials. For this reason, radioactive incidents are rare. The two categories of packaging for shipment are Type A and Type B packaging.

Type A packaging contains low-level commercial radioactive shipments. These containers include cardboard boxes, wooden crates, metal drums, and cylinders for compressed radiological gases such as xenon (Figure 11.31). Measuring devices, such as radiography instruments and soil density meters, contain radioactive materials and technically may be considered Type A packaging. Radiopharmaceuticals, those medicines that contain a radioactive material, are packaged in small quantities and generally limited to air transportation because of their short half-lives (degradation time).

Type B packaging is the strongest packaging and is used for more highly radioactive shipments. These containers include steel reinforced concrete casks, lead pipe, and heavy-gauge metal drums (Figure 11.32). They can survive serious accidents and fire without releasing the radioactive material. Type B containers are designed to carry fissionable materials, high-grade raw materials, nuclear fuels (both new and spent), and highly radioactive metals.

The various model fire codes have provisions for occupancies that store or handle radioactive materials. Most codes require occupancies that contain radioactive materials to have radiation detection equipment to monitor for radiation releases. The alarm panel should be located in a part of the facility where radioactive materials are *not* stored or handled. Radiation warning signs should be posted at the entrances to all areas where radioactive materials are stored and handled. Make sure that all radioactive materials are stored in their appropriate containers when they are not in actual use.

Radioactive materials that are stored outside structures should be in fire-resistant containers. The various fire codes may also specify a minimum distance that containers must be kept from buildings, property lines, public ways, or exit discharges from structures.

CORROSIVE MATERIALS

Corrosives are those chemicals that cause visible destruction or irreversible harm to skin tissue. Liquid corrosives that are leaking also have a severe corrosion rate on steel. Corrosives include inorganic acids (such as hydrochloric acid, nitric

Figure 11.31 A small radioactive Type A container.

Figure 11.32 A common radioactive Type B container.

acid, sulfuric acid, hydrofluoric acid, and perchloric acid), halogens (such as bromine, iodine, fluorine, and chlorine), and several strong bases (such as sodium hydroxide, potassium hydroxide, calcium oxide, calcium hypochlorite, hydraxine, and hydrogen peroxide). Corrosives may also be strong oxidizers; however, they are classified as corrosives because of their hazardous effect when someone inhales, ingests, or touches them.

The principal hazard associated with inorganic acids involves the possibility of leakage. If this occurs, the acids may come in contact with other materials, resulting in a fire or explosion. The halogens are noncombustible, but may support combustion or cause certain materials to spontaneously ignite. Fluorine and iodine may become explosive under certain conditions. Strong bases have oxidizing capabilities that may accelerate the burning of various materials or cause them to burn with explosive force.

Storage and Handling of Corrosives

Corrosives come in a wide variety of containers (Figures 11.33 a through c). The containers range in size from glass/plastic bottles and carboys to plastic drums. Fiberboard drums and multilayered paper bags are used for acid materials and caustics in dry form. Wax coated bottles are used to store hydrofluoric acid because the acid also attacks glass. Intermodal portable tanks, tank trucks, railroad tank cars, barges, and pipelines are used to transport bulk shipments of corrosives. Because of the density of corrosives, storage containers and tanks that are smaller than those used for other types of liquids are used to store and transport corrosives. Corrosives can weigh up to twice that of an equal volume of water.

Most building and fire codes have specific requirements for occupancies that store and handle corrosive materials. The bulk of these requirements are aimed at containing any spill of the corrosive material that may occur. Inside storage areas should have spill control and drainage protection in place, as well as adequate ventilation systems. Tanks that are stored inside structures may be required to be equipped with liquid level alarms that are intended to be triggered when the level of the corrosive in the tank drops.

Figure 11.33a Common corrosives, such as muriatic acid for swimming pools, may be found in simple plastic jugs.

Figure 11.33b Corrosives may come in a variety of containers.

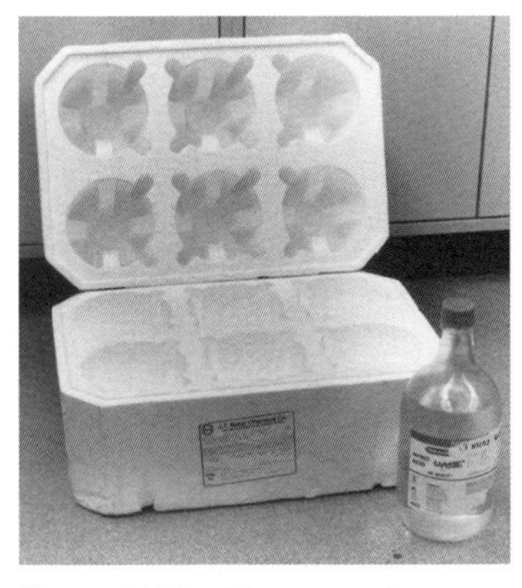

Figure 11.33c Glass corrosives may be packed inside protective cases.

Large tanks that are stored outside of structures should have some type of secondary containment system around them. This is typically in the form of a dike. The various fire codes may also specify a minimum distance that tanks and containers must be kept from buildings, property lines, public ways, or exit discharges from structures.

OTHER REGULATED MATERIALS

Hazardous materials classification systems, such as the UN system described in Chapter 10 of this manual, typically have a catch-all category for hazardous materials that do not fit within any of the other classifications. These include items, such as consumer commodities, that present limited hazards during storage, handling, and transportation due to their form, quantity, and packaging. Dealing with these types of materials requires good judgment and common sense on the part of the inspector. The inspector should try, as much as possible, to fit these materials into one of the previously described categories and apply those regulations. If this is not possible, examine each situation on an individual basis to determine an appropriate level of protection for that occupancy.

Appendices, Glossary, and Index

Appendix A

Summaries of Court Decisions Regarding Liability

Adams v. State, 555 P. 2nd 235 (1976)

A suit against the state was filed following a fire in a motel in which five persons died. The State of Alaska had inspected the motel eight months before the fire and had failed to issue a letter to the owner citing the violations of the State Fire Safety Code, despite the fact that the inspector had indicated to his superior that the motel presented an "extreme life hazard." The Supreme Court of Alaska reversed and remanded for trial a lower court's granting of the state's motion for judgment on the pleading. The court ruled that the statute that immunizes the state from tort claims arising out of failure to perform discretionary functions did not immunize the state from negligent failure to alleviate known fire hazards.

Coffey v. City of Milwaukee, 74 Wis. 2d 526, 247 N.W. 2d 132 (1976)

A tenant in an office building brought suit against the City of Milwaukee following a fire that damaged his tenant space. Despite arrival of the fire department in time to control and extinguish the fire, a defective standpipe was unable to furnish the necessary water to fight the fire. The building had been inspected, but the inspector had failed to detect and/or order replacement of the defective standpipe.

The Supreme Court of Wisconsin affirmed a lower court ruling that overruled the city's and inspector's demurrers. It stated that building inspections do not involve a "quasi-judicial" function within the meaning of governmental tort immunity statute and that the city could not claim it was merely performing a "public duty," since there was no distinction drawn between "public duty" and "special duty" owed to the tenant under the circumstances.

Grogan v. Kentucky, 577 S.W., 2d 4 (1979)

Following the Beverly Hills Supper Club fire (May 28, 1977), suit was filed against the City of Southgate and the Commonwealth of Kentucky for failure to enforce laws dealing with fire safety. The Supreme Court of Kentucky held the city and commonwealth not liable, and that government ought to be free to enact laws without exposing its supporting taxpayers to liability for failures of omission in its attempt to enforce them.

Halvorsen v. Dahl, 89 Wash., 2d 673, 574 P. 2d 1190 (1978)

The Supreme Court for the State of Washington revised and remanded for trial a superior court dismissal charging liability against the City of Seattle following a hotel fire in which a man was killed. The court ruled that the Seattle housing code did impose special duty to those individuals who reside in buildings covered by the code, and that the city had long-term knowledge of violations of that code by the building and had undertaken to force compliance on several occasions but had not followed through.

Gordon v. Holt. 412, N.Y.S., 2d., 534 (1979)

The City of Utica, New York was defendant in a suit filed on behalf of tenants and owners of an apartment house following a fire. They claimed the city was liable because the building department issued a certificate of occupancy when there were known major violations of the building code. The Supreme Court, Appellate Division, Fourth Department, held that the city building code did not create a special duty flowing from the city to specific tenants of a particular building so as to permit those tenants to file suit for breach of duty owed to the public at large to provide adequate police and fire protection.

Cracraft v. City of St. Louis Park, 279 N.W., 2d 801 (1979)

Suit was brought against the City of St. Louis Park, Minnesota following an explosion and fire in a high school involving a 55-gallon (208 L) drum of duplicating fluid. Two students were killed and a third seriously injured. The courts held that the city was not liable as no special circumstances existed so as to create a special duty on the part of the city toward individual members of the public injured in the explosion.

Wilson v. Nepstad, 282 N.W., 2d, 664 (1979)

Following an apartment fire in Des Moines, Iowa that involved deaths and injuries, a district court dismissed a municipal court tort claim action. The Supreme Court of Iowa reversed and remanded the case on the basis that certain statutes were intended to impose municipal tort liability for negligence based on a breach of statutory duty, in this case inspecting the property in a negligent manner and issuing an "inspection certificate" which by implication warranted the premises to be safe for the purpose of human habitation.

Modlin v. City of Miami Beach, Fla., 201 So. 2d 70 (1967)

The City of Miami Beach was the defendant in a case filed on behalf of a woman when a store mezzanine collapsed. The case against the city was based on negligent inspection during construction some five years previous which failed to discover the defect that led to the collapse. The Florida Supreme Court ruled that the building inspector owed no special duty to the woman killed; and therefore, the city was not liable for the negligent inspection by its employee.

Appendix B

Example of a Citation Program

CODE CITATION POLICY AND PROCEDURES

SECTION I. Purpose

1.1. To gain compliance with the Uniform Fire Code, California Administrative Code and Title 19 when all reasonable efforts have been unsuccessful.

1.2 A course of legal action to be taken when a condition exists that causes a threat to life or property from fire or explosion.

SECTION II. Background

2.1 During the year (1997), the Fire Department wrote 2,426 violations, achieved 1,698 corrections and had 728 outstanding violations.

2.2 Our present process of enforcement (City Attorney, District Attorney, Office Hearings, filing of complaint, etc.) does not lend itself to providing uniformity of compliance within the community. The majority of fire violations written are characteristic of the following three conditions:

2.2.1 *Transient* problems such as overcrowding of public entertainment facilities, illegal parking in fire lanes, mischievous fire setting.

2.2.2 *Changeable* or portable situations such as illegal locking devices on public exit doors and obstructions to aisles or exitways.

2.2.3 *Maintenance* of fire extinguishing and alarm systems, portable situations such as electrical violations, housekeeping including outdoor fire hazards.

2.3 Transient violations are specific occurrences which should be acted upon immediately through a citation process. Changeable violations are corrected by the person responsible (in most cases) on a temporary basis, but are changed back after the inspector leaves the premises. The same situations are encountered yearly and are not being permanently abated.

2.4 Citizen awareness of the Fire Department's ability to cite for violations would create an effective deterrent in maintaining corrective abatements on a more permanent basis. The citation process would be used on a discretionary basis and would be very cost effective from the standpoint of available manpower utilization and steadily increasing workload demand.

SECTION III. Policy

3.1 Members of the Fire Prevention Division who are authorized by the Division Chief in charge shall have the authority to issue citations for fire and life safety violations of the Uniform Fire Code and the California Administrative Code, Title 19, Ordinance No. 1088, Resolution No. 1296.7.

3.2 It is the intent of the Fire Prevention Division to achieve compliance of the majority of code violations by traditional means of inspection, notification, the granting of reasonable time limits to comply, and reinspection.

3.3 Citations shall be issued by Fire Prevention Personnel when the following conditions exist:

3.3.1 Failure to gain reasonable compliance for Uniform Fire Code violations.

3.3.2 Deliberate or mischievous fire setting not involving property loss.

3.3.3 Justification is evident that the violation was restored after inspection (Inspection records must verify the facts of violation).

3.3.4 Obstruction of fire lanes.

3.3.5 Upon direction of the City Attorney's office.

SECTION IV. Definitions

4.1 Reasonable time to comply.

4.1.1 Generally means after the third visit or second reinspection, with proper time intervals between visits for responsible party to make corrections of conditions.

4.2 Justification is evident that violation was restored after inspection.

4.2.1 When the same violation has been written two times during any one-year period of a facility inspected two times or more annually.

4.2.2 When the same violation has been written two times during any two-year period of a facility inspected one time per year.

SECTION V. Procedure

5.1 Citations issued under the conditions as set forth in policy should have previous notification history as set forth below.

5.1.1 Exiting or overcrowding of public assembly facilities requires two previous notices of violation.

5.1.2 Engine company referrals to Prevention Division require engine company survey and reinspection, notice of violation written by inspector.

5.1.3 Prevention inspection (originated by Fire Prevention) requires survey report and notice of violation.

5.1.4 Repeated violation which is corrected during inspection process and restored to violation condition afterwards, requires previous history of specific abatement process within the previous two years with the same responsible party.

5.1.5 Illegal controlled burning requires a warning notice of violation issued. Second offense by same party requires citation to be issued.

5.1.6 Deliberate mischievous fire setting by persons of responsible age requires a citation to be issued to person or persons committing act.

5.2 Citing for misdemeanor violations. The citation is a release stating the defendant will appear in court or post bail in lieu of physical arrest. If not sure if you have the owner, check the liquor license and cite whomever is listed. (It may be a corporation). If no liquor license, cite whomever is listed on the business license. If this is not available, check City Business License Division for correct owner.

5.2.1 Try to give citation directly to the owner. You may have to return at a later time to catch him or her in the place of business. If the owner is not available, mail the citation (unsigned and certified mail) to the owner's address listed on the liquor or business license.

5.2.2 The citation must be signed by the responsible party for the premises or the person committing the act in cases of transient fire setting or illegal burning.

5.2.3 All misdemeanors by the bail schedule require a court appearance by the violator. At that time the defendant appears and either (a) pleads guilty, in which case we hear nothing more about it; (b) has matter continued, which is not our concern; or (c) pleads not guilty, in which case the court sets a date for trial and for pretrial conference, and notifies the city attorney's office or the acting agency. It is the latter hearing date, set by the court, that requires the citing officer's appearance.

5.3 Citing for Infractions. The procedure for writing the citation for an infraction is the same as for a misdemeanor violation.

5.3.1 All infractions can be paid at the court clerk's office or by mail (after phoning clerk to find out what their final assessment will be). Do not indicate to the recipient of the citation what you think the fine will be.

5.4 Failure to Appear in Court. When the violator fails to appear in court on the appointed date, the citation is forwarded to the District Attorney's office, which will then notify the agency issuing the citation. A declaration must be filled out by the issuing officer at which time a warrant will be issued for the violator. A maximum of three weeks from the due date on the citation will be allowed for the officer to complete the paper work and the District Attorney to act.

5.5 Failure to Abate Violation. After the date of appearance has expired a reinspection of the violation is warranted. Failure to abate violation requires that a second citation be issued.

5.6 Attach copies of previous notices issued with your copy of citation. Submit citation copy and notices to the Fire Prevention Officer, who will review them and forward to the proper authorities.

5.7 The department citation logs shall be filled out and kept current. Log will be posted on wall behind inspector's desk. The Prevention Officer and/or Division Chief shall be notified the following day of any citations issued by Fire Prevention personnel.

5.8 The Division Chief shall maintain a citation book record indicating which citation book numbers have been assigned to officers.

5.9 Attached is a schedule of recommended bails set for violations. Bail costs are required if violator is physically arrested and may be used by judge to determine first offense fines. (**Note:** The bail schedule is lengthy and is not printed here.)

SECTION VI. Exceptions

6.1 All exceptions or deviations shall be discussed with the Division Chief and City Attorney prior to writing of a citation.

Appendix C

Mount Prospect Fire Department
Fire Prevention Bureau
112 E. Northwest Hwy.
Mount Prospect, Illinois 60056
Phone: (847) 818-5253 Fax: (847) 818-5240

INSPECTION REPORT FORM

Address:	Suite #:	Inspection Date:
Name of Business		Inspector
Business License #	Occupancy Classification	Occupancy ID #:
Strip Mall / Building Complex Name	Contact Person:	Phone #:

TYPE OF INSPECTION

- ☐ Fire Prevention Activity
- ☐ Business License
- ☐ Certificate of Occupancy
- ☐ Fire Prevention Permit
- ☐ Complaint
- ☐ Hydrostatic Test
- ☐ Sprinkler Acceptance Test
- ☐ Fire Alarm Acceptance Test
- ☐ Fire Pump Acceptance Test
- ☐ Construction Site
- ☐ Underground Flushing
- ☐ Flow Test
- ☐ Hood/Suppression System

You are here notified to remedy the conditions stated above within _____ hours/days from the date of this order. Appeal from this order may be made within 10 days from the date of service. Direct such appeal to the Fire Marshal of the Mount Prospect Fire Prevention Bureau by telephone, **(847) 818-5253**, or by writing the Office of the Fire Chief, 112 E. Northwest Hwy., Mount Prospect, Illinois 60056.

Date Complied:______________________ Date Notified:______________________

Inspector:______________________ Received By:______________________

WHITE - OFFICE **YELLOW - FILE** **PINK - SITE**

MOUNT PROSPECT FIRE DEPARTMENT

FIRE PREVENTION BUREAU

112 E NORTHWEST HIGHWAY MOUNT PROSPECT, IL 60056
847-818-5253 FAX: 847-818-5240

ADDRESS **SUITE #** **INSPECTION DATE**

NAME OF BUSINESS **INSPECTOR**

BUSINESS LICENSE # **OCCUPANCY CLASSIFICATION** **SHIFT/BADGE #**

STRIP MALL/BUILDING COMPLEX NAME **BUSINESS HOURS**

OWNER/MANAGER OF PROPERTY **OWNER/MANAGER ADDRESS** **OWNER PHONE #**

EMERGENCY CONTACT **EMERGENCY PHONE #1 (NO PAGERS)** **EMERGENCY PHONE #2**

	YES	NO
FIRE ALARM SYSTEM?	☐	☐
SPRINKLER SYSTEM?	☐	☐
KNOX BOX?	☐	☐
FIRE PUMP?	☐	☐

FIRE ALARM/SUPPRESSION MONITORING COMPANY:__________

PHONE #:__________ **POSITION #:**__________

HAZARD MATERIALS ON SITE? YES ☐ NO ☐

APPROXIMATE BLDG. SQ FOOTAGE:__________

TYPES:__________ **QUANTITIES:**__________

Item	Code		Item	Code		Item	Code	
1. OUTSIDE	**1.00**		**3. LIFE SAFETY**	**3.00**		**5. ELECTRICAL**	**5.00**	
Fire Hydrant	**1.10**		**Exits**	**3.10**		Open Wiring	5.01	
Obstructions/Condition	1.11		Improper Number	3.11		Cover off panel	5.02	
Fire Lane	**1.20**		Blocked/Obstructed	3.12		Open junction boxes/outlets/ switches	5.03	
Improper Markings	1.21		Locks	3.13		Extension cord use	5.04	
Obstructed/Condition	1.22		Wrong door swing	3.14		Panel not accessible	5.05	
Knox Box	**1.30**		Door needs repair	3.15		Broken conduit	5.06	
Address	**1.40**		Arrangement	3.16		Panel not marked	5.07	
			Exit Access	**3.20**		Ground fault	5.08	
2. FIRE PROTECTION SYSTEMS	**2.00**		Blocked/Obstructed	3.21		Explosion proof equipment	5.09	
Sprinkler/Standpipe/Fire Pump	**2.10**		**Exit Enclosures (Stairs, etc.)**	**3.30**		**6. FIRE BARRIERS**	**6.00**	
Annual Test	2.11		Storage/Improper use	3.31		**Fire wall/Partitions**	**6.10**	
Valves open/supervised	2.12		**Exit Signs**	**3.40**		Holes/cracks	6.11	
Valves not labeled	2.13		Needs repair	3.41		Wrong rating	6.12	
Valves not accessible	2.14		Improperly located	3.42		**Fire Doors/Frames**	**6.20**	
No spare sprinkler/wrench	2.15		**Emergency Lights**	**3.50**		Annual test	6.21	
Improper coverage	2.16		Needs repair	3.51		Need fire door	6.22	
No inside bell	2.17		Inadequate Coverage	3.52		No fusible link	6.23	
Fire Dept. Connection	**2.20**					Needs repair	6.24	
Obstructed/Condition	2.21		**4. FLAMMABLE LIQUIDS**	**4.00**		**Ceiling**	**6.30**	
Strobe/Condition	2.22		Improper Storage	4.01		Titles missing	6.31	
Sign	2.23		Improper Dispensing	4.02		Holes/cracks	6.32	
Not painted red	2.24		Ignition Sources	4.03				
Fire Alarm	**2.30**		Posting of "No Smoking" Signs	4.04		**7. GENERAL**	**7.00**	
Annual Test	2.31		Fire Dept. Permit/Displayed	4.05		Housekeeping	7.01	
Panel not accessible	2.32		**Spray Booths**	**4.10**		No posted occupancy limits	7.02	
Panel not labeled/Zone map	2.33		Sprinklers/Residue	4.11		Compressed gas cylinders	7.03	
Improper detector spacing	2.34		Ventilation/Booth	4.12		Kitchen filters	7.04	
Other Fixed Suppression System	**2.40**		Filters/Residue	4.13		Improper storage	7.05	
Semi/Annual Test	2.42					Gas meters	7.06	
Not provided	2.43					Boiler certification	7.07	
Portable Fire Extinguisher	**2.50**					Emergency fuel shutoff (gas station)	7.08	
Annual Inspection	2.51							
Location	2.52							
Access/Obstruction	2.53							

ITEM	REMARKS AND RECOMMENDATIONS FOR CORRECTION

TOTAL VIOLATIONS:__________ REINSPECTION IN:______CALENDAR DAYS DATE OF REINSPECTION:__________

YOU ARE HEREBY NOTIFIED TO CORRECT THE ABOVE VIOLATIONS WITHIN THE TIME FRAME GIVEN. APPEAL FROM THIS ORDER MAY BE MADE WITHIN TEN (10) WORKING DAYS FROM THE DATE OF SERVICE. DIRECT SUCH APPEAL TO THE FIRE MARSHAL OF THE MOUNT PROSPECT FIRE PREVENTION BUREAU BY TELEPHONE: 847-818-5253, OR IN WRITING TO: 112 E NORTHWEST HIGHWAY, MOUNT PROSPECT, IL 60056.

THE UNDERSIGNED HEREBY ACKNOWLEDGES THAT CONSTRUCTION REPAIRS OR ALTERATIONS HAVE NOT TAKEN PLACE WITHOUT A PERMIT FROM THE VILLAGE OF MOUNT PROSPECT. FUTURE OR PAST WORK WITHOUT A PERMIT IS SUBJECT TO ALL PENALTIES AND/OR FINES.

OWNER/AGENT:____________________ TITLE:_______________ DATE:__________

WHITE/OFFICE • YELLOW/FILE OR REINSPECTION • PINK/OWNER OF BUSINESS
SYMBOLS: X = VIOLATION O = NO VIOLATION / = DOES NOT PERTAIN

Appendix D

Standard Map Symbols

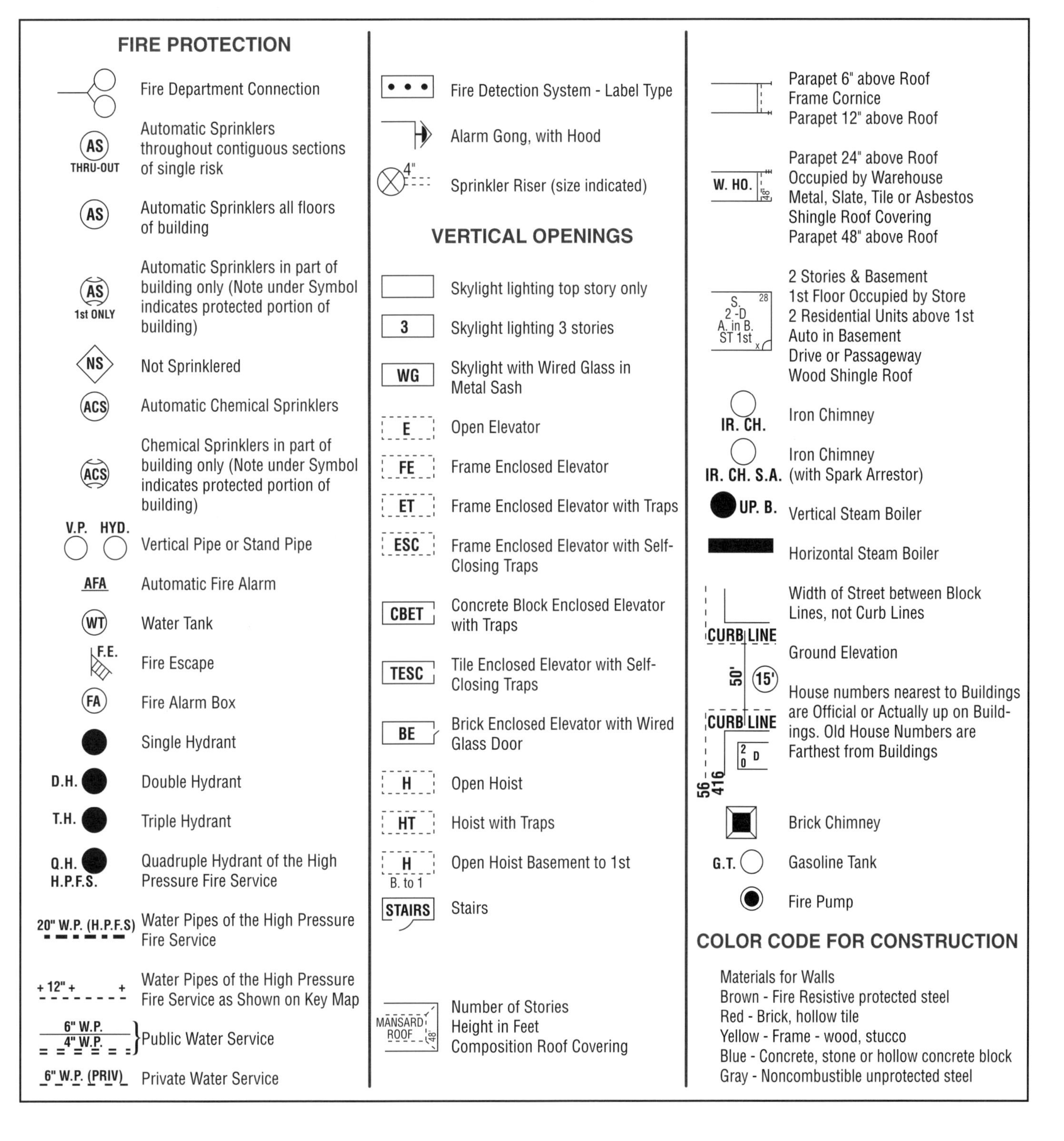

Appendix E

Model Code Organizations

MODEL FIRE PREVENTION CODES

NFPA 1, *Fire Prevention Code®*
National Fire Protection Association

The BOCA National Fire Prevention Code®
Building Officials & Code Administrators International, Inc.

The Uniform Fire Code™
International Conference of Building Officials
5360 South Workman Mill Road
Whittier, CA 90601

Standard Fire Prevention Code©
Southern Building Code Congress International, Inc.

National Fire Code of Canada
National Research Council Canada
Ottawa, Ontario
Canada K1A OR6

MODEL BUILDING CODES

The Uniform Building Code™
International Conference of Building Officials
5360 South Workman Mill Road
Whittier, California 90601

The BOCA National Building Code®
Building Officials & Code Administrators International, Inc.

The Standard Building Code©
Southern Building Code Congress International, Inc.

National Building Code of Canada
National Research Council Canada
Ottawa, Ontario
Canada K1A OR6

MODEL CODE ORGANIZATION INFORMATION

This section explains model code organizations and the ways in which they provide a national forum for presenting and discussing proposed code revisions. It is important that fire inspectors understand this process so that they can actively participate in code reform and revision.

These nonprofit organizations use a democratic committee system of active members to update and produce code revisions. Any individual, organization, or professional group can become a member and participate in the code revision process; however, designated individuals vote for their assigned jurisdiction. State and local officials are encouraged to participate in model code organizations because they are responsible for code enforcement in their respective jurisdictions.

NATIONAL FIRE PROTECTION ASSOCIATION (NFPA)

One Batterymarch Park
Quincy, MA 02269
(617) 770-3000

Publishes the *National Fire Codes*, a compilation of standards, recommended practices, manuals, guides, and model laws. The NFPA also publishes various fire-related textbooks, instructional aids, and produces audiovisual materials. The basis for NFPA is the dissemination of timely standards developed by more than 210 technical committees. The NFPA consists of over 60,000 members from fire departments, business, industry, health care, architecture, engineering, manufacturing, local and state government, and other professional trade areas and organizations. In addition to the *Fire Prevention Code®* noted above, the NFPA has a wide variety of other standards that fire inspectors may use. These include NFPA 30, *Flammable and Combustible Liquids Code*; NFPA 70, National Electrical Code®, NFPA 101, *Life Safety Code®*, and a wide variety of fire protection equipment and systems standards.

Services provided by the NFPA include:
Code maintenance through changes
Code and standards revision or confirmation every three to five years

Formal interpretation procedures
Engineering advisory service

Additional NFPA publications that may be of assistance to the fire inspector include:

Fire Protection Handbook - A comprehensive reference that covers almost every aspect of fire protection engineering. It is often referred to as the "bible" of fire protection.

NFPA Inspection Manual - A pocket-size publication written primarily for private fire inspection; however, most of the information contained in this guide is helpful to fire inspectors. This publication explains various types of hazards encountered during inspections.

Conducting Fire Inspections: A Guide For Field Use - A pocket-size publication that identifies individual classes of occupancies. This guide highlights special areas to inspect and includes checklists for maintaining accurate inspections.

Life Safety Code Handbook - A publication that explains and illustrates NFPA 101, Life Safety Code in greater detail.

National Electrical Code Handbook - A publication that explains and illustrates NFPA 70, *National Electrical Code* in greater detail.

Building Officials & Code Administrators International, Inc. (BOCA)

4051 West Flossmoor Road
Country Club Hills, IL 60477
(708) 799-2300

Publishes the BOCA *Plumbing Code®*, BOCA *National Building Code®*, BOCA *Mechanical Code®*, BOCA *National Fire Prevention Code®*, BOCA *Property Maintenance Code®*, and BOCA *Energy ConservationCode®*, *The Building Official and Code Administration (BOCA) Bulletin*, research reports, and Professional Development Series.

Note: These codes were formerly known as the *Basic Plumbing Code, Basic Building Code, Basic Mechanical Code*, and so forth.

Services provided by BOCA include:

Code maintenance through annual changes
Examination planning
Consultation
Administrative studies of local government code enforcement agencies

Southern Building Code Congress International, Inc. (SBCCI)

900 Montclair Rd.
Birmingham, AL 35213-1206
(205) 591-1853

Publishes the *Standard Building Code© Standard Fire Prevention Code©*, and related construction codes.

Services provided by SBCCI include:

Code maintenance through annual changes
Code interpretations
Consultation

International Conference of Building Officials (ICBO)

5360 South Workman Mill Road
Whittier, California 90601
(562) 699-0124

This organization, in conjunction with the Western Fire Chiefs Association, publishes the *Uniform Building Code™*, the *Uniform Fire Code™*, *Building Standards* magazine, and related construction codes.

Services provided by ICBO include:

Code maintenance through annual changes
Examination planning
Code interpretation
Consultation
Voluntary certification of members

OTHER ORGANIZATIONS RELATED TO FIRE INSPECTION

In addition to the above, many organizations and agencies actively promote various phases of fire protection and prevention:

Alliance of American Insurance
20 North Wacker Drive
Chicago, Illinois 60606

American Insurance Association (AIA)
85 John Street
New York, New York 10038

America Institute of Architects
1735 New York Avenue N.W.
Washington, D.C. 20006

American Planning Association
1776 Massachusetts Avenue N.W.
Washington D.C. 20005

Association of Major City Building Officials (AMCBO)
c/o NCSBCS
481 Carlisle Drive
Herndon, Virginia 22070

Building Research Advisory Board
2101 Constitution Avenue N.W.
Washington, D.C. 20418

Bureau of Explosives
Association of American Railroads
1920 L Street, N.W.
Washington, D.C. 20036

Council of American Building Officials (CABO)
5201 Leesburg Pike
Falls Church, Virginia 22041

Factory Mutual System
1151 Boston-Providence Turnpike
Norwood, Massachusetts 02062

Industrial Risk Insurance (IRI)
85 Woodland Street
Hartford, Connecticut 06102

International Association of Electrical Inspectors (IAEI)
802 Busse Highway
Park Ridge, Illinois 60068

International Association of Fire Chiefs (IAFC)
4025 Fair Ridge Drive
Fairfax, VA 22033-2868

International Association of Fire Fighters (IAFF)
1750 New York Avenue N.W.
Washington, D.C. 20006

International Association of Plumbing and Mechanical Officials (IAPMO)
5032 Alhambra Avenue
Los Angeles, California 90032

National Association of Housing and Redevelopment Officials (NAHRO)
2600 Virginia Avenue NW
Washington, D.C. 20037

National Bureau of Standards (NBS)
United States Department of Commerce
Washington, D.C. 20234

Southwest Research Institute (SWRI)
6220 Culebra Road
San Antonio, Texas 78284

Underwriters Laboratories, Inc.
333 Pfingsten Road
Northbrook, Illinois 60062

United States Department of Agriculture
Information Division
Washington, D.C. 20250

United States Department of Interior
Safety Division
Bureau of Mines
Washington D.C. 20240

United States Department of Treasury
Bureau of Alcohol, Tobacco, and Firearms
1200 Pennsylvania Avenue, N.W.
Washington, D.C. 20226

United States Fire Administration
National Fire Academy
16825 South Seton Avenue
Emmitsburg, MD 21727

Appendix F

Fire Protection Systems Test Checklists

Colorado Springs Fire Department
Fire Alarm System
Installer Certification

Permit #: ______________ **Plan Review #:** ______________

Date: ______________ **City License #:** ______________ **Exp. Date:** ______________

	Owner/Occupant	System Installer	System Supplier
Business Name:			
Address:			
Contact:			
Telephone #:			

Approved plans on site? ☐ **Yes** ☐ **No**

Alarm System is: ☐ **Manual** ☐ **Manual and Automatic** ☐ **Automatic**

Alarm Classification: ☐ **Local** ☐ **Emergency Voice** ☐ **Remote Station** ☐ **Proprietary** ☐ **Central Station**

NFPA 72 Record of Completion (Fig. 1-7.2.1 - '96 Ed.) attached? **Yes** ☐ **No** ☐

Yes	No	N/A	
☐	☐	☐	Loop resistance is within specifications (test may be required if system wiring changed from plans) Verify loop resistance
☐	☐	☐	1. Conductor pair short-circuited at far end
☐	☐	☐	2. Measure and record resistance of each circuit
☐	☐	☐	3. Verify loop resistance does not exceed manufacturer's specified limits
☐	☐	☐	Verify stray voltage does not exist between installation conductors and ground or between installation conductors

Certification of System Installation:

Complete this section after system is installed but prior to conducting operational acceptance tests. Check wiring for open shorts, ground faults, and improper branching. This system installation was inspected by ______________ (person) on ______________ and was found to comply with the installation requirements of:

- ☐ NFPA 72
- ☐ Article 760 of NEC
- ☐ Manufacturer's Instruction/ Specifications (provide)
- ☐ Other (specify & provide) ______________

Print Name: ______________ Signed: ______________

Organization: ______________ Date: ______________

Certification of System Operation:

All operations features and functions of this system were tested by ______________ (person) on ______________ and found to be operating properly in accordance with the requirements of:

- ☐ NFPA 72
- ☐ Job Specs
- ☐ Manufacturer's Instructions (provide)
- ☐ Other (specify & provide) ______________

Print Name: ______________ Signed: ______________

Organization: ______________ Date: ______________

Colorado Springs Fire Department
Non-Approved, Non-Required Alarm Systems
Acceptance Test
NFPA 72

CSFD Project #	Date:	Inspector:
CSFD Permit #		Project Name:
Installation Company:		Building ID:
Installer Name:		Project Address:
Installer Phone:		Responsible Party's Phone:
Monitoring service:	Monitoring service contact number:	
Location of alarm control panel:		
Alarm make and model:		
Type of alarm system: ☐central ☐proprietary	☐remote station	☐local (constantly attended location)

NON-REQUIRED, NON-APPROVED SYSTEMS ONLY

Yes	No	N/A	
☐	☐	☐	Keys to locked areas labeled and placed in Knox Box
☐	☐	☐	Panel location is readily accessible and located as approved by CSFD

Comments: ______________________________

System tested by:	Print name here	Sign here

☐ **System Approved**
☐ **System Disapproved** Date: __________ Inspector's Initials __________

Colorado Springs Fire Department
Residential Fire Alarm Acceptance Test
NFPA 72, UFC Sections 1001 and 1007

CSFD Project #	Date:	Inspector:
CSFD Permit #		Project Name:
Installation Company:		Building ID:
Installer Name:		Project Address:
Installer Phone:		Responsible Party's Phone:
Monitoring service:	Monitoring service contact number:	
Location of alarm control panel:		
Alarm make and model:		
Type of alarm system: ☐ central ☐ proprietary ☐ remote station ☐ local (constantly attended location)		

DOCUMENTATION

Yes	No	N/A	
☐	☐	☐	"Installer Certification" form provided to CSFD
☐	☐	☐	Installer has permit and plans approved by CSFD on site
☐	☐	☐	All components located same as approved drawing and specifications
☐	☐	☐	Copy of system instruction manuals on site
☐	☐	☐	Contractor dB pretest list provided for spot check verification

INSTALLATION

Yes	No	N/A	
☐	☐	☐	Devices are located in all areas required by the code
☐	☐	☐	Visual warning devices (strobes) provided in handicap accessible areas as per plans
☐	☐	☐	Fire alarms, supervisory signals and trouble alarms distinctly different and descriptively annunciated (labeled lights or LEDs)
☐	☐	☐	Auto dial devices programmed to dial supervising station

OPERATIONAL TESTS

Yes	No	N/A	
☐	☐	☐	Notify fire department and alarm monitoring agency, prior to testing, that testing will take place and an emergency response is not desired, unless 911 call is placed
☐	☐	☐	Notify on-site personnel, prior to testing, that testing will take place, and signal arranged in case of actual fire

Perform the following tests under both primary and secondary power

Pass	Fail	N/A	
☐	☐	☐	Power light on, in normal condition
☐	☐	☐	A second initiating device overrode silence switch
☐	☐	☐	Alarm initiating devices actuated
☐	☐	☐	Fire alarm equipment not concealed, obstructed, or impaired

OPERATIONAL TESTS (continued)

Pass	Fail	N/A	
			Control Panel
☐	☐	☐	Alarm signal received at off site location showing correct zone
☐	☐	☐	All lamps and LEDs illuminate via panel lamp test switch
			Power Supply Supervision
☐	☐	☐	Loss of AC power
☐	☐	☐	Fire alarm circuit breaker labeled “Fire Alarm Control Circuit” in red
☐	☐	☐	Verified dedicated 120 Volt AC branch circuit
			Batteries
☐	☐	☐	Batteries match approved plans, specifications, and calculations
☐	☐	☐	No leakage
☐	☐	☐	Connections tight
☐	☐	☐	If lead-acid batteries, electrolyte level full
			Transient Suppressors
☐	☐	☐	Lightning protection equipment installed

Control Panel Trouble Signals

Pass	Fail	N/A	
☐	☐	☐	Trouble signals and panel light operated for each circuit tested
☐	☐	☐	All panel trouble signals operate properly
☐	☐	☐	Trouble signals received at off site location when supervised function is disconnected
			Signal Silencing/Reset
☐	☐	☐	Audible and visual supervisory/trouble silencing switch
☐	☐	☐	Alarm signal silencing switch
☐	☐	☐	Alarm signal reset switch

Detection And Initiating Devices

Pass	Fail	N/A	
			Fire Extinguishing or Suppression System (Including Sprinklers) Alarm Switch
☐	☐	☐	Operates audible evacuation devices and signal received at FACP
			Fire Gas and Other Gas Detectors
☐	☐	☐	Tested as per manufacturer's listing
			Heat Detectors - Avoid damage to non-restorable fixed temperature elements!
☐	☐	☐	Verify temperature rating vs. ambient temperature of location
☐	☐	☐	Temperature classification identified on device
			Fixed Temperature, Rate of Rise
☐	☐	☐	Heat tested with heat source
			Fixed Temperature, Non-Restorable, Line Type
☐	☐	☐	Tested mechanically and electrically for function
			Fixed Temperature, Non-Restorable, Spot Type
☐	☐	☐	Tested electrically for function
			Restorable, Line Type, Pneumatic Tube Only
☐	☐	☐	Heat source or pressure pump

Residential Fire Alarm Systems
NFPA 72, UFC Sections 1001 & 1007
and UFC 10-3 &10-4

OPERATIONAL TESTS (continued)

Pass	Fail	N/A	
			Smoke Detectors
			System Type
☐	☐	☐	Test in place with listed aerosol smoke into sensing chamber or by other approved method
☐	☐	☐	Alarm responded
☐	☐	☐	Device is listed and marked with sensitivity range
			Air Sampling and Duct Type
☐	☐	☐	Test as per manufacturer's recommended test method
			Projected Beam
☐	☐	☐	Introduce aerosol smoke or optical filter in beam path
			Smoke Detector with Thermal Element
☐	☐	☐	Operate each part independently as described for respective devices
			Radiant Energy Sensing Fire Detectors
☐	☐	☐	Test only per manufacturer's prescribed test method
			Water Flow Switch
☐	☐	☐	Flow though inspector's test valve with orifice equal to that of smallest sprinkler orifice

Alarm Notification Devices

Pass	Fail	N/A	
☐	☐	☐	Audible and visual alarm devices operated properly, and signal throughout building
			Audible Alarms
☐	☐	☐	Measure and record dBs
☐	☐	☐	Audible levels of fire alarm warning device exceeds 15 dBa (decibels above ambient) or 5 dBs above maximum sound level with 30 seconds duration
☐	☐	☐	Alarm evacuation signal under 120 dBa maximum noise level
			Visible
☐	☐	☐	Verify approved installation layout matches approved plans

Supervising Station Fire Alarm Systems

Pass	Fail	N/A	
☐	☐	☐	24 hour monitoring agency has the appropriate telephone number for emergency response in protected property's community
			All Equipment
☐	☐	☐	Test all system functions as per manufacturer recommended test procedure
☐	☐	☐	Actuated initiating device, and signal received within 90 seconds
☐	☐	☐	If test jacks are used, first and last tests performed without use of test jacks

Pass	Fail	N/A	*DACT (Digital Alarm Communicator Transmitter)*
☐	☐	☐	DACT seizes primary line while primary line is in use and signals received at supervising station
☐	☐	☐	Transmission attempt completed within 90 seconds from off-hook to on-hook.
☐	☐	☐	DACT trouble indicated at both supervising station and protected property within four minutes of fault detection

OPERATIONAL TESTS *(continued)*

Pass	Fail	N/A	
			DACT (Digital Alarm Communicator Transmitter) (continued)
☐	☐	☐	Cause DACT to transmit signal to Supervising Station DACR (Digital alarm communications receiver) while a fault in primary line is simulated, and DACT used secondary telephone line to complete transmission to DACR
			DART (Digital Alarm Radio Transmitter)
☐	☐	☐	DART transmits a trouble signal to supervising station within four minutes
			McCulloh Transmitter
☐	☐	☐	McCulloh transmitter produces not less than three complete rounds of no less than three complete signal impulses
			RAT (Radio Alarm Transmitter)
☐	☐	☐	Indication of fault at protected premises or trouble signal transmitted to supervising station

Special Procedures

Pass	Fail	N/A	
			Alarm Verification
☐	☐	☐	Verify alarm response for smoke detector circuits having alarm verification under 150 seconds (60 for verification + 90 for normal signal transmission)
☐	☐	☐	***Upon completion of test, system restored to normal***

Comments: ______________________________

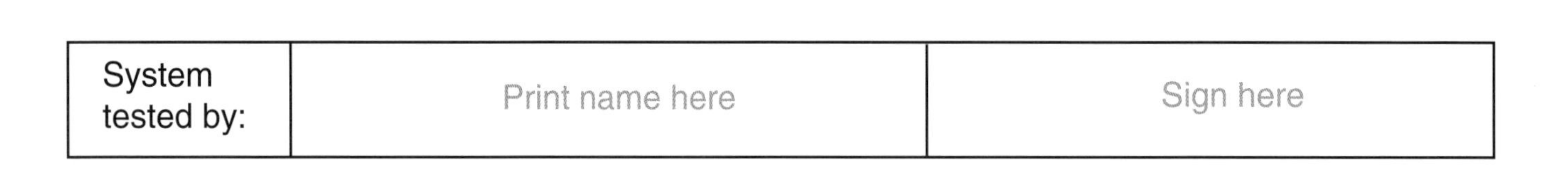

System tested by:	Print name here	Sign here

☐ **System Approved**
☐ **System Disapproved** Date: __________ Inspector's Initials __________

Residential Fire Alarm Systems
NFPA 72, UFC Sections 1001 & 1007
and UFC 10-3 & 10-4

Colorado Springs Fire Department
Fire Alarm Acceptance Test
NFPA 72, UFC Sections 1001 and 1007

CSFD Project #	Date:	Inspector:
CSFD Permit #		Project Name:
Installation Company:		Building ID:
Installer Name:		Project Address:
Installer Phone:		Responsible Party's Phone:
Monitoring service:	Monitoring service contact number:	
Location of alarm control panel:		
Alarm make and model:		
Type of alarm system: ☐central ☐proprietary ☐remote station ☐local (constantly attended location)		

DOCUMENTATION

Yes	No	N/A	
☐	☐	☐	"Installer Certification" form provided to CSFD
☐	☐	☐	Installer has permit and plans approved by CSFD on site
☐	☐	☐	All components located same as approved drawing and specifications
☐	☐	☐	Copy of system instruction manuals on site
☐	☐	☐	Contractor dB pretest list provided for spot check verification

INSTALLATION

Yes	No	N/A	
☐	☐	☐	Devices are located in all areas required by the code
☐	☐	☐	Visual warning devices (strobes) provided in handicap accessible areas as per plans
☐	☐	☐	Fire alarms, supervisory signals and trouble alarms distinctly different and descriptively annunciated (labeled lights or LEDs)
☐	☐	☐	Manual fire alarm (pull) boxes mounted at proper height and location, readily accessible, unobstructed, and along normal path of egress at each exit from each level with additional boxes when travel distance is greater than 200'
☐	☐	☐	Auto dial devices programmed to dial supervising station
☐	☐	☐	Knox Box contains all necessary keys for access and operation of fire alarm system

OPERATIONAL TESTS

Yes	No	N/A	
☐	☐	☐	Notify fire department and alarm monitoring agency, prior to testing, that testing will take place and an emergency response is not desired, unless 911 call is placed
☐	☐	☐	Notify on-site personnel, prior to testing, that testing will take place, and signal arranged in case of actual fire

Perform the following tests under both primary and secondary power

Pass	Fail	N/A	
☐	☐	☐	Power light on, in normal condition
☐	☐	☐	A second initiating device overrode silence switch

OPERATIONAL TESTS *(continued)*

Pass	Fail	N/A	
☐	☐	☐	Alarm initiating devices actuated
			Fire/Life safety features operated properly:
☐	☐	☐	Magnetic door releasing hardware
☐	☐	☐	Ventilation shutdown
☐	☐	☐	Smoke control systems
☐	☐	☐	Emergency lighting
☐	☐	☐	Elevator recall to designated floors
☐	☐	☐	Egress systems
☐	☐	☐	Fire alarm fire emergency phone jacks operate
☐	☐	☐	Fire alarm equipment not concealed, obstructed, or impaired
☐	☐	☐	Verify isolation from ground (ground faults)
☐	☐	☐	Verify conductor-to-conductor isolation (short circuit faults)

Control Panel

☐	☐	☐	Secondary means of alarm transmission provided
☐	☐	☐	Field zones properly identified on panel and match labeled panel zones & drawing
☐	☐	☐	Alarm signal received at off site location showing correct zone
☐	☐	☐	All lamps and LEDs illuminate via panel lamp test switch
			Fuses
☐	☐	☐	Verify rating
☐	☐	☐	Verify supervision trouble signal when fuses are removed
			Interfaced Equipment
☐	☐	☐	Integrity verified
☐	☐	☐	Compatibility of equipment verified
☐	☐	☐	Trouble signal verified at FACP

Power Supplies

Pass	Fail	N/A	
☐	☐	☐	Disconnection of Secondary batteries causes trouble signal
☐	☐	☐	If provided, test UPS (Uninterrupted Power Supply)
☐	☐	☐	Fire alarm circuit breaker labeled "Fire Alarm Control Circuit" in red
☐	☐	☐	Verified dedicated 120 Volt AC branch circuit
			Batteries
☐	☐	☐	Batteries match approved plans, specifications, and calculations
☐	☐	☐	No leakage
☐	☐	☐	Connections tight
☐	☐	☐	If lead-acid batteries, electrolyte level full
			Transient Suppressors
☐	☐	☐	Lightning protection equipment installed

OPERATIONAL TESTS *(continued)*

Control Panel Trouble Signals

Pass	Fail	N/A	
☐	☐	☐	Trouble signals and panel light operated for each circuit tested
☐	☐	☐	All panel trouble signals operate properly
☐	☐	☐	Trouble signals received at off site location when supervised function is disconnected

Signal Silencing/Reset

Pass	Fail	N/A	
☐	☐	☐	Audible and visual supervisory/trouble silencing means
☐	☐	☐	Alarm signal silencing means
☐	☐	☐	Alarm signal reset means

Supervisory Initiating Devices

Pass	Fail	N/A	
☐	☐	☐	Supervisory device actuated and signal verified
☐	☐	☐	Visible audible signal only at annunciator and control panel
			Control Valve Switch
☐	☐	☐	Within first two revolution, 1/5 travel distance, or manufacturer's specification
☐	☐	☐	Device indicating pin is not in the stem's threads
☐	☐	☐	Restoration to normal
			Air Pressure Switch
☐	☐	☐	Variance of + or - 10psi from required
☐	☐	☐	Restoration to normal

Detection And Initiating Devices

Pass	Fail	N/A	
			Fire Extinguishing or Suppression System (Including Sprinklers) Alarm Switch
☐	☐	☐	Operates audible evacuation devices and signal received at FACP
			Fire Gas and Other Gas Detectors
☐	☐	☐	Tested as per manufacturer's listing
			Heat Detectors - Avoid damage to non-restorable fixed temperature elements!
☐	☐	☐	Verify temperature rating vs. ambient temperature of location
☐	☐	☐	Temperature classification identified on device
			Fixed Temperature, Rate of Rise
☐	☐	☐	Heat tested with heat source
			Fixed Temperature, Non-Restorable, Line Type
☐	☐	☐	Tested mechanically and electrically for function
			Fixed Temperature, Non-Restorable, Spot Type
☐	☐	☐	Tested electrically for function
			Restorable, Line Type, Pneumatic Tube Only
☐	☐	☐	Heat source or pressure pump

OPERATIONAL TESTS *(continued)*
Detection And Initiating Devices *(continued)*

Pass	Fail	N/A	
			Smoke Detectors
			System Type
☐	☐	☐	Test in place with listed aerosol smoke into sensing chamber or by other approved method
☐	☐	☐	Alarm responded
☐	☐	☐	Device is listed and marked with sensitivity range
			Air Sampling and Duct Type
☐	☐	☐	Test as per manufacturer's recommended test method
			Projected Beam
☐	☐	☐	Introduce aerosol smoke or optical filter in beam path
			Smoke Detector with Thermal Element
☐	☐	☐	Operate each part independently as described for respective devices
			Radiant Energy Sensing Fire Detectors
☐	☐	☐	Test only per manufacturer's prescribed test method
			Fire Alarm Boxes
☐	☐	☐	For key operating signal, test both pre-signal and general alarm circuits
			Water Flow Switch
☐	☐	☐	For wet systems: Flow though inspector's test valve with orifice equal to that of smallest sprinkler orifice
☐	☐	☐	For dry, pre-action, and deluge systems: Flow through alarm test bypass

Alarm Notification Devices

Pass	Fail	N/A	
☐	☐	☐	Audible and visual alarm devices operated properly, and signal throughout building
			Audible Alarms
☐	☐	☐	Measure and record dBas
☐	☐	☐	Audible levels of fire alarm warning device exceeds 15 dBa (decibels above ambient) or 5 dBs above maximum sound level with 30 seconds duration
☐	☐	☐	Alarm evacuation signal under 120 dBa maximum noise level
			Speakers
☐	☐	☐	Measure and record dBas
☐	☐	☐	Voice alarm signal is tone signal for 1/2 to one second "on" and one second "off" for three cycles, then voice alarm
☐	☐	☐	Voices are clearly understood and message makes sense
			Visible
☐	☐	☐	Verify approved installation layout matches approved plans

OPERATIONAL TESTS *(continued)*

Supervising Station Fire Alarm Systems

Pass	Fail	N/A	
☐	☐	☐	24 hour monitoring agency has the appropriate telephone number for emergency response in protected property's community
			All Equipment
☐	☐	☐	Test all system functions as per manufacturer recommended test procedure
☐	☐	☐	Actuated initiating device, and signal received within 90 seconds
☐	☐	☐	If test jacks are used, first and last tests performed without use of test jacks
			DACT (Digital Alarm Communicator Transmitter)
☐	☐	☐	DACT seizes primary line while primary line is in use and signals received at supervising station
☐	☐	☐	Transmission attempt completed within 90 seconds from off-hook to on-hook.
☐	☐	☐	DACT trouble indicated at both supervising station and protected property within four minutes of fault detection
☐	☐	☐	Cause DACT to transmit signal to Supervising Station DACR (Digital alarm communications receiver) while a fault in primary line is simulated, and DACT used secondary telephone line to complete transmission to DACR
			DART (Digital Alarm Radio Transmitter)
☐	☐	☐	DART transmits a trouble signal to supervising station within four minutes
			McCulloh Transmitter
☐	☐	☐	McCulloh transmitter produces not less than three complete rounds of no less than three complete signal impulses
			RAT (Radio Alarm Transmitter)
☐	☐	☐	Indication of fault at protected premises or trouble signal transmitted to supervising station

Emergency Communications Equipment

Pass	Fail	N/A	
			Amplifier/Tone generators
☐	☐	☐	Verify proper switching and operation of back-up equipment
☐	☐	☐	Operate function
☐	☐	☐	Verify receipt of proper audible and visible signals at FACP
			Phone Jacks
☐	☐	☐	Initiate communication through jack
			Phone Sets
☐	☐	☐	Verify proper operation
			System Performance
☐	☐	☐	Verify that open or shorted telephone zones do not interfere with system performance
☐	☐	☐	Verify that speakers do not cause audio feedback when microphone is used

Special Procedures

Pass	Fail	N/A	
			Alarm Verification
☐	☐	☐	Verify alarm response for smoke detector circuits having alarm verification under 150 seconds (60 for verification + 90 for normal signal transmission)

☐ ☐ ☐ ***Upon completion of test, system restored to normal***

Comments: ____________________

System tested by:	Print name here	Sign here

☐ **System Approved**
☐ **System Disapproved** Date: __________ Inspector's Initials __________

Fire Alarm Systems
NFPA 72, UFC Sections 1001 & 1007
and UFC 10-3 & 10-4

Colorado Springs Fire Department
Automatic Sprinkler System
Installer Certification

Permit #: ____________ Plan Review #: ____________

Date: ____________ City License #: ____________ Exp. Date: ____________

	Owner/Occupant	System Installer	System Supplier
Business Name:			
Address:			
Contact:			
Telephone #:			

Approved plans on site? ☐ **Yes** ☐ **No**

Contractor's Material and Test Certificate for Aboveground Piping attached? ☐ **Yes** ☐ **No**

Certification of System Installation:

Complete this section after system is installed but prior to conducting operational acceptance tests. This system installation was inspected by ____________ (person) on ____________ and was found to comply with the installation requirements of:

- ☐ Please Circle: NFPA 13 13R 13D 25 231C
- ☐ UBC Standard
- ☐ Manufacturer's Instruction/ Specifications (provide)
- ☐ Other (specify & provide) ____________

Print Name: ____________ Signed: ____________

Organization: ____________ Date: ____________

Certification of System Operation:

All operations features and functions of this system were tested by ____________ (person) on ____________ and found to be operating properly in accordance with the requirements of:

- ☐ Please Circle: NFPA 13 25
- ☐ UFC and UBC Standards
- ☐ Manufacturer's Instructions (provide)
- ☐ Other (specify & provide) ____________

Print Name: ____________ Signed: ____________

Organization: ____________ Date: ____________

AUTOMATIC SPRINKLER SYSTEMS

FORM 2-J
(Page 1 of 2)

CONTRACTOR'S MATERIAL AND TEST CERTIFICATE FOR ABOVEGROUND PIPING

PROCEDURE
Upon completion of work, inspection and tests shall be made by the contractor's representative and witnessed by an owner's representative. All defects shall be corrected and system left in service before contractor's personnel finally leave the job.

A certificate shall be filled out and signed by both representatives. Copies shall be prepared for approving authorities, owners and contractor. It is understood the owner's representative's signature in no way prejudices any claim against contractor for faulty material, poor workmanship, or failure to comply with approving authority's requirements or local ordinances.

PROPERTY NAME	DATE
PROPERTY ADDRESS	

PLANS	ACCEPTED BY APPROVING AUTHORITY('S) NAMES		
	ADDRESS		
		☐ YES	☐ NO
		☐ YES	☐ NO
INSTRUCTIONS	HAS PERSON IN CHARGE OF FIRE EQUIPMENT BEEN INSTRUCTED AS TO LOCATION OF CONTROL VALVES AND CARE AND MAINTENANCE OF THIS NEW EQUIPMENT? IF NO, EXPLAIN	☐ YES	☐ NO
	HAVE COPIES OF APPROPRIATE INSTRUCTIONS AND CARE AND MAINTENANCE CHARTS AND NFPA 13A BEEN LEFT ON PREMISES? IF NO, EXPLAIN	☐ YES	☐ NO
LOCATION OF SYSTEM	SUPPLIES BLDGS.		

	MAKE	MODEL	YEAR OF MANUFACTURE	ORIFICE SIZE	QUANTITY	TEMPERATURE RATING
SPRINKLERS						

PIPE AND FITTINGS	PIPE CONFORMS TO_____________ STANDARD	☐ YES	☐ NO
	FITTINGS CONFORM TO_____________ STANDARD IF NO, EXPLAIN	☐ YES	☐ NO

	ALARM DEVICE			MAXIMUM TIME TO OPERATE THROUGH TEST PIPE	
ALARM VALVE OR FLOW INDICATOR					

	DRY VALVE			Q.O.D.		
DRY PIPE OPERATING TEST						

	TIME TO TRIP THRU TEST PIPE*		WATER PRESSURE	AIR PRESSURE	TRIP POINT AIR PRESSURE	TIME WATER REACHED TEST OUTLET*		ALARM OPERATED PROPERLY	
	MIN.	SEC.	PSI	PSI	PSI	MIN.	SEC.	YES	NO
Without Q.O.D.									
With Q.O.D.									

IF NO, EXPLAIN

*MEASURED FROM TIME INSPECTOR'S TEST PIPE IS OPENED.

AUTOMATIC SPRINKLER SYSTEMS

FORM 2-J
(Page 2 of 2)

DELUGE & PREACTION VALVES

OPERATION ☐ PNEUMATIC ☐ ELECTRIC ☐ HYDRAULIC

PIPING SUPERVISED ☐ YES ☐ NO | DETECTING MEDIA SUPERVISED ☐ YES ☐ NO

DOES VALVE OPERATE FROM THE MANUAL TRIP AND/OR REMOTE CONTROL STATIONS ☐ YES ☐ NO

IS THERE AN ACCESSIBLE FACILITY IN EACH CIRCUIT FOR TESTING ☐ YES ☐ NO | IF NO, EXPLAIN

MAKE	MODEL	DOES EACH CIRCUIT OPERATE SUPERVISION LOSS ALARM		DOES EACH CIRCUIT OPERATE VALVE RELEASE		MAXIMUM TIME TO OPERATE RELEASE	
		YES	NO	YES	NO	MIN.	SEC.

TEST DESCRIPTION

HYDROSTATIC: Hydrostatic tests shall be made at not less than 200 psi (13.6 bars) for two hours or 50 psi (3.4 bars) above static pressure in excess of 150 psi (10.2 bars) for two hours. Differential dry-pipe valve clappers shall be left open during test to prevent damage. All aboveground piping leakage should be stopped.
FLUSHING: Flow the required rate until water is clear as indicated by no collection of foreign material in burlap bags at outlets such as hydrants and blow-offs. Flush at flows not less than 400 GPM (1514 L/min) for 4-inch pipe, 600 GPM (2271 L/min) for 5-inch pipe, 750 GPM (2839 L/min) for 6-inch pipe, 1000 GPM (3785 L/min) for 8-inch pipe, 1500 GPM (5678 L/min) for 10-inch pipe and 2000 GPM (7570 L/min) for 12-inch pipe. When supply cannot produce stipulated flow rates, obtain maximum available.
PNEUMATIC: Establish 40 psi (2.7 bars) air pressure and measure drop which shall not exceed 11/2 psi (0.1 bars) in 24 hours. Test pressure tanks at normal water level and air pressure and measure air pressure drop which shall not exceed 11/2 psi (0.1 bars) in 24 hours.

TESTS

ALL PIPING HYDROSTATICALLY TESTED AT ______ PSI FOR ______ HRS.
DRY PIPING PNEUMATICALLY TESTED ☐ YES ☐ NO
EQUIPMENT OPERATES PROPERLY ☐ YES ☐ NO
IF NO, STATE REASON

DRAIN TEST | READING OF GAGE LOCATED NEAR WATER SUPPLY TEST PIPE: STATIC PRESSURE: ______ PSI | RESIDUAL PRESSURE WITH VALVE IN TEST PIPE OPEN WIDE ______ PSI

Underground mains and lead in connections to system risers flushed before connection made to sprinkler piping.

VERIFIED BY COPY OF THE U FORM NO. 85B ☐ YES ☐ NO | OTHER EXPLAIN

FLUSHED BY INSTALLER OF UNDERGROUND SPRINKLER PIPING ☐ YES ☐ NO

BLANK TESTING GASKETS

NUMBER USED	LOCATIONS	NUMBER REMOVED

WELDING

WELDED PIPING ☐ YES ☐ NO

IF YES...

DO YOU CERTIFY AS THE SPRINKLER CONTRACTOR THAT WELDING PROCEDURES COMPLY WITH THE REQUIREMENTS OF AT LEAST AWS D10.9, LEVEL AR-3 ☐ YES ☐ NO

DO YOU CERTIFY THAT THE WELDING WAS PERFORMED BY WELDERS QUALIFIED IN COMPLIANCE WITH THE REQUIREMENTS OF AT LEAST AWS D10.9, LEVEL AR-3 ☐ YES ☐ NO

DO YOU CERTIFY THAT WELDING WAS CARRIED OUT IN COMPLIANCE WITH A DOCUMENTED QUALITY CONTROL PROCEDURE TO INSURE THAT ALL DISCS ARE RETRIEVED, THAT OPENINGS IN PIPING ARE SMOOTH, THAT SLAG AND OTHER WELDING RESIDUE ARE REMOVED, AND THAT THE INTERNAL DIAMETERS OF PIPING ARE NOT PENETRATED ☐ YES ☐ NO

HYDRAULIC DATA NAMEPLATE

NAMEPLATE PROVIDED ☐ YES ☐ NO | IF NO, EXPLAIN

REMARKS

DATE LEFT IN SERVICE WITH ALL CONTROL VALVES OPEN:

SIGNATURES

NAME OF SPRINKLER CONTRACTOR

TESTS WITNESSED BY

FOR PROPERTY OWNER (SIGNED)	TITLE	DATE
FOR SPRINKLER CONTRACTOR (SIGNED)	TITLE	DATE

ADDITIONAL EXPLANATION AND NOTES

Colorado Springs Fire Department
Automatic Sprinkler Systems Acceptance Test
NFPA 13 and UFC Section 1003

CSFD Project #	Date:
CSFD Permit #	Inspector:
Installation Company:	Project Name:
Installer Name:	Building ID:
Installer Phone:	Project Address:
Installer License #:	Responsible Party's Phone:
System Location:	
Location of main sprinkler riser:	
FDC location:	
Water supply test results gpm@ psi	
Type of system: ☐ wet ☐ dry ☐ preaction ☐ deluge ☐ water spray ☐ water mist	
System additives: ☐ antifreeze ☐ anti-corrosive ☐ wetting agent ☐ other ______	
Make and model of back flow preventer:	
Protected area (square foot per riser):	
If dry pipe or preaction, system volume: (gallons)	
Fire pump present: ☐ yes ☐ no	

DOCUMENTATION

Yes	No	N/A	
☐	☐	☐	Installer's certificate on file with CSFD
☐	☐	☐	Certified installer has permit and plans approved by CSFD on site
☐	☐	☐	All components located same as approved drawings or "As Builts" submitted
☐	☐	☐	Copy of system instruction manuals on site
☐	☐	☐	Owner has a current copy of NFPA 25 on site or copy of service contract

COMPONENTS

Yes	No	N/A	
☐	☐	☐	Inspector's test valve is smoothbore, corrosion resistant orifice same diameter as smallest sprinkler head
☐	☐	☐	Relief valve is present if pressure is 175 psi or greater, or 10 psi above maximum system pressure in a gridded wet pipe system
☐	☐	☐	Double check valve or back flow prevention valve installed
☐	☐	☐	FDC piping sized to system according to approved plans
☐	☐	☐	FDC has one hose connection per 250 gpm flow required
☐	☐	☐	FDC has national standard hose thread internal threaded swivel fittings
☐	☐	☐	Hose connections are 2 1/2" diameter unless otherwise approved
☐	☐	☐	Hose valves are installed on a 4" riser minimum
☐	☐	☐	Pressure gauges on both sides of alarm check valve or back flow preventer
☐	☐	☐	Valve room light and heat provided
☐	☐	☐	Extra heads and correct head wrench for each sprinkler head type in cabinet in riser room as specified in plan review

Automatic Sprinkler Systems
NFPA 13 and UFC Section 1003

COMPONENTS *(continued)*

Yes	No	N/A	
☐	☐	☐	Extra sprinkler heads are of proper type and temperature
☐	☐	☐	If pressure at any hose connection outlet is greater than 100 psi, pressure reducing valves present to reduce the pressure to 100 psi

INSTALLATION

Yes	No	N/A	
☐	☐	☐	*Permanent system identification signs for each:*
☐	☐	☐	Control valve (if multiple valves present)
☐	☐	☐	Riser valves
☐	☐	☐	Hydraulic calculation label attached to riser
☐	☐	☐	FDC
☐	☐	☐	Water supply valves, test valves, and flow switch electrically monitored for I-1 occupancies (20 heads or greater) and all others (100 or greater)
☐	☐	☐	Water supply valves are of indicating type and supervised by one of four approved means: ____ sealed (small systems only) ____ locked (small systems only) ____ tamper switch - local alarm ____ tamper switch -off site alarm
☐	☐	☐	Main drain to exterior with turned down elbow
☐	☐	☐	Riser supported by hanger or attachment, according to plans
☐	☐	☐	Provisions for flushing pipe (i.e., end caps)
☐	☐	☐	Entire system able to be drained
☐	☐	☐	Pipe protected from damage (freezing and corrosion)
☐	☐	☐	Horn and strobe water flow alarm is located above the FDC (or within 20' on same side of building)
☐	☐	☐	FDC readily accessible and free from obstructions
☐	☐	☐	All FDC caps securely in place
☐	☐	☐	FDC has check valve and drip valve
☐	☐	☐	4' of pipe installed between FDC and check valve provided in a heated area
☐	☐	☐	FDC properly connected to riser: • wet, single riser system connects to system side of control valve • wet, multi-riser system connects to supply side of control valve • dry system connects between indicating control valve and dry-pipe valve
☐	☐	☐	Maximum distance between hangers complies with NFPA 13
☐	☐	☐	Sprinkler heads are properly spaced (15' maximum or manufacturer's specifications)
☐	☐	☐	If subject to damage, sprinkler heads have guards in place
☐	☐	☐	Sprinkler heads free from coatings not applied by manufacturer
☐	☐	☐	Sprinkler heads parallel to slope of ceiling, roof, and /or stairs
☐	☐	☐	Piping penetrations have proper clearance: 1" clearance for 1" to 3 1/2" pipe 2" clearance for 4" and larger pipe
☐	☐	☐	Escutcheon plates installed
☐	☐	☐	In systems over 100 psi, armovers are secured from vertical movement, or 12" maximum (steel) unsupported length

OPERATIONAL TESTS

Pass	Fail	N/A	
☐	☐	☐	FDC clapper operates easily
☐	☐	☐	Hydrostatic testing of system @ 200 psi for 2 hours (50 psi above static, if static is above 150 psi) including FDC
☐	☐	☐	Main drain flow test completed per flow curve
☐	☐	☐	Exterior flow test completed
☐	☐	☐	Flow test through inspector's test valve
☐	☐	☐	Activation of flow alarm for all types of systems within 90 seconds per NFPA 72

DRY & PRE-ACTION SYSTEMS ONLY

Yes	No	N/A	
☐	☐	☐	If system volume is 500 gallons or greater, then a quick opening device is required
☐	☐	☐	If system volume is greater than 750 gallons, then multiple dry pipe valves are required, unless system can provide water at the most remote sprinkler in less than 60 seconds
☐	☐	☐	Paddle type water flow device has not been installed in dry, preaction or deluge systems
☐	☐	☐	Dry system compressor with minimum 1/2" fill line and pressure gauges
☐	☐	☐	Air pressure maintained above the water pressure as per manufacturer's specifications
☐	☐	☐	Auxiliary drains (drum drip valve) at lowest point in the system from inspector's test valve

Pass	Fail	N/A	
☐	☐	☐	Dry system relief valve functions automatically
☐	☐	☐	Low air alarm monitored
☐	☐	☐	System to refill with air from compressor within 30 minutes
☐	☐	☐	Operational test of dry pipe valve performed
☐	☐	☐	Quick opening device tested
☐	☐	☐	Air pressure test @ 40 psi for 24 hours (1 1/2 psi loss allowed)

Comments: ______________________________

____

____

____

____

____

____

____

____

System tested by:	Print name here	Sign here

☐ **System Approved**

☐ **System Disapproved** Date: __________ Inspector's Initials __________

Automatic Sprinkler Systems
NFPA 13 and UFC Section 1003

Colorado Springs Fire Department
Residential Sprinkler Systems
(Within Dwelling Unit Only)
Acceptance Test
NFPA 13D and 13R

CSFD Project #	Date:
CSFD Permit #	Inspector:
Installation Company:	Project Name:
Installer Name:	Building ID:
Installer Phone:	Project Address:
Installer License #:	Responsible Party's Phone:
System Location:	
Location of main sprinkler riser:	
FDC location:	
Water supply test results: gpm@	psi
Type of system: ☐ wet ☐ dry ☐ preaction	☐ deluge
System additives: ☐ antifreeze ☐ anti-corrosive	☐ wetting agent ☐ other ________
Total # of residential sprinkler heads: ______ # normal	______ # intermediate
Type of heads: ______ # upright ______ # pendent	______ # sidewall
Make and model of back flow preventer:	
Protected area (square foot per riser):	
If dry pipe or preaction, system volume: (gallons)	
Fire pump present: ☐ yes ☐ no	

DOCUMENTATION

Yes	No	N/A	
☐	☐	☐	Installer's certificate on file with CSFD
☐	☐	☐	Certified installer has permit and plans approved by CSFD on site
☐	☐	☐	All components located same as approved drawings or "as builts" submitted
☐	☐	☐	Copy of system instruction manuals on site
☐	☐	☐	Owner has a current copy of NFPA 25 on site or service/maintenance agreement

COMPONENTS

Yes	No	N/A	
☐	☐	☐	Piping materials listed and specified in NFPA 13D and 13R
☐	☐	☐	Only new sprinklers used in the installation of new systems
☐	☐	☐	At least three sprinklers of each type are kept on the premises

COMPONENTS *(continued)*

Yes	No	N/A	
☐	☐	☐	Minimum of one source of automatic water supply
☐	☐	☐	Single, listed control valve to shut off both the domestic and sprinkler system
☐	☐	☐	Separate valve provided for domestic water shut off
☐	☐	☐	1” or larger drain and test connection valve on the system side of the control valve
☐	☐	☐	Minimum of one 1 1/2” or 2 1/2” fire department connection provided
☐	☐	☐	Pressure gauges provided to indicate pressures on both sides of main check valves
☐	☐	☐	All sprinklers within a compartment have the same temperature classification
☐	☐	☐	Standard or quick-response used outside the dwelling unit
☐	☐	☐	Non-metallic escutcheons are listed
☐	☐	☐	Only manufacturer applied coatings may be on sprinklers

INSTALLATION

Yes	No	N/A	
☐	☐	☐	13D and 13R systems installed in areas not subject to freezing
☐	☐	☐	13D and 13R system properly insulated
☐	☐	☐	No system additives in 13D or 13R systems
☐	☐	☐	Pipe hanger, bracing and support same as NFPA 13 or manufacturer’s specifications
☐	☐	☐	Listed residential sprinklers used inside the dwelling unit, and directly adjacent corridors if ceilings are flat, smooth and less than 10’ in height
☐	☐	☐	Intermediate temperature rated RS used where: 1. Maximum ambient ceiling temps are between 101°F and 150°F 2. Under glass or plastic skylights 3. In unventilated concealed space under un-insulated roof or attic 4. Near specific heat sources listed in NFPA 13D and 13R
☐	☐	☐	Maximum area protected by a single sprinkler is 144 ft^2 or less, or manufacturer’s specifications
☐	☐	☐	Maximum distance between sprinklers does not exceed 12’, or manufacturer’s specifications
☐	☐	☐	Maximum distance to a wall partition does not exceed 6’, or manufacturer’s specifications
☐	☐	☐	Minimum distance between sprinklers is 8’, or manufacturer’s specifications
☐	☐	☐	Deflectors of pendent and upright sprinklers are 1” to 4” below the ceiling, or manufacturer’s specifications
☐	☐	☐	Deflectors of sidewall sprinklers are between 4” to 6” below the ceiling, or manufacturer’s specifications
☐	☐	☐	Sprinklers positioned so that they are not adversely affected by obstructions

OPERATIONAL TESTS

Pass	Fail	N/A	
☐	☐	☐	Hydrostatic testing: Wet @ 200 psi, or 50 psi above maximum; if maximum is 150 psi or greater for 2 hours including FDC (failure is any leakage or loss of pressure)
☐	☐	☐	Bucket Test: Most remote head(s) will flow 18 gallons per minute into a container graduated in gallons

Comments: ____

____

____

____

____

____

____

____

____

____

System tested by:	Print name here	Sign here

☐ **System Approved**
☐ **System Disapproved** Date: __________ Inspector's Initials __________

Colorado Springs Fire Department
Standpipe System
Installer Certification

Permit #: ______________________ **Plan Review #:** ______________

Date: ____________ **City License #:** ______________ **Exp. Date:** ______________

	Owner/Occupant	System Installer	System Supplier
Business Name:			
Address:			
Contact:			
Telephone #:			

Approved plans on site? ☐ **Yes** ☐ **No**

Standpipe System is: ☐ **Wet** ☐ **Dry**

Contractor's Material and Test Certificate for Aboveground Piping attached? ☐ **Yes** ☐ **No**

Certification of System Installation:

Complete this section after system is installed but prior to conducting operational acceptance tests. This system installation was inspected by ______________________ (person) on ______________ and was found to comply with the installation requirements of:

- ☐ NFPA 14
- ☐ UFC and UBC
- ☐ Manufacturer's Instruction/ Specifications (provide)
- ☐ Other (specify & provide) ______________________

Print Name: ______________________ Signed: ______________________

Organization: ______________________ Date: ______________________

Certification of System Operation:

All operations features and functions of this system were tested by ______________________ (person) on ______________ and found to be operating properly in accordance with the requirements of:

- ☐ NFPA 14
- ☐ Job Specs
- ☐ UFC and UBC
- ☐ Manufacturer's Instructions (provide)
- ☐ Other (specify & provide) ______________________

Print Name: ______________________ Signed: ______________________

Organization: ______________________ Date: ______________________

Colorado Springs Fire Department
Standpipe Systems Acceptance Test
NFPA 14 and UFC Section 1004

CSFD Project #	Date:
CSFD Permit #	Inspector:
Installation Company:	Project Name:
Installer Name:	Building ID:
Installer Phone:	Project Address:
Installer License #:	Responsible Party's Phone:
Location of control valve:	
Water supply test results: gpm@	psi
FDC location:	
Type of system: ______Class I (2.5" Only)	______Class III (1.5" And 2.5")
Fire pump present: ☐ Yes ☐ No	

DOCUMENTATION

Yes	No	N/A	
☐	☐	☐	Installer's certificate provided to CSFD
☐	☐	☐	Certified installer has permit and plans approved by CSFD on site
☐	☐	☐	All components located same as approved drawings
☐	☐	☐	Copy of system instruction manuals on site
☐	☐	☐	Owner has a current copy of NFPA 25 on site, or copy of service contract

COMPONENTS

Yes	No	N/A	
☐	☐	☐	All components installed as per approved drawings
☐	☐	☐	Minimum of one roof outlet
☐	☐	☐	Check valve near FDC system connection
☐	☐	☐	Pipe between FDC and check valve has automatic drip valve
☐	☐	☐	Each floor has water flow device with a test connection
☐	☐	☐	Separate standpipes provided in each required exit stairway
☐	☐	☐	"Standpipe" sign located above standpipe FDC, also indicates required pressure for demand
☐	☐	☐	Connection to the water supply has an OS&Y or other listed indicating-type valve and a check valve located near supply

INSTALLATION

Yes	No	N/A	
☐	☐	☐	Standpipe components installed as per approved plans
☐	☐	☐	Each floor connected to fire alarm system

INSTALLATION *(continued)*

Yes	No	N/A	
☐	☐	☐	Multiple standpipes interconnected at bottom
☐	☐	☐	FDC not more than 100' from hydrant
☐	☐	☐	FDC 18" to 48" above adjoining grade surface
☐	☐	☐	No shut off in FDC piping
☐	☐	☐	Approved pressure reducing valves where static pressure exceeds 100 psi
☐	☐	☐	Standpipes are located in noncombustible stair enclosures and protected from freezing
☐	☐	☐	Hose connections are readily accessible and not more than 6' from the floor
☐	☐	☐	Riser supports are provided at the lowest level, alternate levels and at the top
☐	☐	☐	Lateral runs from standpipe to the hose valve over 18" are provided with hangers
☐	☐	☐	Horizontal standpipe runs detail hangers at a maximum of 15' spacing

OPERATION

Yes	No	N/A	
☐	☐	☐	FDC clapper valves operate
☐	☐	☐	Each outlet valve functional and hose threads in good condition
☐	☐	☐	Each valve operates smoothly throughout full range
☐	☐	☐	Tighten hose caps to prevent leaking during test
☐	☐	☐	Hydrostatically test all systems @ 200 psi (or 50 psi above maximum, if maximum is 150 psi or greater) for two hours including FDC (failure is any leakage or loss of pressure)
☐	☐	☐	Flow tested at hydraulically most remote hose connection
☐	☐	☐	Automatic and semi dry systems flow a minimum of 250 gpm at the uppermost hose connection within three minutes of system actuation
☐	☐	☐	Hydraulically test at low point of system • place gauge at low point • place gauge at the top level • observe and record results
☐	☐	☐	Main drain flow test successfully completed using test curve
☐	☐	☐	Manual standpipe - FD Pumper was used to verify system design
☐	☐	☐	**Class I Test: per 1991 UFC Appendix III-C** • air tested at 25 psi to check for leaks • flow test: at 500 gpm to roof outlet with maximum of 15 psi pressure loss
☐	☐	☐	**Class III Test: per UFC** • flow test: at 500 gpm through each riser from uppermost outlet and maintain a residual pressure of 65 psi
☐	☐	☐	If listed pressure regulating device present, flow test to verify psi setting
☐	☐	☐	If pump installed on system, tests are conducted with pump operating
☐	☐	☐	Hose valve caps removed after test to drain water

Standpipe Systems
NFPA 14 and UFC Section 1004

Comments: ____________________

System tested by:	Print name here	Sign here

☐ **System Approved**
☐ **System Disapproved** Date: __________ Inspector's Initials __________

Colorado Springs Fire Department
Fire Pump System
Installer Certification

Permit #: ______________________ **Plan Review #:** ______________

Date: ____________ **City License #:** ______________ **Exp. Date:** ______________

	Owner/Occupant	System Installer	System Supplier

Approved plans on site? ☐ **Yes** ☐ **No**

Certification of System Installation:

Complete this section after system is installed but prior to conducting operational acceptance tests. Check wiring for open shorts, ground faults, and improper branching. This system installation was inspected by ____________ (person) ____________ on ____________ and was found to comply with the installation requirements of:

- ☐ NFPA 20
- ☐ Manufacturer's Instruction/ Specifications (provide)
- ☐ Other (specify & provide) ____________________

Print Name: ______________________ Signed: ______________________

Organization: ______________________ Date: ______________________

Certification of System Operation:

All operations features and functions of this system were tested by ____________ (person) on ____________ and found to be operating properly in accordance with the requirements of:

- ☐ Please circle: NFPA 20 25
- ☐ Job Specs
- ☐ Manufacturer's Instructions (provide)
- ☐ Other (specify & provide) ____________________

Print Name: ______________________ Signed: ______________________

Organization: ______________________ Date: ______________________

Colorado Springs Fire Department
Fire Pump Acceptance Test
NFPA 20

CSFD Project #		Date:
CSFD Permit #		Inspector:
Installation Company:		Project Name:
Installer Name:		Building ID:
Installer Phone:		Project Address:
Installer License #:		Responsible Party's Phone:
Location of fire pump:		
Location of sprinkler valve:		
FDC location:		
Water supply test results:	gpm@	psi
Make and model of fire pump:		
Make and model of controller:		
Type of driver: ☐ electric	☐ diesel	☐ combination

DOCUMENTATION

Yes	No	N/A	
☐	☐	☐	Certified Installer has permit and approved plans on site
☐	☐	☐	Copy of NFPA 25 or maintenance agreement on site
☐	☐	☐	Copy of all operation and maintenance information on site
☐	☐	☐	Installer's certification on file with CSFD, including the following 1. Hydrostatic testing documentation; Tested @ 200 psi (or 50 psi above static) for 2 hours including FDC (failure is any leakage or loss of pressure) 2. Manufacturer's pump test characteristic curve, and 3. Electrical certification by electrical contractor
☐	☐	☐	All components located match approved plans

COMPONENTS

Yes	No	N/A	
☐	☐	☐	Supervised fire alarm
☐	☐	☐	All components listed for fire service use
☐	☐	☐	Nameplate affixed to pump
☐	☐	☐	Pressure gauge (minimum 3 1/2" diameter) connected near discharge to indicate two times working pressure (minimum 200 psi) in psi, bars, or both
☐	☐	☐	Compound gauge (minimum 3 1/2" diameter) connected near suction pipe to indicate two times working pressure (minimum 100 psi) in inches of Hg, psi, or both
☐	☐	☐	Only steel pipe is used above ground
☐	☐	☐	Pump is enclosed in sheltered area
☐	☐	☐	Pump house has electrical primary lighting, emergency lighting, heat, pitched floors with floor drain, and (if diesel driver present) ventilated

COMPONENTS *(continued)*

Yes	No	N/A	
☐	☐	☐	Low-water pressure or low-water level alarm provided
☐	☐	☐	Valves are of indicating type and electrically supervised by one of two approved methods: ☐ tamper switch - local alarm ☐ tamper switch - off site alarm
☐	☐	☐	One inch or greater clearance around all pipes that pass through walls or floors
☐	☐	☐	Hose valves mounted on hose valve header according to NFPA 20 (see chart in standard)
☐	☐	☐	Listed indicating (or butterfly) valve and drain (or ball drip) valve installed in pipe line to the hose valve header close to the pump
☐	☐	☐	Pump bypass provided

RELIEF VALVE ONLY

Yes	No	N/A	
☐	☐	☐	Relief valve between pump and pump discharge attached so that it can be removed for repairs without disturbing piping
☐	☐	☐	Relief valve discharges into open pipe or into a cone with an observation window secured to
☐	☐	☐	outlet of the relief valve
☐	☐	☐	Relief valve not piped to pump suction or supply connection
☐	☐	☐	A shut off valve is not installed in the relief valve supply or discharge piping

INSTALLATION

Yes	No	N/A	
☐	☐	☐	All components installed according to listing
☐	☐	☐	Circulation relief valve is set below shut off pressure at minimum expected suction pressure
☐	☐	☐	Pump components protected from damage (mechanical, thermal, & chemical)
☐	☐	☐	Guards are installed for flexible connecting shafts
☐	☐	☐	All joints are welded or secured with approved fittings
☐	☐	☐	Butterfly valves installed 50' or more upstream from pump suction flange
☐	☐	☐	Suction pipe is installed below frost-line or in frost-proof casings
☐	☐	☐	Elbows installed at least ten pipe diameters away from pump on suction side only
☐	☐	☐	Eccentric tapered reducers or increases installed level side up
☐	☐	☐	Pump and suction are on the same foundation unless strain relief (flexible couplings) provided
☐	☐	☐	If water is taken from an open source, suction screens installed
☐	☐	☐	If water is taken from a stored supply, vortex plate installed
☐	☐	☐	If pumps are installed in series, butterfly valves not installed between pumps
☐	☐	☐	No device(s) present in suction pipe which will stop, restrict starting, or restrict discharge of a fire pump or driver

OPERATIONAL TESTS

Yes	No	N/A	
☐	☐	☐	Alarm signals received at constantly attended location or 24 hour monitoring service
☐	☐	☐	Suction pipe free from air leaks and/or air pockets
☐	☐	☐	No excessive vibrations, heat build up, or noise from operating fire pump
☐	☐	☐	Volts, amps, and speed measured and recorded

OPERATIONAL TESTS *(continued)*

Pass	Fail	N/A	
☐	☐	☐	Flow test completed: ☐ Minimum ☐ Rated ☐ Peak
☐	☐	☐	Load start test: started and ran to discharge equal to peak load
☐	☐	☐	Test performed to ensure no phase reversal in primary or secondary power supply
☐	☐	☐	No less than six automatic and six manual operations on normal power (let run for five minutes each time)
☐	☐	☐	Automatic operation starts pump from all starting features (e.g., pressure switches & remote starting signals)
☐	☐	☐	Pump started once from each power source after manual operation
☐	☐	☐	Pump started from emergency power, if provided, three automatic and three manual operations
☐	☐	☐	Fire pump ran for at least one hour during all of the preceding tests
☐	☐	☐	*CSFD Stationary Fire Pump Field Acceptance Form* completed

Comments: ____

System tested by:	Print name here	Sign here

☐ **System Approved**
☐ **System Disapproved** Date: __________ Inspector's Initials __________

Colorado Springs Fire Department
Stationary Fire Pump
Field Acceptance Form

Fire Department Official:	Date:
Business Name:	Address:
Installing Contractor & Representative:	
Insurance Co. & Representative:	

GENERAL

Pump	Manufacturer:	Model:	Serial No.:
	Rated GPM:	Rated PSI/Head:	Rated RPM:
Driver	Manufacturer:	Model:	Serial No.:
Controller	Manufacturer:	Model:	Serial No.:
Motor	Rated HP:	Rated RPM:	Drive Type:
	If Electric:Rated Voltage	Rated Amperage:	
	If Diesel: Fuel supply 8 hours (gal/HP)		Fuel ______ Gallons
Fire Pump:	Start ___ PSI	Stop ___ PSI	City Static: ____ PSI
Jockey Pump:	Start ___ PSI	Stop ___ PSI	

WATER FLOWED

Capacity	Hose Streams				GPM	RPM	Suction PSI	Discharge PSI	Net Increase	Amp Leg (Electric Drivers Only)			Voltage	Run Time
	#	Hose Size	Expected Performance	Pitot PSI						1	2	3		
Churn														
100%														
150%														

OPERATIONAL TESTS

		Yes	No	
	Pump Room has adequate:	☐	☐	Pump starts on water flow from roof
☐	Drain	☐	☐	Pump starts on pressure drop
☐	Heat	☐	☐	All fire pumps tested
☐	Light	☐	☐	Pressure relief valve verified and operable
☐	Ventilation (all except electric)	☐	☐	Piping has adequate support
☐	6 manual starts		in.	Hose size
☐	6 automatic starts		in.	Tip size
☐	No excessive vibration		ft.	Hose length
☐	No overheating		psi	Pitot reading
☐	Pump operates with simulated alarm condition		gpm	Calculated Flow Rate

Stationary Fire Pump - Field Acceptance Form (continued)

OPERATIONAL TEST (continued)				
	Electric Drive		Diesel Drive	
☐	Shut down and restart pump at rated capacity without circuit breaker interruption	☐	Alarms: Overspeed (120%), low oil pressure, high temperature, battery failure(s), failure to start (6 cycles, # per battery), charger failure(s)	
☐	If secondary power provided, switch from normal to standby power at rated capacity without significant interruption to pump, inspect transfer switch	☐	Supervision: Pump *Run, Off or Manual* position, charging system trouble, closed valve trouble	
☐	Emergency start operates	☐	Both battery chargers and amp-meters function	
☐	Trip circuit breaker	☐	Tachometer, oil pressure, H_2O temperature, engine hours, amps gauges are all listed and operational	
☐	Supervised alarms: Motor running, power loss, electric phase reverse	☐	Manual shut off ***or*** automatic shut off after 30 minute minimum run time for fire pump	
		☐	Timer set for 30 minutes/week cycle	
		☐	Governor setting:	RPM
		☐	Emergency trip:	RPM

Problems Found: ______________________________

____

Corrections Made: ______________________________

____

Date Corrected: ____________________ By: ____________________

This report is to certify the fire pump system has been properly tested and inspected for reliability to cover the items listed in this report, is consistent with the manufacturer's requirements, and all corrections have been made.	
Signature of Tester:	Date:
Agency:	Phone:
Fire Department Inspector:	

Guidelines For Conducting Fire Pump Tests

Fire pumps shall be tested annually, in accordance with the following:

1. Record the starting and running amperage on all legs of the controller.
2. Test the fire pump on back up power (if applicable.)
3. Verify the fire pump will start on pressure drop.
4. Verify the back-up pump will operate if lead pump fails (if applicable.)
5. Check mercury pressure switch and timer for proper settings.

The Fire Pump Acceptance Test form is to be completed and provided to the Fire Department.

Note: The Fire Pump Acceptance Test form is designed for multiple use:

- annual testing
- quarterly testing
- new installations

Whichever type of test is done, mail a copy of the completed form to the Colorado Springs Fire Prevention Bureau.

Colorado Springs Fire Prevention Bureau, 101 West Costilla Street, Suite 129, Colorado Springs, Colorado 80903
Phone (719) 578-7040 Fax (719) 578-6029

Fire Pump Testing Procedure

General

This guideline was developed to assist both service and fire personnel with the procedures and methods of performing and documenting the testing of fire pumps.

Fire pumps are required to be tested upon installation, and annually (quarterly for high-rise structures.) The performance evaluation test is per the requirements of NFPA 20 and 25, and the Uniform Fire Code.

Definitions:

Acceptance Test: A fire pump test made after a new installation and where major pump repairs have been performed. Tests shall be witnessed by Fire Department staff. Installers of the system will conduct the test.

Annual or Quarterly Confidence Test: An abbreviated version of the fire pump test may not involve the system installer but may involve the insurance carrier or another service contractor.

Test Criteria

NFPA Standard Number 20 has three specific requirements that must be satisfied before a fire pump system can be approved as follows:

1. At churn, shut-off, or zero flow, the net pressure should not exceed 140% of the rated pressure of horizontal split-case pumps. The net pressure should not exceed 140% of the rated pressure for end-suction and vertical shaft turbine-type pumps.

2. At 100% of the rated pump capacity, the net pump pressure should be 100% of the rated pressure.

3. At 150% of the rated pump capacity, the net pump pressure should be at least 65% of the rated pump pressure.

Test Equipment

Acceptance or Annual/Quarterly Confidence Test

1. Enough three inch diameter fire hose to reach from the pump test header to either the playpipe nozzles.
2. Secure method of holding hose. *MANUAL METHODS ARE NOT ACCEPTABLE!*
3. One underwriter style playpipe for each 250 gpm of pump capacity
4. One pitot tube with a 150 psi pressure gauge
5. Two calibrated pressure gauges incremented for a 200-300 PSI range
6. One hand-held tachometer or revolution counter
7. One "amprobe-style" voltmeter
8. Barrier guard fireline tape
9. Two portable radios

Pre-Test Activities - Field Performance and Annual Tests

1. Complete the "general information" section of the Fire Pump Acceptance Test form.

2. The fire pump's information is stamped on the pump's nameplate. This information, in addition to the pump's certification performance curve, is required to perform the test.

3. Pump drive information will be available on the driver nameplate.

4. The controller information should be posted inside its cabinet door.

5. The fire pump and pressure maintenance (jockey) pump starting pressures will be posted or can be read directly from the mercury-type limit switches. The city static pressure can be found on the pump intake pressure gage.

Field Performance Fire Pump Test

1. Inspect the room or building housing the fire pump to ensure compliance with the pertinent codes and plan design.

2. Inspect the fire pump, driver, piping, jockey pump and the controller to ensure compliance with the pertinent codes and plan design.

3. Survey the site of the pump test to determine how the pump discharge will drain or be disposed of. The discharge for the smallest fire pump can approach 750 gpm and contain rocks and debris. Secure the discharge area from vehicular or pedestrian traffic.

4. If the pump is monitored, notify the monitoring agency that a fire pump test is being conducted and that an alarm condition will be transmitted. During the pump test, verify that an alarm condition is being received.

Field Performance Fire Pump Test (continued)

5. Check the pump for excessive noise, vibration and leaks. Other areas to check include:
   - operation of the circulating relief valve
   - pump packing is emitting one drop of water per second

6. Isolate the fire pump from the fire protection system by closing the discharge OS&Y valve, and opening either the indicating control valve to either the pump test header or pump flow loop, depending on the type of installation.

7. If a test header is used, connect the playpipes with nozzles and hose lines to the pump test header. The playpipes should be secured to a test bracket near the test header. If a permanent playpipe's bracket is not provided, a portable playpipe bracket can be used.

8. Replace the pump's suction and discharge gages with **calibrated test gages**.Teflon tape should be wrapped around the gage threads to ensure an adequate seal.

9. If the pump utilizes an electric driver, a volt-ammeter will be used to measure each phase of the driver. The leads are located in the pump controller.

10. Depending on the type of equipment used, access to the driver end plate may be required to measure the motor's rpm. This is especially important on diesel drive fire pumps because the original equipment tachometers are not calibrated.

11. Allow approximately 30 seconds between each successive start. During this portion of the test, verify that an alarm condition is being transmitted.

12. To begin the automatic start portion of the test, open the fire protection system control valve slowly allowing a drop in the system pressure. The pump should automatically start.

13. While operating, record the pressure registered on both the discharge and intake gages, the rpm's, and, if applicable, the amperage measurements. Record the date in the NO FLOW section of the test form.

14. Connect the hoselines, playpipes, and gate valves to the fire department connection. When discharging water, attempt to direct the water discharge toward the roof drains or toward an open perimeter area. If discharge is above the building, secure the discharge area from vehicles and pedestrians.

15. If applicable, open the indicating control valve to the floor test loop. This is located between the suction and discharge piping below the pump bypass. Activate the flow meter; do not attempt to calibrate or adjust the device.

Field Performance Fire Pump Test (continued)

16. Open the fire protection system indicating control valve to charge the system. Fully open a 2 1/2 inch gage valve slowly and pitot the flow. Notify the Control Officer of the pitot measurement. During this period the drive rpm, discharge pressure, suction pressure, flow meter measurement, and amperage (if applicable) should be recorded. Open each successive hoseline as instructed by the Control Officer and measure the flow using the pitot tube. After opening each hoseline measure the flow from each hoseline previously opened. EXAMPLE: If opening hoseline #3; pitot hoselines #1 and #2 after measuring the flow from hoseline #3. Record the pitot measurements and interpret the flow information.

17. Review the discharge information of hoseline #1 in comparison to the flow meter reading. Evaluate the two measurements and, if necessary, have the test site representative calibrate the flow meter.

18. Determine the net pump pressure from the flow measurements.

19. Adjust the gate valve on each hose connection (outlet, not FDC!) so that the total flow equals 100% of the pump's rated capacity. Adjust the individual gage valves so that the pitot measurements equals the data calculated by the Control Officer.

20. Once the specified pitot pressure is achieved, notify the Control Officer. Measure the driver rpm, discharge and suction pressures, flow meter reading, and amperage (if applicable).

21. Once the necessary measurements are recorded and the field calculated results satisfy the Control Officer, shut down the roof hoseline and remove all of the equipment.

22. Remove the calibrated test gages and the volt-ammeter after closing the indicating control valve to the flow test loop. Ensure that the fire protection system control valves are open and supervised, either electronically or by padlock and chain. Place the fire pump controller in the automatic mode and re-inspect the entire installation to verify that the system is fully operative and will function in the event of fire.

23. Notify the property representatives that the test is complete and that the test results will be forwarded. Notify the alarm monitoring agency.

Colorado Springs Fire Department
Foam Fire Protection Systems
Installer Certification

Permit #: ______________________ **Plan Review #** ______________

Date: ____________ **City License #:** ______________ **Exp. Date:** ______________

	Owner/Occupant	System Installer	System Supplier
Business Name:			
Address:			
Contact:			
Telephone #:			

Approved plans on site? ☐ **Yes** ☐ **No**

Certification of System Installation:

Complete this section after system is installed but prior to conducting operational acceptance tests. This system installation was inspected by ______________________ (person) on ______________ and was found to comply with the installation requirements of:

- ☐ Please circle: NFPA 11 11A 16 16A
- ☐ UFC Article 79 Section 7902.2.4.2 and UFC Standard 79-1
- ☐ Manufacturer's Instruction/ Specifications (provide)
- ☐ Other (specify & provide) ______________________

Print Name: ______________________ Signed: ______________________

Organization: ______________________ Date: ______________________

Certification of System Operation:

All operations features and functions of this system were tested by ______________________ (person) on ______________ and found to be operating properly in accordance with the requirements of:

- ☐ Please circle: NFPA 11 11A 16 16A 25
- ☐ Job Specs
- ☐ Manufacturer's Instructions (provide)
- ☐ Other (specify & provide) ______________________

Print Name: ______________________ Signed: ______________________

Organization: ______________________ Date: ______________________

Colorado Springs Fire Department
Foam Fire Protection Systems Acceptance Test
NFPA 11, 11A, 16, 16A, and UFC Article 79-1

CSFD Project #	Date:
CSFD Permit #	Inspector:
Installation Company:	Project Name:
Installer Name:	Building ID:
Installer Phone:	Project Address:
Installer License #:	Responsible Party's Phone:
System Location:	
Location of main sprinkler riser:	
FDC location:	
Water supply test results: gpm@ psi	
Type of system: ☐ wet ☐ dry ☐ preaction ☐ deluge	
System additives: ☐ antifreeze ☐ anti-corrosive ☐ wetting agent ☐ other________	
Make and model of back flow preventer:	
Protected area (square foot per riser):	
If dry pipe or preaction, system volume: (gallons)	
Fire pump present: ☐ yes ☐ no	

DOCUMENTATION

Yes	No	N/A	
☐	☐	☐	Installer's certificate on file with CSFD
☐	☐	☐	Certified installer has permit and plans and calculations approved by CSFD on site
☐	☐	☐	All components located same as approved drawings
☐	☐	☐	Copy of system instruction manuals on site
☐	☐	☐	Owner has a current copy of NFPA 25 or service/maintenance agreement on site

COMPONENTS

Yes	No	N/A	
☐	☐	☐	Self-contained breathing apparatus provided with lifeline attached
☐	☐	☐	All components are listed
☐	☐	☐	Automatic detection with manual actuation means
☐	☐	☐	Reserve foam concentrate on site
☐	☐	☐	Supervised fire alarm

OPERATIONAL TESTS

Pass	Fail	N/A	
☐	☐	☐	Foam concentrate stored between 35°F and 100°F
☐	☐	☐	Strainers inspected and cleaned after each use
☐	☐	☐	Operating instructions posted at control stations
☐	☐	☐	Discharge test meets approved specifications
☐	☐	☐	Flush system with water after foam discharge

Comments: ______________________________

System tested by:	Print name here	Sign here

☐ **System Approved**
☐ **System Disapproved** Date: __________ Inspector's Initials __________

FIRE COLORADO SPRINGS DEPARTMENT

Colorado Springs Fire Department
Water Spray System
Installer Certification

Permit #: ______________________ **Plan Review #** ______________

Date: __________ **City License #:** ______________ **Exp. Date:** ______________

	Owner/Occupant	System Installer	System Supplier
Business Name:			
Address:			
Contact:			
Telephone #:			

Approved plans on site? ☐ **Yes** ☐ **No**

Certification of System Installation:

Complete this section after system is installed but prior to conducting operational acceptance tests. This system installation was inspected by ______________ (person) on __________ and was found to comply with the installation requirements of:

- ☐ Please circle: NFPA 10 25
- ☐ Manufacturer's Instruction/ Specifications (provide)
- ☐ Other (specify & provide) ______________

Print Name: ______________ Signed: ______________

Organization: ______________ Date: ______________

Certification of System Operation:

All operations features and functions of this system were tested by ______________ (person) on __________ and found to be operating properly in accordance with the requirements of:

- ☐ Please circle: NFPA 10 25
- ☐ Job Specs
- ☐ Manufacturer's Instructions (provide)
- ☐ Other (specify & provide) ______________

Print Name: ______________ Signed: ______________

Organization: ______________ Date: ______________

Colorado Springs Fire Department
Water Spray Systems Acceptance Test
NFPA 15, UFC Standard 79-2

CSFD Project #	Date:
CSFD Permit #	Inspector:
Installation Company:	Project Name:
Installer Name:	Building ID:
Installer Phone:	Project Address:
Installer License #:	Responsible Party's Phone:
System Location:	
Location of main riser:	
FDC location:	
Water supply test results: gpm@ psi	
System additives: ☐ antifreeze ☐ anti-corrosive ☐ wetting agent ☐other ________	
Make and model of back flow preventer:	
Protected area (square foot per riser):	
Hazard protected:	

DOCUMENTATION

Yes	No	N/A	
☐	☐	☐	Installer's certificate on file with CSFD
☐	☐	☐	Certified installer has permit and plans approved by CSFD on site
☐	☐	☐	All components located same as approved drawings and specifications
☐	☐	☐	Copy of system instruction manuals on site
☐	☐	☐	Owner has current copy of NFPA 25 on site or service/maintenance agreement
			Power light on, lit normal condition

COMPONENTS

Yes	No	N/A	
☐	☐	☐	All system components are listed (except drain valves and signs)
☐	☐	☐	All system components are installed as shown in approved plans
☐	☐	☐	System actuation valve has both automatic and manual actuation means
☐	☐	☐	Required warning signs provided
☐	☐	☐	At least one manual actuation device, each independent of the manual actuation device at the system actuation valve, is conspicuously located, readily accessible during the emergency, and properly identified as to the system controlled
☐	☐	☐	One or more FDC's provided

INSTALLATION

Yes	No	N/A	
☐	☐	☐	Extra heavy pattern fittings where pressure exceeds 175 psi
☐	☐	☐	Water supply control valve remains readily accessible during a fire situation
☐	☐	☐	Water supply control valves are of indicating type and electrically supervised by one of two approved means: _______ tamper switch - local alarm _______ tamper switch - off site alarm
☐	☐	☐	System actuation valves are as close to the protected hazard as accessibility during the emergency will permit
☐	☐	☐	Drain not directly connected to a sewer system
☐	☐	☐	All pipe and riser assembly adequately supported and protected from damage where explosion potential exists
☐	☐	☐	Gauges permit easy removal
☐	☐	☐	Gauges protected from damage (including freezing and mechanical)
☐	☐	☐	Automatic detection equipment protected from damage
☐	☐	☐	Automatic flame detectors located so that no portion of the protected hazard extends beyond the perimeter line of detectors
☐	☐	☐	System arranged for automatic operation with support of manual actuation means
☐	☐	☐	Nozzles spacing maximum 10' vertical or horizontal
☐	☐	☐	If runoff may contain burning, flowing liquid, it will not expose firefighters, critical equipment and piping, other important structures, or property of others

OPERATIONAL TESTS

Pass	Fail	N/A	
☐	☐	☐	Automatically in 40 seconds or less
☐	☐	☐	Manually at each device
☐	☐	☐	Discharge patterns not impeded by obstructions
☐	☐	☐	Audible alarm within 90 seconds after flow begins
☐	☐	☐	If separate detection system activates the water spray system, the separate system shall initiate an alarm independent of water flow

ULTRA HIGH SPEED WATER SPRAY SYSTEMS ONLY

Application: ____local ____ area ____dual ____ personnel

Yes	No	N/A	
☐	☐	☐	Response time less than 100 milliseconds
☐	☐	☐	Wet system only
☐	☐	☐	Minimum 25 gpm/nozzle and .50 gpm/sq. ft. verified
☐	☐	☐	Minimum pressure at most remote nozzle is 50 psi or greater
☐	☐	☐	System is smaller than 500 gallon capacity
☐	☐	☐	Protection time duration of greater than 15 minutes
☐	☐	☐	No fire hose connections allowed
☐	☐	☐	Automatic with supplemental manual operation means
☐	☐	☐	No timers or delays in system
☐	☐	☐	System actuation valves are protected from physical damage

Water Spray Systems
NFPA 15 and UFC Section 79-2

ULTRA HIGH SPEED WATER SPRAY SYSTEMS ONLY (continued)

Yes	No	N/A	
☐	☐	☐	Air bleeder valves shall be placed at all piping high points to bleed air in the system
☐	☐	☐	Strainers in place
☐	☐	☐	Radiant energy sensing devices or equivalent
☐	☐	☐	Control panel located in area protected from physical injury and electromagnetic energy which could induce false activation

Comments: ______________________________

____

____

____

____

____

____

____

____

System tested by:	Print name here	Sign here

☐ **System Approved**
☐ **System Disapproved** Date: __________ Inspector's Initials __________

Colorado Springs Fire Department
CO_2/Clean Agent System
Installer Certification

Permit #: ______________________ **Plan Review #** ______________

Date: ______________ **City License #:** ______________ **Exp. Date:** ______________

	Owner/Occupant	System Installer	System Supplier

Approved plans on site? ☐ **Yes** ☐ **No**

Certification of System Installation:

Complete this section after system is installed but prior to conducting operational acceptance tests. This system installation was inspected by ______________ (person) on ______________ and was found to comply with the installation requirements of:

☐ Please circle: NFPA 12 12A 2001

☐ Manufacturer's Instruction/ Specifications (provide)

☐ Other (specify & provide) ______________

Print Name: ______________ Signed: ______________

Organization: ______________ Date: ______________

Certification of System Operation:

All operations features and functions of this system were tested by ______________ (person) on ______________ and found to be operating properly in accordance with the requirements of:

☐ Please circle: NFPA 12 12A 2001

☐ Job Specs

☐ Manufacturer's Instructions (provide)

☐ Other (specify & provide) ______________

Print Name: ______________ Signed: ______________

Organization: ______________ Date: ______________

Colorado Springs Fire Department
Gaseous Inerting and Extinguishing
Systems Acceptance Test
NFPA 12 and 2001

CSFD Project #	Date:
CSFD Permit #	Inspector:
Installation Company:	Project Name:
Installer Name:	Building ID:
Installer Phone:	Project Address:
Installer License #:	Responsible Party's Phone:
System Location:	
Type of extinguishing agent: ☐CO_2 ☐ FM-200 ☐Other ________	

All Gaseous Inerting and Extinguishing Systems

DOCUMENTATION

Yes	No	N/A	
☐	☐	☐	Installer's certificate provided to CSFD
☐	☐	☐	Certified installer has permit and CSFD approved plans and calculations on site
☐	☐	☐	All components located same as approved drawings
☐	☐	☐	Copy of system instruction manuals on site
☐	☐	☐	Owner has a current copy of maintenance agreement/contract on site
☐	☐	☐	Configuration of the hazard is the same as specified in the original plans

COMPONENTS

Yes	No	N/A	
☐	☐	☐	SCBA provided for rescue purposes
☐	☐	☐	Adequate and reliable primary and 24 hour minimum standby power sources
☐	☐	☐	Dedicated 120 Volt AC branch circuit and properly labeled
☐	☐	☐	All automatic detection devices listed and used in a listed or approved method for detecting hazardous conditions
☐	☐	☐	Alarm to indicate failure of supervised device, distinct from operation or hazardous condition alarms
☐	☐	☐	Manual release means provided
☐	☐	☐	1) Mechanical manual release, or
☐	☐	☐	2) Electrical manual release when control equipment monitors battery level and provides a low battery signal
☐	☐	☐	Warning and instruction signs at entrances and inside protected areas
☐	☐	☐	Cylinder provided with a pressure relief safety device (e.g. frangible disc)
☐	☐	☐	Pre-discharge alarms provided
☐	☐	☐	Zones identified on panel

INSTALLATION

Yes	No	N/A	
☐	☐	☐	All system components maintain minimum clearances from energized electrical equipment
☐	☐	☐	Manual control located not more than 48" above floor level, accessible during a fire, and identified as to the protected hazard, function performed, and method of operation
☐	☐	☐	Protected area is properly sealed with a minimum of non-closable areas
☐	☐	☐	Openings sealed or provided with automatic closures
☐	☐	☐	Actual room volume matches the volume specified on approved plans

OPERATIONAL TESTS

Pass	Fail	N/A	
☐	☐	☐	CSFD Dispatch and Alarm Receiving Office notified: 1. Fire system tested is being conducted, and 2. Emergency response from personnel not desired
☐	☐	☐	Non-destructive operational test performed on all devices pertinent to system function
☐	☐	☐	Tests performed under primary power then secondary power
☐	☐	☐	System disarmed
☐	☐	☐	System activated fire alarm
☐	☐	☐	Pre-discharge alarms operate upon system actuation
☐	☐	☐	24 hour monitoring service received signals and identified each time and type of signal
☐	☐	☐	All automatic valves tested, unless testing will release agent and/or damage the valve, (i.e., no destructive testing)
☐	☐	☐	Release circuit tested at the storage container
☐	☐	☐	Devices labeled with proper designations and instructions
☐	☐	☐	Energy sources shut down in release area upon system activation
☐	☐	☐	Auxiliary life safety functions operate upon system operation (egress, ventilation shut down, emergency lighting, etc.)
☐	☐	☐	Manual releases clearly identified, at proper height, along egress path, accessible during a fire
☐	☐	☐	Manual releases require two separate and distinct actions for operation
☐	☐	☐	All manual actuators are in place and tamper seals are intact
☐	☐	☐	Nameplate data is the same as the specifications
☐	☐	☐	Storage containers securely held in position
☐	☐	☐	Piping securely supported
☐	☐	☐	Piping has no corrosion, kinks, or other damage
☐	☐	☐	Pressure test for 10 minutes at 150 psig (Pressure drop of <10% allowed)
☐	☐	☐	Nozzles are clean and secure
☐	☐	☐	Nozzles orientation and orifice size unchanged from original design
☐	☐	☐	Nozzles have seals where needed
☐	☐	☐	Metal nozzles grounded for electrostatic protection
☐	☐	☐	Discharge nozzles protected by frangible discs or blow-off caps
☐	☐	☐	**System returned to normal**

Gaseous Inerting and Extinguishing Systems
CO_2 and Clean Agent Systems - NFPA 12 and 2001

CO_2 SYSTEMS ONLY

Pass	Fail	N/A	
☐	☐	☐	Abort switches not used on carbon dioxide systems
☐	☐	☐	CO_2 storage is connected to discharge piping and actuators
☐	☐	☐	Objects exposed to CO_2 discharge properly grounded

CLEAN AGENT SYSTEMS ONLY

Yes	No	N/A	
☐	☐	☐	Automatic detection and automatic actuation used
☐	☐	☐	Abort switches clearly identified
☐	☐	☐	Abort switches are of the dead man type (require constant manual pressure)
☐	☐	☐	Abort switch operated; audible and visual supervisory signals received at fire alarm control panel
☐	☐	☐	No nozzles located at head level or below
☐	☐	☐	Disable agent storage container release mechanism; replace with electrically actuated mechanism (e.g. light bulb)
☐	☐	☐	The discharge time for halocarbons does not exceed 10 seconds
☐	☐	☐	The discharge time for inert gas agents does not exceed 60 seconds to achieve 95% of the design concentration

Comments: ____

System tested by:	Print name here	Sign here

☐ **System Approved**

☐ **System Disapproved** Date: ____________ Inspector's Initials ____________

FIRE COLORADO SPRINGS DEPARTMENT

Colorado Springs Fire Department
Dry/Wet Chemical & Powder System
Installer Certification

Permit #: ____________________ **Plan Review #** ____________

Date: __________ **City License #:** ______________ **Exp. Date:** ______________

	Owner/Occupant	System Installer	System Supplier
Business Name:			
Address:			
Contact:			
Telephone #:			

Approved plans on site? ☐ **Yes** ☐ **No**

Certification of System Installation:

Complete this section after system is installed but prior to conducting operational acceptance tests. This system installation was inspected by ____________ (person) on ____________ and was found to comply with the installation requirements of:

- ☐ Please circle: NFPA 17 17A 96
- ☐ UFC Article 10
- ☐ Manufacturer's Instruction/ Specifications (provide)
- ☐ Other (specify & provide) ____________

Print Name: ____________ Signed: ____________

Organization: ____________ Date: ____________

Certification of System Operation:

All operations features and functions of this system were tested by ____________ (person) on ____________ and found to be operating properly in accordance with the requirements of:

- ☐ Please circle: NFPA 17 17A 96
- ☐ UFC Article 10
- ☐ Job Specs
- ☐ Manufacturer's Instructions (provide)
- ☐ Other (specify & provide) ____________

Print Name: ____________ Signed: ____________

Organization: ____________ Date: ____________

Colorado Springs Fire Department
Fixed Industrial Fire Protection Systems
Acceptance Test
NFPA 17 and 17A, UFC Section 1006

CSFD Project #	Date:
CSFD Permit #	Inspector:
Installation Company:	Project Name:
Installer Name:	Building ID:
Installer Phone:	Project Address:
Installer License #:	Responsible Party's Phone:
System Location:	
Type of Extinguishing Agent: ☐ Wet Chemical ☐ Dry Chemical ☐Other ______	
Name of Extniguishing Agent:	
Equipment Protected:	
Extinguishing chemical cylinder	Expellant gas cartridge provided? ☐ yes ☐ no
Normal pressure: psi	If yes,
Manufacturer's minimum pressure: psi	Normal weight (if carbon dioxide): pounds
Normal weight: pounds	Normal pressure (if nitrogen): psi
Manufacturer's minimum weight: pounds	
Automatic shutdown of power sources: ☐ yes ☐ no	If yes, ☐ gas ☐ electric

DOCUMENTATION

Yes	No	N/A	
☐	☐	☐	Installer's certificate provided to CSFD
☐	☐	☐	Certified installer has permit and plans approved by CSFD on site
☐	☐	☐	All components located same as approved drawings
☐	☐	☐	Instruction manual on site
☐	☐	☐	Current maintenance tag/paperwork in place on site
☐	☐	☐	Proof of hydrostatic testing of wet chemical containers, auxiliary pressure containers, and hose assemblies

COMPONENTS

Yes	No	N/A	
☐	☐	☐	All components are listed and installed according to manufacturer's specifications and approved plans
☐	☐	☐	Both automatic and manual means of operation

COMPONENTS *(continued)*

Yes	No	N/A	
☐	☐	☐	System electrically supervised by one of two means: ____local ____off site
☐	☐	☐	Pressure gauge on pressurized containers
☐	☐	☐	Failure/ trouble alarms distinctive from operation or hazardous condition alarms
☐	☐	☐	Portable fire extinguisher of rating not less than 40-B Sodium Bicarbonate, Potassium
☐	☐	☐	Bicarbonate, or Ammonium Phosphate provided within 30' of protected hazard along egress path near exit.

INSTALLATION

Yes	No	N/A	
☐	☐	☐	Manual pull station readily accessible and 3' to 5' above floor
☐	☐	☐	Remote manual actuators identified as to hazard protected
☐	☐	☐	Power and/or fuel supply shut off provided
☐	☐	☐	Supervision of trouble in auto detect system, electrical actuation circuit, & electrical power supply
☐	☐	☐	If alarm system present, it is connected to extinguishing system
☐	☐	☐	Piping system securely supported
☐	☐	☐	Agent main containers and expellant gas assemblies located near protected hazard(s), but protected from exposure to damage, easily accessible, and not above ceiling, ceiling tiles, nor sitting on floor
☐	☐	☐	Discharge nozzles connected and supported so that they are not readily misaligned
☐	☐	☐	A fusible link or heat detector provided above each appliance or group of appliances protected by a single nozzle

TESTING

Pass	Fail	N/A	
☐	☐	☐	Manual pull station activates system
☐	☐	☐	Operation of fusible test link activates system
☐	☐	☐	Dump test with test cartridge
☐	☐	☐	Nozzle blow off caps in place
☐	☐	☐	Pressure gauges in operable range
☐	☐	☐	Automatically shuts off energy sources
☐	☐	☐	Building fire alarm sounds upon system activation
☐	☐	☐	Special devices (water shut off valves, pressure switches, test valves, etc.) function per design of manufacturer

Fixed Fire Protection Systems
Non-Cooking Wet/Dry Chemical
NFPA 17 and 17A

Comments: ______________________________

System tested by:	Print name here	Sign here

☐ **System Approved**
☐ **System Disapproved** Date: __________ Inspector's Initials __________

Colorado Springs Fire Department
Commercial Cooking Fire Protection Systems
Acceptance Test
NFPA 96, UFC Section 1006

CSFD Project #	Date:
CSFD Permit #	Inspector:
Installation Company:	Project Name:
Installer Name:	Building ID:
Installer Phone:	Project Address:
Installer License #:	Responsible Party's Phone:
System Location:	
Type of Extinguishing Agent: ☐ Wet Chemical ☐ Other ________	
Name of Extniguishing Agent:	
Equipment Protected:	
Extinguishing chemical cylinder	Expellant gas cartridge provided? ☐ yes ☐ no
Normal pressure: psi	If yes,
Manufacturer's minimum pressure: psi	Normal weight (if carbon dioxide): pounds
Normal weight: pounds	Normal pressure (if nitrogen): psi
Manufacturer's minimum weight: pounds	
Automatic shutdown of power sources: ☐ yes ☐ no	If yes, ☐ gas ☐ electric

DOCUMENTATION

Yes	No	N/A	
☐	☐	☐	Installer's certificate provided to CSFD
☐	☐	☐	Certified installer has permit and plans approved by CSFD on site
☐	☐	☐	All components located same as approved drawings
☐	☐	☐	Owner has a copy of service agreement on site
☐	☐	☐	Instruction manual on site
☐	☐	☐	Proof of hydrostatic testing of wet chemical containers, auxiliary pressure containers, and hose assemblies

COMPONENTS

Yes	No	N/A	
☐	☐	☐	Connected to fire alarm system
☐	☐	☐	Readily accessible manual actuation device(s)
☐	☐	☐	If manual actuation device electrical, standby power must be supplied and visual means to show system energized (e.g., an indicating light)
☐	☐	☐	Manual pull station easily accessible and not more than 48" from floor along egress path
☐	☐	☐	Instructions posted adjacent to manual actuation device
☐	☐	☐	All manual activation devices located in the same area are labeled as to hazard protected

WET CHEMICAL ONLY

Yes	No	N/A	
☐	☐	☐	Fusible links located per approved plans
☐	☐	☐	Fusible link or heat detector located at each branch duct-to-common duct connection (actuates system and shuts off energy to appliance)
☐	☐	☐	Fusible link or heat detector above cooking appliance or group of cooking appliances protected with a single nozzle
☐	☐	☐	Piping and nozzles are secured
☐	☐	☐	Chemical container easily accessible, not above ceiling or sitting on floor
☐	☐	☐	Each cooking appliance, individual hood and branch exhaust duct protected by a single system or by multiple systems designed for simultaneous operation

SPRINKLERS & WATER SPRAY ONLY

Yes	No	N/A	
☐	☐	☐	Operation of a sprinkler automatically shuts off energy sources under the hood
☐	☐	☐	Sprinklers listed for protecting the deep fat fryer(s) are provided per the approved plans
☐	☐	☐	A listed indicating control valve for the water supply is provided
☐	☐	☐	One sprinkler (325-375°F) is at top of each vertical duct and at midpoint of each offset
☐	☐	☐	One sprinkler (325-375°F) is at entrance of the horizontal duct and at 10' centers starting 5' from duct entrance
☐	☐	☐	One sprinkler (325-375°F) is provided 1"-12" above duct collar
☐	☐	☐	One sprinkler is centered per each 10' length of plenum chamber
☐	☐	☐	Sprinklers in ducts are accessible for maintenance

Hood/Duct Only

Yes	No	N/A	
☐	☐	☐	Airflow data provided to CSFD
☐	☐	☐	Listed hood is of appropriate size
☐	☐	☐	Fire actuated damper in supply air plenum
☐	☐	☐	Grease extracting filter listed, tight fitting, and firmly held in place
☐	☐	☐	Grease extracting filter easily accessible and removable
☐	☐	☐	Grease extracting filter installed in vertical manner or to manufacturer's specifications
☐	☐	☐	16" separation between deep fat-fryer and open flame cooking appliance or a 12" high baffle plate is provided between the appliances from front to back
☐	☐	☐	Listed grease filters are in place

Solid Fuel Only

Yes	No	N/A	
☐	☐	☐	Heavy metal container for ashes
☐	☐	☐	Long-handled tools provided for fuel position adjustment and ash removal
☐	☐	☐	Solid fuel unit three feet or more from deep fat fryers
☐	☐	☐	If make-up air is through hood, then make-up air shuts down upon system actuation and exhaust continues to run

OPERATIONAL TESTS

Pass	Fail	N/A	
☐	☐	☐	Pressure gauges in operable range
☐	☐	☐	Nozzle blow off caps are in place
☐	☐	☐	Both manual pull system and operation of fusible test link activates system
☐	☐	☐	Building fire alarm sounds upon system activation
☐	☐	☐	Energy shutdown devices operate on system activation and **all** equipment under the hood shuts down when the system activates

Comments: ______________________________

System tested by:	Print name here	Sign here

☐ **System Approved**
☐ **System Disapproved** Date: __________ Inspector's Initials __________

Colorado Springs Fire Department
Paint Spray Booths and Rooms
Fire Protection Systems Acceptance Test
NFPA 33 and UFC Article 45

CSFD Project #	Date:
CSFD Permit #	Inspector:
Installation Company:	Project Name:
Installer Name:	Building ID:
Installer Phone:	Project Address:
Installer License #:	Responsible Party's Phone:
Type of system: ☐ Sprinkler ☐ Fixed (Dry/Wet Chemical) ☐ CO_2 ☐ Other ________	
Location of cabinet:	
Location of main riser:	
Location of control valve:	
Location of alarm control panel:	
Monitoring service:	Responsible Party:
Monitoring service contact number(s)	

DOCUMENTATION

Yes	No	N/A	
☐	☐	☐	Installer's certificate provided to CSFD
☐	☐	☐	Certified installer has permit and plans approved by CSFD
☐	☐	☐	All components located same as approved drawings
☐	☐	☐	Copy of system instruction manuals on site
☐	☐	☐	Written program concerning filters to include:
☐	☐	☐	Daily disposal procedures
☐	☐	☐	Changing filters when spraying job completed
☐	☐	☐	Changing filters when spraying with another substance

COMPONENTS

Yes	No	N/A	
☐	☐	☐	"No Smoking" signs posted
☐	☐	☐	Hot work, cutting, or welding warning signs posted
☐	☐	☐	Storage, use, and handling of flammable materials in accordance with UFC Article 79
☐	☐	☐	A, E, I, or R occupancies; spraying room is equipped with an automatic sprinkler system and separated vertically and horizontally from other areas
☐	☐	☐	In other occupancies, spray booth, spray area, or spray room approved for such use
☐	☐	☐	If booth has frontal area more than 9 sq. ft., and not equipped with doors, then a metal deflector or fire curtain not less than 4 1/2" deep, installed at the upper outer edge of the booth over the opening
☐	☐	☐	Baffles are non-combustible, readily removable, and prevent the deposit of overspray before it enters the exhaust duct

COMPONENTS *(continued)*

Yes	No	N/A	
☐	☐	☐	Floors are non-combustible, non-sparking material
☐	☐	☐	Exit doors from pre-manufactured spray booths not less than 2'6" x 6'8"
☐	☐	☐	Filters are non-combustible or made of listed material
☐	☐	☐	Aluminum components are not used in the construction of the spraying area, booth, or room
☐	☐	☐	Minimum of one 40 BC portable fire extinguisher provided within 30' along each egress path near exit
☐	☐	☐	Approved metal waste containers provided for discarding saturated rags and towels

INSTALLATION

Yes	No	N/A	
☐	☐	☐	Interior surfaces smooth and continuous without edges
☐	☐	☐	Interior surfaces allow free passage of exhaust air
☐	☐	☐	Separations no less than 3' or a greater distance approved by CSFD, by a wall, or by a partition
☐	☐	☐	If illuminated, fixed lighting units have heat treated or hammered wire glass, and designed so that the glass does not become hot
☐	☐	☐	All electrically conductive components are bonded and grounded
☐	☐	☐	Flexible power cords are installed according to listing

AUTOMATIC SPRINKLER SYSTEM ONLY

Yes	No	N/A	
☐	☐	☐	Copy of NFPA 25 or service agreement on site
☐	☐	☐	Wet pipe system only
☐	☐	☐	Sprinkler heads are protected with thin paper or plastic bags and bags are changed frequently
			Upon sprinkler activation system performs the following:
☐	☐	☐	Shuts down all energy supplies (electrical and compressed air)
☐	☐	☐	Activates local alarm and facility's alarm
☐	☐	☐	Shuts down coating material delivery system
☐	☐	☐	Terminates all spray application operations
☐	☐	☐	Stop any conveyors into and out of spray area
☐	☐	☐	Any make-up or exhaust is not interfaced with the fire alarm system and ventilation remains functional throughout fire alarm condition
☐	☐	☐	Complete *Automatic Sprinkler System Acceptance* Test form

CARBON DIOXIDE OR CLEAN AGENT SYSTEMS ONLY

Yes	No	N/A	
			Upon activation, the system performs the following functions:
☐	☐	☐	Ventilation shutdown
☐	☐	☐	Dampers closed
☐	☐	☐	Complete *Gaseous Inerting and Extinguishing Systems Acceptance Test* form

Paint Spray Booths and Rooms
Fire Protection Systems
NFPA 33 and UFC Article 45

LIQUID ELECTROSTATIC SPRAY APPLICATION ONLY

Yes	No	N/A	
			Supervised flame detection apparatus reacts in 1/2 second or less and accomplishes the following:
☐	☐	☐	Activates local alarm and facility's alarm
☐	☐	☐	Shuts down coating material delivery system
☐	☐	☐	Terminates all spray application operations
☐	☐	☐	Stop any conveyors into and out of spray area
☐	☐	☐	Disconnects power to high voltage elements in spray area and de-energizes the system

POWDER SPRAY APPLICATION ONLY

Yes	No	N/A	
			Upon sprinkler activation system performs the following:
☐	☐	☐	Shuts down all energy supplies (electrical and compressed air) to conveyor, ventilation, application, and powder collection equipment
☐	☐	☐	Activates local alarm and facility's alarm
☐	☐	☐	Closes segregation dampers

OPERATIONAL TESTS

Pass	Fail	N/A	
☐	☐	☐	All spray booths readily accessible for cleaning
☐	☐	☐	All spray booths have a clear space of 3 feet which must be kept clear of storage or combustible materials (painted lines on floor)
			Dry filters:
☐	☐	☐	Filter roll automatically advances when air velocity is less than 100 lineal feet
☐	☐	☐	Shuts down spraying operations if filter roll fails to automatically advance
☐	☐	☐	Visible gauges or audible alarm to indicate that required air is not maintained
☐	☐	☐	Not use with materials known to be prone to spontaneous heating

Comments: ____

____

____

____

____

____

____

____

____

System tested by:	Print name here	Sign here

☐ **System Approved**
☐ **System Disapproved** Date: __________ Inspector's Initials __________

Paint Spray Booths and Rooms
Fire Protection Systems
NFPA 33 and UFC Article 45

Appendix G

Plano Fire Department Approving Authority Checklist for Fire Sprinkler Systems

1. Name of Facility: ____________________
2. Facility Physical Add.: ____________________
3. Municipality & Zip: ____________________
4. Sprinkler Contractor: ____________________
5. Contractor: Physical Add.: ____________________
6. Cont. Municipality & Zip: ____________________
7. Cont. Registration No.: ____________________
8. Contractor Cont. No.: ____________________
9. Name of Responsible Layout Tech.: ____________________

10.0 Authority Hazard Classification:

10.1 Total Area: ____________________ sq. ft.
10.2 Light Hazard Area: ____________________ sq. ft.
10.3 Ordinary Hazard, Gp. 1 Area: ____________________ sq. ft.
10.4 Ordinary Hazard, Gp. 2 Area: ____________________ sq. ft.
10.5 Ordinary Hazard, Gp. 3 Area: ____________________ sq. ft.
10.6 Extra Hazard, Gp. 1 Area: ____________________ sq. ft.
10.7 Extra Hazard, Gp. 2 Area: ____________________ sq. ft.
10.8 High Piled Storage Area: (>12') ____________________ sq. ft.

11.0 Environmental Treatment:

11.1 Area Wet: ____________________ sq. ft.
11.2 Area Dry: ____________________ sq. ft.
11.3 Area Dry Pendent: ____________________ sq. ft.
11.4 Area Pre-Action: ____________________ sq. ft.
11.5 Area Deluge: ____________________ sq. ft.
11.6 Area Anti-freeze: ____________________ sq. ft.

12.0 Minimum Number of Risers: (Calculate for largest floor in multi-floor structures. Round up to the next highest whole number. See NFPA 13, 3-3.1.)

12.1 General:
(Sum of areas)
(————)
(max. area)

$$\frac{\text{"10.2"} + \text{"10.3"} + \text{"10.4"} + \text{"10.5"}}{52{,}000} = \square$$

12.2 Extra Hazard:

$$\frac{\text{"10.6"} + \text{"10.7"}}{25{,}000} = \square$$

12.3 High Piled Storage:

$$\frac{\text{"10.8"}}{40{,}000} = \square$$

12.4 Total = $\square$

NOTE: Single systems serving high piled storage and ordinary hazard areas shall not exceed a total of 52,000 square feet for combined area and not more than 40,000 square feet for high piled storage (NFPA 13, 3-3.1)

			Y	N	
12.5	Are there enough risers?		❑	❑	
13.0	**The sprinkler system is:**		**Y**	**N**	
	Hydraulically calculated		❑	❑	
	Pipe Schedule		❑	❑	
14.0	**Sprinkler Spacing:**				
	Are sprinkler locations compatible with the per-sprinkler maximum area limitations?				
14.1	**Light Hazard:** (NFPA 13, 4-2.2.1)		**Y**	**N**	**NA**
14.1.1	Smooth Ceiling and Beam and Girder.				
14.1.1.1	Calculated	—225 sq. ft.	❑	❑	❑
14.1.1.2	Schedule	—200 sq. ft.	❑	❑	❑
14.1.2	Open Wood Joists				
	Schedule or Calc.	—130 sq. ft.	❑	❑	❑
14.1.3	Other Construction				
	Schedule or Calc.	—130 sq. ft.	❑	❑	❑
14.2	**Ordinary Hazard** (NFPA 13, 4-2.2.2)				
14.2.1	No high piled storage.				
	All configurations	—130 sq. ft.	❑	❑	❑
14.2.2	High piled storage. (NFPA 13, 4-2.2.4)				
14.2.2.1	Schedule, all const.	—100 sq. ft	❑	❑	❑
14.2.2.2	Calc., all const.	—130 sq. ft.	❑	❑	❑
14.3	**Extra Hazard:** (NFPA 13, 4-2.2.3)				
14.3.1	Schedule, all const.	—90 sq. ft.	❑	❑	❑
14.3.2	Calc., all const.	—100 sq. ft.	❑	❑	❑
14.4	**Spacing Between Upright and Pendent Heads:**				
14.4.1	Maximum of 7½' from any wall. (Small room rule is not valid in Delaware.)		❑	❑	❑
14.4.2	Maximum of 15' between sprinklers (NFPA 13, 4-2.1.5.1)		❑	❑	❑
14.4.3	Minimum of 4" from a wall. (NFPA 13, 4-2.1.5.2)		❑	❑	❑
14.4.4	Minimum of 6' from each other? (Residential minimum is 8')		❑	❑	❑

NOTE: On line items 14.4.1 and 14.4.2, some residential sprinklers have been listed for greater spacings. Check the listings.

		Y	N	NA
14.5	**Sidewall Sprinkler Spacing:**			
14.5.1	Are the specified sprinklers listed for this hazard?	❑	❑	❑

NOTE: Sidewall sprinklers are special application devices. Most are listed only for light hazard. Check the listing.

14.5.2	Are the specified sprinklers within their listed coverage?	❑	❑	❑

15.0 **Obstructions:** Have obstructions such as columns, firdowns, beams, offsets, etc., been considered so that every square foot of floor area is covered by sprinklers? ❑ ❑ ❑

16.0 **Temperature Ratings** (Degrees Fahrenheit) (NFPA 13, Table 3-16.6.1)

	Classification	Max. Clg. Temp.	Quantity
16.1	Ordinary (135-170)	100	______
16.2	Intermediate (175-225)	150	______
16.3	High (250-300)	225	______
16.4	Extra High (325-375)	300	______
16.5	Very Extra High (400-475)	375	______
16.6	Ultra High (500-575)	475	______
16.7	Ultra High II (650)	625	______

		Y	N
16.8	Do these fit the expected ceiling temperatures?	❑	❑
16.9	Are these compatible with NFPA 13, Figure 3-16.6.3 (a) Heater and Danger Zones at Unit Heaters, and Table 3-16.6.3 (A), Distance of Sprinklers from Heat Sources?	❑	❑

17.0	**For areas not involving high piled storage:**	**Y**	**N**	**NA**
17.1	Are the contractor-selected hazard classifications correct?	❑	❑	❑
17.2	Does the hydraulically most remote area size-and-density-point fall on or to the right of the required density curve of NFPA 13 Figure 2-2.1 (B)?	❑	❑	❑
17.3	Is the selected remote area truly the hydraulically most remote?	❑	❑	❑
17.4	Review the hydraulic computations?	❑	❑	❑
17.4.1	Has the minimum hose allowance per NFPA 13, Table 2-2.1 (B) been included at the base of the riser?	❑	❑	❑
17.4.2	Is the total water demand greater than the remote area times density (A x d) + the minimum hose allowance of NFPA 13, Table 2-2.1 (B)? (It should be a minimum of 5% to 10% greater)	❑	❑	❑
17.4.3	Have all the fittings, valves and special devices been included?	❑	❑	❑
17.4.4	Does the Supply-Demand Curve show adequate safety factor?	❑	❑	❑
17.4.5	Has the static head loss been included in the computations?	❑	❑	❑
17.4.6	Is the plotted water supply consistent with municipal records?	❑	❑	❑

		Y	N	NA
18.0	**For Schedule Pipe Systems:**			
18.1	Are the applied sizes of pipe in accordance with the hazard classification and schedules of NFPA 13, paragraphs 3-5 through 3-7?	❑	❑	❑
18.2	Calculate the static head loss for the highest sprinkler heads as P = h x .434 __________			
18.3	Is this value + 15 psi less than the pressure available at the base of the riser at the flows required for a schedule system per NFPA 13, Table 2-2.1 (A)?	❑	❑	❑
19.0	**Dry Systems:**			
19.1	Is (Are) the systems divided into volumes not more than 500 gallons for gridded systems or 750 gallons for non-gridded systems? (NFPA 13, 5-2.3.1)?	❑	❑	❑
19.2	Is (Are) systems with check valve subdivision divided into branches with not more than 400 gallons for gridded systems and 600 gallons for non-gridded systems? (See NFPA 13, 5-2.3.1)	❑	❑	❑
19.3	Is the dry pipe valve located in a heated and lighted room?	❑	❑	❑
19.4	If the system capacity exceeds 350 gallons for gridded systems or 500 gallons for non-gridded systems, have quick opening devices been provided?	❑	❑	❑
20.0	**Anti-freeze Systems:**			
20.1	Is the system size less than 40 gallons? (See NFPA 13, A-5-5-2)	❑	❑	❑
20.2	Is the proper anti-freeze solution specified on the drawing? (NFPA 13, 5-5.3)	❑	❑	❑
21.0	**Pre-Action Systems:**			
21.1	Are there less than 1000 closed sprinklers on a system?	❑	❑	❑
22.0	**High Piled Storage:** Have the proper standards been applied?			
22.1	Indoor General Storage — NFPA 231	❑	❑	❑
22.2	Rack Storage of Materials — NFPA 231C	❑	❑	❑
22.3	Storage of Rubber Tires — NFPA 231D	❑	❑	❑
22.4	Storage of Baled Cotton — NFPA 231E	❑	❑	❑
22.5	Storage of Rolled Paper — NFPA 231F	❑	❑	❑
23.0	**Additional Provisions:** Does the design detail the following:			
23.1	Alarm provisions: Water motor gong or electric bell?	❑	❑	❑
23.2	Fire Department Connection?	❑	❑	❑
23.3	Inspector's Test?	❑	❑	❑
23.4	Supervisory Provisions?	❑	❑	❑
23.4.1	Valves? (NFPA 13, 3-14.2.3)	❑	❑	❑
23.4.2	Pre-Action Systems? (NFPA 13, 5-3.5.3)	❑	❑	❑

		Y	N	NA
23.4.3	Deluge Systems? (NFPA 13, 5-3.6)	❑	❑	❑
23.5	Are hanger details correctly specified?	❑	❑	❑
23.6	Are elevations of piping indicated?	❑	❑	❑
23.7	Is the drawing to a standard scale?	❑	❑	❑
23.8	Is a riser diagram provided?	❑	❑	❑
23.9	Are the canopies sprinklered?	❑	❑	❑

24.0 **Design** Accepted ❑
Rejected ❑ With the following comments:

1. ______________________________
2. ______________________________
3. ______________________________
4. ______________________________
5. ______________________________

Name of Reviewer: ______________________________

Date: ______________ Signature: ______________________________

Glossary

A

Accessibility — The ability of fire apparatus to get close enough to a building to conduct emergency operations.

Acute — Severe, rapid onset, usually of short duration.

Alarm-Initiating Device — A mechanical or electrical device that activates an alarm system. There are three basic types of alarm-initiating devices: manual, products-of-combustion detectors, and extinguishing system activation devices.

Ampere — (1) The amount of current sent by one volt through one ohm of resistance; (2) Unit of measurement of electrical current.

Arson — The willful and malicious burning of property.

Authority — Relates to the empowered duties of an official, in this case the fire inspector. The level of an inspector's authority is commensurate with the enforcement obligations of the governing body.

Automatic Sprinkler System — A system of water pipes, discharge nozzles, and control valves designed to activate during fires by automatically discharging enough water to control or extinguish a fire.

B

Backdraft — Instantaneous combustion that occurs when oxygen is introduced into a smoldering fire. The stalled combustion resumes with explosive force.

Basement Plans — Drawings showing the belowground view of a building. The thickness and external dimensions of the basement walls are given, as are floor joist locations, strip footings, and other attached foundations.

Bearing Walls — Walls of a building that by design carry at least some part of the structural load of the building.

BLEVE (Boiling Liquid Expanding Vapor Explosion) — The failure of a closed container as a result of over pressurization caused by an external heat source.

Board of Appeals — A group of five to seven people with experience in fire prevention and code enforcement who arbitrate differences in opinion between fire inspectors and property owners or occupants.

Boiling Point — The temperature at which the vapor pressure of a liquid is equal to the external pressure applied to it.

Bonding — The connection of two objects with a metal chain or strap in order to neutralize the static electrical charge between the two.

Boyle's Law — This law states that the volume of a gas varies inversely with the applied pressure. The formula is: $P_1V_1=P_2V_2$ where

P_1 = original pressure

V_1 = original volume

P_2 = final pressure

V_2 = final volume

Brands — (1) Large, flying, burning embers that are lifted by a fire's thermal column and carried away with the wind. (2) Small burning pieces of wood or charcoal used to test the fire resistance of roof coverings and roof deck assemblies.

British Thermal Unit (BTU) — The amount of heat required to raise the temperature of one pound (0.45 kg) of water one degree Fahrenheit.

Bureau of Alcohol, Tobacco, and Firearms (BATF) — Division of the U.S. Department of Treasury; regulates the storage, handling, and transportation of explosives.

C

Calorie — The amount of heat required to raise the temperature of one gram (0.035 oz.) of water one degree centigrade.

Carbon Dioxide (CO_2) — A gas heavier than air used to extinguish Class B or C fires by smothering or displacing the oxygen.

Centigrade (Celsius) — A temperature scale in which the boiling point of water is 100°C (212°F) and the freezing point is 0°C (32°F) at normal atmospheric pressure.

Charles' Law — Scientific law that says the increase or decrease of pressure in a constant volume of gas is directly proportional to corresponding increase or decrease of temperature. The formula is stated: $P_1T_1 = P_2T_2$ where

P_1 = original pressure

T_1 = original temperature

P_2 = final pressure

T_2 = final temperature

CHEMTREC — The Manufacturing Chemists Association's name for its Chemical Transportation Emergency Center. The center provides immediate information on how to handle hazardous materials incidents. The toll-free phone number is 1-800-424-9300.

Chronic — Of long duration (opposite of acute).

Circuit — The complete path of an electric current.

Citation — A legal reprimand for failure to comply with existing laws or regulations.

Clear Width — The actual unobstructed opening size of an exit.

Codes — Rules or laws used to enforce requirements for fire protection, life safety, or building construction.

Combustible Liquid — Any liquid having a flash point at or above 100°F (37.8°C) and below 200°F (93.3°C).

Combustion — The self-sustaining process of rapid oxidation of a material that produces heat and light.

Common Hazard — A condition likely to be found in almost all occupancies and generally not associated with a specific occupancy or activity.

Common Path of Travel — The route of travel used to determine measured egress distances in code enforcement. The common path of travel is considered to be down the center of a straight corridor and a 1-foot (0.3 m) radius around each corner. Also called the *normal path of travel.*

Complaint — An objection to existing conditions that is brought to the attention of the fire inspection bureau.

Compliance — Meeting the minimum standards set forth by applicable codes or regulations.

Compressed Gas — Gas that, at normal temperature, exists solely as a gas when pressurized in a container.

Conduction — The transfer of heat energy from one body to another through a solid medium.

Conductor — A substance or material that transmits electrical or heat energy.

Construction Classification — The rating given to a particular building based on the materials and methods used to construct it and their ability to resist the effects of a fire situation.

Convection — The transfer of heat energy by the movement of air or liquid.

Corrosives — Those materials that cause harm to living organisms by destroying body tissue.

Cryogenics — Gases that are converted into liquids by being cooled below -150°F (-101°C).

D

Dead-end Corridor — A corridor in which egress is possible in only one direction.

Detailed View — Additional, close-up information shown on a particular section of a larger drawing.

Dikes — Temporary or permanent barriers that prevent liquids from flowing into certain areas or that direct the flow as desired.

Dimensioning — A drawing that places a building on a site plan to clearly show its size and arrangement relative to existing conditions.

Draft Curtains — Dividers hung from the ceiling in large open areas that are designed to minimize the mushrooming effect of heat and smoke.

Dry Chemical — Any one of a number of powdery extinguishing agents used to extinguish fires.

Dry Standpipe System — A standpipe system that either has water supply valves closed or that has no fixed water supply to it.

Ducts — Hollow pathways used to move air from one area to another in ventilation systems.

E

Electric Shock — Injury caused by electricity passing through the body. Severity of injury depends upon the path the current takes through the body, the amount of current, and the resistance of the skin.

Electrical Systems— Those wiring systems designed to distribute electricity throughout a building.

Electron — A minute component of an atom that possesses a negative charge.

Elevation View — An architectural drawing used to show the number of floors of a building and the grade of surrounding ground.

Equivalency — Alternative practices that are acceptable for meeting a minimum level of fire protection.

EX — Rating symbol used on lift trucks that are safe for use in atmospheres containing flammable vapors or dusts.

Exhaust System — A ventilation system designed to remove stale air, smoke, vapors, or other airborne contaminants from an area.

Exit — That portion of a means of egress that is separated from all other spaces of the building structure by construction or equipment and that provides a protected way of travel to the exit discharge.

Exit Access — The portion of a means of egress that leads to the exit. Hallways, corridors, and aisles are examples of exit access.

Exit Capacity — The maximum number of people who can discharge through a particular exit.

Exit Discharge — That portion of a means of egress that is between the exit and a public way.

Exit Stairs — Stairs that are used as part of a means of egress. The stairs may be part of either the exit access or the exit discharge when conforming to requirements in the *Life Safety Code*.

Expellant Gas — Any of a number of inert gases that are compressed and used to force extinguishing agents from a portable fire extinguisher. Nitrogen is the most commonly used expellant gas.

Explosive — Any material or mixture that will undergo an extremely fast self-propagation reaction when subjected to some form of energy.

Extinguishing Agent — Any substance used for the purpose of controlling or extinguishing a fire.

F

Factory Mutual System (FM) — Fire research and testing laboratory that provides loss control information for the Factory Mutual System and anyone else who may find it useful.

Fahrenheit — Temperature scale in which the boiling point of water is 212°F (100°C) and the freezing point is 32°F (0°C) at normal atmospheric pressure.

Field Sketch — A rough drawing of an occupancy that is made during an inspection. The field sketch is used to make a final inspection drawing.

Fire Alarm System — (1) A system of alerting devices that takes a signal from fire detection or extinguishing equipment and alerts building occupants or proper authorities of a fire condition. (2) A system used to dispatch fire department personnel and apparatus to emergency incidents.

Fire Cause — The combination of fuel supply, heat source, and a hazardous act that results in a fire.

Fire Cause Determination — The process of establishing the cause of a fire incident through careful investigation and analysis of the available evidence.

Fire Damper — A device that automatically interrupts air flow through all or part of an air handling system, thereby restricting the passage of heat and the spread of fire.

Fire Department Connection (FDC) — An inlet appliance that has two or more 2½-inch (65 mm) connections or one large-diameter (4-inch [100 mm] or larger) connection through which fire apparatus can boost the pressure or amount of water flowing through a sprinkler or standpipe system.

Fire Detection System — A system of detection devices, wiring, and supervisory equipment used for detecting fire or products of combustion and then signaling that these elements are present.

Fire Door — A specially constructed, tested, and approved door installed to prevent fire spread.

Fire Drill — A training exercise to ensure that the occupants of a building can exit the building in a quick and orderly manner in case of fire.

Fire Extinguisher — A portable fire fighting device designed to combat incipient fires.

Fire Hazard — Any material, condition, or act that contributes to the start of a fire or that increases the extent or severity of fire.

Fire Load — The maximum amount of heat that can be produced if all the combustible materials in a given area burn.

Fire Partition — A fire barrier that extends from one floor to the bottom of the floor above or to the underside of a fire-rated ceiling assembly. A fire partition provides a lower level of protection than a fire wall.

Fire Point — The temperature at which a liquid fuel produces sufficient vapors to support combustion once the fuel is ignited. The fire point is usually a few degrees above the flash point.

Fire Prevention Code — A law enacted for the purpose of enforcing fire prevention and safety regulations.

Fire Resistance Rating — The amount of time a material or assembly of materials will resist a typical fire as measured on a standard time-temperature curve.

Fire Retardant — A chemical applied to material or another substance that is designed to retard ignition or the spread of fire.

Fire Risk — The probability that a fire will occur and the potential for harm it will create.

Fire Stop — Materials used to prevent or limit the spread of fire in hollow walls or floors, above false ceilings, in penetrations for plumbing or electrical installations, or in cocklofts and crawl spaces.

Fire Wall — A wall with a specified degree of fire resistance that is designed to prevent the spread of fire within a structure or between adjacent structures.

Flameover — Condition that occurs when a portion of the fire gases trapped at the upper level of a room ignite, spreading flame across the ceiling of the room.

Flame Spread Rating — A numerical rating assigned to a material based on the speed and extent to which flame travels over its surface.

Flame Test — A test designed to determine the flame spread characteristics of structural components or interior finishes.

Flammable and Explosive Limits — The upper and lower concentrations of a vapor expressed in percent mixture with an oxidizer that will produce a flame at a given temperature and pressure.

Flammable Liquid — Any liquid having a flash point below 100°F (37.8°C) and having a vapor pressure not exceeding 40 psi absolute (276 kPa).

Flashover — The stage of a fire at which all combustibles are heated to their ignition temperatures and the area becomes fully involved in fire.

Flash Point — The lowest temperature at which sufficient vapors are produced to form an ignitable mixture.

Floor Plan — An architectural drawing showing the layout of a floor within a building as seen from above. It outlines where each room is and what the function of the room is.

Fluoronated Surfactants — Chemicals that lower the surface tension of a liquid, in this case fire fighting foams.

Foam — An extinguishing agent produced by mixing a foam-producing compound with water and aerating the solution for expansion. These agents are primarily used for extinguishing Class B fires, but in some cases may be used on Class A fires as well.

G

Gas — A compressible substance, with no specific volume, that tends to assume the shape of a container. Molecules move about most rapidly in this state.

Grounding — Reducing the difference in electrical potential between an object and the ground by the use of various conductors.

H

Half-Life — (1) Time required for half of something to undergo a process. (2) Time required for half the amount of a substance in or introduced into a living system or ecosystem to be eliminated or disintegrated by natural processes. (3) Period of time required for any radioactive substance to lose half of its strength or reduce by one-half its total present energy.

Halogenated Agents — Chemical compounds (halogenated hydrocarbons) that contain carbon plus one or more elements from the halogen series. Halon 1301 and Halon 1211 are most commonly used as extinguishing agents for Class B and C fires.

Hazardous Material — Any material that poses an unreasonable risk to the health and safety of persons and/or the environment if it is not properly controlled during handling, storage, manufacture, processing, packaging, use, disposal, or transportation.

Heat — A form of energy that is proportional to molecular movement. To signify its intensity, it is measured in degrees of temperature.

Heat Transfer — The flow of heat from a hot substance to a cold substance. This may be accomplished by convection, conduction, or radiation.

Hoseline — A section of flexible conduit that is connected to a water supply source for the purpose of delivering water onto a fire.

HVAC System (Heating, Ventilating, and Air Conditioning System) — The mechanical system used to provide environmental control within a structure.

Hydrostatic Test — A testing method used to check the integrity of pressure vessels.

I

Ignition Source — A method (either wanted or unwanted) that provides a means for the initiation of self-sustained combustion.

Ignition Temperature — The minimum temperature at which a fuel other than a liquid will continue to burn once it is ignited.

Immunity — Freedom from legal liability for an act or physical condition.

Incipient Phase — The first stage of the burning process where the substance being oxidized is producing some heat, but the heat has not spread to other substances nearby.

Inspection — A formal examination of an occupancy and its associated uses or processes to determine its compliance with the fire and life safety codes and standards.

Ionization — The process by which an object or substance gains or loses electrons, thus changing its electrical charge.

L

Legend — An explanatory list of symbols on a map or diagram.

Liability — To be legally obligated or responsible for an act or physical condition.

***Life Safety Code* (NFPA 101)** — A building standard designed to protect lives in the event of a fire.

Liquefied Gas — A confined gas that at normal temperatures exists in both liquid and gaseous states.

Liquid — An incompressible substance that assumes the shape of its container. The molecules flow freely, but substantial cohesion prevents them from expanding as a gas would.

Live Load — The force placed upon a structure by the addition of people, objects, or weather.

M

Magazine — A storage facility approved by the Bureau of Alcohol, Tobacco, and Firearms (BATF) for the storage of explosives.

Means of Egress — A safe and continuous path of travel from any point in a structure leading to a public way. The means of egress is comprised of three parts: the exit access, the exit, and the exit discharge.

Mechanical Systems — Large equipment systems within a building that may include, but are not limited to, climate control systems; smoke, dust, and vapor removal systems; trash collection systems; and automated mail systems. These do not include general utility systems such as electric, gas, and water.

N

National Fire Protection Association (NFPA) — A nonprofit educational and technical association dedicated to protecting life and property from fire by developing fire protection standards and educating the public.

National Response Center — A federal organization charged with coordinating the response of numerous agencies to emergency incidents involving the release of significant amounts of hazardous materials.

Neutron — A part of the nucleus of an atom that has a neutral electrical charge.

NFPA 704 Labeling System — A system for identifying hazardous materials in fixed facilities. The placard is divided into sections that identify the degree of hazard with respect to health, flammability, and reactivity, as well as special hazards.

Noncombustible — Incapable of supporting combustion under normal circumstances.

Normal Operating Pressure — The normal amount of pressure that is expected to be available from a hydrant, prior to pumping.

North American Emergency Response Guidebook (NAERG) — A manual that aids emergency response and inspection personnel in identifying hazardous materials placards. It also gives guidelines for initial actions to be taken at hazardous materials incidents.

O

Occupancy Classification — The classifications given to structures by the model code used in that jurisdiction based on the intended use for the structure.

Occupational Health and Safety Administration (OSHA) — A United States federal agency that develops and enforces standards and regulations for occupational health and safety in the workplace.

Occupant Load — The total number of people who may occupy a building or portion of a building at any one given time.

Ohms — Units of measurement of electrical resistance.

Ordinance — A law set forth by a government agency, usually at the local municipal level.

Oxidizer — Any material that provides oxygen for combustion.

P

Panic Hardware — Hardware mounted on exit doors in public buildings that enables doors to be opened when pressure is applied.

Parapet — A portion of a wall that extends above the level of the roof.

Pitot Tube — An instrument containing a Bourdon Tube that is inserted into a stream of water to measure the velocity pressure of the stream. The gauge reads in units of pounds per square inch (psi [kPa]).

Placard — A diamond-shaped sign that is affixed to all sides of a vehicle transporting hazardous materials. The placard indicates the primary class of the material, and in some cases the exact material being transported.

Plans Review — The process of reviewing building plans and specifications to determine the safety characteristics of the proposed building. This is generally done before permission is granted to begin construction.

Plan View — A drawing containing the two-dimensional view of a building as seen from directly above the area.

Plot Plan — An architectural drawing showing the layout of buildings and landscape features for a given plot of land. The view is from directly above.

Point of Origin — The exact location at which a particular fire started.

Police Power — The authority that may be given to an inspector to arrest, issue summons, or issue citations for fire code violations.

Predischarge Alarm — An alarm that sounds before a total flooding fire extinguishing system is about to discharge, thus giving occupants the opportunity to leave the area.

Pre-Fire Planning — Advance planning of fire fighting operations at a particular location, taking into account all factors that will influence fire fighting tactics.

Pressure-Reducing Valve — A valve installed at standpipe connections that is designed to reduce the amount of water pressure at that discharge to a specific pressure, usually 100 psi (700 kPa).

Pressure Relief Device — An automatic device designed to release excess pressure from a container.

Proton — A part of an atom that possesses a positive charge.

Pyrolysis — The chemical decomposition of a substance through the action of heat.

Public Way — A parcel of land such as a street or sidewalk that is essentially open to the outside and is used by the public for moving from one location to another.

R

Radiation — Transfer of heat energy by electromagnetic waves.

Radioactive Material — A material whose atomic nucleus spontaneously decays or disintegrates, emitting radiation.

Radioactive Particle — Particles emitted during the process of radioactive decay. There are three types of radioactive particles: alpha, beta, and gamma particles.

Reactive Material — A material that will react violently when combined with air or water.

Reactivity — The ability of two or more chemicals to react and release energy and the ease with which this reaction takes place.

Refrigerant — A substance used within a refrigeration system to provide the cooling action.

Residual Pressure — The pressure remaining at a given point in a water supply system while water is flowing.

Response Time Index (RTI) — A numerical value representing the speed and sensitivity with which a heat responsive fire protection device (like a fusible link) responds.

Responsibility — An act or duty for which someone is clearly accountable.

Right of Entry — The rights set forth by the administrative powers that allow the inspector to inspect buildings to ensure compliance with applicable codes.

Riser — A vertical water pipe used to carry water for fire protection systems above ground, such as a standpipe or sprinkler riser.

Roof Covering — The final outside cover that is placed on top of a roof deck assembly. Common roof coverings include composite or wood shake shingles, tile, slate, tin, or asphaltic tar paper.

Roof Deck — The bottom components of the roof assembly that support the roof covering. The roof deck may be constructed of such components as plywood, wood studs (2 inches x 4 inches [50mm x 100 mm]) or larger, lathe strips, and so on.

S

Salamander — A portable heating device generally found on construction sites.

Sanction — A notice or punishment attached to a violation for the purpose of enforcing a law or regulation.

Sectional View — A vertical view of a building as if it were cut into two parts. The purpose of a sectional view is to show the internal construction of each assembly.

Semiconductor — Material that is neither a good conductor nor a good insulator, and therefore may be used as either in some applications.

Site Plan — A drawing that provides a view of the proposed construction in relation to existing conditions. It is generally the first sheet on a set of drawings.

Smoke Damper — A device that restricts the flow of smoke through an air handling system.

Smokeproof Enclosures — Stairways that are designed to limit the penetration of smoke, heat, and toxic gases and that serve as part of a means of egress.

Solid — A substance that has a definite shape and size. The molecules of a solid generally have very little mobility.

Special Fire Hazard — A fire hazard arising from the processes or operations that are peculiar to the individual occupancy.

Specific Gravity — The ratio of the weight of a liquid to an equal volume of water.

Sprinkler — The waterflow discharge device in a sprinkler system. The sprinkler consists of a threaded intake nipple, a discharge orifice, and a deflector to create an effective fire stream pattern.

Standard — A document containing requirements and specifications outlining minimum levels of performance, protection, or construction.

Standpipe System — Fire department hose outlets installed on different levels of a building to be used by firefighters and/or building occupants.

Static Electricity — An accumulation of electrical charges on opposing surfaces created by the separation of unlike materials or by the movement of surfaces.

Static Pressure — The pressure at a given point in a water system when no water is flowing.

Steiner Tunnel — A horizontal furnace 25 feet (7.6 m) long, 17½ inches (445 mm) wide, and 12 inches (305 mm) high. A 5,000 BTU (5 270 kj) flame is produced in the tunnel and the extent of flame travel across the surface of the test material is observed through ports in the side of the furnace.

Steiner Tunnel Test — A test used to determine the flame spread ratings of various materials.

Storage Tanks — Storage vessels that are larger than 60 gallons (227 L) and are located in a fixed location.

Structural Abuse — Using or changing a building beyond its original designed capabilities.

Supervisory Circuit — An electronic circuit within a fire protection system that monitors the system's readiness and transmits a signal when there is a problem with the system.

T

Temperature — The measurement of the intensity of the heating of a material.

Threshold Limit Value (TLV) — The concentration of a given material that may be tolerated for an 8-hour exposure during a regular work week without ill effects.

Title Block — A small information section on the face of every plan drawing. The title block contains such information as name of project, title of the particular drawing, the scale used, and date of drawing and/or revisions.

Tort — A wrongful act (except for breach of contract) for which a civil action will lie.

Toxicity — The ability of a substance to do harm within the body.

Toxic Material — Any material classified as a poison, asphyxiant, irritant, or anesthetic.

Travel Distance — The distance from any given area in a structure to the nearest exit or to a fire extinguisher.

U

Underwriters Laboratories, Inc. (UL) — An independent fire research and testing laboratory.

United States Department of Transportation (DOT) — The federal organization that regulates the transportation of hazardous materials.

Unstable Material — Material that is capable of undergoing chemical changes or decomposition with or without a catalyst.

V

Vapor Density — The weight of a gas as compared to the weight of air.

Vaporization — The passage from a liquid to a gaseous state. Rate of vaporization depends upon the substance involved, heat, and pressure.

Vapor Recovery System (VRS) — A system that recovers gasoline vapors emitted from a vehicle's gasoline tank during product dispensing.

Ventilation — The systematic removal of heated air, smoke, and gases from a structure and replacing them with cooler air to reduce damage and facilitate fire fighting operations.

Violation — An infringement of existing rules, codes, or laws.

Voltage — Units of electric potential.

W

Wet Standpipe System — A standpipe system that has water supply valves open and maintains water in the system at all times.

Index

G

Index by Kari Bero, Bero-West Indexing Service, Seattle, WA.

FIRE INSPECTION AND CODE INFORCEMENT
6th EDITION
1st PRINTING, 4/98

COMMENT SHEET

DATE ________________ NAME ______________________________

ADDRESS ____

ORGANIZATION REPRESENTED ______________________________

CHAPTER TITLE ___________________________ NUMBER ________

SECTION/PARAGRAPH/FIGURE _______________________ PAGE ________

1. Proposal (include proposed wording, or identification of wording to be deleted), OR PROPOSED FIGURE:

2. Statement of Problem and Substantiation for Proposal:

RETURN TO: IFSTA Editor
Fire Protection Publications
Oklahoma State University
930 N. Willis
Stillwater, OK 74078-8045

SIGNATURE ______________________

Use this sheet to make any suggestions, recommendations, or comments. We need your input to make the manuals as up to date as possible. Your help is appreciated. Use additional pages if necessary.